ASSESSMENT OF CROP LOSS
FROM AIR POLLUTANTS

Proceedings of an International Conference
Raleigh, North Carolina, USA
October 25–29, 1987

Organized by the Research Management Committee of the
National Crop Loss Assessment Network (NCLAN)

for the

US Environmental Protection Agency
Environmental Research Laboratory
Corvallis, Oregon, USA

ASSESSMENT OF CROP LOSS FROM AIR POLLUTANTS

Edited by

WALTER W. HECK

USDA-Agricultural Research Service, North Carolina State University, Raleigh, North Carolina, USA

O. CLIFTON TAYLOR

University of California, Riverside, California, USA

and

DAVID T. TINGEY

US Environmental Protection Agency, Corvallis, Oregon, USA

ELSEVIER APPLIED SCIENCE
LONDON and NEW YORK

ELSEVIER SCIENCE PUBLISHERS LTD
Crown House, Linton Road, Barking, Essex IG11 8JU, England

Sole Distributor in the USA and Canada
ELSEVIER SCIENCE PUBLISHING CO., INC.
52 Vanderbilt Avenue, New York, NY 10017, USA

WITH 53 TABLES AND 127 ILLUSTRATIONS

© ELSEVIER SCIENCE PUBLISHERS LTD 1988
(except Chapters 3 and 12)
© CENTRAL ELECTRICITY GENERATING BOARD—Chapter 8
© CROWN COPYRIGHT—Chapter 22

Softcover reprint of the hardcover 1st edition 1988

British Library Cataloguing in Publication Data

Assessment of crop loss from air pollutants.
1. Crops. Effects of pollutants of atmosphere
I. Heck, Walter W.   II. Taylor, O. Clifton.
III. Tingey, David T.   IV. National Crop
Loss Assessment Network
632'.19

ISBN-13: 978-94-010-7109-3   e-ISBN-13: 978-94-009-1367-7
DOI: 10.1007/978-94-009-1367-7

Library of Congress Cataloging in Publication Data

Assessment of crop loss from air pollutants: proceedings of an
international conference. Raleigh, NC, October 25–29, 1987/edited
by Walter W. Heck, O. Clifton Taylor, David T. Tingey.
        p.   cm.
"Organized by the Research Management Committee of the National
Crop Loss Assessment Network (NCLAN) for the U.S. Environmental
Protection Agency, Environmental Research Laboratory, Corvallis, OR."
    Includes bibliographies and index.

    1. Plants, Effect of air pollution on—Congresses. 2. Crops
losses—Congresses. I. Heck, Walter Webb, 1926–   . II. Taylor,
O. C. (Oliver Clifton), 1918–   . III. Tingey, David T.
IV. National Crop Loss Assessment Network (U.S.). Research
Management Committee. V. Corvallis Environmental Research
Laboratory.
SB745.A88 1988
338.1'4—dc 19   88-16980
                                                        CIP

Typeset and Printed by The Universities Press (Belfast) Ltd.

PREFACE

During late 1985, the Research Management Committee (RMC) of the National Crop Loss Assessment Network (NCLAN) decided the most appropriate way to bring the NCLAN program to a successful conclusion was to hold an international conference. It was envisaged as an opportunity to present an overview of results from the NCLAN program and as a chance to view the results in the context of ongoing research by members of the international community.* Although we wanted the Conference to have an assessment orientation, it was also intended for the Conference to focus on current state-of-knowledge. The Conference was designed to overview the needs of crop loss assessment, current approaches to assessment, progress in the development of predictive models, the use of the information for economic predictions, and the application of the data in policy decisions. Every effort was made to assure a broad representation of ideas.

The Conference program was developed to evaluate major issues that address regional/national assessments of impacts of atmospheric pollutants on agricultural production. Sessions were structured to address specific issues by invited speakers, and by contributed papers and posters. First, background needs for doing loss assessment research including specific approaches and a rather detailed review of the NCLAN program were addressed (Session I). Session II addressed the needs for defining the exposure environment (e.g. extrapolating to regional concentrations and exposure characterization). Field approaches for determining crop loss were reviewed in Session III. Session IV addressed the importance of an understanding of the physiological processes in

---

* This volume includes a list of NCLAN publications published or in press by March 1988.

assessment programs for use in process oriented models. Session V acknow-ledged a lack of basic understanding of the importance of pollutant mixtures on crop response, and on information relating to how abiotic and biotic factors affect the response of crops to gaseous pollutants. Session VI highlighted statistical and process level modeling and their use in assessment programs. Economic assessments and policy implications were addressed in Session VII. Sessions II through VII included contributed papers and posters which appear in a special issue of *Environmental Pollution* (Volume 53, 1988).

We developed and implemented a strong peer review process for invited papers. The process was a cooperative venture between the session leaders and the Conference Committee. Drafts were requested well before the Conference so the general approach of each paper could be approved. The session chairmen and co-chairmen who identified three peer reviewers, were respon-sible for collating and commenting on the reviews, and dealing with the authors.

Two of the invited papers were not submitted due to time constraints on the original authors. To maintain the integrity of this volume, we included information relative to the subject matter of these two papers. We appreciate Thomas J. Moser and co-authors permitting us to include their contributed paper as Chapter 14 in this volume and Adams and Crocker (Chapter 19), who were asked at a late date to develop a short paper on model requirements for an economic evaluation. We realize these papers do not fully address the original intent of the invited papers but they do cover primary ingredients of the original themes.

The thirty international attendees represented 15 countries. Participation from the European Community reflected their interest in European assessment activities. The Conference reflected current concepts associated with assess-ment efforts within the scientific community.

The session chairmen and co-chairmen represent a cross section of interna-tional research workers in the field. We recognize the importance of these individuals in the organization of the Conference, their willingness to spend time coordinating the peer review process, the excellent job they did in moderating their sessions, and their overall support. In appreciation of their tireless efforts, we recognize:

Session I: The Need for Crop Loss Assessment	Richard M. Adams, Chairman
Session II: Meteorology, Atmospheric Chemistry and Regional Monitoring—Extrapolation	Bruce B. Hicks, Chairman Allen S. Lefohn, Co-Chairman
Session III: Yield Assessment Using Field Approaches for Measuring Crop Loss	Michael H. Unsworth, Chairman Howard E. Heggestad, Co-Chairman
Session IV: The Value of Physiological Understanding in Crop Loss Assessment	David T. Tingey, Chairman Robert G. Admundson, Co-Chairman

Session V: Abiotic and Biotic Interactive Stress Factors	Leonard H. Weinstein, Chairman Richard A. Reinert, Co-Chairman
Session VI: Statistical and Simulated Modeling Approaches	Victor C. Runeckles, Chairman Eric M. Preston, Co-Chairman
Session VII: Economic Considerations and Policy Implications	O. Clifton Taylor, Chairman Patricia M. Irving, Co-Chairman
Poster Session	Robert J. Kohut, Co-Chairman Lance W. Kress, Co-Chairman

We thank all NCLAN participants for their unqualified support of the Conference and all organizations which gave financial assistance.

Finally, we thank our spouses for patience beyond the call of duty.

Walter W. Heck
O. Clifton Taylor
David T. Tingey

ACKNOWLEDGEMENTS

The success of the Conference was due to a large degree to the experience and efficiency of our Conference Coordinator, Janet McFayden of North Carolina State University. Janet's unfailing eye for detail as well as for the long-term goals of the Conference are gratefully acknowledged by the Conference Chairmen. Janet managed all details and problems at the Conference with grace and poise and left us to enjoy the Conference and interact with the participants. We also recognize the support of Clara Edwards, Mable Bullock, Jody Castleberry, Ramona Logan, Debby Cross, and Jeanie Hartman who carried out many of the details necessary to the successful operation of the Conference and preparing the proceedings; Steve Vozzo served as Conference photographer.

The Committee acknowledges the tireless effort of the editorial group at Northrop Services under Helen Mathews, including Jaynie Allen, Janice Braswell, Mike Clark, Linda Cooper, and Deborah Ussery-Baumrucker for the excellent editorial work they did on both the early drafts of manuscripts and on the final copy prior to sending to the publisher. These efforts have improved the quality of all the manuscripts.

The Conference Chairmen appreciate the financial assistance from the US Environmental Protection Agency (EPA) and the USDA's Agricultural Research Service. Support was also provided by North Carolina State University, Carolina Power and Light Company, Southern California Edison Company, and the Air Pollution Control Association.

The following organizations supported a Trade Exhibition during the Conference; we appreciate their support:

Raleigh Valve and Fitting Company, Raleigh, NC, USA
Elsevier Applied Science, Ltd, London, UK

Air Pollution Control Association, Pittsburgh, PA, USA
Thermo Environmental Instruments Company, Franklin, MA, USA

We also acknowledge the close working relationship we have had with Elsevier Applied Science and their help and encouragement in the preparation of this publication.

The use of trade names in any paper within this volume does not imply endorsement by any of the funding sources of the products named, nor criticism of similar ones not mentioned.

All papers included in this volume meet the EPA peer review requirements of three reviews in addition to the editorial reviews. We recognize the contributions of those who peer reviewed the papers for this publication. The volume has not been subjected to EPA review and therefore does not necessarily reflect the views of the Agency and no official endorsement should be inferred.

CONTENTS

VI. STATISTICAL AND SIMULATED MODELING APPROACHES

VII. ECONOMIC CONSIDERATIONS AND POLICY IMPLICATIONS

INTRODUCTION

RAYMOND G. WILHOUR

Environmental Protection Agency, Gulf Breeze, Florida, USA

The Clean Air Act of 1970 and related amendments established the legislative mandate and defined the process by which National Ambient Air Quality Standards (NAAQS) are established for the purpose of protecting human health (primary standard) and public welfare (secondary standard) against "any known or anticipated adverse effects" from criteria air pollutants. Public welfare refers to all non-health effects and the criteria pollutants include ozone (O_3), carbon monoxide, particulate matter, sulfur dioxide (SO_2), nitrogen dioxide (NO_2), hydrocarbons, and lead.

The Environmental Protection Agency's (EPA) process for setting NAAQS includes development of Ambient Air Quality Criteria Documents for each criteria pollutant. The documents summarize and interpret the state of science concerning, among other subjects, effects of a particular criteria pollutant (e.g., O_3) on agricultural crops. From this, the Office of Air Quality Planning and Standards (OAQPS) develops a staff paper which describes the key scientific studies, identifies important factors for consideration in establishing a NAAQS, and recommends a range of standards that appear reasonable to protect public welfare (secondary standard), given the current state of scientific knowledge. Following extensive reviews of the document by the public and the Clean Air Science Advisory Committee (CASAC), and after appropriate revisions, the document is provided to the Administrator of EPA for executive decisions regarding a final NAAQS.

The first secondary standards for the above criteria pollutants were promulgated in 1971. In recognition of the need to periodically review the standards as new information becomes available, EPA developed a plan in 1975 to review NAAQS at 5-year intervals, beginning with photochemical oxidants. The review process is very similar to that described above for establishing NAAQS and was initiated with the preparation of the 1978 Air Quality Criteria Document for Ozone and Other Photochemical Oxidants. This review determined that the 1971 standard for photochemical oxidants (1-h average of 0.08 ppm O_3) could not be scientifically supported. This position was taken because scientific information was insufficient to reliably determine a national perspective of the influence of various O_3 exposure scenarios on human welfare. The standard was changed to an O_3 standard and the secondary standard was set to equal the primary standard (1-h average of 0.12 ppm). Likewise, O_3 exposure–response data were not available for commercially important vegetation and data and procedures for extrapolating economically important response data to a regional scale were not developed. Thus, the primary standard was determined to be an acceptable surrogate for the secondary standard. This observation was generally shared by both regulatory officials and the scientific community.

A growing concern for an economic analysis of the effects of air pollutants on human welfare peaked during the 1978 review of the O_3 standard and remains an important concern to this day. The concern was initiated by a series of Presidential Executive Orders to conduct cost impact analyses for all major regulations. Clearly, during the 1978 review, there were insufficient data to estimate, with any reasonable level of certainty, the expected economic consequences of O_3 at ambient, or at alternative concentrations, on the economy.

It is important to note that, while there were limitations, a number of very important conclusions were derived from the air pollution plant effects data base at that time. Air pollutants were capable of impairing physiological functions, causing foliar injury, and reducing growth and yield of vegetation at ambient concentrations. Nationally, O_3 was the pollutant likely to have the most serious effect on vegetation and economic losses were probably greatest on agricultural crops. Species and varieties/cultivars of vegetation varied greatly in sensitivity to air pollutants, and phenological and environmental factors affected the nature and magnitude of response.

The concern for developing credible estimates of the effects of air pollutants on public welfare (agriculture), within a reasonably short period of time (5 years), led EPA managers and scientists to examine various options for coordinating a highly focused research program among EPA, other federal agencies such as the US Department of Agriculture (USDA), and private institutions such as the Electric Power Research Institute (EPRI). From these interactions emerged the concept of the National Crop Loss Assessment Network (NCLAN).

The original, overall objective around which NCLAN was formulated was the development of credible economic assessments of the effects of the air pollutants O_3, SO_2, and NO_2, alone and in combination, on agriculture. Prior to initiation of NCLAN, it became clear that the original intent to obtain major resources from sources other than EPA and USDA, especially for SO_2 and NO_2 studies, was not viable. Additionally, to meet EPA's priority in welfare effects studies, it was necessary to have a credible evaluation of the effects of O_3 on the agricultural economy by 1986, in time for the second review of the NAAQS for O_3. Thus, the principal research objective for the NCLAN program was the development of a credible evaluation of the economic effects of ambient O_3 on US agriculture.

The national assessment perspective required that the NCLAN research program focus its attention on major agricultural crops and major agricultural producing regions of the US. The economy of funding and timing of a final assessment required that scientists with appropriate expertise and research facilities be incorporated into the research program. Further, since research success is better assured when participants are committed to the research objectives and approach, it was decided to organize a Research Management Committee (RMC) that would develop a research plan as its first responsibility. The original RMC members were selected with two criteria in mind: (1) experience in conducting air pollution studies in the field and (2) location in one of five defined major agricultural regions. The following investigators and regions were represented in the original RMC: Walter W. Heck (Chairman, Southeast), Leonard H. Weinstein (Northeast), Joseph E. Miller (Central States), O. Clifton Taylor (Co-Chairman, Southwest), and Eric M. Preston (later David T. Tingey) (Northwest).

During the initial planning/organizational phase, it became apparent that three important supporting functions were needed to meet the goals of the NCLAN program: economic assessment, air quality support and statistical consulting. Development and use of economic procedures to translate farm-level crop loss data to estimate the economic effects of O_3 on the national economy were the main goals of the economic assessment component. Definition of O_3 air quality in agricultural areas of the US was the major goal of the air quality component. Guidance in the development of statistically reliable O_3 dose–plant yield response models was the primary role of the statistical consultant. These supporting functions were led by Richard M. Adams (Oregon State University), James Reagan (EPA), and John O. Rawlings (North Carolina State University). Adams became a full member of the RMC and Rawlings became the official statistical consultant to NCLAN and the RMC.

The RMC was tasked with preparing a 5-year research strategy for defining the economic effects of ambient O_3 concentrations on the agricultural productivity of the US and a detailed research plan for the first year of NCLAN. A number of research approaches were debated and an empirical approach emerged as most likely to meet the objectives within the time and budget limitations. This approach required that (1) O_3 exposure–response functions be developed for major crop species and cultivars, (2) crops be grown under normal agricultural practices and exposed to O_3 in regions where they are economically important, and (3) common experimental methodologies and quality assurance procedures be employed at all research sites in all regions.

The above requirements were used by the RMC as guiding principles in developing the 5-year NCLAN research strategy and the detailed plans for the

Fiscal Year 1980 research activities. The strategy and first year plans were reviewed by representatives of EPA's OAQPS. This office is responsible for recommending to the EPA Administrator (1) whether or not NAAQS require modification and (2) alternative forms of the standard. This review was to confirm that the research objectives and outputs were compatible with the policy concerns of EPA. An expert peer scientific review panel was constituted and was charged with reviewing the validity and efficiency of the scientific approach in meeting the stated research objectives. An annual cycle of planning and peer scientific review was instituted to confirm that the research objectives and accomplishments were relevant to policy needs of EPA and that the research approach was scientifically sound.

Frequent and open communication both within the NCLAN program and the scientific community was accomplished by (1) frequent conference calls between the NCLAN principals, (2) quarterly NCLAN newsletters defining status and progress, (3) annual planning meetings and peer scientific reviews, (4) annual NCLAN progress reports, (5) scientific articles, and (6) special sessions at national scientific conferences (e.g., Air Pollution Workshop). Internal communications were important for research coordination but the two-way exchange between the NCLAN program and the scientific community was also very important. This communication drew upon the experiences and knowledge of a broad base of scientists to guide the direction of the NCLAN program. In turn, the experiences and findings of this national research program (NCLAN) were used to guide the direction of the much larger and broader international research program on the effects of air pollutants on agriculture.

Following a series of reviews including administrative and scientific, the NCLAN research program was undertaken in 1980. A series of regionally distributed field studies were initiated with the objective of defining O_3 exposure–yield response functions for major agricultural crops grown in major agricultural areas of the US. These studies constituted the backbone of the NCLAN research program; however, special exposure–response studies (e.g., genotypic variation, 7- versus 12-h annual seasonal mean, peak versus mean O_3 concentrations, moisture and SO_2 interactions, etc.), physiological investigations, and mechanistic simulation modeling were also conducted for use in interpreting and extrapolating of the empirically generated exposure–response functions. These latter studies aided in estimating the level of confidence associated with regional and national yield estimates and economic losses associated with alternative O_3 air quality scenarios. The results of these studies constitute the most comprehensive exposure–response data base ever produced by a research program on the effect of air pollutants on agricultural crops.

A credible regional or national evaluation of crop losses expected from ambient O_3 and selected air quality scenarios requires a reasonable estimate of ambient O_3 conditions prevalent in agricultural areas of the US. Since these air quality data were not available, NCLAN pioneered the use of kriging (a spatial statistical methodology) to estimate ambient O_3 conditions over broad agricultural regions of the US. The procedure was used to estimate average 7- and 12-h seasonal mean O_3 concentrations on a statewide basis during 1981–83, the base period for the revised NCLAN economic assessment.

The O_3 exposure–response functions plus the O_3 exposure data provided the necessary information for conducting a credible regional/national assessment of the effects of O_3 on agriculture. The last major accomplishment of the NCLAN program was the incorporation of small scale biological models (exposure–response functions) into a large scale economic model to calculate the national effects of alternative O_3 levels on crop productivity and associated producer and consumer costs. Using this economic model, the final NCLAN economic evaluation estimated the benefits of a 10, 25 or 40% reduction of O_3 to be 0·8, 1·9 or 2·8 billion dollars, respectively (in 1982 dollars).

The NCLAN program was a highly successful research venture attaining every research goal that was central to the program. The nature and quality of scientific accomplishments were viewed as outstanding by the scientific community, federal research managers and policy makers who provided financial support to the program. The program should be viewed as a model in planning, management, and integration of very complex environmental research issues. The success of the NCLAN program is a tribute to the dedication, absolute cooperation (even in difficult budget situations) and hard work of the program managers, scientists, and technicians of the NCLAN program. The establishment of clear research objectives that persisted throughout the program, a relatively firm but long-term funding obligation by EPA and USDA, and a relatively consistent group of participating research managers, scientists and technicians were cementing agents that added stability to the research program.

For every noteworthy accomplishment, there will always be those that applaud and those that criticize the achievement. With respect to NCLAN, both opinions share a common element: the program was successful in producing a credible assessment of the economic effects of O_3 on agriculture but additional research is needed to define and reduce the uncertainty surrounding such assessments. Major sources of uncertainty include the effects of genetic composition, developmental stage, environmental factors, associated pollutants, and exposure dynamics on crop response to O_3. Better estimates of O_3 exposure statistics prevalent in agricultural regions of the US would also improve crop loss estimates. Finally, refinements to economic assessment procedures would probably better translate farm-level crop losses to economic consequences at a national scale.

The goals of the International Conference on Assessment of Crop Loss from Air Pollutants were to (1) describe the NCLAN program and present the final results, (2) summarize the current state-of-science in assessing the effects of air pollutants on crops, and (3) present an international perspective on research needed to reduce the uncertainties in future assessments.

Session I

THE NEED FOR
CROP LOSS ASSESSMENT

(Richard M. Adams, *Chairman*)

1

CROP ASSESSMENT: INTERNATIONAL NEEDS AND OPPORTUNITIES

DONALD A. HOLT

Illinois Agricultural Experiment Station, Urbana, Illinois, USA

1.1 INTRODUCTION

In biblical times, prophets predicted crop status and recommended measures by which famine could be averted. The economic, social, political and environmental importance of crop production still dictate the need for large area crop assessment systems. To the extent that levels of crop production can be predicted with some degree of accuracy, appropriate actions can be taken by individuals and groups to buy and sell, to replenish or reduce reserves, to redistribute supplies to areas of shortage, to utilize alternative crop products for specific purposes, and to negotiate appropriate prices.

In the past few decades, crop assessment has become a tool of policymakers and a basis for regulatory activity. The National Crop Loss Assessment Network (NCLAN) project was a pioneering effort to assess the large-scale effects of air pollutants on US crop production. The objectives were to obtain information upon which useful cost-benefit analyses could be based, to conduct such analyses, and thus, to provide a basis for sound regulatory decisions. The project accomplished these objectives and became a prototype for programs required to accomplish these objectives in the future.

In this chapter, I will present a conceptual framework for large area crop production forecasting to show the relationships between assessing large-scale air quality effects and assessing other large-scale effects on crop performance. I will then discuss the NCLAN project as a prototype system for generating the information needed in assessing air quality and other effects on crops. Next, the need to continue and improve air quality research will be discussed. Finally, I will describe a sophisticated, institutionalized, decision support system for air quality regulatory activities.

1.2 LARGE AREA PRODUCTION FORECASTING

Since air quality is only one of many factors that affect crop production, and since air quality interacts with other factors to influence crop quantity and quality, air quality effects should be assessed in the context of an overall crop assessment system. In order to provide useful information, such a system should depict the effects and interactions of all major determinants of crop yield and quality, including air quality. Of course, such a system would be enormously complex. In practice, crop assessors make simplifying assumptions and considerable extrapolations, and learn to live with much uncertainty.

Most modern large area crop assessment systems have been developed to support the grain trade. With the emergence of the global agricultural economy, it has become economically advantageous to grain traders to gain as much information as possible about crop status over the entire world. The sums of money exchanged in grain trading are very large, and there is a tremendous demand for information on the current and future status of grain crops as they are growing in the field.

To obtain this information, market participants rely on reporting services, such as those provided by the United States Department of Agriculture-Economic Statistics and Cooperative Service (formerly the Statistical Reporting Service) and other similar government agencies at national and state or provincial levels, and by private firms or individuals who gather such information and sell it or use it in internal operations.

The methods used to gather grain production information include extensive mail surveys of farmers; observations and measurements by trained enumerators; mathematical predictions based on weather information; subjective, usually unofficial, "roadside" surveys by trained or untrained observers; and, in some cases, highly subjective techniques, based on astrological or other metaphysical phenomena. The general philosophy seems to be "some information is better than none".

1.3 A CONCEPTUAL MODEL FOR LARGE AREA CROP ASSESSMENT

In the course of a research program sponsored by Control Data Corporation (CDC), my former colleagues at Purdue University and I developed a large area crop assessment system based on weather-driven simulation models. The system was implemented by CDC and its results were marketed as one of CDC's AGSERV services. The users of this service were primarily large grain-trading or other agribusiness firms. Some of the information generated in this project was used to assess the effect of the interaction of O_3 and moisture stress on soybean production (King and Nelson, 1987).

Drawing upon experience gained in the CDC project, we conceptualized a general large area crop assessment system. I find it useful to think about assessing air quality effects in the larger context of assessing environmental and other effects on crop production.

1.3.1 Characteristics desired of large area production forecasting models

Our objective during the Purdue effort was to develop a modeling approach to large area production forecasting that would have the following characteristics:

1. The functional relations between causal factors, such as weather and yield, should be broadly applicable, e.g., the coefficients should be the same for a given crop; at least over areas where similar cultivars are grown.
2. The influences of the inherent productivity of specific areas and other causal factors on crop yields should be clearly and logically separated in the model, so that they may be evaluated separately.
3. The validation of the model should not require extensive historical data from each area for which predictions will be made.

1.3.2 Theoretical background

The general theory is that there is a certain amount of energy that could be fixed into economic biomass, but various factors cause this amount of energy to be reduced, so that the amount actually fixed is much less than that which impinges on the earth's atmosphere. Within this framework, the estimated production of a crop over a given area, hereinafter called a cell, can be computed using the following equations:

$$P = A \times Y, \quad \text{and}$$

$$Y = M \times W \times E \times B \times Q, \quad \text{where}$$

P = production of the crop (economic biomass) in a specific cell (units of weight or volume);

A = land area on which the crop is growing within the cell (units of area);

Y = yield of the crop per unit land area (units of weight or volume per unit area);

M = maximum average yield per unit land area that would be achieved in a cell if no limitations were imposed by weather, pests, economic conditions, or air quality (units of weight or volume per unit area);

W = weather factor, depicting the limiting effect of weather;

E = pest factor, depicting the limiting effects of weeds, diseases, and insects;

B = management factor, depicting large-scale economic limitations on crop yield; and

Q = air quality factor, depicting crop yield limitations imposed by air pollutants.

Factors W, E, B, and Q are dimensionless variables varying from 0·0 to 1·0. Of course, each of these major yield-limiting factors would have to be evaluated with a detailed algorithm, model, or group of models. Such models could take many forms, but would have to generate final values varying from 0·0 to 1·0, so that the degree of yield limitation caused by each factor would be reflected in the final yield and production estimates.

1.3.3 The concept of maximum yield

Maximum yield (M) is a key concept in this approach to production forecasting. It represents the average yield per unit area that would be obtained in a cell if weather, pests, and management were optimal for the production of that crop throughout a particular season, and if there were no pollutant effects. Differences in M between cells are due to differences in inherent productivity of the land in the cells. Inherent productivity, by this definition, is a function of soil characteristics, topography, germplasm utilization, and the average level of management that is employed in the cell.

If historical data on yields, weather conditions, pests, economic conditions, and air quality are available for a cell, the M for each cell can be evaluated as follows:

$$M = Y/(W \times E \times B \times Q)$$

By definition, M is a theoretically possible yield, but is practically unattainable. It should be a very useful number by which to represent and compare the inherent productivity of cells because, by definition, it is independent of year-to-year variations in weather, pests, short-term management decisions, and air quality. It should be expected to change only gradually with progressive deterioration in or improvement of soils in the cell, changes in the productivity of the crop germplasm utilized, and/or improvement in the average level of technology employed. Only catastrophic natural phenomena or revolutionary, immediately adopted technological developments should change M drastically.

Trends in M are assumed to be approximately linear over periods of several years. The long-term trends in M should be more consistent than historic trends in average yields, because the latter are influenced by weather cycles of perhaps 10 to 20 years and perhaps by other long-period phenomena.

M has some additional characteristics that should be noted. Theoretically, if W, E, B, and Q are evaluated with accurate deterministic models, the M of a particular cell could be estimated from one year of historical weather data from that cell. Two years would be required to estimate the long-term trend in M. Since there are inevitably some weaknesses in the various models and errors in the data bases, it is most likely necessary to evaluate M from more than 2 years of data, but not from the 30 or so that are necessary for the regression approach.

M will be relatively stable and increase gradually over a period of years if the models generating W, E, B, and Q are accurate and complete. This assumption rests on the likelihood that crop genetic potential will be continually improved. Thus, the stability of M serves as a test of the system. It is interesting to note that a graph of M over time has a steeper slope than a graph of yield over time due to the multiplicative and relative nature of the yield expression. If our theoretical development of this yield expression is correct, potential crop yields are increasing more rapidly than actual yields (Fig. 1.1). On the surface at least, this would seem to be a very important deduction, although we have not yet thought through its implications.

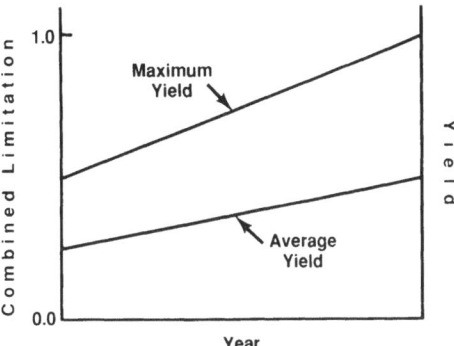

Fig. 1.1. Maximum yield, as defined for the proposed large area production forecasting system, increases more rapidly over time than expected average yield.

It is also interesting to note that M will be underestimated if an important term is omitted from any of the models generating W, E, B, or Q, but if the missing factor does not vary much between years, yield estimates will still be accurate. In fact, in our application of this theoretical approach to an actual production forecasting system, we found that useful production estimates could be made using only weather information, without actually evaluating E, B or Q. Of course, if the model is to be used to evaluate the effects of air quality on crops, there must be some way to depict those effects in the model.

1.3.4 Other major factors affecting yield

The weather factor (W), within this conceptual framework, represents the integrated, quantitative effect on yield of weather conditions that prevailed during an entire growing season. W is generated by mathematical expressions depicting the effects of moisture, temperature, photoperiod, solar radiation, and weather episodes, such as hail or high winds, on estimated yield.

While we believe that physiology-based simulation models will be most appropriate for estimating W, it is conceivable that W could be computed with regression models. Equations for calculating similar weather indices have been obtained in the past by regression analysis (Shaw and Durost, 1965). Weather catastrophes that totally destroy crops are not included in W, because they affect production estimates by reducing harvested area.

The pest factor (E) represents, by our definition, the limiting influence of diseases, insects, and weeds on the crop. E presumably can be estimated by utilizing survey information and remote sensing in conjunction with appropriate quantitative models or equations.

The management factor (B) depicts the limitations placed on yields by suboptimal fertilization, cultural practices, etc. This factor, in our system, contributes to variations in yield-influencing management between successive seasons. Such variations might be due to economic conditions or to government programs. For example, unusually high fertilizer prices might cause a general reduction in fertilizer use in a particular year, thus reducing yields. A distinction must be made between the year-to-year variations in management due to B and the long-term trends in technology adaptation, which are accounted for by projecting a trend in M.

1.3.5 Depicting the effect of "normal weather"

Our theoretical approach should eliminate most of the need for "normal" weather files in production forecasting. M can be evaluated using records of actual weather, pests, etc. When information on pests, management, and air quality is lacking, the yield equation can be reduced to

$$Y = M \times W$$

Using $\bar{Y}$ (the historic average yield for a cell, and a number that may be easier to obtain than historic weather data) and $\bar{M}$ (the average maximum yield for a cell, computed using data from the M trend line), $\bar{W}$ (the average weather factor) can be evaluated easily by this simple manipulation:

$$\bar{W} = \bar{Y}/\bar{M}$$

Thus, one number ($\bar{W}$) can be used to represent the average limiting effect of weather on crop yield in that cell, averaged over all years for which there are historic yield data. To obtain this value by running the weather-driven portions of crop models using historic daily weather data would be virtually impossible for many cells, and would be a costly, time-consuming undertaking, at best.

Once $\bar{W}$ is computed for a cell, it can be used in place of an entire "normal" daily weather file. After initializing W at $\bar{W}$, models can be run to the current date with actual weather data, and the seasonal weather factor ($\hat{W}$) obtained by weighted averaging of W and $\bar{W}$. W should be weighted in proportion to the fraction of total, already accumulated, heat units required for the crop to mature. This is tremendously less complicated than the current procedure and requires much less computer time. For all practical purposes, it means that the models need be run only once to provide sequential yield predictions for a cell throughout a growing season.

1.3.6 Using the model

We believe that as experience is gained in using this theoretical approach to production forecasting, the yield expression will become a very important analytic tool. It can be used in either a predictive or retrospective mode. For example, one could average M, W, E, B and Q of a given cell over several years and find out which factor was most limiting in that cell. A comparison of these limitations between counties, states and nations would be extremely interesting and useful in determining research and extension priorities. The retrospective use of the model would be particularly useful for estimating the crop losses caused by air pollutants.

The system described here would presumably be operated by constructing a two-dimensional array of values for M and for each dimensionless factor. Each value in each array would represent the situation existing at one point on a map (pixel) of M or of a specific factor. The number of pixels required per unit land area would be determined by the degree of geographical resolution required to depict any factor accurately. The geographical data bases generated in this manner would be extremely valuable for cost-benefit analyses of air quality effects and related regulatory changes.

Spatial and temporal scales are important considerations in any modeling effort (LeDuc and Holt, 1988). It will be neither necessary nor possible to achieve the same degree of spatial or temporal resolution for each factor (Fig. 1.2). The weather factor, for example, could probably be mapped with greater accuracy than the air-quality factor, because weather stations are distributed more densely in rural areas than are air-sampling sites. The values of some factors would be generated each day by appropriate models. Other factors, such as the economic factor, would be evaluated less frequently, perhaps only once per season, depending on the frequency of surveys or other means by which information on driving variables is collected.

To generate a forecast or after-the-fact estimate, a seasonal value for each pixel would be obtained by averaging the daily or less frequent values for that pixel over the season; the maps (data arrays) would be overlaid; and the overall multiplicative expression would be evaluated for each pixel on the combined map, thus generating an estimated yield for each pixel (Fig. 1.3). The yield estimates would then be aggregated to provide production estimates for each cell.

This entire system can be computerized, using models driven primarily by weather and air quality data, for which a large, rapid collection system exists. Thus, it is technically feasible to produce frequent large area crop production forecasts, rapidly assessing the potential effects of large scale weather events and other changes in driving variables.

The yield predictions will improve as the season progresses and the models

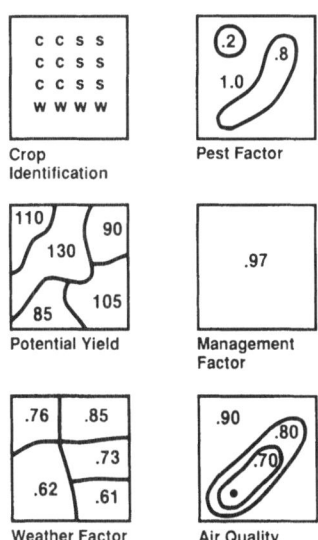

Crop
Identification

Pest Factor

Potential Yield

Management
Factor

Weather Factor

Air Quality
Factor

Fig. 1.2. The pattern and degree of resolution on maps of various factors used to predict yield and production in a large area production forecasting system usually vary with the density of measuring sites and the nature of the underlying phenomena.

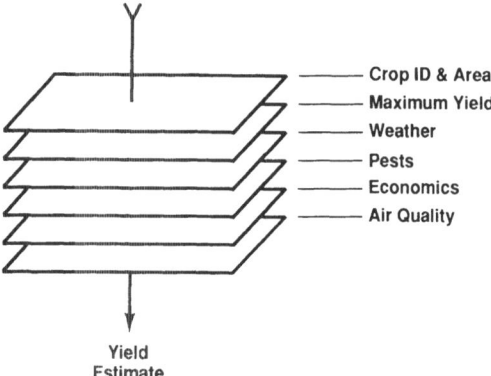

Yield
Estimate

Fig. 1.3. Maps of factors used to predict yield are overlaid, a computation is made for each pixel, and the resulting yield estimates are aggregated to a large area production forecast.

will depict the effects of more actual and less "normal" weather. The forecast information can be distributed, via computer or other networks, to all interested parties in a timely manner. Among other advantages, this rapid dissemination of crop status information can tend to equalize the commodity bargaining positions of all market participants, both large and small. The retrospective information generated by the models should also be made available over computerized telecommunication networks.

It is important to keep in mind that models are theoretical representations of the function of real systems. If a model (theory) is to be useful for predicting future events, the existence and detailed nature of the functional relationships in the model must be confirmed by experimentation. NCLAN was a project that evolved to develop a model, provide functional relationships for the model, estimate the parameters of the functional relationships, and refine the estimates of these parameters, through field and laboratory experimentation. Thus, NCLAN is a prototype of a research system developed to support an institutionalized, national crop assessment system.

1.4 A NATIONAL CROP LOSS ASSESSMENT SYSTEM

It is now technically feasible to create a nationally coordinated, model-driven, large area crop assessment system. The system should take the form of an expert system, designed to answer the question, "What quantity and quality of crops will be or have been produced this year in a specific area"? Of course, a number of other more specific questions could be answered by such a system. A large area crop assessment capability would greatly facilitate a decision-support system for air quality regulation and for many other federal, state, and private activities (Fig. 1.4). General specifications of the system should be as

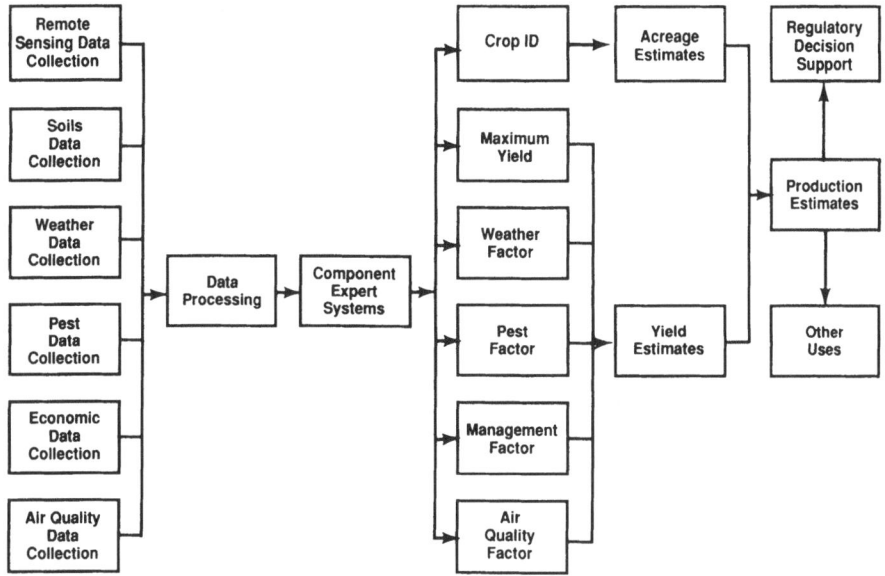

Fig. 1.4. Data collection and processing capabilities are linked to large area production forecasting, regulatory decision support, and other uses through component expert systems within which the models are imbedded.

follows:

1. The system should be supported by an ongoing research and development program, which should be considered an integral part of the crop assessment program.
2. The models involved should depict the effects and interactions of all major factors affecting crop yield and quality.
3. The expert system should be modular, such that geographically separated but well-coordinated groups can develop, operate, maintain, and refine different model components.
4. The modules and submodules should be interfaced to permit information exchange and aggregate analyses.
5. Each module should encompass a major factor affecting crop yield and quality, such as air quality.
6. Each module should take the form of an expert system, or a package of expert systems, so that it can include both qualitative and quantitative information and utilize the subjective knowledge of experts.
7. The groups responsible for developing, operating, refining, and maintaining the system, as well as the users of the system, should be linked by an effective, computerized telecommunications network that permits ready exchange of data and messages among computers.

1.5   A DECISION SUPPORT SYSTEM FOR
AIR QUALITY REGULATION

1.5.1   NCLAN as a prototype

The NCLAN project is an excellent prototype for a decision support system underlying an air quality regulatory program. Its principal objective was to develop a base of information for air quality regulatory decisions. By accomplishing this, assessment of crop loss in the United States due to air pollutants was enabled on a national scale.

The essential characteristics of the NCLAN project included the following.

1. It involved a number of well-equipped and well-supported centers, where field and laboratory research was conducted, plus mechanisms for funding specific, related research projects at other locations. The centers were located in and were representative of major agricultural regions of the United States and provided data for the modeling effort.
2. It involved a modeling effort, through which a number of biological, statistical and economic models were developed or adapted, tested, refined and integrated into an overall model by which producer and consumer losses due to the effect of air pollutants on crops could be estimated on a national scale.
3. It had a well-organized coordinating system, consisting of a project management committee that was broadly representative of the scientists involved in the project. The project management committee, along with an annual, intensive, week-long workshop involving all participants, served to network the participants effectively and resulted in a project that was extremely well-coordinated nationally, yet benefited from specialized, decentralized management.
4. There were formalized and standardized quality control mechanisms, including persons with direct responsibility for devising, implementing, monitoring and continuously reviewing instrument calibration procedures and other research protocols, and a peer-review panel for annual evaluation of the project.

1.5.2   Major accomplishments of NCLAN

NCLAN researchers faced a difficult research challenge, namely, that of designing and conducting meaningful experiments in the field environment that involve air quality factors. Open top chambers equipped with elaborate apparatus for infusing, monitoring and controlling the levels of various air constituents enabled the researchers to meet this challenge. Researchers at the various locations rigorously adhered to mutually agreed upon quality control protocols, thus greatly increasing the value and broad applicability of the information generated in the experiments.

Experiments were carefully designed to provide the most sensitive statistical tests that could be achieved within resource constraints. The result was a number of statistically valid biological dose–response functions representing the anticipated response of crops to various levels of air pollutants under several different soil and climatic regimes. NCLAN researchers obtained some information on the interactions among pollutants, and particularly, on the interactions between pollutants and water stress. This information was used in the national assessments.

One of the major accomplishments of the NCLAN project was successfully integrating small-scale biological models (dose–response functions) into a large-scale economic model, thus enabling a meaningful national assessment of the impact of air pollutants on crops and on associated producer and consumer costs. The sensitivity of the overall model to differences in the dose–response functions and to the presence or absence of information regarding the pollutant/water stress interaction was measured.

The sensitivity analyses provided information that was valuable not only for regulatory decisions, but also for planning ongoing research. For example, one of the sensitivity analyses revealed that the overall cost assessment was more sensitive to uncertainty about O_3/water stress interactions than to uncertainty about differences among crop cultivars regarding their tolerance of O_3 stress (Adams and McCarl, 1985). As a result, the limited resources were focused on studies of the O_3/water stress interaction.

1.6 CONTINUING NEEDS FOR AIR QUALITY RESEARCH

1.6.1 Need for more information
There will always be a need for crop-related air quality research. While the NCLAN project and related research efforts provided information on which to base a preliminary national assessment of pollutant-induced crop losses, that information will not adequately support future regulatory activity. The value of the information generated in those efforts will depreciate with time, probably at an accelerating rate.

The differences in pollutant effects caused by differences in daily exposure regime have not been fully characterized. More information is needed regarding the variability among crop cultivars to resist or tolerate effects of air pollutants. While cultivar differences were not as important as the O_3/water stress interaction, according to NCLAN estimates, they were nevertheless important, when averaged over the entire nation. Interactions among pollutants need more study. The complex interactions between air quality and many other factors affecting crop yield and quality, including weather, insect, and disease stress, are not well understood.

The nature of the interactions among air pollutants and insect, disease, and weed pests of crops will change, as the pests change in response to natural selection, pesticides, changes in production systems, and other causes. These interactions could be of great importance, particularly in perennial crops such as forests and forage crops.

New crop cultivars with different levels of resistance to and/or tolerance of pollutants will be developed. Biotechnology will hasten this process. New crop production techniques will interact differently with pollutants. In parallel with changes in technology and regulations, there will be changes in the nature and levels of pollutants generated by industry, automobiles and other sources.

1.6.2 Changes in research techniques, simulation techniques, and computers

There already have been, and will continue to be, changes in air quality research techniques. There are increased efforts to develop more advanced biosensors, by which the physiological status of crops can be more easily and accurately measured. Some of these will permit remote sensing of biological variables, thus enabling less intrusive research procedures. The ability to sense and record biological phenomena on a larger scale and in greater detail will be very important in future air quality research.

Among the most important changes in air quality research are new techniques for dealing with the great complexity of models needed to depict the function of the crop–soil–air system accurately. The crop–soil–air system is multidimensional. When this system is studied as a component of an overall socioeconomic system, the complexity increases considerably.

Attempts to simulate the behavior of the system are necessary if concepts of system behavior are to be evaluated fully and accurately. The necessary simulation models, even those that depict the function of only portions of the system, are extremely complex and computationally intensive. They suffer what some have described as the "curse of dimensionality".

Current NCLAN models do not depict changes in plant susceptibility to pollutant exposure that are associated with advancing maturity. Endowing the models with this additional realism will require generating separate dose–response functions for each stage of maturity and programming models to track stages of maturity as the season advances. This will add another dimension to models that are already quite complex.

The addition of each new factor known to influence the system, plus its interactions with other factors, adds dimensions to the model. If a model is to depict the function of the real system accurately, it must be realistically complex, or at least approach the complexity of the real system. Each developmental step toward that level of complexity increases the computational requirements of the model.

At best, simulation models will not predict system function with complete accuracy. Inevitably, inadequate information on, or errors in, measuring or estimating driving variables, extraneous effects not depicted in the model, and/or errors in estimating parameters in the model will induce errors in the predictions. To be useful, the models should reflect this uncertainty. Thus, at least some of the variables and coefficients should be entered as probability distributions, rather than as single values, thus greatly increasing computational requirements.

The technique of stochastic-dynamic programming, coupled with the enormous computing power of supercomputers, is a way of dealing with the dimensionality problem (Taylor, 1987). Crop loss projections generated by stochastic-dynamic models are accompanied by estimates of the uncertainty associated with the estimates. These uncertainty measurements will provide the basis for risk assessment and help to ensure that the potential effects of regulatory or other changes are fully explored before implementation. In addition to serving as a basis for risk management, stochastic-dynamic models should permit estimates of how much model complexity and prediction accuracy is sufficient to provide adequate information for regulatory or other decisions.

Interpretation of model output will be greatly facilitated by innovative use of the very sophisticated graphics capabilities that can now be interfaced with and implemented by supercomputers. Graphics have been used by Onstad and Ruesink (1986) to depict the ecological relationship among corn plants growing in typical corn monoculture, corn borers attacking and infesting these plants, and a protozoan pest of corn borers that is infecting the insects.

A detailed simulation model of this ecological relationship generates millions of numbers per second. The numbers are color-coded and displayed graphically in a remarkable visualization that simulates the phenomena. Variables representing the status of each species involved are represented as colors, the intensity of which varies in proportion to the magnitude of the variable. Mixtures of the colors, which are carefully chosen so that the mixtures are readily distinguished from any of the pure colors, depict the interactions among the organisms.

One can envision a model to predict pollutant concentrations over large areas, in which model output is represented as colors on maps of the areas in question. Concentrations of pollutants and degree of pollutant damage to crops could be coded as different colors or as different color intensities. Various interactions could be represented by colors or mixtures of colors. Experience suggests that this capacity to visualize changes propagated by complex biological and ecological processes provides a degree of understanding that is not otherwise attainable.

1.6.3 Linking regulatory risks to the design of field experiments

We need to design field experiments with use of the information they generate in mind. Carmer and Walker (1988) published a landmark article in the *Journal of Production Agriculture,* a new publication of the American Society of Agronomy. In this article, the authors outline a procedure for improving the design of field experiments and selecting optimum significance levels for statistical tests, based on estimates of the economic, social, political and/or environmental consequences of making type 1, type 2, or type 3 errors of inference.

The approach suggested by Carmer and Walker addresses relationships among the standardized true differences that can be detected in an experiment; the probabilities of Type 1, Type 2, and Type 3 errors; and the relative losses and risks associated with the errors. Since the magnitude of the standardized true difference that can be detected is at least in part a function of the design and conduct of the experiment, researchers are provided with a tool for tailoring their field experiments and their statistical tests to the large-scale consequences of failing to draw correct inferences from the experiment.

While Carmer and Walker discussed their approach in the context of individual experiments and individual farmer decisions, it seems likely that the associated philosophy and procedures can be extended to larger scale situations involving several related experiments, large agricultural areas, and far-reaching regulatory decisions. Their approach would add a new dimension of decision support for regulatory activity or other large-scale decision making. The following brief explanation of the proposed approach to selecting significance levels is meant only to provide the reader with some feel for the potential of this approach.

Briefly, the procedure consists of:

1. estimating the relative loss associated with making a wrong decision (Types 1 and 3 errors) versus failing to detect a true difference (Type 2 error);
2. deciding how the relative loss varies with changes in the magnitude of the standardized true difference that can be detected in the experiment, i.e., establishing a relative loss curve;
3. combining knowledge of the three error rates with the relative loss curve to obtain a relative risk function;
4. computing the weighted-average risk of making the three types of errors for all the comparisons in the experiment and for varying levels of the standardized true difference to be detected; and
5. selecting a significance level for statistical tests that minimizes this weighted-average risk.

When weighted-average risk is plotted against significance level for various levels of relative risk, a family of curves is generated with minimums at different levels of significance. By visually observing such graphs or by computation, one can select an optimum level of significance for statistical tests in particular experiments. The optimum level is that at which the weighted-average risk of economic, social, political and/or environmental losses associated with Type 1, Type 2, and Type 3 errors is minimized. Carmer and Walker describe situations in which the often arbitrarily chosen Type 1 error probability levels of 0·05 and 0·01 for tests of significance are not appropriate.

The approach articulated by Carmer and Walker provides a way to link large-scale air quality and other cost-benefit analyses more directly to underlying field research (Fig. 1.5). To realize the full potential of this approach, researchers must continue to improve the ways in which relative losses are evaluated. Large-scale models of the type employed in the NCLAN project should be very useful tools for this purpose. The stochastic-dynamic programming models of the future will be even more useful for evaluating the risks associated with inaccurate information. Establishing the appropriate relative loss curve, which is probably not linear in every situation, will be a challenging task.

The Carmer and Walker approach to risk management in field experiments

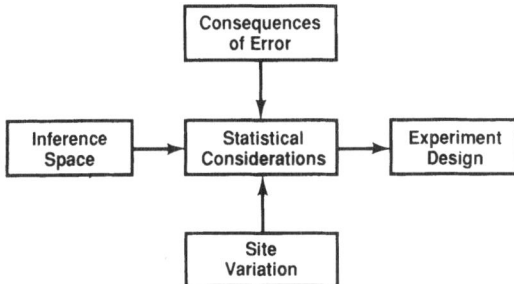

Fig. 1.5. Experiment designs should reflect natural variation within the site of the experiment, variation within the desired inference space, and the economic, environmental, social, and/or political consequences of making Type 1, Type 2 and Type 3 errors.

also provides a way to assess the adequacy of specific experiment designs, layouts, and procedures for providing sound information in support of specific regulatory decisions. The cost-benefit ratios of the experiments themselves, as components of the total decision support system, can be evaluated. Thus, it should be possible to optimize experiment designs to fit both regulatory goals and budget limitations. This should be particularly valuable to those who solicit and those who allocate funds for field and laboratory research on crop responses to air pollutants. A more systematic and quantitative method of measuring the value of the information generated by such experiments is necessary.

1.7 CREATING THE AIR QUALITY REGULATORY EXPERT SYSTEM

The models developed to support air quality regulatory decisions can be used much more effectively if they are embedded in expert systems (Fig. 1.6). Expert systems are a type of decision-aid computer software that employ artificial intelligence science and technology. Artificial intelligence is the science of programming computers to perform tasks requiring such human

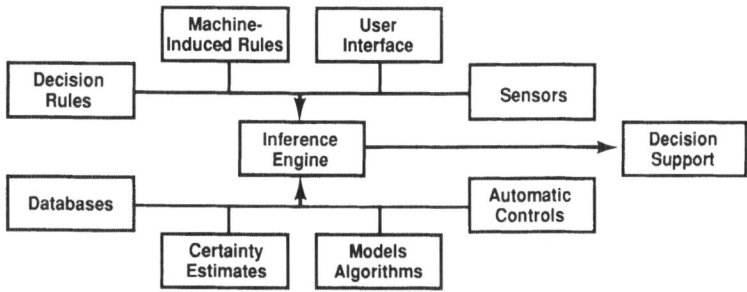

Fig. 1.6. Through the control structure (inference engine) of expert systems, several useful components are linked to decision support.

skills as reasoning, sound judgement, learning from experience, communicating in natural language, and interpreting images. Expert systems incorporate the judgement and experience of experts. When a model or models are embedded in an expert system, the expert system plays two roles. It manages the operation of the models by properly configuring the input data and summarizing and interpreting the output data. It also provides a means by which the more subjective knowledge of experts can be incorporated into the decision-support system.

In its simplest form, an expert system contains the rules of thumb, generalizations, and other decision rules by which experts make decisions in a specific subject matter domain. These rules usually take the form of "IF-THEN" statements. A simple expert system will ordinarily also contain a database of factual information about the subject matter in question. The rules serve as means by which the expert system selects and analyzes data to obtain answers to specific questions and to generate recommendations for action.

More sophisticated expert systems contain mechanisms whereby new rules and generalizations are generated from an analysis of case histories and observations. These inference mechanisms are the real contribution of artificial intelligence to expert systems. At a higher level of sophistication, models or other complex algorithms are embedded in expert systems. The models actually provide an opportunity to incorporate very complex IF-THEN rules into expert systems.

These complex rules take the general form "IF—certain conditions exist at the beginning of a time period, AND IF—certain conditions prevail during the time period, THEN—the condition of the system at the end of the time period is as predicted by the model". Embedding models in expert systems permits users to explore the outcome of a number of very complex scenarios, drawing on the subjective expertise of experts and the quantitative expertise embodied in models.

At a higher level of development, expert systems are connected directly to various sensors, which capture information about the system in question. If the expert system is also connected to and implements action through automatic control mechanisms, the entire decision-making loop is completed. Whether or not this total automation would be of value in air quality regulation remains to be seen. It might, however, be useful in controlling specific sources of air pollutants.

In general, an expert system is a working model of the decision-making process in a specific subject matter domain. Thus it should be possible to develop and maintain an expert system or package of expert systems to answer the specific question: What regulatory action is appropriate in this particular air quality situation? The expert system, incorporating the expertise of scientists and regulatory officials with the quantitative information generated by field experiments and represented in models, should be the core of a decision support system for air quality regulation.

Such an expert system should not be considered a final product. Its development will of necessity be an open-ended process. Just as human experts must continue to accumulate experience and refine concepts, so an expert system must continue to be improved and refined, in response to changing needs for information and to changes in knowledge about the biological system in question. Thus, an expert system must be sustained by an ongoing research and development effort.

1.8 CONCLUSIONS

The decision-support system for air quality regulation should be institutionalized. The system needs continuity and an infrastructure of people, facilities, equipment, cooperative relationships, and accumulated experience that is sustained indefinitely. While the components of this system must evolve with time, the evolution should be carefully managed so as not to disrupt the continuity of research efforts, weaken the infrastructure, or disperse the people, facilities, equipment, and, especially, the accumulated experience that enable the work to be accomplished. Having a well-maintained, decision support infrastructure in place will assure that the air quality knowledge base continues to grow and improve, specific air quality problems can be solved, and individual air quality questions can be answered promptly and efficiently, as they arise.

ACKNOWLEDGEMENTS

The author wishes to acknowledge the essential contributions of C. S. T. Daughtry, S. E. Hollinger, W. L. Nelson, H. F. Reetz and R. Stuff to the conceptual development of the large-area crop assessment system proposed in this paper.

REFERENCES

Adams, R. M. and B. A. McCarl. (1985). Assessing the benefits of alternative oxidant standards on agriculture: the role of response information. *J. Environ. Econ. Managmt*, **12**, 264–76.

Carmer, S. G. and W. M. Walker. (1988). Significance from a statisticians viewpoint. *J. Prod. Agr.* (in press).

King, D. A. and W. L. Nelson. (1987). Assessing the impacts of soil moisture stress on regional soybean yield and its sensitivity to ozone. *Agric. Ecosys. Environ.*, **20**, 23–35.

LeDuc, S. K. and D. A. Holt. (1988). The scale problem: modeling plant yield over time and space. In *Plant growth modeling for resource management, Vol. 1, Current Models and Methods*, ed. by K. Wisiol and J. D. Hesketh. Boca Raton, FL, CRC Press (in press).

Onstad, D. W. and W. G. Ruesink. (1986). Simulation model of the epizootiology of a microsporidium infecting an insect. In *Fundamental and applied aspects of invertebrate pathology*, ed. by R. A. Sampson, J. M. Flak, and D. Peters,

Wageningen, The Netherlands, The Foundation of the Fourth International Colloquium of Invertebrate Pathology, 576–9.

Shaw, L. H. and D. D. Durost. (1965). The effect of weather and technology on corn yields in the corn belt, 1929–62. *Agricultural Economic Report No. 80.* United States Department of Agriculture, Economic Research Service.

Taylor, C. R. (1987). The fading curse of dimensionality. *University of Illinois Agricultural Economics Staff Paper No. 87 E-385.*

2

RESEARCH APPROACHES TO POLLUTANT CROP LOSS FUNCTIONS

D. P. ORMROD, B. A. MARIE, and O. B. ALLEN
University of Guelph, Guelph, Ontario, Canada

2.1 INTRODUCTION

The emphasis of this paper will be on the various approaches to experimental determination of crop loss functions, rather than on the functions that now exist. The crop loss functions currently in use have been thoroughly reviewed by Krupa and Kickert (1987a) in a journal paper and in the form of a report to The Acid Deposition Research Program (Krupa and Kickert, 1987b). Pollutant crop loss functions must be based on credible research, since the research forms the basis for function development. The experimental design, data collection, and statistical analyses all must stand the scrutiny of peer review. There are many alternative research approaches, but only one sequence can be carried out in a particular project. The alternatives must be discussed, considered, and decided upon. The National Crop Loss Assessment Network (NCLAN) project was an outstanding example of this type of process, and the results demonstrated the merit of an orderly approach.

The determination of crop loss functions is at the heart of the current objectives of most air pollution studies. Much work has already been done toward understanding the relationship between plant structure and function and exposure to pollutants in various combinations and durations. Many mechanisms of pollutant injury are well understood. However, the challenge of the broader question still remains: how to best quantify the effect of pollutants on crop yield. A knowledge base addressing this question exists and arises particularly from NCLAN, but also from a host of other completed projects. We recognize both the strengths and the weaknesses of this knowledge base, and there are many aspects of the effect of pollutants on crop yield that deserve further study.

There are three concerns to address in the planning stage of any research project that aspires to estimate crop loss due to air pollutants. The first concern is with the type of questions we are asking. The specific definition of the objectives of a study is critical when choosing the best approach for collecting and interpreting valid data. The second concern involves the design of a study plan to achieve these objectives. This consideration includes the choice of factors and levels for experimentation and the precision required to detect the treatment effects, the experimental design, and the various approaches to analysis and interpretation of the data. The third concern addresses the characteristics of the exposure facilities to be used. This paper will discuss the diversity of exposure systems, both chambered and chamberless. Finally, there will be a recommendation for integrated approaches. Figure 2.1 is a conceptual depiction of the research approaches for determining crop loss functions.

2.2 DEFINITION OF THE QUESTIONS

The research approach to estimating crop loss depends to a great extent on the specific question that is being asked. In many cases, the objectives are not well defined, and a nonfocused approach is taken, one which often does not make the best use of the facilities and the investigator's time. The most fundamental and most common question, one which is almost certainly too broad, is "Does air pollution have an effect on crop yield?" That question has been answered in the affirmative and refined to "What is the magnitude of the effect that we can expect pollutant exposure to have on crop yield?" This question can be approached by many types of both field and laboratory experiments, as well as by epidemiological studies. The approach taken defines the applicability and precision of the generated data. In experimental approaches, the scope of the question is narrowed to focus on a type of crop, growth stage, or set of environmental conditions.

An example of a specific question that is frequently asked is "What effect can we expect pollutant exposure, in conjunction with other environmental stresses, to have on crop yield?" This is critical to the extrapolation of results from experiments to the farmer's field, because pollutant stresses operate in conjunction with environmental stresses in the ambient environment. "Environmental stresses" may be applied in several ways, each approach responding to a different question. Outdoor exposures to ambient air pollutants repeated over time are a kind of unplanned, multifactor, experimental approach to the interaction of pollutant and other environmental stresses. The stresses differ from year to year, but over a long period of time, they approximate the average effect. There are often conflicting results from year to year, because the interacting environmental effects are not constant. The alternative approach is to isolate one type of environmental stress and apply it in a controlled fashion at multiple levels, in conjunction with air pollutant stresses. These two approaches may be thought of as epidemiological and

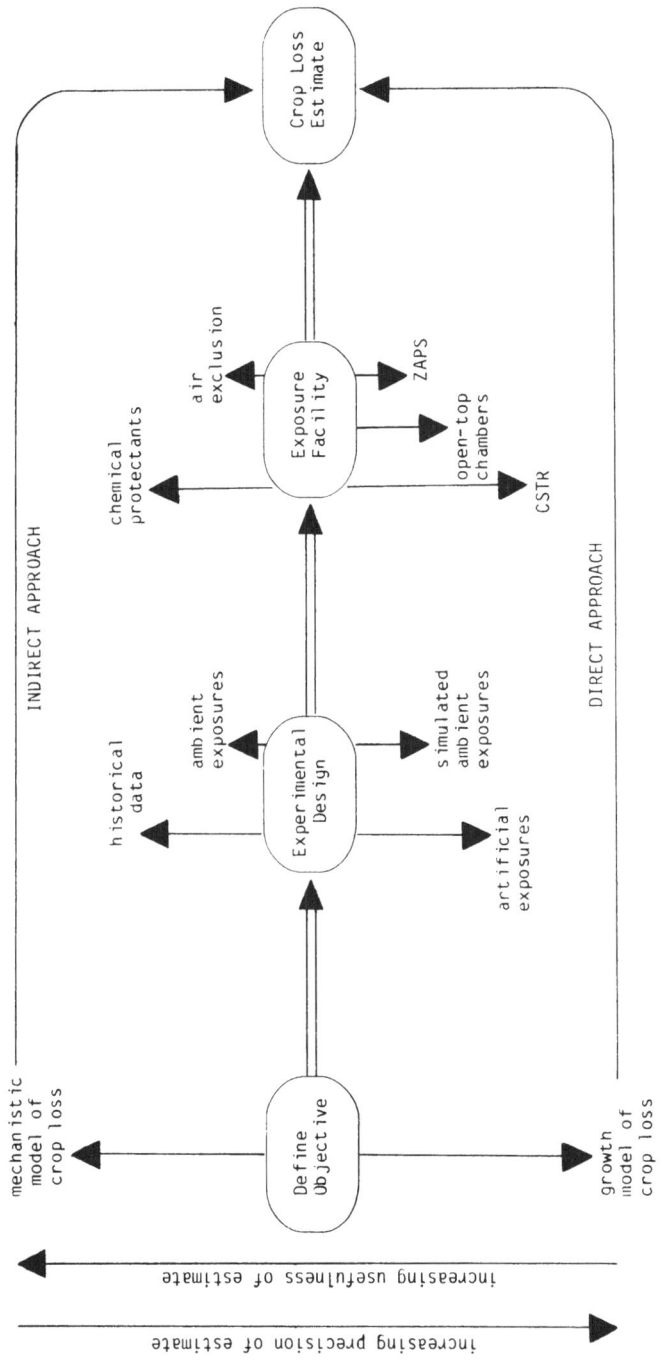

Fig. 2.1. Conceptual research approaches to crop loss assessment.

experimental, respectively. It may be useful to think of two kinds of factors in a designed experiment: (1) those that will be varied, and (2) those that will be held constant. The former certainly should encompass ambient conditions, but may also include extremes (in excess of ambient levels) in order to increase precision (Allen *et al.*, 1988). It is desirable that those factors to be held constant be as close as possible to ambient levels. Of course, it is not possible to examine the interaction of those factors to be varied with those to be held constant.

Another question that should be asked is "How can the knowledge of these responses and interactions be used for predictive or explanatory purposes?" The answer to this question is very complex and relates to how close to ambient the exposure was, as well as how realistic the other environmental conditions were under which the exposure occurred. In the field, the environmental conditions change over time in a complex way, so it is difficult to even characterize these fluctuations, let alone to duplicate them in a laboratory situation. Hence, one is always extrapolating when applying laboratory results to the field. If it is highly desirable that the results be suitable for predictive modeling, then aspects of the experimental design and exposure system must mimic ambient conditions as closely as possible. An exception to this occurs in the selection of the upper limits of an experimental factor; the inclusion of levels in excess of those observed in the ambient environment is desirable to increase experimental precision. The difficulty associated with decreasing the artificial boundaries of an experimental setup is usually a corresponding decrease in precision. In many cases, the changes in growth under ambient conditions are subtle, and decreased precision reduces the probability of detecting treatment effects. The experimental protocol must be designed to minimize the coefficient of variation.

2.3   WHAT CONSIDERATIONS MUST BE MADE WITH RESPECT TO EXPERIMENTAL APPROACHES TO ANSWERING THESE QUESTIONS?

2.3.1   Treatment selection

Selection of experimental treatments leading to the pollutant crop loss functions has been a particularly vexing problem. The temporal concentration pattern of pollutants in the ambient environment is influenced by many variables. Concentrations are determined by meteorological conditions, presence of pollutant precursors, and other environmental conditions. Episodes with relatively high concentrations may occur at various times of the day and vary in duration. There has not been a clear definition of the components of an exposure, e.g., concentration, duration, that are most important in causing crop responses. The characterization and representation of plant exposure to air pollutants is a problem that includes both interpretation of ambient episodes and choice of experimental treatment regimes. A simple summary statistic from one situation cannot be converted easily to another situation if it involves changes in the concentration patterns. In crop response studies, it is particularly difficult to extrapolate from short experimental exposures to season-long responses.

Research designed to define pollutant crop loss functions must characterize and represent plant exposures that incorporate the time element. When plant yield is considered, the effect on yield depends on the integrated impact of the pollutant exposures during the growth of the plant, that is, the growing season (or seasons, in the case of perennial species), which varies with location, species and cultivar. Crop sensitivity may vary over the season with some growth stages being more sensitive than others. Treatment regimes must characterize the effects of pollutant episodes at specific periods of plant growth. It is probably impossible to select an exposure summary statistic that adequately characterizes the effect of all combinations of ambient air pollutant concentration and exposure duration. The lack of an appropriate statistic does result in the use of many different statistics. These statistics cannot be readily compared directly without the expensive task of returning to the original data.

An alternative approach to describing pollutant exposure is foliar flux density. It is the amount of pollutant entering the plant, rather than the concentration around the foliage, that represents the exposure of the leaf cells. This approach is supported by a large body of data relating increased stomatal aperture to increased response of plants to gaseous pollutants (Thomas and Hill, 1935; Ting and Dugger, 1971; Fletcher *et al.*, 1972; Tingey *et al.*, 1982; Olszyk and Tingey, 1984). The conclusion from these data is that the effect of O_3 on plants is to a greater or lesser degree dependent on the amount of gas that enters the leaf. However, it has not been well documented that a strong general relationship exists between the O_3 flux to the leaf and the amount of injury, or yield loss. There are probably other resistances in the pathway between ambient injury and the actual site of impact, so that neither canopy nor stomatal flux may be the correct indicator of the dose of pollutants to the cellular target (Taylor *et al.*, 1982). In addition, it is difficult to measure or quantify, over an extended period, the relationship between ambient pollutant concentration and pollutant flux to plants because of the interactive effects of environmental and biological variables. It is difficult to imagine how foliar flux could be measured over a season as a reliable description of dose.

A mean, based on various averaging times, is the most common statistic used to characterize and represent plant exposures to air pollutants. One can view exposure as a function of time given by some functional form, $f(t)$. Any summary statistic is a reduction of this function to a number (technically a functional), for example: total dose is $\int_0^T f(t)\,dt$, and average concentration is $1/T \int_0^T f(t)\,dt$; alternatively, the height of the highest peak is $0 \leqslant t \leqslant T f(t)$. But a mean value treats all concentrations as being equally effective in causing a plant response and largely ignores the importance of peak concentrations. The total dose (mean concentration times duration) has also been used to describe plant exposure, but it has the same deficiencies as the mean. It is the peak concentrations that are most likely to have some impact. The dynamics of the exposure are important as well, that is, whether the exposure concentration is constant or variable. This leads to questions about the adequacy of simple summary statistics to characterize seasonal exposures in crop loss studies, because they do not consider exposure dynamics within the season.

The two types of exposure regimes currently used to develop pollutant crop loss functions have been referred to as the modified ambient and simulated ambient regimes. In the modified ambient, the exposure represents the current temporal variation in ambient concentrations. This approach is suitable for long-term studies: exposures track the environment, and a range of concentrations can be created above and below ambient levels. However the exposures cannot be truly replicated in time or space. In the simulated ambient approach, the exposures represent the temporal variation in ambient concentrations, but not the current ambient levels; the exposure regimes are preplanned to be independent of the instantaneous ambient levels. In this case, exposures can be replicated in time and space, a range of concentrations can be created, and hypotheses about the effects of various characteristics of the exposure can be tested. The primary limitation is that the exposure pattern is arrived at somewhat artificially. In general, exposure regimes for the development of crop loss functions should reflect the episodic (seasonal and diurnal) nature of the pollutant occurrence, represent the study area, follow ambient air quality, and utilize a range of pollutant(s) concentrations above and below the current ambient concentrations.

Until very recently, relatively little attention had been focused on the development of suitable exposure regimes as an important component of experimental design. The methods that have been used include consideration of the characterization of ambient air quality by site and region, and the construction of recommended experimental fumigation regimes (Garsed and Rutter, 1984; Lefohn *et al.*, 1986).

2.3.2 Experimental design

The experimental design focuses a research program on its specific objectives. Selecting the appropriate design is a critical step in ensuring the success of the study and the applicability of the results. The experimental design includes the number and kind of factors controlled, the patterns of randomization, and the number of replicates to be used. These factors will determine what treatment comparisons may be made, whether dose–response curves can be fitted, the precision of estimates, and the range of conditions over which inferences can be made.

Most commonly, air pollution experiments are factorials, consisting of two continuous variables, or one continuous and one discontinuous variable. Examples of the former type are exposures of plants to various concentrations of SO_2 and O_3, or air temperature and O_3. An example of the second type is exposure of several cultivars or species to various concentrations of pollutant gas. More ambitious studies may consist of three or more factors, arranged in some kind of nested design. The levels of each factor in the experiment are governed by the end use of the data: the question that is being asked and, to a lesser extent, the nature of the factor. The most common approach is for the levels to be evenly spaced, a statistical requirement for the analysis of the data by orthogonal polynomial contrasts. This requirement does not apply in analysis of data by polynomial regression. The latter approach works better with more treatments than the number required for orthogonal polynomial contrasts, but it is more flexible, in that it allows a disproportionately large number of treatments in the area of interest, as well as some outside the area of interest, for increased precision.

Determining how to arrange these factorial treatments in a design is challenging for air pollution studies, because the nature of the treatments, i.e., gases, dictates that experimental units be fixed and tend to be limited in number. In many laboratories, they are also blocked in small numbers, due to physical limitations. Pollutant studies are therefore excellent candidates for efficient experimental approaches, such as incomplete block or incomplete factorial designs. In our laboratory, an incomplete block design for six treatments in four chambers was recently demonstrated to be up to three times as efficient as the comparable randomized complete block design. This efficiency term means that to obtain the same information from a randomized complete block design, up to three times as many replicates would be required. Our laboratory has also used incomplete factorial designs, replicated in incomplete blocks, to estimate response surfaces for plant growth following exposure to mixtures of pollutants (Ormrod *et al.*, 1984; Marie and Ormrod, 1987). It is not necessary for every treatment combination to be represented in the experimental design. The response surface for a three × five factorial can be estimated using only seven of the fifteen treatments, plus three or four replications of the central treatment for each replicate of the other treatments. These approaches represent a significant saving in time and effort.

2.3.3 Precision

The concern for an appropriate level of precision in a study has already been touched upon; this is largely determined by the degree to which the experimenter wishes to detect a treatment effect, and the size of the error variance. The error variance is a measure of the effect of all factors on the experimental units other than those factors which are under investigation; for example, treatments, blocks and replicates. It follows then, that the closer an experimental situation is to an actual field situation, the less control the experimenter is likely to have over the multitude of other inputs. One technique for improving precision, which is not in widespread use, is analysis of covariance. This involves the nondestructive measurement and recording of a covariate such as planar area, plastochron index, or height prior to experimentation. The post-treatment data are then adjusted to the mean covariate value, so that treatment effects are not diminished by plant-to-plant variation.

A second concern with respect to precision is the repeatability of the treatments, i.e., how well the replicates truly replicate each other. Treatments such as dynamic exposures which simulate ambient exposures may have a higher degree of variability from run to run. Those exposures that actually track the ambient concentrations are not replicable. When the treatments of a factorial are not closely replicated, and are assigned their expected level rather than the observed level (e.g., O_3 concentrations), the result is that a portion of the total sum of squares that belongs in the "treatment" source of variation ends up in the "error" term instead. This, in turn, inflates the error mean square, leading to a loss of precision.

2.3.4 Extrapolation

Extrapolation in this context describes the use of the observed experimental values to approximate the field values. The accuracy of this approach is very difficult to quantify. It has already been qualitatively assessed as being poor for experiments conducted in continuous stirred tank reactors or indoor exposure chambers. This was determined by a comparison of yield loss statistics from experiments in these facilities with results from outdoor studies. A comparison of these two types of exposure is essentially not a valid comparison, because the differences in growth and exposure conditions between the two contribute to the yield parameters, but it does effectively demonstrate the limitation of indoor studies. However different the absolute results of the two exposures are, the more interesting comparison is whether the relationship among the treatments remains the same between indoor and outdoor types of experiments. In many comparisons, the relationship is not maintained, suggesting that there are interactions between pollutants and other environmental factors that affect pollutant impact in positive or negative ways.

Both the need to extrapolate data to the field and the need for precision touch on the final consideration in choosing an experimental approach. That is, how applicable are the experimental results to crops grown under normal cultivation practices? This consideration is central to the goal of crop loss assessment experiments and is dependent on how well the experiment simulates ambient exposure.

2.4 WHAT ARE THE TYPES OF APPROACHES, AND HOW DO THEY SATISFY THE CONSIDERATIONS PREVIOUSLY DESCRIBED?

This section will discuss the variety of exposure situations that lead to the development of crop loss functions involving air pollutants. The approaches utilize laboratory chambers (cubes or continuous stirred tank reactors (CSTRs)), field chambers (open-top chambers), and open field systems (zonal air pollution, chemical protectants, air exclusion); statistical/epidemiological methods may also be used to assess losses. These four primary approaches, in the order described, have decreasing perturbation, usually accompanied by decreased precision, of the growing environment of the crop.

The assessment of the impact of air pollutants on crop productivity requires that the ambient environment not be altered in significant ways by the experimental conditions and that the growing conditions of the crop be as natural as possible. It is necessary to exclude ambient pollutants and include treatment pollutants, and a means to accomplish this is a chamber or enclosure. Exposure systems should share many common features. Each system must have a monitoring system that measures pollutant concentration continuously or frequently in a time-sharing system that sequentially measures concentrations in chambers or at field sites. The systems should employ inert tubing for sampling lines, and there should be continuous air flow in all lines to reduce time lags. The monitoring system and its calibration should conform to approved methods. In most exposure systems, the air pollutants are generated artificially and are dispensed to exposure chambers or field plots. In some systems, proportional air filtration is used to provide different levels of ambient pollutants. Table 2.1 summarizes the more common exposure situations, with respect to the strengths, limitations, and type of data generated by each.

2.4.1 Laboratory exposures: CSTRs or cubes
The use of indoor exposure systems, such as continuous stirred tank reactors or exposure cubes, to determine crop loss responses to long-term pollutant stress has been largely discredited due to the differences between the environmental conditions of the indoor chambers relative to field studies. Furthermore, comparisons between crop loss functions determined from indoor and outdoor studies have not revealed a high correlation. The ability to define the dynamic exposure patterns that occur in ambient situations may ameliorate this problem, by allowing laboratory experiments to simulate the ambient environment more closely.

2.4.2 Open-top field chambers
Field chamber systems are adaptations of greenhouse and laboratory chamber designs. The open-top chamber system is the most widely used field chamber at present. These upright cylinders are covered on the sides with clear film and have the advantage of portability, moderate cost, and ease of maintenance. The size and shape of the chambers may be modified for different plant statures, although a standardized version is widely used.

Outdoor open-top chambers have the advantages of widespread use with well-defined operational characteristics and pollutant concentrations. There is only moderate environmental modification compared with closed chambers, a wide range of treatments can be imposed, and the effects of air filtration and gradients are well-defined (Heck *et al.*, 1984*a, b*; Heagle *et al.*, 1986; Heggestad *et al.*, 1986). Limitations of open-top chambers include limitations on the number and size of plants that can be accommodated, differences in microclimate from ambient air, and lack of filtration systems for some pollutants. There is now a good knowledge base on these chamber effects, based on years of experience with the open top chamber now in use in crop exposure studies.

The open-top chamber system has a high-volume flow of filtered air introduced around the inner lower perimeter of the cylinder to reduce ambient pollutant influx through the open top. The cylinders can be used as ambient air

TABLE 2.1

Summary of experimental approaches to determination of crop loss functions

	Chamberless systems		Chambered systems		Exposure regimes	
	Air exclusion chemical exclusion	ZAPS plume	Open-top chambers	Continuous stirred tank reactors	Ambient or modified ambient	Artificial
Pros	No chamber effects, so limited environmental modification Adaptable to range of plant sizes	No chamber effects, so no environmental modification Treatments may be applied	Wide range of treatments possible Cost effective Filtration and gradients well defined Pollutants defined	Wide range of pollutant treatments and wide range of treatments of other environmental stresses possible, because chambers are well controlled environmentally	Exposures are a more realistic representation of field stress Ambient type exposure is coupled to environment	Suitable for an exposure system of limited sophistication Exposures are easily characterized in terms of duration and concentration Suitable for screening studies

Cons	Confounding factors, e.g., fungicidal effect Environmental modification, e.g., canopy air movement Treatments dependent on ambient pollutants, so not replicable in time or space	Treatment integrity depends on environmental conditions Space required for separation or replications Treatments confounded with ambient pollutants	Moderate environmental modification Plant size is limited Microclimatic modification due to chamber effects	Artificial environment (i.e., pots, lights) Applicability to field crop loss assessment may be limited	Ambient and modified ambient are not replicable in time or space Simulated ambient not coupled to environment	Not suitable for long-term studies Difficult to extrapolate to field crop loss
Type of data	Indirect estimates of plant growth before pollutants	Dose–response functions for growth effects of pollutants confounded to some degree with other environmental stresses	Dose–response functions for growth effects confounded to some degree with other environmental stresses	Dose–response functions for growth effects and mechanistic effects such as CO_2 fixation, stomatal conductance, and interaction studies with other stresses	Dose–response functions from simulated and ambient exposures Modified dose–response functions from ambient	Dose–response functions for growth and mechanistic effects, such as stomatal conductance, CO_2 fixation, pollutant flux, and interaction with other environmental stresses

exclusion systems to test the difference in plant growth between ambient air and filtered air, or gaseous pollutants can be added to the incoming air stream to provide a range of concentrations among cylinders in an experiment. The rate of pollutant addition is adjusted to control the pollutant concentration hourly or daily, according to the experimental design.

There are many relative strengths of the open-top chamber system. The portability of open-top chambers and the ease with which covering films can be added or removed facilitates storage and maintenance and allows standard crop production practices to be followed during land preparation, seeding and early crop growth before the chambers are set in place and/or covered with clear film. Open-top chambers reduce temperature deviation between chamber air and ambient air and allow sufficient control of gas concentrations for single or mixed-gas exposures. The cylinders can be placed in various designs in the field to optimize proximity to the monitoring and control equipment in relation to soil conditions.

The weaknesses of open-top chambers relate mainly to microclimatological characteristics. Air flowing from the lower inner perimeter of the chamber out through the open top reduces the intrusion of outside air; consequently the air flow pattern is different from that in the open field. Air temperatures may be higher as a result of trapped long wave radiation, and rainfall is not always evenly distributed within a chamber, so supplemental irrigation may be necessary. Photosynthetically active radiation can be reduced by poorly maintained plastic film and neighboring chambers. These differences between chamber and field conditions may affect crop responses to air pollutants (Roberts, 1984).

Concerns about open-top chambers also include the suspicion of nonspecific filtration. Charcoal filter systems in open-top designs may remove some ambient gases and particles from the air other than the pollutant under study. Open-top chambers do not have the equivalent of ambient vertical gradients of pollutant concentration.

It is usually more challenging to maintain desired pollutant concentrations in open-top chambers (Heagle *et al.*, 1979). Some intrusion of outside air through the open chamber top is bound to occur, and this outside air may dilute the pollutant within the chamber (Unsworth *et al.*, 1984*a, b*). The amount of intrusion varies with wind speed, but can be minimized by the use of innovative design. The addition of a truncated cone (frustum) to the top of open-top chambers can greatly reduce the intrusion of ambient air and improve the uniformity of the environment and pollutant concentration within the chambers, yet at the same time reduce ambient light and rainfall.

There are several variations of the open-top design. The chamber developed by Heagle *et al.* (1973) has been used widely for many crop species, and forms the basis for the NCLAN program. Various versions have improved air flow characteristics and other special features (Kats *et al.*, 1976; 1985). Very large open-top chambers developed for use with grapevines may prove useful for other large stature species. Open-top chambers may be fitted with portable tops for rain exclusion or application studies (Hogsett *et al.*, 1985).

Many other field chambers of various designs have been used for crop response studies (Ashenden *et al.*, 1982; Whitmore and Mansfield, 1983). They all are essentially fully enclosed by transparent film. These chambers rely on high air flow-through rates to minimize temperature increases compared with ambient air, an approach which may increase water use by the plants. Many of these field chambers evolved from greenhouse exposure systems. Chamber shapes have ranged from square or rectangular to cylindrical or geodesic dome.

The environmental modifications imposed by open-top chambers are relatively slight compared to other types of field chambers, and have been documented so the chamber effect on crop response can be considered in response evaluations.

2.4.3 Chamberless exposure: ZAPS

Chamber-free field exposure systems were developed to provide a method for exposing large field plots, thereby increasing the size of experimental units, and to prevent the environmental modification caused by enclosing plants in open-top or other field chambers. Such field exposure systems are typically a series of tubes with calibrated orifices, which are distributed over a field to emit gaseous pollutants (Lee *et al.*, 1978; Laurence *et al.*, 1982). The primary advantage of such systems is that plants are exposed to pollutants under ambient conditions. The disadvantages relate to the diminished control over the level of pollutants and the nature of the exposure. These systems are greatly affected by wind speed and direction, and plants are subjected to ambient air pollutants, as well. Brookhaven National Laboratory (Evans, personal communication) is testing a circular system of exposure pipes, which, under computer control, reacts to wind direction by opening or closing gas delivery ports on the circle. This is modelled after a European system and should increase the accuracy with which treatments are applied.

A major limitation of plume-type non-chambered exposure systems is that only the addition of pollutant is possible. A pollutant-free environment is required to achieve a full range of treatments or the effects of ambient pollutant conditions cannot be ascertained. Replication of treatments requires a large investment in space and equipment, as well.

2.4.4 Chamberless exposure: air exclusion

Another nonchambered approach is the air exclusion system, which has a high air flow and well-defined exclusion characteristics that allow it to be used in both ambient air and pollutant addition studies. These systems impose only minor environmental modification during the periods of duct inflation (Kuja *et al.*, 1986; Olszyk *et al.*, 1986*a*, *b*). A concern with the air exclusion system is the atypical air flow across canopies. The final nonchambered research approach to establishing dose–response models is the use of naturally occurring or artificially induced pollutant gradients in field situations (Oshima *et al.*, 1976; Laurence *et al.*, 1982).

Nonchambered field systems have the advantages of minimal environmental modification, greater area and plant number, and greater maximum plant size. Nonchambered systems function best with homogeneous canopies. The disadvantages include the need for further development of mechanisms for monitoring and control and for increasing uniformity within a treatment, the intrusion of the ambient polluted air into treatments, and the large space requirements for a range of treatments and replications.

Nonchambered exposure systems are attractive because of the lack of structure-induced artifacts in the exposure studies and the minimum alteration of the environment. Both types of nonchambered systems, the plume and the air exclusion systems, can be useful depending on the specific experimental objective. They can provide supplementary experimental approaches to studies involving open-top chambers. Large circular or square plume systems encircle the entire exposure plot to provide large-scale exposure with no environmental modification. The dispersion characteristics for the systems have been well-designed based on computer modeling and field testing. The distance between emitters and receptor plants has been determined to provide uniformity in pollutant concentrations under ambient wind conditions. Development of computer-based pollutant dispensing, monitoring, and control systems, and environmental monitoring has permitted continuous use of such systems and characterization of the pollutant exposures within experimental plots (Greenwood *et al.*, 1982; McLeod *et al.*, 1985).

2.4.5 Chamberless exposure: chemical protectants

The use of chemical protectants to exclude air pollutants, particularly O_3, from foliar surfaces has been extensively utilized over a long period of time. Such fungicides as Benlate, ethylene diurea, and polyamines have been applied both as foliar spray and soil drench to prevent O_3 effects on crop plants (Hofstra *et al.*, 1978; Beckerson and Ormrod, 1986). This is in effect a form of chamberless exposure; however, not without limitations. The chemical protectants used were not developed initially as O_3 protectants; the discovery that they often protect against O_3 injury is a serendipitous aside to their primary function. These primary modes of action tend to confound the results; it is difficult to partition the effect on crop yield between fungicidal and antiozonant.

2.4.6 Indirect measurement

Estimations of crop yield loss may be made indirectly by measuring crop metabolism during the application of pollutant stress, and extrapolating changes in metabolic activity to alterations in yield. Photosynthesis rates and stomatal or leaf diffusive resistance are the response parameters that are most commonly correlated with gaseous pollutant stress. Their usefulness as predictors of stress-related yield loss is limited by the fact that the relationship between carbon dioxide uptake or carbon fixation and eventual yield is not direct. Some plant species have demonstrated a capacity for recovery from pollutant stress, which may be difficult to include or account for in a model that relies on changes in physiological parameters to predict crop yield loss. Many of these physiological parameters are sensitive to other environmental stresses, which can confound field observations. This is why much indoor experimentation has been devoted to establishing the relationship between physiological parameters and yield loss, in response to pollutant stress. To date, there is little evidence in the literature of the successful estimation of field loss from physiological measurements obtained in the laboratory.

2.4.7 Epidemiological estimation

The use of an epidemiological approach to estimate crop loss due to air pollutants has had limited application (Leung *et al.*, 1982; Moskowitz *et al.*, 1982), and relies on two estimates: the empirical relationship between pollutant dose and crop yield, and seasonal pollutant concentrations in agricultural areas. The dose–response relationship has received a good deal of experimental attention and has been discussed already in this presentation. The latter estimate presents several challenges. The first is that little agricultural land in either Canada or the United States is well monitored for air pollutants. This means that in many cases, seasonal pollutant concentrations for large areas are extrapolated from meagre data bases. The second challenge is the distillation of this seasonal data into a statistic(s) that can be utilized in an empirically derived equation. Early attempts (Oshima, 1974) at deriving this statistic included the concept of dose, which is defined as $ppm\,h^{-1}$ and could be mathematically thought of as the area under the curve describing pollutant concentration, integrated over time. An alternate statistic is the 7-h daylight growth season average (Heagle *et al.*, 1979). Both of these approaches have limitations: the former requires access to the original data base, the latter equates chronic and acute exposures that have the same dose. Lefohn and Benedict (1982) identified these weaknesses and developed an alternate statistic, which utilizes readily available data, separates chronic and acute exposures of the same dose, and provides a mathematical description of an exposure in terms of concentration, duration and frequency. The NCLAN program has estimated 7-h mean O_3 concentrations for each of the major US crop growing areas, using the kriging technique, to estimate crop loss to O_3 (Lefohn *et al.*, 1987). There is still much discussion concerning the O_3 statistic that is best related to crop yield loss.

2.5 SUMMARY

There are many research approaches to determining crop loss functions. They all have merits and disadvantages that must be weighed against the objectives of a particular study. Regardless of the approach, however, an integrated study should have the following characteristics: a clear definition of the question at hand, a rigorous experimental design and data analysis plan, an experimental procedure that includes quality assurance and quality control processes, and an acceptance of the limitations of the chosen system. The NCLAN project is an excellent example of this process: a thorough, rigorous integrated investigation of a relatively simple question. It is an example well worth emulation.

REFERENCES

Allen, O. B., B. A. Marie, and D. P. Ormrod. (1988). Relative efficiency of factorial designs for estimating response surfaces with reference to gaseous pollutant mixtures. *J. Environ. Qual.* (in press).

Ashenden, T. W., P. W. Tabner, P. Williams, M. E. Whitmore, and T. Mansfield. (1982). A large-scale system for fumigating plants with SO_2 and NO_2. *Environ. Pollut. Ser. B,* **3,** 21–26.

Beckerson, D. W. and D. P. Ormrod. (1986). Polyamines as antiozonants for tomato. *HortSci.,* **21,** 1070–1.

Fletcher, R. A., N. O. Adedipe, and D. P. Ormrod. (1972). Abscisic acid protects bean leaves from ozone-induced phytoxicity. *Can. J. Bot.,* **50,** 2389–91.

Garsed, S. G. and A. J. Rutter. (1984). The effects of fluctuating concentrations of sulphur dioxide on the growth of *Pinus sylvestris* L. and *Picea sitchensis* (Bong.) Carr. *New Phytol.,* **97,** 175–95.

Greenwood, P., A. Greenhalgh, C. Baker, and M. Unsworth. (1982). A computer-controlled system for exposing field crops to gaseous air pollutants. *Atmos. Environ.,* **16,** 2261–6.

Heagle, A. S., D. E. Body, and W. W. Heck. (1973). An open-top field chamber to assess the impact of air pollution on plants. *J. Environ. Qual.,* **2,** 365–8.

Heagle, A. S., R. B. Philbeck, H. H. Rogers, and M. B. Letchworth. (1979). Dispensing and monitoring ozone in open-top field chambers for plant effects studies. *Phytopathology,* **69,** 15–20.

Heagle, A. S., W. W. Heck, V. M. Lesser, J. O. Rawlings, and F. L. Mawry. (1986). Injury and yield response of cotton to chronic doses of ozone and sulfur dioxide. *J. Environ. Anal.,* **15,** 375–82.

Heck, W. W., W. W. Cure, J. O. Rawlings, L. J. Zaragoza, S. A. Heagle, H. E. Heggestad, R. J. Kohut, L. W. Kress, and P. J. Temple. (1984*a*). Assessing impacts of ozone on agricultural crops: 1. Overview. *J. Air Pollut. Control Assoc.,* **34,** 729–35.

Heck, W. W., W. W. Cure, J. V. Rawlings, L. J. Zaragoza, A. S. Heagle, H. E. Heggestad, R. J. Kohut, L. W. Kress, and P. J. Temple. (1984*b*). Assessing impacts of ozone on agricultural crops: II. Crop yield functions and alternative exposure statistics. *J. Air Pollut. Control Assoc.,* **34,** 810–17.

Heggestad, H. E., J. H. Bennett, E. H. Lee, and L. W. Douglass. (1986). Effects of increasing doses of sulfur dioxide and ambient ozone on tomatoes: plant growth, leaf injury, elemental composition, fruit yields, and quality. *Phytopathology,* **76,** 1338–44.

Hofstra, G., D. A. Littlejohns, and R. T. Wukasch. (1978). The efficacy of the antioxidant ethylene-diurea (EDU) compared to carboxin and benomyl in reducing yield losses from ozone in navy bean. *Plant Dis. Rep.,* **62,** 350–52.

Hogsett, W. E., D. T. Tingey, and S. R. Holman. (1985). A programmable exposure control system for determination of the effects of pollutant exposure regimes on plant growth. *Atmos. Environ.,* **19,** 1135–45.

Hogsett, W. E., D. Olszyk, D. P. Ormrod, G. E. Taylor, Jr, and D. T. Tingey. (1987). Air pollution exposure systems and experimental protocols: Volume 1: Review and evaluation of performance. U.S. EPA Report No. 900/3-87/037a, Environmental Research Laboratory, Corvallis, OR.

Kats, G., C. R. Thompson, and W. C. Kuby. (1976). Improved ventilation of open-top greenhouses. *J. Air Pollut. Control Assoc.,* **26,** 1089–90.

Kats, G., P. J. Dawson, A. Bytnerowicz, J. W. Wolf, C. R. Thompson, and D. M. Olszyk. (1985). Effects of ozone or sulfur dioxide on growth and yield of rice. *Agric. Ecos. Environ.,* **14,** 103–17.

Krupa, S. and R. N. Kickert. (1987*a*). An analysis of numerical models of air pollutant exposure and vegetation response. *Environ. Pollut.,* **44,** 127–58.

Krupa, S. and R. N. Kickert. (1987*b*). An analysis of numerical models of air pollutant exposure and vegetation response. *The Acid Deposition Research Program*, **10**, 113.

Kuja, A., R. Jones, and A. Enyedi. (1986). A mobile rain exclusion canopy and gaseous pollutant reduction system to determine dose–response relationships between simulated acid precipitation and yield of field grown crops. *Water Air Soil Pollut.*, **31**, 307–15.

Laurence, J. A., D. C. Maclean, R. H. Mandl, R. E. Schneider, and K. S. Hansen. (1982). Field tests of a linear gradient system for exposure of row crops to SO_2 and HF. *Water Air Soil Pollut.*, **17**, 399–407.

Lee, J. J., R. A. Lewis, and D. E. Body. (1978). A field experimental system for the evaluation of the bioenvironmental effects of sulfur dioxide. *Proceedings Fort Union Coal Field Symposium*, Vol. 5, pp. 608–20. Montana Academy of Science, Billings, Montana.

Lefohn, A. S. and H. M. Benedict. (1982). Development of a mathematical index that describes ozone concentration, frequency and duration. *Atmos. Environ.*, **16**, 2529–32.

Lefohn, A. S., W. E. Hogsett, and D. T. Tingey. (1986). A method for developing ozone exposures that mimic ambient conditions in agricultural areas. *Atmos. Environ.*, **20**, 361–6.

Lefohn, A. S., H. P. Knudsen, J. A. Logan, J. Simpson, and C. Bhumralkar. (1987). An evaluation of the kriging method to predict 7-h seasonal mean ozone concentrations for estimating crop losses. *J. Air Pollut. Control Assoc.*, **37**, 595–602.

Leung, S. K., W. Reed, and S. Geng. (1982). Estimations of ozone damage to selected crops grown in southern California. *J. Air Pollut. Control Assoc.*, **32**, 160–4.

Marie, B. A. and D. P. Ormrod. (1987). Dose response relationships of the growth and injury effects of ozone and sulphur dioxide on Brassicaceae seedlings. *Can. J. Plant Sci.*, **66**, 659–67.

McLeod, A. R., J. E. Fackrell, and K. Alexander. (1985). Open-air fumigation of field crops: criteria and design for a new experimental system. *Atmos. Environ.*, **19**, 1639–49.

Moskowitz, P. D., E. A. Coveney, W. H. Medeiros, and S. C. Morris. (1982). Oxidant air pollution: A model for estimating effects on U.S. vegetation. *J. Air Pollut. Control Assoc.*, **32**, 155–60.

Olszyk, D. M. and D. T. Tingey. (1984). Fusicoccin and air pollutant injury to plants. Evidence for enhancement of SO_2 but not O_3 injury. *Plant Phys.*, **76**, 400–2.

Olszyk, D. M., A. Bytnerowicz, G. Kats, P. J. Dawson, J. Wolf, and C. R. Thompson. (1986*a*). Crop effects from air pollutants in air exclusion systems vs. field chambers. *J. Environ. Qual.*, **15**, 417–22.

Olszyk, D. M., A. Bytnerowicz, G. Kats, P. J. Dawson, J. Wolf, and C. R. Thompson. (1986*b*). Effects of sulfur dioxide and ambient ozone on winter wheat and lettuce. *J. Environ. Qual.*, **15**, 63–9.

Ormrod, D. P., D. T. Tingey, M. L. Gumpertz, and D. M. Olszyk. (1984). Utilization of a response surface technique in the study of plant responses to ozone and sulphur dioxide mixtures. *Plant Physiol.*, **75**, 43–8.

Oshima, R. J. (1974). Development of a system for evaluating and reporting economic crop losses by air pollution in California. II: Yield study, IIA: Prototype ozone dosage-crop loss conversion function. Prepared for the California Air Resources Board, California Department of Food and Agriculture.

Oshima, R. J., M. P. Poe, P. K. Braegelmann, D. W. Baldwin, and V. Vanway. (1976). Ozone dosage-crop loss function for alfalfa: a standardized method for assessing crop losses from air pollutants. *J. Air Pollut. Control Assoc.*, **26**, 861–5.

Roberts, T. M. (1984). Long-term effects of sulphur dioxide on crops: An analysis of dose–response relations. *Phil. Trans. R. Soc. London*, **305**, 299–306.

Taylor, G. E., Jr., D. T. Tingey, and H. C. Ratsch. (1982). Ozone flux in *Glycine max* (L.) Merr.: sites of regulation and relationship to leaf injury. *Oecologia*, **53**, 179–89.

Thomas, M. D. and G. R. Hill, Jr. (1935). Absorption of sulphur dioxide by alfalfa and its relation to leaf injury. *Plant Physiol.*, **10**, 291–307.

Ting, I. P. and W. M. Dugger. (1971). Ozone resistance in tobacco plants: a possible relationship to water balance. *Atmos. Environ.*, **5**, 147–50.

Tingey, D. T., G. L. Thutt, M. L. Gumpertz, and W. E. Hogsett. (1982). Plant water status influences ozone sensitivity of bean plants. *Agric. Environ.*, **7**, 243–54.

Unsworth, M. H., A. S. Heagle, and W. W. Heck. (1984a). Gas exchange in open-top field chambers. I. Measurement and analysis of atmospheric resistance to gas exchange. *Atmos. Environ.*, **18**, 373–80.

Unsworth, M. H., A. S. Heagle, and W. W. Heck. (1984b). Gas exchange in open-top field chambers. II. Resistances to ozone uptake by soybeans. *Atmos. Environ.*, **18**, 381–5.

Whitmore, M. E. and T. A. Mansfield. (1983). Effects of long-term exposures to SO_2 and NO_2 on *Poa pratensis* and other grasses. *Environ. Pollut. Ser. A*, **31**, 217–35.

3

THE NCLAN PROGRAM FOR CROP LOSS ASSESSMENT

ERIC M. PRESTON and DAVID T. TINGEY

US Environmental Protection Agency, Corvallis, Oregon, USA

3.1 INTRODUCTION

The Clean Air Act requires the US Environmental Protection Agency (EPA) to set National Ambient Air Quality Standards (NAAQSs) for "any air pollutant which, if present in the air, may reasonably be anticipated to endanger public health or welfare and whose presence in the air results from numerous or diverse mobile and/or stationary sources" (Padgett and Richmond, 1983). The Agency is responsible for developing and promulgating primary (to protect public health) and secondary (to protect public welfare) NAAQSs. The 1977 amendments to the Clean Air Act require that the criteria (scientific basis) for the standards be periodically reviewed and revised to include new information. In 1978, EPA reviewed the scientific literature to determine the impact of ozone on vegetation and published this analysis in the Air Quality Document for Ozone and Other Photochemical Oxidants (US EPA, 1978). The credibility of the secondary standard for ozone, based in part on these studies, suffered because there were insufficient data to determine reliably either the effects of ozone on the yield of major agronomic crops under field conditions or to determine resultant economic consequences.

Several realities in the late 1970s precluded a credible comprehensive economic assessment.

1. Ozone exposure–response data were not available for the economically important crop species. Relatively few studies measured effects of exposure on growth and yield. In addition, available exposure–response studies were usually not comparable due to important differences in exposure regimes, response variables measured, and exposure methods.

Experimental exposures were often so unrealistic that inferences to the field were inappropriate.

2. Data on the spatial distribution of ozone in agricultural areas were incomplete and not expressed in a form that allowed estimation of potential effects on crops.

3. Important interactions between pollutant exposure and key environmental variables were poorly understood.

On the other hand, advances in experimental exposure technology (Heagle *et al.*, 1973, Lee and Lewis, 1978) and improvements in quality control and data management of EPA's National Aerometric Data Bank (Storage and Retrieval of Aerometric Data (SAROAD) made the time right to generate more realistic exposure–response data and to attempt a comprehensive economic assessment.

In 1980, the EPA initiated a model crop loss assessment program. The effort was launched with a series of meetings with EPA's Office of Air Quality Planning and Standards (OAQPS) and air pollution scientists. Objectives were defined and the merits of alternative methods of achieving these objectives were discussed. The National Crop Loss Assessment Network (NCLAN) was conceived as a multidisciplinary study to produce an economic assessment of the effects of air pollutants on US agriculture based upon state-of-the-science exposure–response data.

NCLAN's primary objectives were to:

1. define the relationships between yields of major agricultural crops and exposure to ozone, sulfur dioxide, nitrogen dioxide, and their mixtures;

2. assess the economic consequences resulting from the exposure of major agricultural crops to ozone, sulfur dioxide, and nitrogen dioxide, and their mixtures; and

3. advance the understanding of the cause and effect relationships that determine crop response to pollutant exposure.

Funding limitations and the resulting priority setting ultimately led to decisions to concentrate field research efforts on developing ozone exposure–response functions (though a few studies using pollutant mixtures were conducted). The process-oriented physiological studies were all but eliminated and the economic assessment was restricted to consider only the effects of ozone. Nevertheless, the NCLAN program approaches the task of developing the "ideal" data set for standard setting on a scale that had never before been attempted.

3.2 RATIONALE FOR THE NCLAN APPROACH TO ECONOMIC ASSESSMENT

Presumably, air pollution exposure imposes costs on production of a given yield. If the costs of producing a given yield under known oxidant exposure are available, a cost function may be estimated econometrically, which defines the

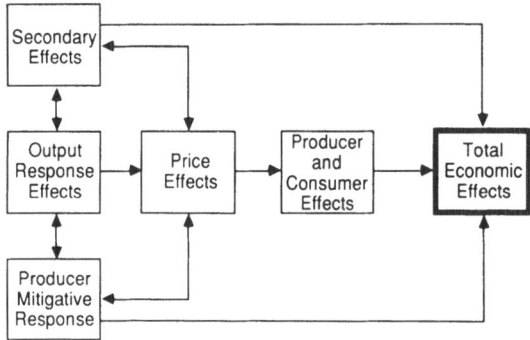

Fig. 3.1. Flow chart illustrating the components of a national economic assessment.

relationship between oxidant exposure and the components of production cost (Mjelde *et al.,* 1984). No explicit crop exposure–response information is required, since this is embedded in the farmers' actual cost and yield data. Unfortunately, there was no national-level database on farm costs that was sufficiently detailed to implement such an analysis. Therefore, other approaches were considered.

These approaches required a knowledge of the biological (yield) response of crops to pollutant exposure and an evaluation of the economic consequence of that biological response (Fig. 3.1). In evaluating the biological response, the interactive effects of air quality, climate, weather, soil fertility, and agronomic practices needed to be considered. In evaluating the economic consequence of the response, allowances had to be made for actions that producers take to mitigate or avoid the effects of air pollution on their crop and for the impact that changes in quantity have had on product price.

The assessment problem was truly interdisciplinary, involving the physical sciences, biological sciences, and economics. The needs of the economic assessment defined the biological information required, and the needs of the biological assessment defined the physical and agronomic information needed. The problem framework had to be structured to ensure that all information-gathering activities were appropriate to the scale of the economic assessment.

3.2.1 Econometric and mathematical programming approaches

Econometric-simulation and mathematical programming approaches appeared to be the most consistent with the conceptual framework for addressing large-scale agricultural assessment problems. Both approaches, in general, build upon crop yield air pollution response functions. Econometric techniques generally begin with multivariate analysis of survey data to define a production function relating the yield of an agricultural crop in a region to variables affecting yield (including air quality). These statistically based yield relationships then become one component of the overall econometric model. Manipulation of the yield relationships can serve to simulate the economic effects of air pollution by tracing their consequences through other relation-

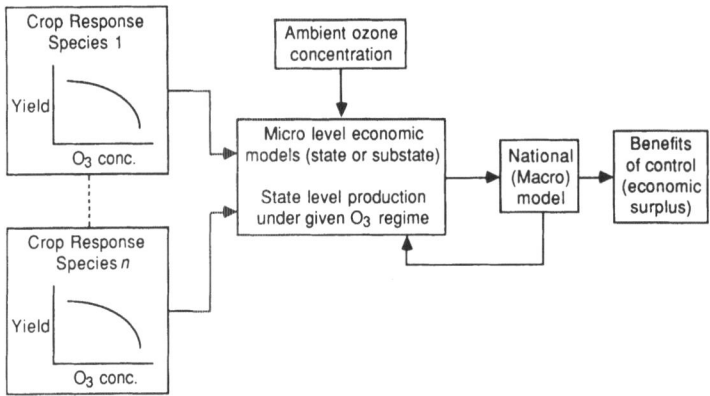

Fig. 3.2. Information flow and logic structure of NCLAN's national economic assessments.

ships in the model. This approach was not adopted because of concerns that the accuracy and resolution of national-level databases on environmental variables were inadequate to separate the relatively small (5–15%) anticipated ozone effect from the effects of other environmental variables on yield. Also, the detection of an ozone effect on yield depends upon a gradient of exposure over the areas sampled. There was concern that the ozone gradient was not steep enough over the major crop growing areas of the United States to permit reliable detection of ozone-related effects.

Mathematical programming techniques appeared to be the most suitable for the assessment problem. In this approach, normative optimization models integrating crop exposure–response data are used to estimate the impacts of varying input parameters, such as air quality, on economic response of the agricultural system. The flow of logic and data requirements of NCLAN economic assessments are illustrated in Fig. 3.2.

3.2.2 Approaches for estimating crop loss functions

Several approaches could be used to estimate crop loss functions. All have their strengths and limitations. Three primary approaches were considered.

The first approach utilizes multivariate statistical techniques to estimate the production function from a regression of spatial yield data on variables suspected to affect yield. Independent variables would have to be randomly sampled in space over a gradient of conditions in the region of interest. Oshima *et al.* (1976) developed response functions for alfalfa in California using this method. It has also been successfully applied to evaluating pest and disease effects on crop yield (James and Teng, 1979). It was successful because the impact of pests or disease was substantial and localized within areas having relatively similar environment and agricultural practices. However, the approach was not considered viable for the NCLAN assessments. Air pollution impacts on crops can be subtle and may be modified as environmental variables change over large crop growing regions. Consequently, a dense sampling pattern over wide geographic areas would be necessary to detect the

5 to 15% yield loss that appears to be typical of many experimental chronic pollutant exposures. In addition, it was considered unlikely that the steep ozone gradient available to Oshima *et al.* (1976) was present in the major crop growing areas of the Midwest. This would make it difficult to distinguish ozone effects from those of other environmental variables.

The second approach utilizes crop simulation models to estimate the response of crops to important inputs (including air pollution). Using available crop simulation models as a starting point, effects of pollutant exposure on yield are modeled to conform to results of laboratory and field experiments. Behavior of the models is then validated against independent experimental field data generated over the range of conditions for which extrapolation is desired. The models are then exercised with available regional input data to generate regional estimates of yield loss. The economic consequences of these loss estimates can then be evaluated.

The strength of this approach was that the process of constructing the models and generating the appropriate data to run them leads to greater understanding of the underlying mechanisms causing the observed yield responses. Such understanding adds credibility to the assessment and aids in setting reasonable limits on extrapolation of results.

However, there were also serious limitations to this approach. Most crop growth models available at the time had high temporal resolution (hourly time step) but were spatially lumped (representing growth effects on a small plot or plant basis). Therefore, to estimate yield effects for an entire crop growing region, the models would have had to be run under a large variety of ozone-environment regimes from a random sample of locations within the region. Such data, with sufficient temporal resolution, for driving variables were generally not available and would be prohibitively expensive to gather. Also, the costs and lead time required to develop and validate the crop models made this approach infeasible as the primary method for producing the exposure–response functions in NCLAN. However, simulation modeling has played a crucial secondary role in modeling the effects of water stress on plant response to ozone and applying this interaction to the economic model (e.g., King, 1987; King and Nelson, 1987).

The third approach utilizes carefully designed and standardized field experiments to generate representative exposure–response functions for major crop species. With adequate replication of field experiments in time and space, empirically based exposure–response functions can be generated that are representative of the crop growing region. The advantages of this approach are that, in the semi-controlled field exposures, yield effects could be definitely associated with pollutant exposure, yield response could be ascertained for exposures at and above ambient conditions (useful for evaluating the consequences of alternative regulatory scenarios in assessments), and the accuracy and precision of the individual response functions can be controlled (through experimental design) and quantified. However, these field experiments are expensive and require a full field season to generate a single replicate of an exposure–response function. Because a large number of these functions are required to provide adequate spatial and temporal replication, this approach suffers some of the same limitations of the first two. Nevertheless, NCLAN scientists, peer reviewers, and the OAQPS staff agreed that, considering the pros and cons of the options available, this approach would provide the most useful information within time and budget constraints.

3.3 THE NCLAN APPROACH

The field experimental approach was chosen as the primary method for generating exposure–response information. Physiological studies and mechanistic simulation modeling were used to provide supplementary information useful in interpreting and extrapolating the empirically generated functions.

A series of nationally coordinated field studies was conducted to provide crop exposure–response data representative of the areas in which they were conducted. A network of field sites was established in regions where ten field crops (corn, soybeans, wheat, hay, tobacco, sorghum, cotton, barley, peanuts and dry beans), accounting for roughly 85% of US acreage, are planted (Fig. 3.3). Ozone exposure–response experiments were conducted for each. When funding permitted, additional independent variables were included in the field experiments (e.g., water stress or other pollutants). Physiological measurements were taken during the growing season when possible, to aid in interpreting results and modeling interactions of independent variables. Response functions so generated and interpreted were used in an economic assessment of the consequences of ozone exposure to US field crops.

Fig. 3.3. The National Crop Loss Assessment Network, Regional Boundaries and Location of the Regional Crop Loss Assessment Projects (RCLAP). Box = auxiliary field site, solid circle = administrative center and field site.

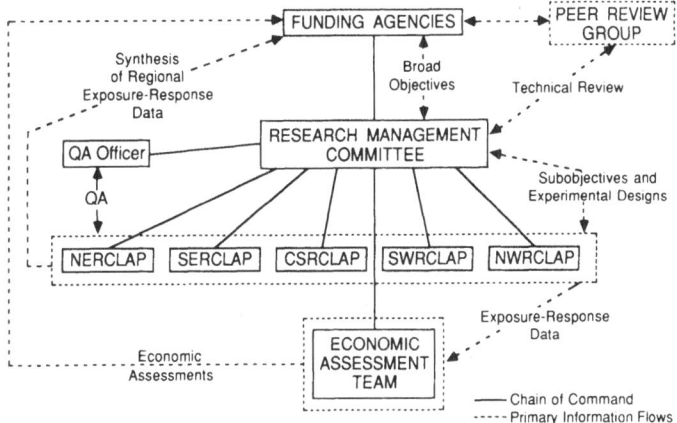

Fig. 3.4. NCLAN organisation and management.

The interdisciplinary nature and widely dispersed locations of NCLAN research made carefully coordinated action and regular, effective communication prerequisites for success. A participatory management structure was implemented to facilitate these goals (Fig. 3.4). The principal investigators from participating research institutions and the EPA program manager constituted a governing Research Management Committee (RMC). The RMC was broadly responsible for translating articulated Agency needs into appropriate research activities within budget constraints. The RMC provided a forum for program planning, tracking, and problem resolution. The RMC was the primary focus for interproject coordination and was primarily responsible for ensuring the comparability of methods used in field research.

Field research was undertaken within the five Regional Crop Loss Assessment Projects (RCLAPs). The regional projects were responsible for research conducted within their region and were accountable for the resources assigned to those projects, following guidelines approved by the RMC.

NCLAN research was quality controlled through two independent processes. An interdisciplinary peer review group from outside NCLAN reviewed research progress annually and recommended project improvements, terminations, or changes in direction. Approximately 50% of the peer review group was replaced every 2 years. This allowed reasonable continuity and infusion of new ideas, but prevented the gradual development of conflicts of interest that might prejudice the group's objectivity. Secondly, the process of generating and analysing research data was subjected to a strict quality assurance program (Coffey *et al.*, 1988). Each research project was subjected to an annual quality assurance audit to ensure that data were of known quality, accuracy, precision, and completeness and were being generated with approved protocols. Audits often detected operational problems before they became serious and prevented loss of data.

3.4 NCLAN ACCOMPLISHMENTS

In its 8-year history, NCLAN has maintained a remarkable record of accomplishing what it set out to do on time and within budget. The majority of the program's funding and research efforts have been devoted to developing crop exposure–response functions of known and documented quality using generally comparable experimental protocols. Exposure–response functions have been developed for the original target agricultural species (see Table 3.1 for selected examples). For many species, the response functions were derived from experiments replicated in at least two growing seasons on more than one cultivar. In some cases, where the crop is normally grown in a large area with wide variability in environmental conditions (e.g., soybeans and wheat), experiments were conducted at more than one site. The exposure–response data set is by far the most comprehensive available for ozone effects on vegetation. The electronically stored database is available to the scientific community for further analysis and scrutiny from the US Environmental Protection Agency, Corvallis, OR.

A second significant accomplishment is the completion of economic assessments fully utilizing the NCLAN exposure–response database. This required conversion of SAROAD air-quality data into an appropriate form and development of an appropriate model of the national farm economy. NCLAN pioneered the use of kriging (a statistical spatial interpolation methodology) to estimate ambient ozone exposures from point sample data in the SAROAD database (Lefohn et al., 1987b). The economic model developed for NCLAN by Adams et al. (1984) was the most comprehensive and sophisticated of any used to date in air pollution assessments.

In 1984, a preliminary assessment was completed (Adams et al., 1984) utilizing exposure–response functions available at that time for six of the major

TABLE 3.1
Examples of exposure–response functions developed by the NCLAN program*

Species "cultivar"	Weibull parameter estimates			Relative losses at three 7-h/seasonal means (ppm)		
	$\hat{\alpha}$	$\hat{\sigma}$	$\hat{c}$	0·04	0·05	0·06
Barley "Poco"	1·988	0·205	4·278	0·1	0·2	0·5
Bean, kidney	2878	0·12	1·171	11·0	18·1	24·8
Corn "Pioneer 3780"	12533	0·155	3·091	1·2	2·6	4·8
Cotton "Acala SJ-2"	5546	0·199	1·228	5·9	10·0	14·0
Peanut "NC-6"	7485	0·111	2·249	6·4	12·3	19·4
Sorghum "Dekalb-28"	8137	0·296	2·217	0·8	1·5	2·5
Soybean "Corsoy"	2785	0·133	1·952	5·6	10·4	15·9
Wheat "Roland"	5479	0·113	1·633	9·4	16·4	23·7

* The Weibull parameters and estimated relative yield loss data are from Heck et al., 1984.

agricultural crops (corn, soybean, wheat, cotton, sorghum and barley) that account for approximately 70% of the total acreage of US field crops. This assessment estimated that reductions in ambient ozone concentrations of 10, 25 and 40% would yield annual economic surpluses of 0·67, 1·71 and 2·52 billion 1980 dollars, respectively. The final assessment (Adams *et al.*, 1988) is significantly more expensive.

Finally, NCLAN Program participants demonstrated that a long-term coordinated, multidisciplinary air pollution study is possible. The successful partnership of EPA, academia and other agencies was facilitated by a well-considered management structure, but it was the dedication, patience and personal motivation of principal investigators that ensured the program's viability and ultimate success in achieving overall goals.

3.5 IMPLICATIONS FOR FUTURE RESEARCH

Future research needs can be assessed from several perspectives. We have structured our recommendations upon three primary end uses of crop effects data: (1) economic assessment studies, (2) development of ambient air-quality standards, and (3) advancement of scientific understanding of pollutant effects on vegetation. We recognize that each of these uses may require different kinds and amounts (intensity) of research effort. We have attempted to be objective in developing these recommendations, but are unapologetic for any biases that remain.

3.5.1 Economic assessment

A substantial amount of exposure-response data has been collected (Heck *et al.*, 1983, 1984) to support an economic assessment of ozone effects on agriculture (Adams *et al.*, 1984, 1988). The results of NCLAN economic assessments clearly show that ozone has a significant impact on the agricultural economy. It is now appropriate to ask, how much improvement in the accuracy and precision of these estimates is justified in support of standard setting, and what additional data are needed to provide this improvement?

3.5.1.1 Do we need more and better national-level assessments?

The initial economic assessment (Adams *et al.*, 1984) was based on the six crop species that are grown on approximately 70% of the crop acreage in the U.S. The final economic estimate (Adams *et al.*, 1988) will be based on eight crops representing approximately 90% of the total US crop acreage. The addition of more crop species is unlikely to greatly influence the magnitude of economic benefit estimates in future national-level assessments.

As a part of the preliminary economic assessment, Adams *et al.* (1984) conducted a series of sensitivity analyses to estimate the influence of various factors such as cultivar selection (differential cultivar sensitivity), ambient ozone concentration, and soil moisture status on the economic assessment. Individually, these factors did influence the magnitude of the monetary losses up to approximately 70%. However, several of the sources of error introduced by uncertainty in the crop less estimates tended to cancel each other out. It is not clear that additional accuracy in the crop loss functions would significantly improve the precision of future economic assessments.

Empirical studies of Adams *et al.* (1982) show that the quality of the economic and biological response information contributes equally to the reliability of the final economic assessment. Consequently, improvements in economic estimates will require not only improvements in the representativeness and precision of the ozone/crop loss functions but also characterization of the economic processes on which the assessment is based.

Finally, there are sources of uncertainty associated with the estimation of the ambient ozone concentrations as well as with the economic model and with the crop-response data. A continued reduction in uncertainty in only one of these areas will not necessarily increase the credibility of the resulting national economic assessment.

3.5.1.2 *Regional crop loss assessments: many unanswered questions*

Regional crop loss assessments, on the other hand, could be improved substantially by more accurate and more complete crop loss data. The NCLAN crop species were selected to represent the major fraction of US agriculture with the minimum number of field experiments. The crop/cultivar selection did not reflect the range and diversity of crops grown in all regions of the country.

These data deficiencies can significantly influence the results of regional crop loss assessments. For example, a recent EPA study that attempted to use data from the NCLAN program for assessing the impacts of ozone on agriculture in the Northeast found that the species for which NCLAN had developed exposure–response functions accounted for less than 50% of the agricultural value of the area. Clearly, in this case, data on additional crops are needed to reduce the errors in the regional economic assessment.

We suggest that future economic assessments focus on specific crop-growing regions. By limiting the scope of such studies to a specific region, additional effort can be placed upon improving the resolution of ambient ozone exposure data, utilizing region-specific experimental exposure regimes, and testing the crop species/cultivars relevant to the region.

Ecoregions (e.g., Omernik, 1987) identify discrete areas of the country that are relatively homogeneous in climatic and edaphic factors. Ambient ozone and environmental data typical of these ecoregions could be used as treatment levels in studies to determine the variation in plant response to ozone within and among ecoregions and to identify the key factors that cause or control the variation. These improvements in data specificity should improve resulting economic assessments.

3.5.1.3 *So many cultivars, so little time*

Commercial cultivars continue to be introduced that have not been and cannot be studied as extensively as those studied in the NCLAN program. It will never be possible to study all regionally important crops or crop cultivars or to account for the ever-changing mix of cultivars. Consequently, methods are needed for rapidly screening cultivar response to ozone and for making inferences about the ozone response of untested species or cultivars from the response of tested species or cultivars.

3.5.1.4 Alternative approaches to economic assessments should be tested

Most previous economic assessments have been based on exposure–response functions (Adams *et al.*, 1984; Kopp *et al.*, 1985; Shortle *et al.*, 1986). However, the use of alternative methods with different inherent assumptions would provide a good check on the results of exposure–response-based assessments. Mjelde *et al.* (1984) used the duality approach to assess the impacts of ozone on grain production in Illinois. Their estimates of the monetary impact were similar to those estimated using the exposure–response function approach. In this case, the duality method was an inexpensive means of producing economic benefit estimates that were consistent with those produced using crop-response functions. However, to date, the duality approach has not been applied on a regional or national basis.

3.5.2 Ambient air-quality standards

Answers to several questions are needed to support standard setting. Relevant questions include:

1. What is an adverse effect?
2. At what levels of pollution do adverse effects occur?
3. What is the magnitude of variation in adverse effects among species and plant cultivars?
4. What is the appropriate air-quality measure (e.g., averaging the time of the exposure that relates plant response to ambient ozone concentrations)?

To provide the necessary scientific facts to answer these and related questions, data from an array of scientific studies are needed.

3.5.2.1 What aspect of pollutant exposure causes the problem?

The temporal dynamics of pollutant exposure are thought to influence plant response. The relative influence of the components of exposure dynamics (exposure concentrations, durations, frequencies, and respite intervals) on plant response is not well understood. Several lines of evidence suggest that peak pollutant concentrations have a greater role in determining plant response than long term average concentrations (US EPA, 1986).

Various air-quality summary statistics have been used to characterize pollutant exposure. The difficulty in selecting the most appropriate air-quality measure has been summarized by Heagle and Heck (1980). Ambient and experimental ozone exposures have been presented as seasonal, monthly, weekly, or daily means; peak hour means; number of hours above a selected concentration; or the numbers of hours above selected concentration intervals. However, none of these statistics adequately characterizes the relationships among ozone concentration, exposure duration, interval between exposures (respite), and plant response. A recent analysis illustrated that no one measure was best for all crop species analysed (Lee *et al.*, 1988). However, measures of air quality that emphasized the importance of peaks and accumulated the impacts of the exposure over the season performed better than other measures of exposure.

3.5.2.2 *How reliable are the NCLAN crop response functions?*

NCLAN was designed, within the necessary time and financial constraints, to provide national assessments of agricultural crop losses likely to occur from ozone exposure. The experimental design was formulated to address the major sources of variation that were originally thought to have the greatest influence on ozone-induced crop losses. Within the Program's resource constraints, it was not possible to determine the representativeness of the NCLAN exposure–response functions. To lend credibility to the NCLAN exposure–response data and the resultant economic assessment, the reproducibility (over time and space) of the exposure–response functions must be established.

Several experimental approaches could be used to validate the exposure–response functions. For example, open-top chamber studies could be conducted at one or more field locations, using a range of simulated ambient ozone exposure regimes (that mimic the temporal variations, as well as concentration ranges in ambient air). Thus, the growing environment would be the only uncontrolled variable. Because simulated ambient exposure can be replicated from year to year and across sites, the only uncontrolled source of variation is that attributable to environmental conditions. Although this research will do little to reduce possible sources of error in the response functions, it will at least permit an estimate of the possible errors that result from varying environmental conditions.

Consideration should also be given to using chemical protectants to validate the exposure–response functions over a range of conditions. Initial studies should be conducted in open-top chambers to determine if the chemical protectant has a direct effect on the crop in the absence of ozone. Subsequent studies should be conducted using open-top chambers at one location and studies with chemical protectants at several locations in the geographic region of interest. The results from the chemical protectant studies would be compared with those predicted from open-top chamber studies as a means for validating the response functions and determining possible biases associated with either method.

3.5.2.3 *Are interactions between pollutants important?*

The study of the effects of pollutant combinations on plants is based on the premise that pollutants co-occur in the atmosphere and that together they may induce more (or less) injury than that induced by the additive effects of the individual pollutants. Components of ambient air, such as sulfur dioxide or nitrogen dioxide may moderate or change plant response to ozone (e.g., Ormrod, 1982). The magnitude of the potential modifications varies with plant species, cultivar, pollutant concentration, duration and frequency of exposure, and the environmental and edaphic conditions in which plants are grown.

However, as future studies of pollutant combinations are planned, greater consideration must be given to the pollutant exposure regimes. The regimes should reflect the concentrations and frequency of pollutant occurrence typical of the area of study or for which inferences will be made. It is not sufficient to simply assume that the gases must co-exist in the atmosphere because they are emitted into the atmosphere.

This simplistic approach fails to account for the temporal and spatial variation in pollutants and their levels of occurrence. Co-occurrence of sulfur dioxide, nitrogen dioxide, and ozone at phytotoxic levels is relatively rare (Lefohn and Tingey, 1984; Lefohn *et al.*, 1987*a*). Also, though most previous studies of plant effects have considered only the simultaneous occurrence of pollutants, most co-occurrences are sequential rather than simultaneous. Because ozone does not frequently co-occur with sulfur dioxide and nitrogen dioxide, greater consideration should be given to studying the interaction of ozone with other regionally distributed pollutants such as acidic precipitation and acidic fog.

In future studies, consideration should also be given to simulating the temporal patterns of pollutant occurrence. In ambient air, nitrogen dioxide concentrations frequently peak and then decline due to photolysis preceding the elevated levels of ozone. Runeckles and Palmer (1987) exposed plants sequentially to nitrogen dioxide and then ozone. Nitrogen dioxide pretreatment altered the plants' subsequent response to ozone.

3.5.2.4 *How do environmental variables influence ozone response?*

It is well established that environmental factors such as light, air temperature, relative humidity, and soil moisture significantly affect the response(s) of plants to ozone and other air pollutants (e.g., US EPA, 1986). Because these factors vary greatly both temporally and spatially, their effects on ozone-induced yield loss in crops require better explanation. A greater understanding of the effects of environmental factors, such as relative humidity, on ozone-induced injury or yield loss is also important for establishing a process-based, theoretical approach for predicting such injury or crop impacts across different locations and years.

The multitude of factors (environmental, nutritional, edaphic, etc.) and subsequent interactions that may influence plant response to ozone are too numerous to analyse effectively and study individually in a reasonable period of time. Consequently, we must consider alternative approaches to the sequential study of the individual factors.

For example, it may be possible to rank the importance of the various factors based on our knowledge of their physiological and biochemical influences on plants. Their relative importance in altering the ozone impact could then be inferred. Alternatively, principal components analysis (PCA) could be applied to experimental data to determine which combination of factors is most important in altering the plant response. Once the "short" list of important factors is known for a particular regional environment, future studies could elucidate their interrelationships.

3.5.3 Advancement of scientific understanding

Because of limitations in funding and the need to provide the essential data for the economic assessment, much of the NCLAN research has been limited to an empirical approach. This technique provided large amounts of plant response data but contributed relatively little to understanding the mechanisms responsible for plant response. To ultimately understand how ozone impairs plant growth and yield and to use this knowledge in a predictive mode, an improvement in our understanding of the basic mechanisms of ozone phyto-toxicity is needed.

Although there have been numerous biochemical and physiological studies to investigate the fundamental nature of ozone phytotoxicity, most have been limited in scope and have focused on a single level of study. For example, the majority of studies of biochemical and/or cellular processes have not con-sidered ozone uptake. Consequently, we frequently know that different clones or cultivars exhibit different biochemical responses to ozone, but we do not know if these differences are the result of different amounts of ozone flux into the plant or if the differences actually represent different biochemical sensitivities to ozone. Unless there is more integration and coordination among the various studies, this limitation will persist.

Future studies should be initiated to develop a conceptual framework of how ozone affects plant processes. From a comprehensive conceptual framework, the data and information necessary to advance scientific understanding could be identified. Given the numerous processes that could be investigated, a conceptual road map is needed to ensure that the limited financial resources are directed toward the most important or promising areas.

An "expert system" may provide a suitable format for a comprehensive conceptual framework. Expert systems are computer programs that suggest solutions to complex problems by mimicking human reasoning processes and employing a knowledge base developed by human experts (Stone *et al.*, 1986). All expert systems contain a data or knowledge base and an executive program that interprets the knowledge base to answer questions posed.

The knowledge base is a collection of facts, generalities, opinions, and empirical information that represents the state of the world, encoded so that the relevant relationships among facts and generalities can be understood. The executive program controls the system's reasoning process. It determines the sequence of logical steps that leads to solution of the problem. Expert systems provide an accessible and powerful means to transfer existing expert knowl-edge and experience to a non-expert.

The expert system would constitute a dynamic decision-making model depicting our developing understanding of the effects of ozone and other stresses on the quantity and quality of crops and the ramifications of these effects in the agricultural infrastructure. The knowledge base could function as an "electronic air-quality criteria document" that is continually updated. The executive program could incorporate the necessary scientific reasoning to apply the growing knowledge base to various Agency regulatory problems.

In the process of developing and implementing an expert system, types of data necessary to exercise the system in a problem-solving mode would become clear. Research could then be designed to provide the data efficiently and systematically.

3.6 SUMMARY

In 1980, the US EPA undertook initiation of a crop loss assessment program. The NCLAN was conceived as a multidisciplinary study to produce an economic assessment of the effects of air pollutants on US agriculture based upon state-of-the-science exposure–response data. A network of field sites was established in regions where ten field crops accounting for roughly 85% of US acreage are planted. Replicated ozone exposure–response experiments were conducted for each of the crops. The resulting exposure–response data set is by far the most comprehensive data set for ozone effects on vegetation available.

National economic assessments have been completed fully utilizing the NCLAN exposure–response database. This required conversion of SAROAD air-quality data into an appropriate form and development of an appropriate model of the national farm economy. NCLAN pioneered the use of kriging to estimate ambient ozone exposures from point sample data in the SAROAD database (Lefohn *et al.*, 1987*b*). The farm economy model developed by Adams *et al.* (1984) was the most comprehensive and sophisticated of any air pollution assessment to date.

A preliminary assessment (Adams *et al.*, 1984) in 1984 estimated that reductions in ambient ozone concentrations of 10, 25 and 40% would yield annual economic surpluses of 0·67, 1·71, and 2·52 billion 1980 dollars, respectively. The results of the final assessment (Adams *et al.*, 1988) significantly improve these estimates.

Several sources of uncertainty in the estimates of economic loss remain that point out future research directions. Given that ozone has an economic impact on agriculture, it is appropriate to ask, how much improvement in the accuracy and precision of these estimates is justified in support of standard setting and what additional data are needed to provide this improvement?

We conclude that significantly improving the national economic assessments will be costly. There are many independent areas of uncertainty in the analyses, and a continued reduction in uncertainty in one or a few of these areas will not necessarily increase the credibility of the resulting national economic assessment.

We suggest that future economic assessments focus on specific crop growing regions. Exposure–response data are needed for important crops in several regions of the United States. By limiting the scope of such studies to a specific region, additional effort can be placed upon improving the resolution of ambient ozone exposure data, utilizing region-specific experimental exposure regimes, and testing the crop species/cultivars relevant to the region.

Methods are needed for rapidly determining the ozone response and for making inferences from the response of tested species or cultivars about the response of untested species or cultivars. In addition, alternative approaches to economic assessments should be tested.

Research has not yet determined unambiguously the relative importance of the components of pollutant exposure in inducing plant response. Both field and controlled-environment studies are needed to determine the relationships between crop response and the statistics used to characterize ozone exposure.

Research is also needed to determine the reliability of the NCLAN exposure–response functions and to determine how important interactions with other pollutants and other environmental variables are in determining plant response to ozone exposure.

Future research could be conducted more efficiently if a general conceptual framework relating pollutant exposure to plant response were available. From such a conceptual framework, it would be possible to identify the data and information necessary to advance scientific understanding. Given the numerous processes that could be investigated, a road map is needed to ensure that the limited financial resources are directed toward the most important areas.

An expert system might provide a useful format for the conceptual framework. Expert systems are computer programs that suggest solutions to complex problems by mimicking human reasoning processes and employing a knowledge base developed by human experts (Stone *et al.*, 1986). All expert systems contain a data or knowledge base and an executive program specifying rules for use of the knowledge base.

The process of developing and implementing an expert system would make evident the types of data necessary to exercise the system in problem solving. Research could then be designed to efficiently and systematically provide those data.

The expert system would also provide a means to continually incorporate improved scientific understanding into regulatory decision-making. The knowledge base could function as an "electronic air-quality criteria document" that is continually updated and the executive program could incorporate the necessary scientific reasoning to apply the growing knowledge bases to suggest solutions to various Agency control scenarios.

REFERENCES

Adams, R. M., T. D. Crocker, and N. Thanavibulchai. (1982). An economic assessment of air pollution damages to selected annual crops in southern California. *J. Environ. Econ. Managemt,* **9,** 42–58.

Adams, R. M., S. A. Hamilton, and B. A. McCarl. (1984). The economic effects of ozone on agriculture. Corvallis, OR, US Environmental Protection Agency, EPA-600/3-84-090.

Adams, R. M., J. D. Glyer, and B. A. McCarl. (1988). The NCLAN economic assessment: Approach, findings and implications. In *Assessment of crop loss from air pollutants*. Proceedings of the international conference, Raleigh, N.C., USA, ed. by W. W. Heck, O. C. Taylor, and D. T. Tingey, 473–504. London, Elsevier Applied Science.

Coffey, D. S., S. C. Sprenger, D. T. Tingey, G. E. Neely, and J. C. McCarty. (1988). National crop loss assessment network: Quality assurance program. *Environ. Pollut.*, **53**, 89–98.

Heagle, A. S., and W. W. Heck. (1980). Field methods to assess crop losses due to oxidant air pollutants. In *Crop Loss Assessment*. Proc. E. C. Stakman Commemorative Symposium, St. Paul, Minn., ed. by P. S. Teng and S. V. Krupa, 296–305. University of Minnesota, Miscellaneous Publication No. 7.

Heagle, A. S., D. E. Body, and W. W. Heck. (1973). An open-top field chamber to assess the impact of air pollution on plants. *J. Environ. Qual.*, **2**, 365–8.

Heck, W. W., R. M. Adams, W. W. Cure, R. J. Kohut, L. W. Kress, and P. J. Temple. (1983). A reassessment of crop loss from ozone. *Environ. Sci. Technol.*, **17**, 572A-81A.

Heck, W. W., W. W. Cure, J. O. Rawlings, L. J. Zaragoza, A. S. Heagle, H. E. Heggestad, R. J. Kohut, L. W. Kress, and P. J. Temple. (1984). Assessing impacts of ozone on agricultural crops: II. Crop yield functions and alternative exposure statistics. *J. Air Pollut. Control. Assoc.*, **34**, 810–17.

James, W. C. and P. S. Teng. (1979). The quantification of production constraints associated with plant diseases. In Applied Biology Volume IV, ed. by T. H. Coaker, 201–67. London, Academic Press.

King, D. A. (1987). A model for predicting the influence of moisture stress on ozone-caused crop losses. *Ecol. Model*, **35**, 29–44.

King, D. A. and W. L. Nelson. (1987). Assessing the impacts of soil moisture stress on regional soybean yield and its sensitivity to ozone. *Agric. Ecosys. Environ.*, **20**, 23–35.

Kopp, R. J., W. J. Vaughn, M. Hazilla, and R. Carson. (1985). Implications of environmental policy for US agriculture: The case of ambient ozone standards. *J. Environ. Managemt*, **20**, 321–31.

Lee, J. H. and R. A. Lewis. (1978). Zonal air pollution system, design and performance. In *The bioenvironmental impact of coal-fired power plants*, 3rd Interim Report, ed. by E. M. Preston and R. A. Lewis, 332–44. Corvallis, OR, US Environmental Protection Agency, EPA-600/3-79-021.

Lee, E. H., D. T. Tingey, and W. E. Hogsett. (1988). Evaluation of ozone exposure indices in exposure–response modeling. *Environ. Pollut.*, **53**, 43–62.

Lefohn, A. S. and D. T. Tingey. (1984). The co-occurrence of potentially phytotoxic concentrations of various gaseous air pollutants. *Atmos. Environ.*, **18**, 2521–6.

Lefohn, A. S., C. E. Davis, C. K. Jones, D. T. Tingey, and W. E. Hogsett. (1987*a*). Co-occurrence patterns of gaseous air pollutant pairs at different minimum concentrations in the United States. *Atmos. Environ.*, **21**, 2435–44.

Lefohn, A. S., H. P. Knudsen, J. Logan, J. Simpson, and C. Bhumralkar. (1987*b*). An evaluation of the kriging method, as applied by NCLAN, to predict 7-h seasonal ozone concentrations. *J. Air Pollut. Control Assoc.*, **37**, 595–602.

Mjelde, J. W., R. M. Adams, B. L. Dixon, and P. Garcia. (1984). Using farmers' actions to measure crop loss due to air pollution. *J. Air Pollut. Control Assoc.*, **31**, 360–4.

Omernik, J. M. (1987). Ecoregions of the conterminous United States. *Ann. Assoc. Am. Geol.*, **77**, 118–25.

Ormrod, D. P. (1982) Air pollutant interactions in mixtures. In *Effects of gaseous air pollutants in agriculture and horticulture*, ed. by M. H. Unsworth, and D. P. Ormrod, 307–31. London, Butterworths Scientific.

Oshima, R. J., M. P. Poe, P. K. Braegelmann, D. W. Baldwin, and V. Van Way. (1976). Ozone dosage–crop loss function for alfalfa: A standardized method for assessing crop losses from air pollutants. *J. Air Pollut. Control Assoc.*, **26**, 861–5.

Padgett, J. and H. Richmond. (1983). The process of establishing and revising national ambient air quality standards. *J. Air Pollut. Control Assoc.*, **33**, 13–16.

Runeckles, V. C. and K. Palmer. (1987). Pretreatment with nitrogen dioxide modifies plant response to ozone. *Atmos. Environ.*, **21**, 717–19.

Shortle, J. S., J. W. Dunn, and M. Phillips. (1986). *Economic assessment of crop damage due to air pollution: the role of quality effects*, Staff Paper No. 118, Department of Agricultural Economics, Pennsylvania State University, College Station.

Stone, N. D., R. N. Coulson, R. E. Frisbie, and D. K. Loh. (1986). Expert systems in entomology: Three approaches to problem solving. *Bull. Entomol. Soc. Am.*, Fall, 161–6.

US Environmental Protection Agency. (1978). Air-quality criteria for ozone and other photochemical oxidants. Research Triangle Park, NC, U.S. Environmental Protection Agency, EPA-600/8-78-004.

US Environmental Protection Agency. (1986). Air-quality criteria for ozone and other photochemical oxidants. Volume III. Research Triangle Park, NC, U.S. Environmental Protection Agency, EPA-600/8-84/020cF.

Session II

METEOROLOGY, ATMOSPHERIC CHEMISTRY AND REGIONAL MONITORING—EXTRAPOLATION

(Bruce B. Hicks, *Chairman*; Allen S. Lefohn, *Co-Chairman*)

4

METEOROLOGY—ATMOSPHERIC CHEMISTRY AND LONG RANGE TRANSPORT

A. P. ALTSHULLER

US Environmental Protection Agency, Research Triangle Park, North Carolina, USA

4.1 INTRODUCTION

Ozone (O_3) is a secondary pollutant with many distinctive characteristics with respect to its sources and modes of formation with regions of the troposphere and in the stratosphere. While O_3 can be rapidly removed by both chemical and physical processes near the surface, its lifetime is substantially extended aloft. The scales of intermediate and longer range transport influencing the atmospheric distribution of O_3 will be discussed. As a result of these various chemical and meteorological processes, atmospheric O_3 concentrations can vary substantially during each day, from day to day, and with season of the year as well as with geographical location. Although O_3 is a regional-scale pollutant, individual and overlapping plumes, especially urban plumes, can significantly augment the regional background concentrations of O_3.

In contrast to O_3, sulfur dioxide (SO_2) is a primary pollutant. Both the sources of SO_2 and its mechanisms of removal are much different than for O_3. Sulfur dioxide exposures tend to be of concern to agriculture on a local rather than on a regional scale.

4.2 OZONE

4.2.1 Sources of ozone

Ozone is produced within (1) the planetary boundary layer (PBL) extending from the surface to about 2 km; (2) in the free troposphere, the region between the PBL and the stratosphere; and (3) in the stratosphere. In the

1960s and even later it was believed that O_3 measured at rural/remote locations was only of stratospheric origin (Junge, 1963). Subsequent research demonstrated that intermediate and longer range transport of O_3 and its precursors occurs downwind of urban areas and of industrial sources (Altshuller, 1986a; US Environmental Protection Agency, 1978, 1986).

Stratospheric O_3 is produced after photolysis of molecular oxygen by ultraviolet solar radiation penetrating the stratosphere. The atomic oxygen formed reacts with molecular oxygen to form O_3 (in the presence of N_2 and O_2 as third bodies). The O_3 can be photolyzed back to O_2 and atomic oxygen as well as react with atomic oxygen to form molecular oxygen. These reactions are important in determining the equilibrium conditions for O_3 in the stratosphere (Junge, 1963).

In the troposphere, O_3 formation depends on the photolysis of NO_2 as a source of atomic oxygen. The O_3 is formed by the reaction of the atomic oxygen with O_2. The O_3 in the troposphere is consumed primarily by reactions with NO and NO_2. In the free troposphere, the reaction of methane, CO, and the more persistent nonmethane organic compounds with NO_x are the primary source of O_3 (Fishman and Carney, 1984; Fishman et al., 1985).

Within the planetary boundary layer, a number of sources of O_3 precursors capable of forming O_3 during subsequent transport in plumes downwind have been identified. These sources include (1) plumes from urban areas, (2) fossil fuel power plant plumes and (3) industrial plumes such as those from petroleum refineries (Altshuller, 1986a).

Under many circumstances it is difficult to quantify the contributions from various sources to the O_3 concentrations measured at a rural location. To estimate such contributions it is desirable to measure several other chemical species, which can serve as tracers of the sources of the O_3. However, only a limited number of the rural O_3 measurement projects also provide such tracer measurements.

To estimate the contribution of the stratospheric source to ground-level O_3 concentrations, [7]Be and ^{90}Sr concentration measurements have been used. Based on such tracer measurements it appears that during the warmer months of the year, the average stratospheric O_3 contribution to the ground-level O_3 concentration probably does not exceed 10 ppb (Altshuller, 1986a).

The in-situ production of O_3 in the boundary layer and the free troposphere is predicted to occur by various tropospheric photochemical models, and such production is necessary to account for the O_3 budget at mid-latitudes in the northern hemisphere. The progress in developing such models and the limitations of earlier efforts have been reviewed (Fishman, 1984).

When continental background concentrations of precursors are introduced into tropospheric O_3 models under summertime conditions, a midday O_3 concentration of 0·035 to 0·040 ppm (69 to 79 $\mu g\,m^{-3}$) is predicted near the surface during the summer at mid-latitudes in the northern hemisphere (Fishman and Carney, 1984). This result is in reasonably good agreement with the mean O_3 concentrations observed at many rural locations during summer months (Altshuller, 1986a).

When average eastern United States precursor emissions are used in such a tropospheric model and the emission rates assumed to be constant for 3·5 days, the O_3 concentration at the surface is predicted to increase to about 0·070 ppm (Fishman *et al.*, 1985). The O_3 concentration increases slightly during the next day or two and then drops off. In another modeling study, a stagnating high pressure system was assumed to occur for a week over southern central England (Derwent and Hov, 1982). The regionally averaged O_3 concentration is predicted to peak at 0·130 ppm. This value is not unreasonable compared to the concentrations of O_3 observed during the slow passage of high pressure systems (Altshuller 1985, 1986*a*, *b*; Vukovich *et al.*, 1977; Vukovich and Fishman, 1986).

The tropospheric models discussed above are one-dimensional box models. These models are limited in that they cannot include the impacts of individual urban plumes or overlapping plumes on O_3 concentrations in rural areas downwind. Higher O_3 concentrations than those predicted by tropospheric box models are observed in rural areas within the flow of urban plumes (Altshuller, 1986*a*).

One-dimensional tropospheric modeling of O_3 formation between 2 and 8 km predicts about a 0·004 ppm or an 11% increase in average northern hemisphere O_3 between 1966 and 1980 (Dignon and Hameed, 1985). At mid-latitudes in the northern hemisphere the absolute magnitude of the increase between 1966 and 1980 in the O_3 concentration is predicted to be 0·006 to 0·007 ppm. These increases in O_3 concentration are associated with the increase in the anthropogenic NO_x emissions during the same period. Increases in CO and nonmethane hydrocarbon emissions have not been included in these model calculations.

Reviews of available O_3 measurement results indicate that the average O_3 concentrations over continents may have increased substantially with time both at the surface (Logan, 1985; Bojkov, 1986) and aloft (Angell and Korshover, 1983; Dignon and Hameed, 1985). It has been estimated that the means of the daily values of surface O_3 concentrations have at least doubled, particularly in the summers of recent years compared to those observed in the second half of the 19th century (Bojkov, 1986).

Based on such estimates it appears that the background level of O_3 may have been substantially lower, possibly half or less that of current values, early in the period of industrialization. However, even the O_3 values measured during the second half of the 19th century may not correspond to the "natural" background O_3.

4.2.1.1 *Diurnal variations in O_3 concentrations*

The profiles of the diurnal variations in O_3 concentrations are very similar among various studies from a number of lower elevation rural monitoring locations. These profiles demonstrate the following characteristics:

1. The O_3 concentration peaks during the late morning and early afternoon hours.

2. A flat peak is observed with little variation in the O_3 concentration during the afternoon hours.
3. The O_3 concentration rapidly decreases during the evening hours.
4. This rapid decrease is followed by a slower decrease in O_3 concentration during the early morning hours with a minimum at about 0600 hours.
5. A rapid increase in O_3 concentration is observed in the morning between 0800 and 1200 hours.

The minimum value for O_3 concentration early in the morning appears lower during the warmer months of the year because it is followed by a more rapid rise in the O_3 during the late morning hours. During the warmer months of the year, the ratio of the diurnal maximum to minimum O_3 concentrations usually is in the range of $2:1$ to $3:1$ (Decker *et al.*, 1976; Evans *et al.*, 1983, Evans, 1985; Karl, 1978*a*). At a number of higher elevation locations the diurnal variations observed in the O_3 concentrations tend to be smaller than at isolated lower elevation rural locations (Singh *et al.*, 1978; Oltmans, 1981; Fehsenfeld *et al.*, 1983; Evans, 1985).

Examples are shown of diurnal O_3 profiles of rural stations in the St Louis area (Fig. 4.1). When these stations are in upwind flow, flat O_3 profiles are observed during the late morning and the afternoon hours.

The rapid rise in O_3 concentration in the morning begins well before the solar radiation intensity and temperature begin to reach their maximum daily values. At least up to about 1000 h, the increase in O_3 concentrations appears to be explained largely by O_3 mixing down from the layer isolated aloft the

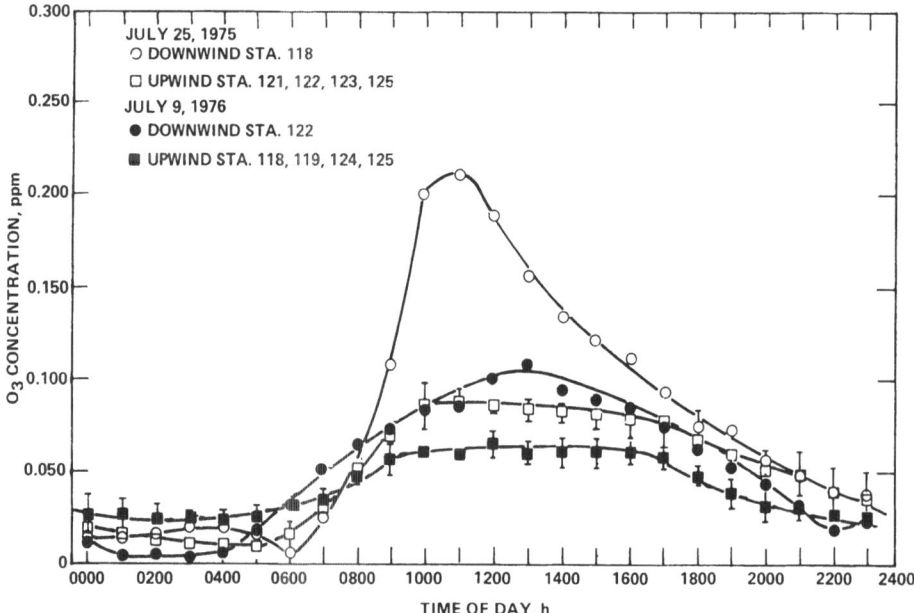

Fig. 4.1. Diurnal profiles of mean upwind of O_3 concentrations and the maximum O_3 concentrations in plumes observed at RAMS stations in St Louis area.

previous evening (Evans, 1979; Kelly *et al.*, 1984*a*; Altshuller, 1986*a*). Any subsequent increase in O_3 concentration late in the morning into the afternoon hours is associated with same day photochemical formation of O_3 within the planetary boundary layer.

As the turbulent mixing due to heating decreases during the evening, the mixed layer collapses, and a shallow nocturnal inversion layer forms at the surface. The O_3 isolated in the layer aloft tends to be stable overnight. Depletion would occur only when the O_3 aloft encounters NO_x within elevated plumes. The O_3 trapped near the surface decreases in concentration overnight as discussed above. The depletion at the surface in rural areas is likely to be controlled by the rate of dry deposition (Garland and Derwent, 1979; Kelly *et al.*, 1984*a*). At mountain locations, excluding mountain valleys, smaller diurnal variations are observed because less dry deposition of O_3 occurs overnight probably because of limited surface area.

4.2.2 Seasonal and shorter time variations in regional O_3 concentrations

The regional O_3 concentrations refer to O_3 generated on larger than an urban scale, which cannot be directly associated with specific plumes. These O_3 concentrations would be those measured at relatively remote locations. However, as discussed previously, these O_3 concentrations do not necessarily approach "natural" background. The regional O_3 concentration varies seasonally and geographically as well as from day to day.

The month-to-month variations in the regional background of O_3 have been based on O_3 concentrations measured during all hours, and O_3 concentrations during midday and on maximum 1-h O_3 concentrations. A plot of the monthly means of the daily averages of the O_3 concentrations at 1200 to 1500 hours Central Standard Time, which were measured in air parcels flowing through rural sites upwind of St Louis during 1975 and 1976 provides an example (Fig. 4.2). Large seasonal variations are apparent with July to January O_3 concentration ratios equal to or exceeding $3:1$ (Altshuller, 1986*b*). The peak 1200 to 1500 hours average monthly O_3 concentrations of 0·070 to 0·080 ppm during 1975 and 1976 around St Louis occur during June, July, or August.

The monthly averages of all hours of O_3 concentrations reported at the nine Sulfate Regional Experiment (SURE) class I sites in the eastern United States in 1978 showed monthly peaks of 0·045 to 0·060 ppm most frequently in June, but monthly peaks occurred earlier in the spring at some of these sites (Mueller *et al.*, 1983*b*). The monthly averages of all hours of O_3 concentrations at three rural sites in Minnesota and North Dakota for the years 1977 to 1981 peaked at 0·045 to 0·050 ppm in May (Pratt *et al.*, 1983).

Peaks in the monthly average O_3 concentrations have been observed at several relatively remote sites, usually higher elevation mountain locations, during the early spring months (Singh *et al.*, 1978). Such early spring peaks have been attributed to but not demonstrated to be associated with stratospheric O_3 injections into the troposphere (Singh *et al.*, 1978).

High O_3 concentrations in the late spring and summer at rural locations are most likely to be associated with photochemical O_3 formation in both the

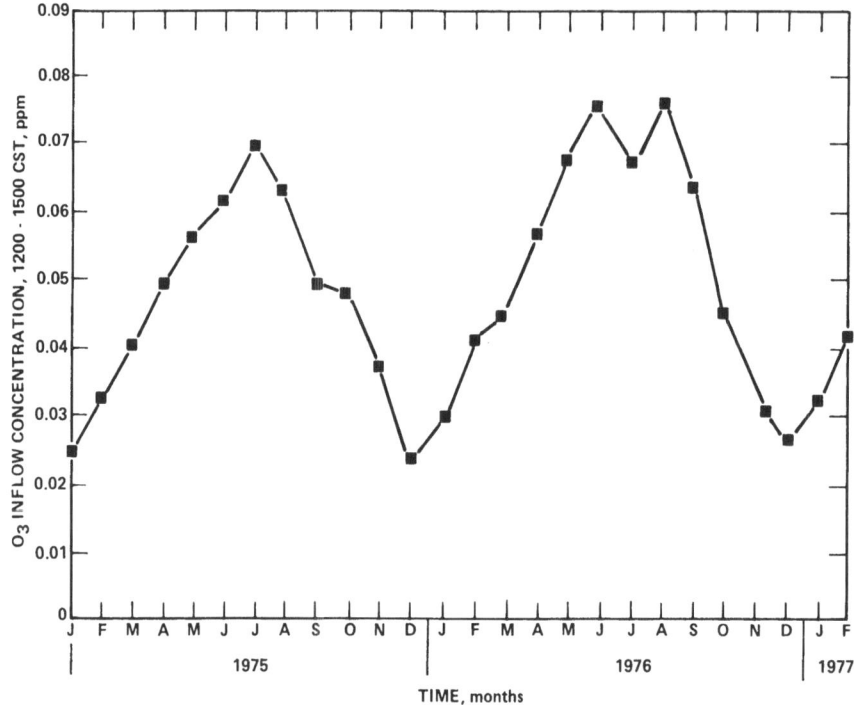

Fig. 4.2. The monthly mean O_3 inflow concentrations, 1200–1500 h, during January 1975 through February 1977 at RAMS stations in St Louis area.

planetary boundary layer and the free troposphere (Altshuller, 1986a). Higher O_3 concentrations observed regionally in the lower midwestern, mid-Atlantic and southern states during the late spring and summer can be associated with augmentation of O_3 that is formed within slowly moving migratory high pressure systems (Vukovich et al., 1977; Altshuller, 1986a,b; Vukovich and Fishman, 1986). Such slowly moving systems occur much less frequently in the western United States, the Upper Midwest, or in New England. Therefore, these systems have less effect on O_3 formation in these regions.

Biogenic emission inventories for isoprene and terpenes predict substantial hydrocarbon loadings in forested areas (Altshuller, 1983). However, the ambient air measurements of hydrocarbons indicate that anthropogenic hydrocarbons usually predominate in rural areas (Altshuller, 1983). In rural farming areas where forests are sparse or nonexistent, biogenic hydrocarbons are unlikely to be a factor in O_3 formation.

In the late spring and summer, substantial day-to-day variations occur in the O_3 concentrations measured at rural locations. Variations in midday O_3 concentrations ranging from 0·035 to 0·040 ppm up to 0·105 to 0·125 ppm have been observed during several summer months in upwind air flow into rural sites around St Louis, MO (Shreffler and Evans, 1982). The higher O_3 concentrations can be associated with the slow passage of high pressure systems (Vukovich et al., 1977; Karl, 1978b; Altshuller, 1985). The lower O_3 concentrations can be associated with passage of cleaner air from the northwest (Altshuller, 1986*b*) or with the lower temperatures and solar radiation intensities associated with rain and/or heavy cloud cover.

4.2.3 Climatological patterns of O_3 distribution during summer months

As noted earlier in the discussion of the 1975 and 1976 monthly mean O_3 concentrations upwind of St Louis, the monthly pattern does change from one year to the next (Altshuller, 1986*b*). A much larger body of O_3 concentrations for the months of July and August for 1977 to 1981 extending from monitoring stations at 100°W longitude eastward have been analysed climatologically (Vukovich and Fishman, 1986). Monthly mean O_3 concentrations of 0·04 to 0·05 ppm are usually observed in the extreme northern United States, in areas west of about 95°W longitude, and in the Gulf Coast states. However, in the lower midwestern and eastern United States, the monthly mean O_3 concentrations can range from 0·06 to 0·09 ppm.

The year-to-year variations in the monthly mean O_3 concentrations differ with geographic location over the 5-year period as follows: Virginia, 0·05 to over 0·09 ppm; central Ohio, 0·06 to 0·08 ppm; eastern Missouri, 0·05 to 0·08 ppm; and central Iowa, less than 0·04 to 0·06 ppm. The shifts in O_3 concentrations from year to year have been attributed to changes in the paths of the migratory high pressure systems (Vukovich and Fishman, 1986).

The monthly mean O_3 concentrations in northeastern Missouri and southeastern Iowa prove to be consistently higher than in northwestern Iowa by as much as 0·03 ppm, but the increment varies considerably from year to year. For the 6 months when concentration differences could be quantified, the average difference in O_3 concentration between northeast Missouri and southeast Iowa was in the range of 0·015 to 0·02 ppm.

When monthly mean O_3 concentrations are compared between central Ohio and the St. Louis area, it is found that for 4 months the O_3 concentration was 0·01 to 0·02 ppm higher in central Ohio, for 3 months the O_3 concentration was 0·005 to 0·010 ppm higher in the St. Louis area, and for 3 months there was no discernible difference in the O_3 concentrations between the two areas. Overall, this comparison indicates a slightly higher O_3 concentration, ≤0·005 ppm, in central Ohio than in the St Louis area. Almost identical average 5-year O_3 values are obtained when comparing central Ohio with either the Washington, DC, area or with Connecticut.

The consistently low monthly mean O_3 concentrations observed in the extreme northern United States can be attributed to less favorable meteorological conditions for O_3 formation in terms of lower temperatures, lower solar radiation intensities, and/or higher wind speeds. Also, the influence of the high pressure systems is minimal for the accumulation of O_3 precursors in these areas. In areas of the western United States with consistently low O_3 concentrations, a minimal influence of high pressure systems on formation of O_3 probably accounts for the low concentrations observed in July and August.

4.2.4 Formation of O_3 in plumes

Measurements of O_3 in the plumes downwind of large cities indicate that the O_3 concentrations often exceed those in the regional background by 0·05 to

0·10 ppm with occasional increments above 0·10 ppm (Spicer *et al.*, 1979; Altshuller, 1986*a*). Smaller increments in O_3 above background O_3 have been observed in the individual plumes of smaller cities along with intermediate incremental O_3 concentrations in overlapping plumes of smaller cities (Spicer *et al.*, 1982; Sexton, 1983). Such results are based both on O_3 measured in aircraft traverses of plumes as well as on surface O_3 measurements downwind of cities (Altshuller, 1986*a*). Excess O_3 concentrations in urban plumes occasionally have been observed up to several hundred kilometers downwind, but most aircraft O_3 measurements have been limited to downwind distances of 50 to 150 km. The extreme range for occurrence of excess O_3 in urban plumes is not well established. The impact of such plumes on crops is dependent on the intensity and duration of the O_3 formation in the plumes as well as the frequency with which the plumes flow over various areas. The impact of urban plumes should be greater in regions with higher densities of cities and towns along with nearby farming areas where O_3 sensitive cultivar varieties are grown. Higher densities of cities and towns can lead to fumigation of a specific rural area by different plumes as wind direction shifts from day to day. Augmentation of the plume of one city or town by that of one or more cities or towns also may occur in such higher density areas.

Ozone formation above regional background also has been observed in plumes from other sources, especially within the plumes of fossil fuel power plants (Altshuller, 1986*a*). However, immediately downwind of such sources the plumes contain large excesses of NO_x as they actually deplete O_3 from the regional background. If these plumes are able to accumulate sufficient hydrocarbon from the background air, an HC-to-NO_x ratio favorable to O_3 formation can result during the daytime. It can take several hours of downwind transport before such a condition is obtained (Gillani *et al.*, 1978; Gillani and Wilson, 1980). With taller stacks, the plume's tendency to fumigate the surface is limited to the portion of the day after the mixing height increases to plume elevations and before the mixed layer collapses below the plume elevations (Gillani and Wilson, 1983; Husar *et al.*, 1978). Therefore, power plant plumes are most likely to have their impact on O_3 at the surface during the afternoon hours. A review of measurements aloft indicates maximum increments over background of 0·02 to 0·05 ppm in the O_3 within power plant plumes (Altshuller, 1986*a*). The net diurnal O_3 within power plant plumes during first day transport actually could be negative, but such an assessment does not appear to have been made on a diurnal basis.

A more detailed evaluation of the impact of O_3 formation in the St Louis plume on rural terrain downwind has been made (Altshuller, 1988). The Regional Air Monitoring System (RAMS) station O_3 measurements upwind and downwind have been used along with aircraft measurements of O_3 in the St Louis plume when available. Examples of diurnal O_3 profiles at downwind sites show late morning or afternoon peaks not observed at upwind sites (Fig. 4.1).

These measurements indicate that during 1975 and 1976 the St Louis urban plume was detectable on 196 days, almost half of the days between April and

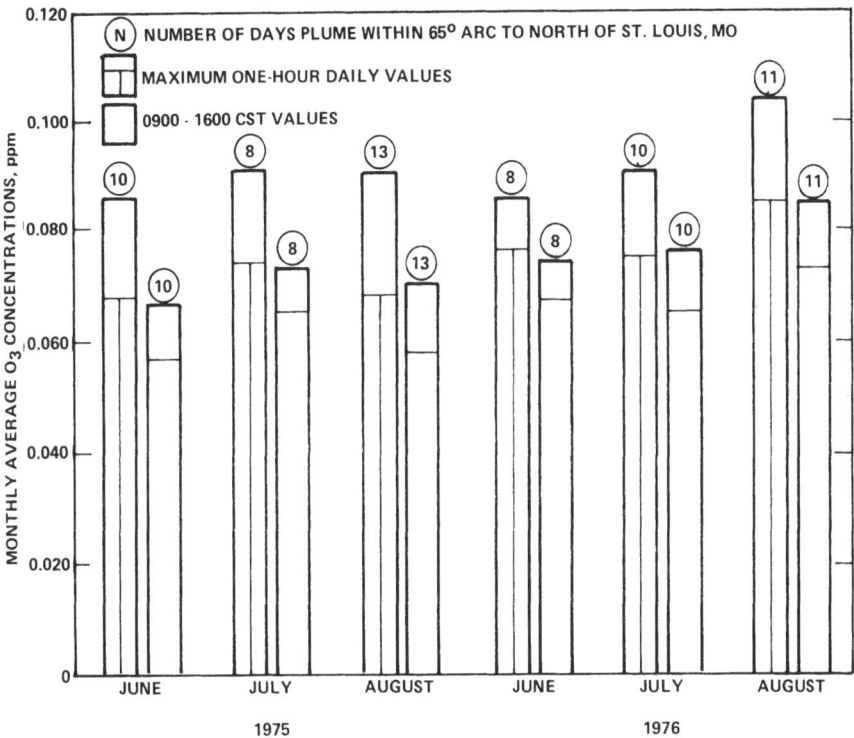

Fig. 4.3. Monthly average O_3 concentrations in June, July, and August 1975 and 1976 within 65° arc north of St Louis, MO. Maximum 1-h concentrations upwind plus plume O_3 values and 0900–1600-h upwind plus plume O_3 values.

October (Altshuller, 1988). On more than half of these days on which plumes form, the prevailing wind is such that the plumes they impact are on an arc from the northwest to northeast of St Louis. Plots of the monthly means for the regional background and the maximum 1-h O_3 concentrations and for the 0900 to 1600-hours daily O_3 concentrations for the regional background and incremental O_3 in the St Louis plume are shown for a 65° arc north of St Louis (Fig. 4.3). The excess O_3 concentrations in the urban plume impact on one or more of the four RAMS stations (114, 115, 121, or 122) to the north of St Louis. The quantity $(\overline{D - U})_{7h}$ is the difference between the mean of the 0900 to 1600-hours O_3 concentrations for the station nearest the center of the plume and the average of means for this time period of the O_3 concentrations at upwind stations. The values of $(\overline{D - U})_{7h}$ (upper part of bars) for months between June and September 1975 and 1976 averages 0·010 ppm and 0·09 ppm, respectively, during the 0900 to 1600-hours time period. The corresponding maximum 1-h O_3 increments were about twice as large. Similar plots for shorter time periods show widely varying urban plume increments, $(\overline{D - U})_{7h}$, from zero to 0·03 to 0·04 ppm (Fig. 4.4). The higher range of values is obtained on several of the 7- to 8-day periods. The corresponding $(\overline{D - U})_{1h}$ values for such shorter periods can range up to 0·05 to 0·07 ppm.

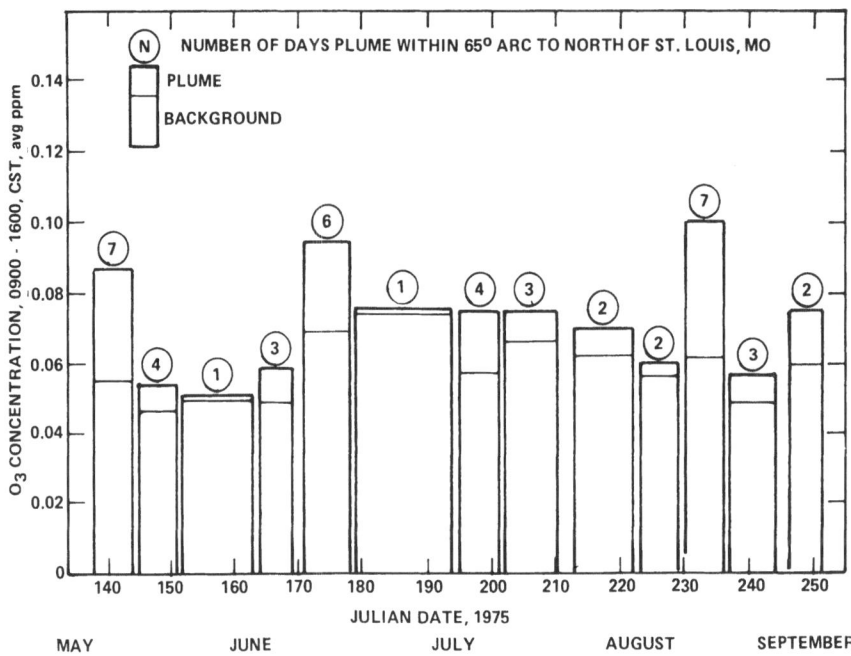

Fig. 4.4. Mean O_3 concentrations during various periods of days in May to September 1975 obtained from daily 0900–1600 h upwind plus plume O_3 concentration values within 65° arc north of St Louis, MO.

In general, significant clustering of higher incremental plume values of O_3 occur over some multiday time periods. Persistence of wind flow in the prevailing wind direction for up to 7 days occurred.

An estimate has been made of the percentage of rural lands impacted on days in late spring and summer by plumes from cities and towns in the region between the Appalachian Mountains and 105°W longitude. Based on reasonably conservative 1-day plume transport distances and plume widths, on any given day, between 5 and 10% of this region is estimated to be impacted by plumes (Altshuller, 1988). The impact of urban plumes would be experienced to a large extent but not exclusively in the prevailing surface wind direction, which tends to be from the south to southeast for plumes in areas to the west of St Louis and south to southwest for plumes in areas to the east of St Louis during the warmer months of the year (Baldwin, 1968). Therefore, most of the impact of plumes in this region will be to the northwest to northeast of urban areas on the downwind rural areas and crop lands.

4.2.5 Longer range transport of O_3

The question arises as to the available evidence demonstrating O_3 transport over longer distances. The evidence for multiday transport of O_3 over ocean areas adjacent to populated land masses is much better than over land (Altshuller, 1986a).

It should not be assumed that the occurrence of high O_3 concentrations as high pressure systems move across the United States from west to east necessarily demonstrate concurrent transport of O_3 over long distances within the planetary boundary layer. The high O_3 concentrations observed may result from favorable meteorological conditions for O_3 formation locally or sub-regionally rather than from O_3 transported from long distances upwind.

High velocity, low-level (nocturnal jet) winds do occur, particularly over the Great Plains and are stronger and more frequent in the summer (Bonner, 1968; Bonner *et al.*, 1968; Evans, 1979; Gillani *et al.*, 1984). A lower frequency of such nocturnal jet winds occurs along the East Coast. Winds can reach 50 km h^{-1}. The prevailing wind direction for these winds aloft during the summer is from the southwest. This means that O_3 and its precursors isolated aloft originating from either urban plumes or power plant plumes can be transported many hundreds of kilometers overnight to the northeast. How-ever, with mixing down to the surface and the deposition of O_3 occurring during the second day and night, it is not evident how much of the original O_3 survives over land beyond the second day. The details of how the NO_x in the narrow nighttime power plant plumes (Smith *et al.*, 1978) can participate in O_3 formation on the subsequent day over the region have not been developed. Nevertheless, the O_3 in the urban and power plant plumes the previous day certainly contributes to the regional background, and this O_3 could originate at least several hundred kilometers upwind of the location of second day measurements.

Ozone and its precursors, CO, hydrocarbons and NO_x, can be transported into the lower free troposphere by updrafts in active cumulus clouds formed during the afternoon hours in the spring and summer months, as well as in the debris from clouds that were active (Ching *et al.*, 1984; Liu *et al.*, 1984*b*; Ching and Alkezweeny, 1985, 1986). In the free troposphere, the residence time of O_3 increases so it can exist well beyond the second day after emission of precursors. The CO, CH_4 and residual nonmethane hydrocarbons, along with available NO_x, can contribute to additional O_3 formation through in-situ formation in the free troposphere as discussed previously. A significant portion of the O_3 and precursors originally present within plumes may be isolated well aloft by these meteorological processes for multiday periods. Subsequently, they will be brought down into the planetary boundary layer by downdraft processes. Consequently, it is likely that a plume can make a longer term contribution to regional O_3 background concentrations after it can no longer be tracked in the planetary boundary layer.

4.2.6 Geographical distribution of higher O_3 concentrations

There is evidence that higher concentrations of O_3 may make a dispropor-tionate contribution to the injury sustained by vegetation exposed to O_3 (US Environmental Protection Agency, 1986). Therefore, several comparisons of rural sites in various geographical areas of the United States have been assembled from the pertinent literature. These comparisons present statistics on the number of hours the 1-h O_3 concentrations exceed 0·08, 0·10, or 0·120 ppm (Tables 4.1 and 4.2). In addition, the circumstances under which the highest O_3 concentrations occur at rural sites are discussed.

TABLE 4.1

Comparisons of peak O_3 concentrations and total hours with O_3 concentrations above 0·10 ppm among rural sites during periods in 1975

Station	Location	Measurement period	Max. 1-h $O_3 \cdot$ conc. (ppm)	Total hours >0·10 ppm	Reference
St Louis RAMS Site 122	45 km N of St Louis, MO	6-27 to 9-30	0·256	101	This work
Bradford, PA	NW PA	6-27 to 9-30	0·124	40	Decker et al., 1976
Poynette, WI	S WI	6-27 to 9-30	0·121	24	Decker et al., 1976
DeRidder, LA	SW LA	6-27 to 10-30	0·128	6	Decker et al., 1976
Lewisburg, WV	SW WV	6-27 to 9-30	0·113	2	Decker et al., 1976
Creston, IA	SW IA	6-27 to 9-30 6-27 to 9-30	0·122	1	Decker et al., 1976
Wolf Point, MT	NE MT		0·064	0	
St. Louis RAMS Site 122	45 km N of St Louis, MO	7-16 to 8-13	0·143	32	This work
Glasgow, IL	104 km NNW of St Louis, MO	7–16 to 8–13	0·130	8	Rasmussen et al. (1977)

Table 4.1 provides the maximum 1-h O_3 concentrations at a number of rural sites during summer periods in 1975 as well as the total number of hours that the hourly O_3 concentrations exceeded 0·10 ppm. Substantially more hours with concentrations above 0·10 ppm were observed at the rural RAMS Site 122° north of St Louis, MO, than at any other rural site listed. This site also is the only location that is frequently within the urban plume. In contrast, very few if any hours with O_3 concentrations above 0·10 ppm occurred at the sites well to the west of the Mississippi River, such as in Creston in southwestern Iowa or Wolf Point in northeastern Montana in 1975 (Decker et al., 1976).

To present a more detailed view of the effect of an urban plume more consideration will be given to RAMS Site 122. At this site, a total of 130 h with concentrations in excess of 0·10 ppm occurred during all of 1975. The RAMS Site 122 was upwind of St Louis for 37 h when the O_3 concentration exceeded 0·1 ppm. On the days in 1975 when RAMS Site 122 was downwind of St Louis with hourly O_3 concentration exceeding 0·10 ppm, the O_3 concentration exceeded 0·10 ppm for only 5 h at the upwind sites. Therefore, about two-thirds of the hours with O_3 concentrations exceeding 0·10 ppm occurred when RAMS Site 122 was impacted by the plume of St Louis. The presence of the urban plume even more strongly influenced the frequency of O_3 concentrations above 0·12 ppm at RAMS Site 122. When this site was within the plume, the O_3 concentrations exceeded 0·12 ppm for 33 h, whereas when the site was in upwind of St Louis the O_3 concentration exceeded 0·12 ppm during 2 h.

During the 1-month period in 1975 when measurements of O_3 were made outside Glasgow, IL (over 100 km north northwest of St Louis) the O_3 concentration exceed 0·10 ppm at Glasgow for 8 h during three days (Ras-

TABLE 4.2

Comparisons of total hours with ozone concentrations above 0·10 ppm and 0·12 ppm and maximum 1-h O₃ concentrations at various rural sites

Station	Location	Measurement period	Max 1-h O₃ conc. ppm	No. of h O₃ conc.			Reference
				>0·08 ppm	>0·10 ppm	>0·12 ppm	
Montague	W MA	Aug to Dec 1977	0·153	60	33	21	Martinez and Singh (1979)
Indian River	S DE	Aug to Dec 1977	0·099	29	0	0	
Research Triangle Park	C NC	Aug to Dec 1977	0·188	80	10	0	
Lewisburg	SE WV	Aug to Dec 1977	0·106	23	3	0	Research Triangle Inst. (1975)
Duncan Falls	SE OH	Aug to Dec 1977	0·107	52	2	0	
Scranton	NE PA	Aug to Dec 1977	0·077	0	0	0	
Fort Wayne	NE IN	Aug to Dec 1977	0·080	0	0	0	
Rockport	NE IN	Aug to Dec 1977	0·099	17	0	0	
Giles Co.	SC TN	Aug to Dec 1977	0·117	63	5	0	
McHenry	W MD	June 14 to August 31, 1974	0·165	262	86	36	
Du Bois	WC PA	June 14 to August 31, 1974	0·200	341	197	77	
McConnelsville	SE OH	June 14 to August 31, 1974	0·165	239	77	14	
Wilmington	SW OH	June 14 to August 31, 1974	0·185	259	142	25	
Wooster	NE OH	June 14 to August 31, 1974	0·165	262	162	39	
Green Mt. NF*	N VT	1979	0·105	63	NA†	0	Evans *et al.* (1982, 1983); Evans (1985)
Croatan NF	NE NC	1979	0·085	7	0	0	
Chequamegon NF	N WI	1979	0·110	53	NA	0	
Mark Twain NF	N MO	1979	0·125	203	NA	NA	
Kisatchie NF	N LA	1979	0·110	5	NA	0	
Custer NF	SE MT	1979	0·070	0	0	0	
La Moure Co.	SE MN	1977 to 1981	NA	NA	30	8	Pratt *et al.* (1983)
Traverse Co.	W MN		NA	NA	18	6	Kelly *et al.* (1982)
Wright Co.	SE ND		NA	NA	19	2	
Pierre, ND	WC ND	July 13 to September 6, 1978	0·056	0	0	0	

* National Forest.
† Not available.

mussen, *et al.*, 1977). This contrasted with the 32 h on 9 days at RAMS Site 122. The lower frequency of elevated O_3 concentrations at Glasgow is likely to result from (1) fewer urban plume fumigations impacting Glasgow than the RAMS Site 122 and (2) depletion of O_3 by dry deposition during the transport of the plume towards Glasgow during the late afternoon or early evening hours.

Measurements were made at a rural site in the Mark Twain National Forest about 125 km south of St Louis during 1979 and 1983 (Evans *et al.*, 1982, 1983; Evans, 1985). A back trajectory analysis for the 1979 episode when the maximum 1-h O_3 concentration reached 0·125 ppm indicated that the site was downwind of the St Louis area (Evans *et al.*, 1982, 1983). An even higher O_3 concentration was observed in 1980, 0·155 ppm, but this episode has not been evaluated (Evans, 1985). Over a 5-year period, 1979 through 1983, the maximum 1-h O_3 concentration at Mark Twain National Forest varied from 0·095 to 0·155 ppm.

Table 4.2 summarizes the total number of hours that hourly O_3 concentrations exceeded 0·08, 0·10, and 0·12 ppm as reported in a number of studies (Research Triangle Institute, 1975; Martinez and Singh, 1979; Evans *et al.*, 1982, 1983; Kelly *et al.*, 1982; Pratt *et al.*, 1983; Evans, 1985).

The Sulfate Regional Experiment (SURE) O_3 measurements in 1977 only began in August so the number of hours is not representative of the entire period that O_3 could have occurred at elevated concentrations in 1977 (Martinez and Singh, 1979). The O_3 measurements made in the SURE study in 1978 did cover the warmer months of the year, but no equivalent statistics for the 1978 results are available. The exception is an 8-day episode in July 1978 (Blumenthal *et al.*, 1981; Mueller *et al.*, 1983a; Liu *et al.*, 1984a). During this episode, the O_3 concentration did exceed 0·10 ppm for a number of hours and peaked at over 0·15 ppm in a few areas. These peak values were attributed in discussion of these aircraft measurements to traverses of urban plumes by the aircraft (Blumenthal *et al.*, 1981).

The slower movement of high pressure systems tends to occur as the systems approach the Appalachian Mountains. Climatologically, high pressure systems stagnating in excess of a week occur approximately once a year over a 40-year period (Korshover, 1975). Therefore, very persistent stagnations with large accumulations of O_3 precursors occur over the Ohio Valley and adjacent areas.

Rural areas well west of the Mississippi River tend to have relatively few if any days per year with hourly O_3 concentrations exceeding 0·10 ppm (Tables 4.1 and 4.2). However, in these western areas, as elsewhere, higher frequencies of elevated O_3 concentrations would be expected to be observed in rural areas within a few hundred kilometers of metropolitan areas.

In the instances where hourly O_3 concentrations exceeding 0·12 ppm are observed, aircraft profiles and back trajectories are usually consistent with downwind transport from a specific urban area (Altshuller, 1988; Blumenthal *et al.*, 1981; Evans *et al.*, 1982, 1983; Martinez and Singh, 1979; Pratt *et al.*, 1983).

It appears that the sites used to construct the isopleths of the monthly means of the daily maximum 1-h O_3 concentrations were selected so as to minimize the influence of urban plumes (Vukovich and Fishman, 1986). If so, it is possible to estimate and superimpose the additive effects of urban plumes onto the regional O_3 concentrations. While different years of O_3 measurements must be utilized, an example illustrating the importance of the contributions from urban plumes is useful to consider.

During the months of July and August 1977 to 1981 the monthly means of the daily 1-h maximum O_3 concentrations were from 0·015 to 0·020 ppm higher in rural areas of northeastern Missouri and southeastern Iowa compared to rural areas of western Iowa (Vukovich and Fishman, 1986). In comparison, the incremental monthly means of the daily 1-h O_3 concentrations within the St Louis plume in the prevailing wind direction to the north were as follows: July 1975, 0·017 ppm; August 1975, 0·022 ppm; July 1976, 0·016 ppm; and August 1976, 0·019 ppm (Altshuller, 1988). These incremental values in this urban plume are in the same range of concentrations as are the differences in the regional background levels between northeastern Missouri and western Iowa. Therefore, the total incremental exposure to O_3 based on this statistic to the rural area north of St Louis would be 0·03 to 0·04 ppm higher than those in rural areas of western Iowa.

4.3 SULFUR DIOXIDE

4.3.1 Sources of SO_2

Anthropogenic sources of gaseous sulfur are likely to dominate emissions in those local areas or regions where crops or forests are grown. Natural gaseous emissions of sulfur for a large region of the eastern United States have been estimated based on 1978 emission inventories to contribute somewhat less than 1% of the total sulfur emissions (Adams *et al.*, 1980, 1981*a*, *b*; Homolya and Robinson, 1984). For the contiguous United States, the natural gaseous emissions of sulfur have been estimated to contribute 4 to 5% of the anthropogenic emissions of gaseous sulfur. The emission rates of natural gaseous sulfur decrease rapidly with increasing latitude (Adams *et al.*, 1981*a*, *b*). Therefore the contributions of natural sulfur emissions to total sulfur emissions over Canada and western Europe also should be small.

In North America, the highest ambient air concentrations of SO_2 in past years have been primarily associated with local SO_2 emissions from smelters, especially in the Sudbury district of Ontario, Canada. The SO_2 emissions during the 1950s and 1960s from the smelters in the Sudbury district have been estimated at over 2 million tons annually, compared to the 3·5 million tons annually for all smelters in the United States (Linzon, 1971). Improvements have occurred in more recent years. In heavily industrialized areas of Europe, regional emissions of SO_2 can also be of concern with respect to possible damage to vegetation (Roberts, 1984).

Worst-case conditions for low-level releases of SO_2 occur with light wind, stable conditions (Turner, 1979). For elevated point sources in relatively level terrain, worst-case conditions can be associated with (1) light wind, unstable conditions for 1 to 3 h and/or (2) moderate wind, neutral conditions with persistent high wind for longer periods. However, this division is not clear cut (Turner, 1979). Under unstable conditions, even with stack heights of 200 to 300 m, maximum SO_2 concentrations are expected to occur within a few kilometers of the source. Under moderate wind and neutral conditions, maximum SO_2 concentrations are less, and maximum impact is located further from the source (Turner, 1979).

Dose–response relations can be obtained from the study made near the iron-ore smelters at Biersdorf, Federal Republic of Germany (Guderian and Stratmann, 1962). The results obtained are consistent with significant effects being associated with fumigations under light wind, stable conditions near the sources (Turner, 1979). The study by Linzon (1971) downwind of the Sudbury smelters involved plots of eastern white pine located from 30 to 176 km from the smelters. In this latter study, damage occurred at greater distances from the sources, in the prevailing wind direction, indicating greater influence from moderate wind.

4.3.2 Distributions of ambient air concentrations of SO_2

Only a limited group of measurements for SO_2 are available at rural locations. The rural sites at which SO_2 measurements have been made were primarily located in the midwestern states east of the Mississippi River, middle Atlantic states, and adjacent states in the southeast (Mueller et al., 1980; 1983b,c; Shaw and Paur, 1983; Altshuller, 1984). The monthly averaged SO_2 concentrations available at the nine SURE Class I sites and the three Ohio River Valley Study sites have been distributed into four applicable concentration ranges in Table 4.3. These distributions are for all months in each study and also for the months between April and September. Even though the sites of Class I stations were selected carefully to avoid local effects, there is a considerable difference in the distributions of values among ranges for these sites. The extremes ranged from the Research Triangle Park, NC, site where 14 of the 15 monthly averaged SO_2 concentration values were at or below 0·005 ppm (13 μg m^{-3}) and none were above 0·010 ppm to the site at Duncan Falls, OH, where only 1 out of the 15 monthly average SO_2 concentration values was at or below 0·005 ppm and 7 of the 15 monthly average values exceeded 0·010 ppm. The sites in New England and in the southeastern states tended to have the lower SO_2 concentration values. During the months of April through September, the SO_2 concentration values occur in the lower concentration ranges with SO_2 concentrations ranging from 0·012 to 0·003 ppm and below.

The results from the three sites in the Ohio River Valley Study (Shaw and Paur, 1983) which define a line from southwest to northeast 590 km in length, indicate that for these sites the SO_2 concentrations tended to have fewer extreme values, than observed at the SURE sites. In general, the SO_2 concentration values at these three sites are closely clustered with the monthly average values during the second and third quarters ranging from 0·04 to 0·07 ppm.

TABLE 4.3

Distribution of monthly average SO_2 values by concentration range from field measurement studies among SO_2 concentration ranges

Site	Measurement periods[a]	Number of months in measurement periods[a] SO_2 concentrations (ppm)				Reference
		<0·005	0·006–0·010	0·011–0·019	0·020–0·025	
Montagne, MA	August 1977 thru October 1978, 8	11, 8	4, 0	0, 0	0, 0	Mueller *et al.* (1983*b*)
Scranton, PA	months between April and October 1977–1978	2, 2	8, 2	4, 4	1, 0	
Indian River, DE		5, 5	5, 2	3, 1	0, 0	
Research Triangle Park, NC		14, 7	1, 1	0, 0	0, 0	
Giles Co, TN		10, 8	2, 0	1, 0	0, 0	
Lewisburg, WV		10, 8	2, 0	2, 0	0, 0	
Duncan Falls, OH		1, 1	7, 5	4, 2	3, 0	
Fort Wayne, IN		8, 7	5, 1	2, 0	0, 0	
Rockport, IN		2, 2	7, 5	6, 1	0, 0	
No. of months		63, 48	41, 16	22, 8	4, 0	
% of total months		48, 67	32, 22	17, 11	3, 0	
Union Co, KT	May 1980 thru August 1981, 8	5, 3	9, 7	2, 0	0, 0	Shaw and Paur (1983)
Franklin Co, IN	months of April –September 1980–81	2, 1	10, 9	4, 0	0, 0	
Ashland Co, OH		6, 5	7, 5	3, 0	0, 0	
No of months		13, 9	26, 21	9, 0	0, 0	
% of total months		27, 30	54, 70	19, 0	0, 0	

[a] all months, April through September.

Isopleths of the monthly average SO_2 values have been constructed for the eastern United States based on the results from the nine SURE Class I stations and the 45 SURE Class II stations (Mueller *et al.*, 1980, 1983*b*). Sulfur dioxide measurements were obtained at all of these stations during August 1977, October 1977, January–February 1978, April 1978, July 1978, and August 1978. The SO_2 isopleths enclosing the areas with the highest monthly average values were in the 0·02 to 0·03 ppm range in the spring, summer, and fall months, and the values were in the 0·03 to 0·04 ppm range in the winter. The maximum monthly averaged SO_2 values at the SURE Class I stations are too small for these stations to be included within the highest SO_2 isopleth areas.

Ten of the 45 SURE Class II stations have been shown to have a higher frequency of SO_2 concentrations above 0·01 ppm than do the other Class II stations or the Class I stations (Mueller *et al.*, 1983*c*). These higher SO_2 concentrations are attributed, at least in part, to local sources of SO_2 near the 10 stations. One or more large point sources of SO_2 are located within 30 km of most of these 10 stations, and point sources are located within 10 km of several of these stations. As a result, the average SO_2 concentrations during the entire study period at stations in this group of 10 stations exceeded the average SO_2 concentrations of 0·009 ppm for the other 35 Class II stations by factors of two to three, and at one site, Brush Valley in western Pennsylvania, the average SO_2 concentration was 0·032 ppm. Examination of the geographical position of these sites with respect to the locations of the areas within the highest SO_2 isopleths indicate that these highest isopleths often are determined by the SO_2 concentrations at the group of 10 stations.

Sulfur dioxide measurements are available at several other rural locations in the eastern United States. Quarterly average SO_2 concentrations have been reported for two locations in upper New England, Coos Co., NH, and Acadia National Park, ME (Altshuller, 1984). The second and third quarter SO_2 values between 1968 and 1973 ranged between 0·002 and 0·005 ppm. Measurements of SO_2 were made at Whiteface Mountain, NY, during July 1982 (Kelly *et al.*, 1984*b*). A filter pack and a flame photometric instrument with enhanced sensitivity that provided a limit of detection of 0·0003 ppm for hourly averaged values were utilized. The monthly mean SO_2 value was 0·0008 ppm with approximate average clean and polluted air SO_2 values of 0·0003 ppm and 0·004 ppm, respectively. At Kejimkujik, Nova Scotia, the geometric mean SO_2 concentration obtained by a filter-pack technique during the period between November 1978 and December 1979 was 0·0003 ppm (Barrie, 1982). At a coastal site 5 km outside of Lewes, DE, the mean SO_2 concentration for August 1982 was 0·0021 ppm, whereas for the period January 25 to February 28, 1983, it was 0·0045 ppm (Wolff *et al.*, 1986). At a rural site in southwestern Louisiana the average SO_2 concentration during a summer period, August 5 to September 9, 1979, was 0·0013 ppm (Kelly *et al.*, 1984*a*).

Although the SO_2 concentrations outside the eastern areas with high area-wide SO_2 emissions are low at rural locations, there are exceptions. The highest SO_2 concentrations reported in 1978 across the United States were in rural areas near western smelters (US Environmental Protection Agency, 1982). The highest annual mean SO_2 concentration was 0·058 ppm at a site in Montana about 3 km northeast of a smelter. Except near smelters, SO_2 concentrations at rural locations in the United States occur at or well below the lowest SO_2 concentrations associated with small yield losses to crops caused by SO_2 (Roberts, 1984).

4.4 CONCURRENT DISTRIBUTIONS OF O_3 AND SO_2 CONCENTRATIONS

There has been interest in the effects on vegetation of concurrent exposures to O_3 and to SO_2 (US Environmental Protection Agency, 1986). Therefore, the concurrent distributions of ambient air O_3 and SO_2 concentrations will be considered.

Ozone and SO_2 exhibit different distributions in the atmosphere both on a seasonal and diurnal basis. The SO_2 concentrations in the first and fourth quarters of the year exceed those in the second and third quarters at rural locations (Mueller *et al.*, 1980, 1983*b*; Shaw and Paur, 1983; Altshuller, 1984).

As discussed previously, O_3 concentrations are higher in spring and summer. Diurnal concentrations of SO_2 during July 1978 averaged over SURE Class I sites showed about a twofold increase in SO_2 concentrations during the morning hours, a small decrease through the remaining daytime hours, and a greater decrease in concentration into the night (Mueller *et al.*, 1983*b*). In contrast, O_3 concentrations usually increase steadily from early morning into the afternoon hours. During a winter period, January to February 1978, almost no diurnal variations were observed in SO_2 concentrations when averaged over SURE Class I sites (Mueller *et al.*, 1983*b*).

Ratios of monthly averages of O_3 to SO_2 at SURE sites are listed in Table 4.4. The O_3-to-SO_2 ratios always equal or exceed 5:1 at the Montague, MA, the Research Triangle Park, NC, the Giles Co., TN, and the Lewisburg, WV, locations. In May, June and July 1978, no O_3-to-SO_2 ratio below 5:1 occurred at eight of the nine sites. At several sites, O_3-to-SO_2 ratios equaled or exceeded 20:1 during some or all of the months between May and September.

Because of a lack of concurrent O_3 and SO_2 measurements and the failure to use the same time-averaging procedures, only estimates of O_3 to SO_2 ratios are possible for the other sites discussed earlier. Based on the available limited concentration measurements for SO_2 in rural areas in upper New England, the southeastern United States and west of the Mississippi River, monthly average SO_2 concentrations are unlikely to exceed 0·003 to 0·005 ppm during the warmer months of the year. Monthly mean O_3 concentrations (all hours) are available at a number of western sites, and these concentrations range between 0·02 and 0·045 ppm between March and October, with the lower O_3 concentrations, 0·02 to 0·03 ppm, occurring in September and October (Viezee *et al.*, 1982; Pratt *et al.*, 1983). Therefore, O_3-to-SO_2 ratios ranging from 5:1 to 10:1 appear as reasonable estimates at western locations during the spring and fall months. Based on the monthly averages of the maximum 1-h O_3 concentrations available, O_3 concentrations in the southeast and upper New England should equal or exceed those west of the Mississippi River (Vukovich and Fishman, 1986). Therefore, the O_3-to-SO_2 ratios in these regions should

TABLE 4.4

Ozone to sulfur dioxide ratios based on monthly average concentration values obtained at SURE Class I monitoring stations during 1978

	March	*April*	*May*	*June*	*July*	*August*	*Sept.*	*Oct.*
Montague, MA	5:1	6:1	14:1	14:1	13:1	≥20:1	10:1	6:1
Scranton, PA	4:1	9:1	8:1	19:1	5:1	4:1	2:1	1:1
Indian River, DE	4:1	4:1	3:1	18:1	≥20:1	16:1	3:1	4:1
Research Triangle Park, NC	13:1	14:1	16:1	17:1	16:1	18:1	11:1	7:1
Giles Co., TN	6:1	≥20:1	≥20:1	≥20:1	≥20:1	≥20:1	≥20:1	7:1
Lewisburg, WV	8:1	≥20:1	≥20:1	≥20:1	≥20:1	≥20:1	≥20:1	17:1
Duncan Falls, OH	ND	ND	ND	6:1	7:1	5:1	3:1	2:1
Fort Wayne, IN	5:1	9:1	≥20:1	>20:1	8:1	4:1	6:1	3:1
Rockport, IN	3:1	5:1	5:1	6:1	5:1	4:1	4:1	3:1

be as high or higher than west of the Mississippi River. The exceptions to these estimates would be near large point sources of SO_2, especially smelters, where much lower O_3-to-SO_2 ratios than those estimated above are likely to occur.

4.5 SUMMARY

Ozone concentrations over rural areas generally tend to decrease substantially from east to west. Significant variations in O_3 concentrations occur at a given rural site from month to month, day to day, and throughout the day.

Monthly means of daily 1-h maximum O_3 concentrations can vary from year to year by 0·02 to 0·04 ppm at any given rural site (Vukovich and Fishman, 1986). However, averaged over 5-year periods, these O_3 concentrations during summer months tend to be within 0·005 ppm of each other at eastern rural sites. In contrast, the monthly means of daily 1-h maximum O_3 concentrations of 0·04 to 0·05 ppm in rural areas several hundred or more kilometers to the west of the Mississippi River are 0·01 to 0·03 ppm lower than in rural areas in the eastern United States.

Mean annual O_3 concentrations based on all hours of the day for 11 rural sites where measurements are available for 4- to 6-year periods show considerable variability from year to year at some of these sites (Pratt et al., 1983; Evans, 1985). Differences in the mean annual O_3 concentrations between high and low concentration years during these 4- to 6-year periods vary from as little as 0·003 ppm to as much as 0·014 ppm. At 4 of the 11 sites, these annual mean differences equaled or exceeded 0·01 ppm.

Unlike the summer monthly means of daily maximum 1-h O_3 concentrations, the mean annual O_3 concentrations did not show a consistent decrease from eastern to western sites in the United States. This latter result may be associated with the use of all hours of measurement as well as by the higher elevations at several of the western sites. Higher elevation sites, outside of mountain valleys, often show essentially flat diurnal profiles (Singh et al., 1978). As discussed earlier, lower elevation rural sites do show appreciable decreases in O_3 concentrations during the night and early morning hours. The differences in these profiles are associated with the substantial rate of dry deposition of O_3 trapped by the shallow nocturnal inversions formed over lower elevation rural sites.

The maximum 1-h O_3 concentration during a year also can vary substantially from year to year as does the number of hours with concentrations exceeding 0·08, 0·10, and/or 0·12 ppm at a site (Pratt et al., 1983; Evans, 1985). Maximum 1-h O_3 concentrations as well as the number of hours at which O_3 concentration is above 0·08, 0·10 and/or 0·12 ppm also show strong geographical corrections. It is unusual for a rural site in the western United States beyond the range of an urban plume to have a maximum 1-h O_3 concentration during a year exceeding 0·08 ppm (Singh et al., 1978; Evans, 1985). In contrast, at rural sites in the eastern United States, the maximum 1-h O_3 concentration during a year frequently exceeds 0·10 ppm.

To the extent that episodic exposures of crops to O_3 concentrations exceeding 0·10 ppm and especially exceeding 0·12 ppm are important in causing crop losses, the impact of urban plumes is very important. Regional concentrations of O_3 can build up to over 0·10 ppm within stagnating high pressure systems. Such systems tend to have their larger impact in portions of the eastern United States (Korshover, 1975). The impact of the incremental O_3 concentration formed in urban plumes is of significance throughout the entire United States downwind of cities (Spicer *et al.*, 1979; Altshuller, 1986*a*).

Higher SO_2 concentrations are observed near large point sources of SO_2, especially near smelters. In most rural areas, the monthly average concentrations of SO_2 are low, usually below 0·01 ppm and often below 0·005 ppm during the warmer months of the year.

Ratios of O_3-to-SO_2 concentrations tend to increase from winter to summer months. In the late spring and summer months, O_3-to-SO_2 ratios often range between 5:1 and 10:1 and equal or exceed 20:1 during some of these months at a number of rural locations.

REFERENCES

Adams, D. F., S. O. Farwell, E. Robinson, and M. R. Pack. (1980). Biogenic sulfur emissions in the SURE region. Final Report by Washington State University to Electric Power Research Institute, 3412 Hillview Ave. EPRI Report No. EA-1516, Palo Alto, CA 94304.

Adams, D. F., S. O. Farwell, M. R. Pack, and E. Robinson. (1981*a*). Biogenic sulfur gas emissions from soils in the eastern and southeastern United States. *J. Air Pollut. Control Assoc.*, **31**, 1083–9.

Adams, D. F., S. O. Farwell, E. Robinson, M. R. Pack, and W. L. Bamesberger. (1981*b*). Biogenic source strengths. *Environ. Sci. Technol.*, **15**, 1493–8.

Altshuller, A. P. (1983). Review: Natural volatile organic substances and their effect on air quality in the United States. *Atmos. Environ.*, **17**, 2131–65.

Altshuller, A. P. (1984). Atmospheric concentrations and distributions of chemical substances. Chapter A-5, Vol. 1. In *The acidic deposition phenomena and its effects: critical assessment review papers*, EPA-600/8-83-016AF 5-1–5-84, ed. by A. P. Altshuller and R. A. Linthurst.

Altshuller, A. P. (1985). Relationships involving fine particle mass, fine particle sulfur and ozone during episodic periods at sites in and around St Louis, MO. *Atmos. Environ.*, **19**, 265–76.

Altshuller, A. P. (1986*a*). The role of nitrogen oxides in nonurban ozone formation in the planetary boundary layer over N. America, W. Europe and adjacent areas of ocean. *Atmos. Environ.*, **20**, 245–68.

Altshuller, A. P. (1986*b*). Relationships between direction of wind flow and ozone inflow concentrations at rural locations outside of St Louis, MO. *Atmos. Environ.*, **20**, 2175–84.

Altshuller, A. P. (1988). Some characteristics of ozone formation in the urban plume of St Louis, MO. *Atmos. Environ.*, **22**, 499–510.

Angell, J. K. and J. Korshover. (1983). Global variation in total ozone and layer-mean ozone: an update through 1981. *J. Clim. Appl. Meteorol.*, **22**, 1611–27.

Baldwin, J. L. (1968). *Climate atlas of the United States.* US Department of Commerce.

Barrie, L. A. (1982). Environment Canada long range transport of atmospheric pollutants program: atmospheric studies. In *Acid precipitation effects on ecological systems,* ed. by F. M. D'Itri, 141–61. Ann Arbor, Ann Arbor Publications.

Blumenthal, D. L., W. S. Keifer, and J. A. McDonald. (1981). Aircraft measurements of pollutants and meteorological parameters during the Sulfate Regional Experiment (SURE) program. Report No. EA-1909. Prepared for Electric Power Research Institute, Palo Alto, CA 94304.

Bojkov, R. D. (1986). Surface ozone during the second half of the nineteenth century. *J. Clim. Appl. Meteorol.,* **25,** 343–52.

Bonner, W. D. (1968). Climatology of the low level jet. *Mon. Weather Rev.,* **96,** 833–50.

Bonner, W. D., S. Esbensen, and R. Greenberg (1968). Kinematics of the low-level jet. *J. Appl. Meteorol.,* **7,** 339–47.

Ching, J. K. S. and A. J. Alkezweeny. (1985). Vertical transport by cumulus clouds. Preprint Volume. Seventh AMS Symposium on Turbulence and Diffusion, 12–15 Nov., Boulder, CO.

Ching, J. K. S. and A. J. Alkezweeny. (1986). Tracer study of vertical exchange by cumulus clouds. *J. Climate Appl. Meteorol.,* **25,** 1702–11.

Ching, J. K. S., S. T. Shipley, E. V. Browell, and D. A. Brewer. (1984). Cumulus cloud venting of mixed layer ozone. *Proceedings of Quadrennial Ozone Symposium, Halkidiki, Greece,* 3–7 Sept. 1974, ed. by C. S. Zerefos and A. Ghazi. Hingham, MA, D. Reidel.

Decker, C. E., L. A. Ripperton, J. J. B. Worth, F. M. Vukovich, W. D. Bach, J. B. Tommerdahl, F. Smith, and D. E. Wagoner. (1976). Formation and transport of oxidant along Gulf Coast and in northern US. EPA-450/3-765-033. Available from National Technical Information Service, 5285 Port Royal Road, Springfield, VA 22161.

Derwent, R. G. and O. Hov. (1982). The potential for secondary pollutant formation in the atmospheric boundary layer in a high pressure situation over England. *Atmos. Environ.,* **16,** 655–65.

Dignon, J. and S. Hameed. (1985). A model investigation of the impact of increases in anthropogenic NO_x emissions between 1967 and 1980 on tropospheric ozone. *J. Atmos. Chem.,* **3,** 491–506.

Evans, R. B. (1979). *The contribution of ozone aloft to surface ozone maxima.* Ph.D. thesis, University of North Carolina at Chapel Hill, School of Public Health, Department of Environmental Science Engineering.

Evans, G. F. (1985). The National Air Pollution Background Network: Final Project Report. Environmental Monitoring Systems Laboratory, US Environmental Protection Agency, Research Triangle Park, NC 27711.

Evans, G., B. Finkelstein, N. Martin, N. Possiel, and M. Graves. (1982). The National Air Pollution Background Network, 1976–1980, EPA-600/4-82-058. US Environmental Protection Agency, Research Triangle Park, NC 27711.

Evans, G., B. Finkelstein, N. Martin, N. Possiel, and M. Graves. (1983). Ozone measurements from a network of remote sites. *J. Air Pollut. Control Assoc.,* **33,** 291–5.

Fehsenfeld, F. C., M. J. Bollinger, S. C. Liu, D. D. Parrish, M. McFarland, M. Trainer, D. Kley, P. C. Murphy, D. L. Albritton, and D. H. Lenshaw. (1983). A study of ozone in the Colorado Mountains. *J. Atmos. Chem.,* **1,** 87–105.

Fishman, J. (1984). Ozone in the troposphere. In *Stratospheric ozone,* Chapter 5, ed. by R. C. Whtiten and S. Prasad. New York, Van Nostrand Reinhold.

Fishman, J. and T. A. Carney. (1984). A one-dimensional photochemical model of the troposphere with planetary boundary-layer parameterization. *J. Atmos. Chem.,* **1,** 351–76.

Fishman, J., F. M. Vukovich, and E. V. Browell. (1985). The photochemistry of synoptic-scale ozone synthesis: Implications for the global ozone budget. *J. Atmos. Chem.,* **3,** 299–320.

Garland, J. A. and R. G. Derwent. (1979). Destruction at the ground and the diurnal cycle of concentration of ozone and other gases. *Q. J. R. Met. Soc.*, **105**, 169–83.

Gillani, N. V. and W. E. Wilson, Jr. (1980). Formation of ozone and aerosols in power plants plumes. *Ann. NY Acad. Sci.*, **338**, 276–96.

Gillani, N. V. and W. E. Wilson, Jr. (1983). Gas-to-particle conversion of sulfur in power plant plumes—II Observations of liquid phase conversions. *Atmos. Environ.*, **17**, 1739–52.

Gillani, N. V., R. B. Husar, J. D. Husar, D. E. Patterson, and W. E. Wilson, Jr. (1978). Project MISTT: Kinetics of particulate sulfur formation in a power plant plume out to 300 km. *Atmos. Environ.*, **12**, 589–98.

Gillani, N. V., D. E. Patterson, and J. D. Shannon (1984). Transport processes. In *The acidity deposition phenomena and its effects: Critical assessment review papers*, Chapter A-3, Vol. 1, ed. by A. P. Altshuller and R. A. Linthurst, 3-1–3-92, EPA-600/8-83-016AF.

Guderian, R. and H. Stratmann. (1962). Freilandversuche zur Ermittlung von Schwelfeldioxidwirkungen auf die Vegetation-I. Teil: Übersicht zur Versuchsmethodik and Versuchanswetung, Koln and Opladen, Westdeutscher Verlag. *Forsch. Ber. d. Landes.* Nordrhein-Westfalen NR. 1118.

Homolya, J. B. and E. Robinson. (1984). Natural and anthropogenic emission sources. In *The acidic deposition phenomena and its effects: critical assessment review papers.* Chapter A-2, Vol. 1, ed. by A. P. Altshuller and R. A. Linthurst. 2-1–2-106, EPA-600/8-83-016AF.

Husar, R. B., D. E. Patterson, J. D. Husar, N. V. Gillani, and W. E. Wilson, Jr. (1978). Sulfur budget of a power plant plume. *Atmos. Environ.*, **12**, 549–68.

Junge, C. E. (1963). *Air chemistry and radioactivity.* New York, Academic Press.

Karl, T. R. (1978a). Day of the week variations of photochemical oxidants in the St Louis area. *Atmos. Environ.*, **12**, 1657–67.

Karl, T. R. (1978b). Ozone transport in the St Louis area. *Atmos. Environ.*, **12**, 1421–31.

Kelly, N. A., G. T. Wolff, and M. A. Ferman. (1982). Background pollutant measurements in air masses affecting the eastern half of the United States. *Atmos. Environ.*, **16**, 1077–88.

Kelly, N. A., G. T. Wolff, and M. A. Ferman, (1984a) Sources and sinks of ozone in rural areas. *Atmos. Environ.*, **18**, 1251–66.

Kelly, T. J., R. L. Tanner, L. Newman, P. J. Galvin, and J. A. Kadlecek. (1984b). Trace gas and aerosol measurements at a remote site in the northeast US. *Atmos. Environ.*, **18**, 2565–76.

Korshover, J. (1975). Climatology of stagnating anticyclones east of the Rocky Mountains in the United States, 1936–1975. National Oceanic and Atmospheric Administrative Technology Memorandum, ERL-ARL-55, US Department of Commerce.

Linzon, S. N. (1971). Economic effects of sulfur dioxide on forest growth. *J. Air Pollut. Control Assoc.*, **21**, 81–6.

Liu, M. K., R. E. Morris, and J. P. Killis. (1984a). Development of a regional oxidant model and application to the northeastern United States. *Atmos. Environ.*, **18**, 1145–61.

Liu, S. C., J. R. McAfee, and R. J. Cicerone. (1984b). Radon 222 and tropospheric vertical transport. *J. Geophys. Res.*, **89**, 7291–7.

Logan, J. A. (1985). Tropospheric ozone: seasonal behavior, trends, and anthropogenic influence. *J. Geophys. Res.*, **90**, 10463–82.

Martinez, J. R. and H. B. Singh. (1979). Survey of the role of NO_x in nonurban ozone formation. Final report on SRI Project 6870-8, prepared for Monitoring and Data Analysis Division, Office of Air Quality Planning and Standards, Research Triangle Park, 27711.

Mueller, P. K., G. M. Hidy, K. Warren, T. F. Lavery, and R. L. Baskett. (1980). The occurrence of atmospheric aerosols in the Northeastern United States. *Ann. NY Acad. Sci.*, **338**, 463–82.

Mueller, P. K., G. M. Hidy, R. L. Baskett, K. K. Fung, R. C. Henry, T. F. Lavery, K. K. Warren, and J. G. Watson. (1983*a*). Analysis of sulfate variability. In *The sulfate regional experiment*: *report of findings*, Section 7, Vol. 3. EA-1901. Prepared for Electric Power Research Institute, 3412 Hillview Avenue, Palo Alto, CA 94304.

Mueller, P. K., G. M. Hidy, R. L. Baskett, K. K. Fung, R. C. Henry, T. F. Lavery, K. K. Warren, and J. G. Watson. (1983*b*). Air quality characteristics. In *The sulfate regional experiment*: *report of findings*, Section 6, Vol. 2. EA-1901. Prepared for Electric Power Research Institute, 3412 Hillview Avenue, Palo Alto, CA 94304.

Mueller, P. K., G. M. Hidy, R. L. Baskett, K. K. Fung, R. C. Henry, T. G. Lavery, K. K. Warren, and J. G. Watson. (1983*c*). Sampling locations. In *The sulfate regional experiment*: *report of findings*, Section 2, Vol. 1. EA-1901. Prepared for Electric Power Research Institute, 3412 Hillview Avenue, Palo Alto, CA 94304.

Oltmans, S. J. (1981). Surface ozone measurements in clean air. *J. Geophys. Res.*, **86**, 1174–80.

Pratt, G. C., R. C. Hendrickson, B. I. Cherone, D. A. Christopherson, M. V. O'Brien, and S. V. Krupa. (1983). Ozone and oxides of nitrogen in the rural upper-midwestern U.S.A. *Atmos. Environ.*, **17**, 2013–33.

Rasmussen, R. A., R. Chatfield, and M. Holdren. (1977). Hydrocarbon and oxidant chemistry observed at a site near St Louis. EPA-600/7-77-056. US Environmental Protection Agency, Research Triangle Park, 27711.

Research Triangle Institute. (1975). Investigation of rural oxidant levels as related to urban hydrocarbon control strategies. EPA-450/3-75-036.

Roberts, T. M. (1984). Effects of air pollutants on agriculture and forestry. *Atmos. Environ.*, **18**, 629–52.

Sexton, K. (1983). Evidence of an additive effect for ozone plumes from small cities. *Environ. Sci. Technol.*, **17**, 402–7.

Shaw, R. W., Jr and R. J. Paur. (1983). Measurements of sulfur in gaseous and particles during sixteen months in the Ohio River Valley. *Atmos. Environ.*, **16**, 1311–21.

Shreffler, J. H. and R. B. Evans. (1982). The surface ozone record from the Regional Air Pollution Study, 1975–1976. *Atmos. Environ.*, **16**, 1311–21.

Singh, H. B., F. L. Ludwig, and W. B. Johnson. (1978). Tropospheric ozone: concentrations and variabilities in clean remote atmospheres. *Atmos. Environ.*, **12**, 2185–96.

Smith, T. B., D. L. Blumenthal, J. A. Anderson, and A. H. Vanderpol. (1978). Transport of SO_2 in power plant plumes: Day and night. *Atmos. Environ.*, **12**, 605–11.

Spicer, C. W., D. W. Joseph, P. R. Sticksel, and G. F. Ward. (1979). Ozone sources and transport in the northeastern United States. *Environ. Sci. Technol.*, **13**, 975–85.

Spicer, C. W., D. W. Joseph, and P. R. Sticksel. (1982). An investigation of the ozone plume from a small city. *J. Air Pollut. Control Assoc.*, **32**, 278–81.

Turner, D. B. (1979). Atmospheric dispersion modeling: A critical review. *J. Air Pollut. Control Assoc.*, **29**, 502–19.

US Environmental Protection Agency (1978). *Air quality criteria for ozone and other photochemical oxidants*, Chapter 4, *Sources and sinks of oxidants*. EPA-600/8-78-004. Available from the Superintendent of Documents. US Printing Office, Washington, DC 20460.

US Environmental Protection Agency. (1982). Experimental concentrations and exposure. In *Air quality criteria for particulate matter and sulfur oxides*, Chapter 5. Vol. II. EPA-600/8-82-0296.

US Environmental Protection Agency. (1986). Air quality criteria for ozone and other photochemical oxidants. In *Properties, chemistry and transport of ozone and other photochemical oxidants and their precursors*. Chapter 3, Vol. II.

EPA/600/84/020bF. Available from Center for Environmental Research Information, Cincinatti, OH 45268.

Viezee, W., H. B. Singh, and H. Shigeishi. (1982). The impact of stratospheric ozone on tropospheric air quality—implications from an analysis of existing field data. Final report SRI Project 1140, CRC Contract no. CAPA-15-76 (1-80). Coordinating Research Council, Inc. 219 Perimeter Center Parkway, Atlanta, GA 30364.

Vukovich, F. M. and J. Fishman. (1986). The climatology of summertime O_3 and SO_2 (1977–1981). *Atmos. Environ.*, **20**, 2423–33.

Vukovich, F. M., W. D. Back, Jr, B. M. Crissman, and W. J. King. (1977). On the relationship between high zone in the rural surface layer and high pressure systems. *Atmos. Environ.*, **11**, 967–83.

Wolff, G. T., N. A. Kelly, M. A. Ferman, M. S. Ruthkosky, D. P. Stroup, and P. E. Korsog. (1986). Measurements of sulfur oxides, haze and fine particles at a rural site on the Atlantic Coast. *J. Air Pollut. Control Assoc.*, **36**, 585–91.

5

THE USE OF GEOSTATISTICS TO CHARACTERIZE REGIONAL OZONE EXPOSURES

H. PETER KNUDSEN

Montana Tech, Butte, Montana, USA

and

ÁLLEN S. LEFOHN

A.S.L. & Associates, Helena, Montana, USA

5.1 INTRODUCTION

One of the key factors in assessing possible crop losses from gaseous air pollutants is the measurement of pollutant concentrations. Ideally, direct measurements of ambient concentration at an agricultural site should be used. However, these data frequently are not available. In the United States, State and Local Air Monitoring Stations (SLAMS) are typically located near urban areas, and thus are not well situated for monitoring air pollutants at many agricultural sites.

To assess agricultural crop losses due to air pollution in the absence of direct measurements, a reliable estimate of air pollutant levels must be found. The National Crop Loss Assessment Network (NCLAN) program has used kriging, a geostatistical interpolation technique, to estimate the seasonal mean ozone (O_3) levels in major crop growing areas of the United States (Heck *et al.*, 1984). The estimates were used in predicting agriculturally related economic benefits anticipated by lowering O_3 levels in the United States (Adams *et al.*, 1984). O_3 monitoring data from SLAMS, archived in the US EPA Storage and Retrieval of Aerometric Data (SAROAD) system, were used as input for the kriging estimates. Lefohn *et al.* (1987) reviewed and evaluated NCLAN's 1984 use of the kriging technique. Based on the Lefohn review, kriging was chosen for the final NCLAN estimates of O_3 in major crop growing areas of the United States (Lefohn and Knudsen, 1987).

Kriging is one of several spatial methods that can be used for estimating air quality. This work discusses kriging and several alternative methods of spatial interpolation, and focuses on the properties and limitations of each method. In addition, this paper discusses the use of kriging to estimate monthly mean O_3 concentrations.

5.2 INTERPOLATION METHODS FOR SPATIAL DATA

Some of the common spatial interpolation methods are triangulation, moving averages, trend surfaces analysis, objective analysis, splines, and kriging. These methods are typically used to make an estimate at a series of points or small areas across a map, usually on a regular grid, such as is shown in Fig. 5.1 for the surface topography of a small area in Wyoming. The grid spacing is a function of the resolution required in the model and the spacing of the data.

5.2.1 Desirable attributes

Each interpolation method has different assumptions and characteristics. Obviously, the selection criteria should address the characteristics of the data and the spatial scale of the model. If the main interest in a model is local prediction accuracy (i.e., accurate estimates at each node or area of the model), then a method designed to estimate local means should be used. If large scale trends or features are to be modeled, a method that smooths out the short-range structures or noise should be used.

The following attributes are used to characterize interpolation methods.

1. Prediction accuracy
   (a) unbiased
   (b) minimum variance
   (c) conditionally unbiased.
2. Quantification of the error in the estimates.
3. Smoothness—esthetics.
4. Long-range trends versus short-range noise.

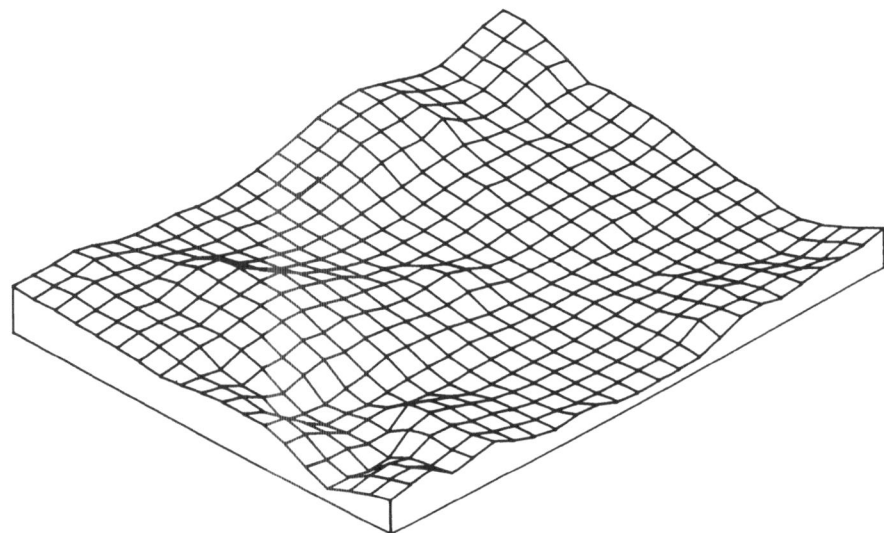

Fig. 5.1. Gridded mesh used in a contouring program.

Prediction accuracy refers to how well the method predicts the true value at each point of the grid. An unbiased estimator is defined as an estimator that does not yield consistently high or low estimates. In other words, an unbiased estimator has zero error on average. Let $Z^*(a)$ be an estimate and $Z(a)$ the true value of a variable at point (a), or within block (a). An unbiased estimator has an expected error of zero, $E[Z(a) - Z^*(a)] = 0$. An estimator that has "good" prediction accuracy must be unbiased and have a small estimation variance (the variance of the point-by-point errors), $\mathrm{Var}[Z(a) - Z^*(a)]$. Another desirable attribute of prediction accuracy is that the estimator be unbiased over any subset of the estimates. For example, the highest 20% estimates should be unbiased. An estimator meeting this criterion is called conditionally unbiased. An example of a conditionally biased estimator would be one that overestimates the high values and underestimates the low values. For air pollution applications, prediction accuracy is an important attribute.

Another important criterion is whether the interpolation method provides some measure of the error in the interpolated values. For example, are the O_3 estimates likely to be within plus or minus 10% of the true values or plus or minus 50%?

Smoothness is basically a criterion of esthetics. Often, contour maps can be described as looking like "whipped cream" and having many peaks and valleys. By altering the interpolation parameters, a map with a much smoother appearance can be created. Maps that are "smooth" also allow the viewer to observe easily the large-scale trends or structures. In some applications, such as mining, the short-range structures or noise are critical and should not be smoothed out.

5.2.2 Triangulation

Triangulation is a method of spatial interpolation used when it is desirable to produce a surface that is continuously differentiable within the triangulation. The first step in triangulation is to compute a set of triangles (line segments connecting data points). A surface is then fitted by least squares to each triangle. Akima's method (1978) uses a quintic surface to achieve the desired continuously differentiable surface within the triangulation surface. Triangulation produces maps with smooth contours and little "whipped cream" effect. However, no measure of prediction accuracy is provided by triangulation.

5.2.3 Moving areas

Moving averages are a simple and flexible way to interpolate spatial data and are widely used in contouring programs. A moving average estimate is simply a weighted average of the nearby data points $Z(x_i)$. The following general formula is used.

$$Z^*(a) = \sum_{i=1}^{n} \lambda_i Z(x_i) \qquad (1)$$

and

$$\lambda_i = d_i^p \Big/ \sum_{j=1}^{n} d_j^p$$

The weights, λ_i, are chosen as a function of the distance, d_i, between the data point and the grid point being estimated. The most common form of weighting is to use the inverse of the distance raised to some power, p. Common powers are in the range of 1 to 3. As the power gets larger, the weights decay faster as d gets larger. The weights are not only a function of the power p, but also a function of the number of data points included in the interpolation. As more data are included, the weights gets smaller; thus, usually the data included in the interpolation are limited to those lying within a specified search radius. Although Equation (1) shows that the sample weighting is a function of the distance and exponent only, the weighting method can easily be altered to include direction of the data values from the grid point (Knudsen and Kim, 1978). This allows the effects of anisotropy to be included in the weighting.

Since the characteristics of the moving average depend upon the interpolation parameters chosen, the only practical way to ensure the best choice of the weighting exponent, the search radius, and the anisotropy factors is to validate the choices. If local prediction accuracy is the desired criterion, a leave-one-out validation (Knudsen and Kim, 1979; Kane *et al.*, 1982) can be used to determine the best parameters. In this approach, a data value is withdrawn from the data set. The remaining data are then used to estimate the value of the withdrawn data value. The estimated value is then compared to the data value and the error computed. This is repeated for each data value. When completed, statistics of the errors are computed. Ideally, the errors will have a zero mean and the variance of the errors should be small. By iteration we can decide whether the exponent should be 1·75, 2·0, or some other value. The parameters giving the best estimation results (i.e., smallest variance of errors) are used.

With careful choice of parameter values, moving averages can provide estimates with nearly the same prediction accuracy as kriging (Knudsen and Kim, 1978; Baafi and Kim, 1983). However, moving averages do not provide an estimate of the magnitude of errors associated with the estimates.

5.2.4 Trend surface analysis

Many contouring programs offer trend surface analysis as a gridding option. The idea is to fit by least squares a mathematical function of the form (Ripley, 1981)

$$f(x, y) = \sum_{r+s} a_{rs} x^r y^s \tag{2}$$

of which the first few functions are

$$a + bx + cy \qquad\qquad \text{linear (plane)}$$
$$a + bx + cx + dx^2 + exy + fy^2 \quad \text{quadratic.}$$

While trend surfaces can be used to model a large-scale trend, they do not model short-scale variations, and thus are of little practical use when local prediction accuracy is important.

5.2.5 Kriging

Although originally developed for ore reserve estimation (Matheron, 1963), kriging has been used in many other spatial estimation applications, such as analyzing and modeling air-quality data (Grivet, 1980) and mapping of sites of toxic contamination (Issacs, 1984).

In kriging, a spatial variable, $Z^*(a)$, is estimated by a linear combination of the neighboring data as follows:

$$Z^*(a) = \sum_{i=1}^{n} \lambda_i Z(x_i) \qquad (3)$$

where

$Z^*(a)$ = estimated value within the area a;
$\quad n$ = number of nearby samples,
$Z(x_i)$ = value of sample at x_i, and
$\quad \lambda_i$ = weight applied to sample i.

The kriging weights, λ_i, are determined by minimizing the estimation variance (a least squares criterion) under the constraint that the estimator is unbiased (the weights must sum to one). The estimation variance to be minimized is expressed in terms of another statistical tool, the variogram. The estimation variance is given by

$$\sigma_\varepsilon^2 = 2 \sum_{i=1}^{n} \lambda_i \gamma(i, a) - \gamma(a, a) - \sum_{i=1}^{n} \sum_{j=1}^{n} \lambda_i \lambda_j \gamma(i, j) \qquad (4)$$

where

$\gamma(i, a)$ = the value of the variance for the distance separating the sample i
$\qquad$ and the area (a) being estimated,
$\gamma(a, a)$ = the average value of the variogram within the area (a), and
$\gamma(i, j)$ = the value of the variogram for the distance separating sample i and
$\qquad$ sample j.

Equation (4) takes into account the main factors affecting the reliability of a spatial estimate. The first term, $\gamma(i, a)$, measures the closeness of the samples to the area being estimated. As the sample distances increase from the area, the term becomes larger. The second term, $\gamma(a, a)$, is a measure of the size of the area being estimated. As the size of the area increases, this term becomes larger. The last term $\gamma(i, j)$ measures the spatial relationship of the samples to each other. If the samples are clustered, this term will be small, and therefore decrease the reliability of the estimate.

The spatial variability of the variable being estimated also affects the reliability of the estimate. This spatial variability is modeled by the variogram $\gamma(h)$. The variogram is the basic tool of spatial statistics and is defined by

$$\gamma(h) = \frac{1}{2n_h} \sum [Z(x + h) - Z(x)]^2 \qquad (5)$$

where

n_h = number of pairs a distance h apart,
$\gamma(h)$ = variogram value for the distance h,
$Z(x)$ = the value of the sample at point x, and
$Z(x+h)$ = the value of the sample at point $x+h$, where h is the distance
between the position x and $x+h$.

The variogram is calculated by including all sample pairs that are a given distance apart and computing the average squared difference for the sample pairs. This calculation is repeated for varying distances and the resulting average squared differences are plotted versus distance to form the variogram.

Using the daily mean of the 7-h O_3 value (0900–1559 hours) calculated for the period May through September 1982, Fig. 5.2 illustrates a typical variogram. In actual kriging calculations, a model is fitted to the experimental variogram, then used in the subsequent kriging calculations. One of the common variogram models, the nested spherical variogram model, is shown as a smooth curve.

Kriging should actually be thought of as a family of estimators. Different estimators are matched with specific assumptions relative to the variable being modeled. The estimator described above is referred to as ordinary kriging and is appropriate for data exhibiting little or no drift, and has an approximately stationary variance (quasi-stationary). If noticeable drift exists in the data, another form of kriging, known as universal kriging, is more appropriate.

Kriging is an estimator that was developed to give unbiased local estimates in the presence of clustered data and estimates that have minimum estimation variance. Because of these two attributes, kriging is the desired estimator when prediction accuracy is important. Kriging is also approximately conditionally

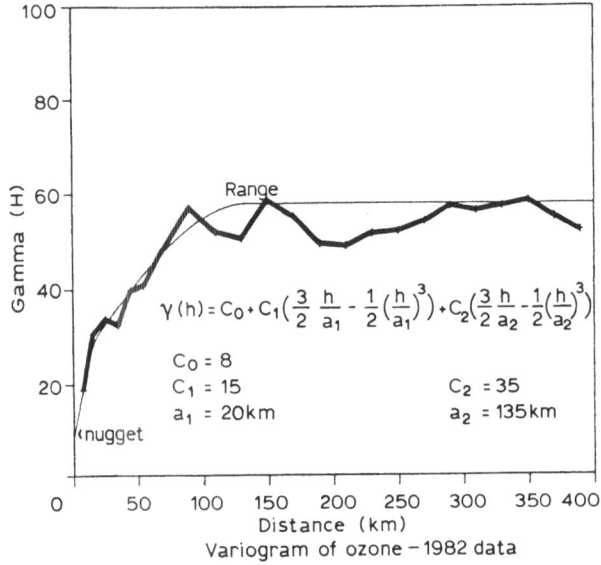

Fig. 5.2. Example of typical variogram (from Lefohn et al., 1987).

unbiased. Therefore, the estimates of high pollutant levels will be unbiased, as will the estimates of low pollutant levels.

For each estimate, kriging also calculates an estimation variance. The distribution of kriging errors is unknown; however, many case studies (Knudsen and Kim, 1978; Baafi and Kim, 1983; and others) have shown that the errors can be reasonably well modeled by a normal distribution.

5.2.6 Objective analysis

In the field of meteorology spatial interpolation is often done by a technique called objective analysis (Gandin, 1963) which is nearly identical to kriging. Both Gandin's techniques of objective analysis and Matheron's (1963) geostatistics were developed at the same time. The basis of Gandin's optimal interpolation is to measure the spatial structure of the meteorological field with a structure function which is defined as the mean-square difference between the values of a random variable at two locations within the field. This is exactly the same as the variogram of geostatistics. Using the structure function, Gandin developed a method of spatial interpolation which is optimal in the sense that the means square interpolation error is minimized. Gandin's method, especially his optimum interpolation with normalization of the weighting factors, is nearly identical to ordinary kriging.

5.2.7 Splines

Spline interpolation and spline smoothing have been used frequently by earth scientists to create maps with smooth contours, much as a draftsman would make manually (Dubrule, 1983). The method of splines involves fitting a function having the shape of a thin plate and forcing it to pass through or near the data points. Spline interpolation forces the spline to pass through or as close as possible to the data points, whereas spline smoothing is used to produce a surface that is smoother and thus may only pass near the data points. The amount of smoothing is a function of a smoothing parameter that is chosen by the user. Splines produce esthetic contour maps, but not necessarily accurate maps. Cross validation techniques have been developed (Wahba, 1979) to determine the smoothing parameter that gives the most accurate map.

Similarities noticed between maps produced by splines and maps produced by kriging led to research by Matheron (1981), Dubrule (1983) and Watson (1984) that showed that for certain covariance functions spline interpolation is equivalent to kriging. Furthermore, they showed that spline interpolation is a special case of kriging and that there is a relationship between the covariance function of kriging and the smoothing parameter of splines.

5.3 ESTIMATION OF THE MONTHLY MEAN DAILY 7-h O_3 CONCENTRATION

In 1982, NCLAN used kriging to estimate the 5-month mean of the daily 7-h (0900–1559 hours) O_3 value in the crop growing regions of the United States.

As a follow-up to this effort, kriging was used to estimate the monthly mean of the maximum daily 7-h value in 68 crop growing regions (selected by EPA) across the United States for the years 1980 to 1984 (Lefohn and Knudsen, 1987). The 68 regions are shown in Figure 5.7. Prior to initiating this work, several questions concerning the validity of applying kriging required an answer. Were O_3 values spatially correlated? If the values were correlated, did O_3 values have an approximately constant variance across the region being studied? This second question is not designed to determine whether the variance is approximately constant in the region being studied, but rather, can regions be found where O_3 displays an approximate constant spatial variance? The final question: did O_3 display drift?

Answers to the above were obtained by performing a statistical study of several months of data for 1980, 1983, and 1984. The following steps were performed in the study.

1. Histograms and maps of the data were plotted.
2. Several partitions of the data were made to define regions that have an approximately constant variance of O_3 values.
3. Variograms were calculated for each of the regions.
4. Steps 1 through 3 were repeated several times until homogeneous regions were found.

Monthly mean O_3 values at a monitoring site vary from month to month, but in general, concentrations are highest in the summer months and lowest in the winter. The variability of O_3 values across monitoring sites is lowest in the winter and highest in the summer. The mean and variance of O_3 values differ across the United States, with the West Coast values having significantly higher variance and higher coefficient of variation than the values for the remainder of the United States. For instance, the variance of O_3 over the entire United States in July 1984 was 171 ppb^2 (Table 5.1); the variance on the West Coast was 359 ppb^2, and 125 ppb^2 for the remainder of the United States. Based on the preliminary kriging analyses, it was decided to partition the United States into only two regions. The West Coast was defined as the area to the west of 114 degrees, and the Eastern United States as the area to the east of 114 degrees. This dividing line was approximately on the border of Arizona and California.

Variograms calculated for July 1984 are shown in Figs 5.3 and 5.4. These

TABLE 5.1
Statistics for maximum monthly 7-h O_3 for July 1984

Entire US	West Coast	East of 114 degrees
Mean = 54 ppb	Mean = 55 ppb	Mean = 54 ppb
Variance = 171 ppb^2	Variance = 359 ppb^1	Variance = 125 ppb^2
Coefficient of	Coefficient of	Coefficient of
variation = 0·24	variation = 0·34	variation = 0·20

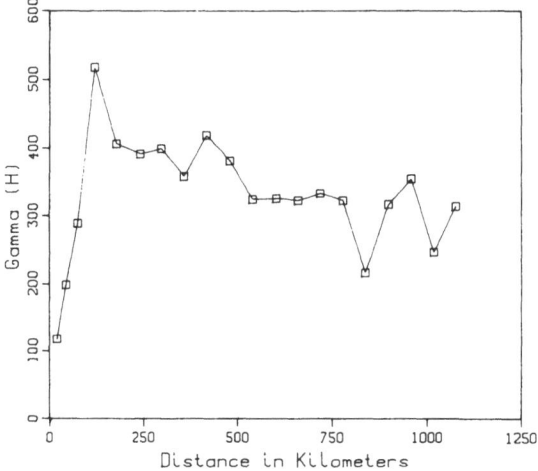

Fig. 5.3. Variogram for max. 7-h O_3 on West Coast for July 1984.

variograms can be described as having average continuity, while showing that O_3 is a spatially correlated variable. The limit of spatial correlation was about 300 km on the West Coast and up to 2000 km for the remainder of the United States. In the summer, the O_3 variograms became strongly anisotropic, with the greatest correlation distances in an east-west direction and the weakest correlation in a north-south direction, as shown in Fig. 5.5.

Using the variograms derived from the analysis, it was concluded that (1) the monthly mean 7-h O_3 value was a spatially correlated variable, and (2) the continental United States should be partitioned into at least two regions in order to meet a quasi-second order (variance) stationary requirement of ordinary kriging. In viewing the various maps showing O_3 values, drift, in general, was not readily observable. Further evidence illustrating a lack of a

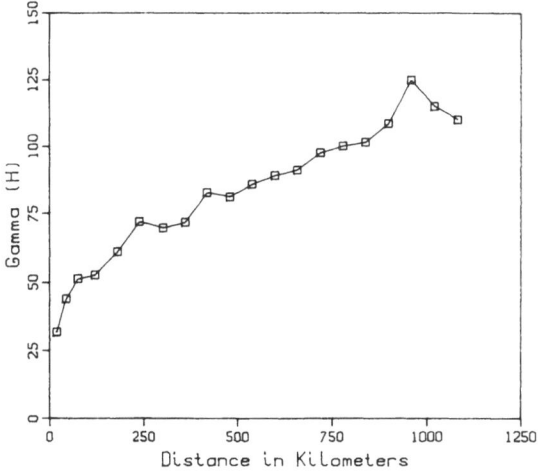

Fig. 5.4. Variogram for max. 7-h O_3 for Eastern US for July 1984.

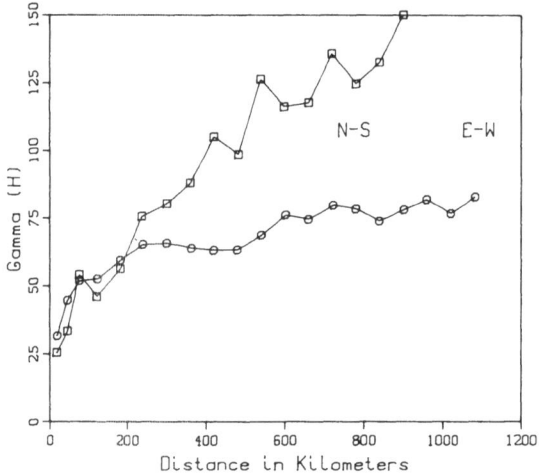

Fig. 5.5. Directional variograms for Eastern US for July 1984.

consistent or general drift was provided by the variograms computed during the study. Of the more than 200 variograms computed, fewer than five showed any evidence of an upward parabolic growth which would indicate the presence of drift. Based on these indications that the kriging assumptions were met, ordinary kriging was used to estimate monthly mean maximum 7-h O_3 levels in the United States for 1980 through 1984.

To create the kriged models of monthly mean values, the following steps were used for the Eastern United States and the West Coast areas.

1. Three variograms were calculated; an omni-directional variogram, an east–west variogram, and a north–south variogram.
2. If the variograms were not appropriate, or if they were greatly different from the variograms for other years, they were inspected to assess whether invalid data had been used.
3. Once acceptable variograms were obtained, a theoretical model was fitted to them.
4. If there was any question about the quality of the fit, the model was validated using the "leave one out" technique.
5. Kriged estimates were made.

Figure 5.6 shows the locations of O_3 monitoring sites where data were used to estimate July 1984 values. The variograms computed from these data are shown in Figs 5.3 and 5.4. The validation results are provided in Table 5.2. As indicated earlier, if the variogram is appropriate, then the errors will have a mean of almost zero and a variance approximately equal to the kriging variance, which is determined from the variogram. The actual variance of prediction errors, $44 \cdot 6 \, ppb^2$, was slightly lower than the kriging variance of $48 \cdot 0 \, ppb^2$ calculated from the variogram. The number of errors within the 95% error bound (2 times the square root of the kriging variance) was $95 \cdot 6\%$. Both values indicated that the variogram model was acceptable.

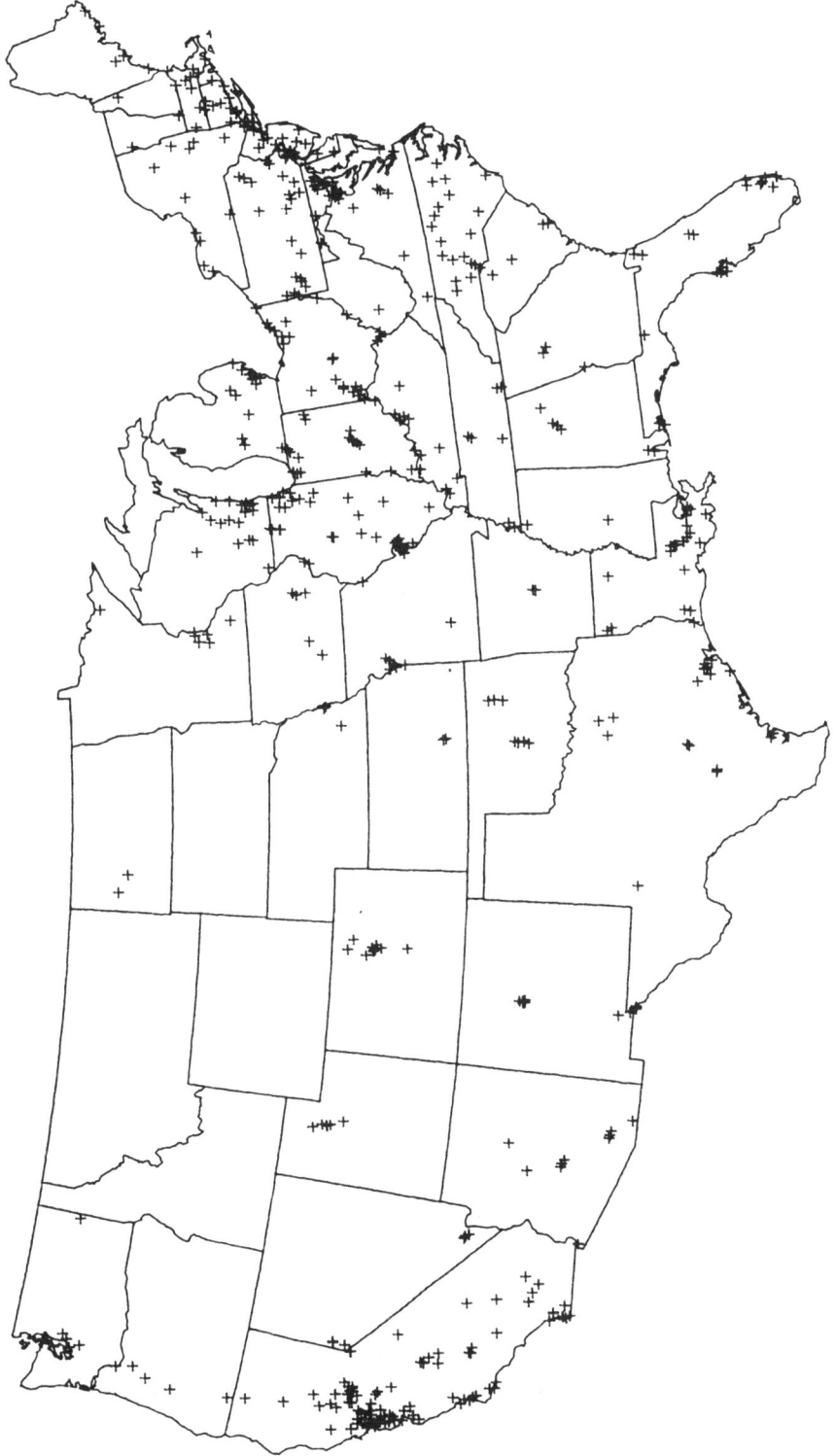

Fig. 5.6. Location of monitoring sites used to estimate monthly maximum 7-h O_3 for July 1984.

TABLE 5.2
Validation results for the July 1984 7th Eastern US
variogram

Statistics of kriging errors	
Mean error	0·17 ppb
Variance of errors	44·6 ppb^2
Predicted kriging variance	48·0 ppb^2
Percent of errors within $2\sigma_k$	95·65%
Number of samples	391

The final kriged estimates are shown in Fig. 5.7. The upper number within each region is the estimated monthly mean of the maximum daily 7-h O_3 value for the entire region, not just the center of the region. The second number is the 95% error bound in ppm on the kriged estimate.

5.4 FUTURE DIRECTIONS

This work has focused on methods to estimate spatial means. As discussed, kriging has been used by several investigators to estimate regional O_3 levels for the 7-h mean parameter. Recent work has demonstrated that it is also possible to use kriging to estimate the cumulative number of hourly occurrences of O_3 equal to or greater than 0·07 ppm (Lefohn et al., 1988a).

Kriging can be used for estimating the spatial distributions of O_3. However, those wishing to use spatial statistical techniques for predicting air quality should be aware that important limitations do éxist. For example, isolated, high elevation sites exhibit O_3 exposure characteristics that are distinct from those observed at lower elevation sites (Barry, 1964; Stasiuk and Coffey, 1974; Mohnen et al., 1977; Lefohn and Jones, 1986; Lefohn and Mohnen, 1986; Miller et al., 1986). The difference in O_3 exposure patterns may make it difficult to predict adequately exposures experienced at high elevation sites using data from lower elevation areas.

Before any mathematical model can be used to estimate pollution levels and the distribution of such pollutants, the investigator must assess whether the pollutant parameter used meets the assumptions associated with the model. For example, SO_2 and NO_2 emissions are primarily associated with point-sources in the United States (Lefohn and Jones, 1986). Thus, SO_2 and NO_2 concentrations would not necessarily be expected to be as spatially correlated as is O_3, or to meet necessarily the quasi-second order stationarity requirement of kriging. Kriging thus may not be an appropriate technique that would be useful for estimating SO_2 or NO_2 concentrations.

Spatial statistical methods can provide useful information to those who wish to predict air quality for those regions where insufficient monitoring data exist. However, the future application of these methods for predicting biological

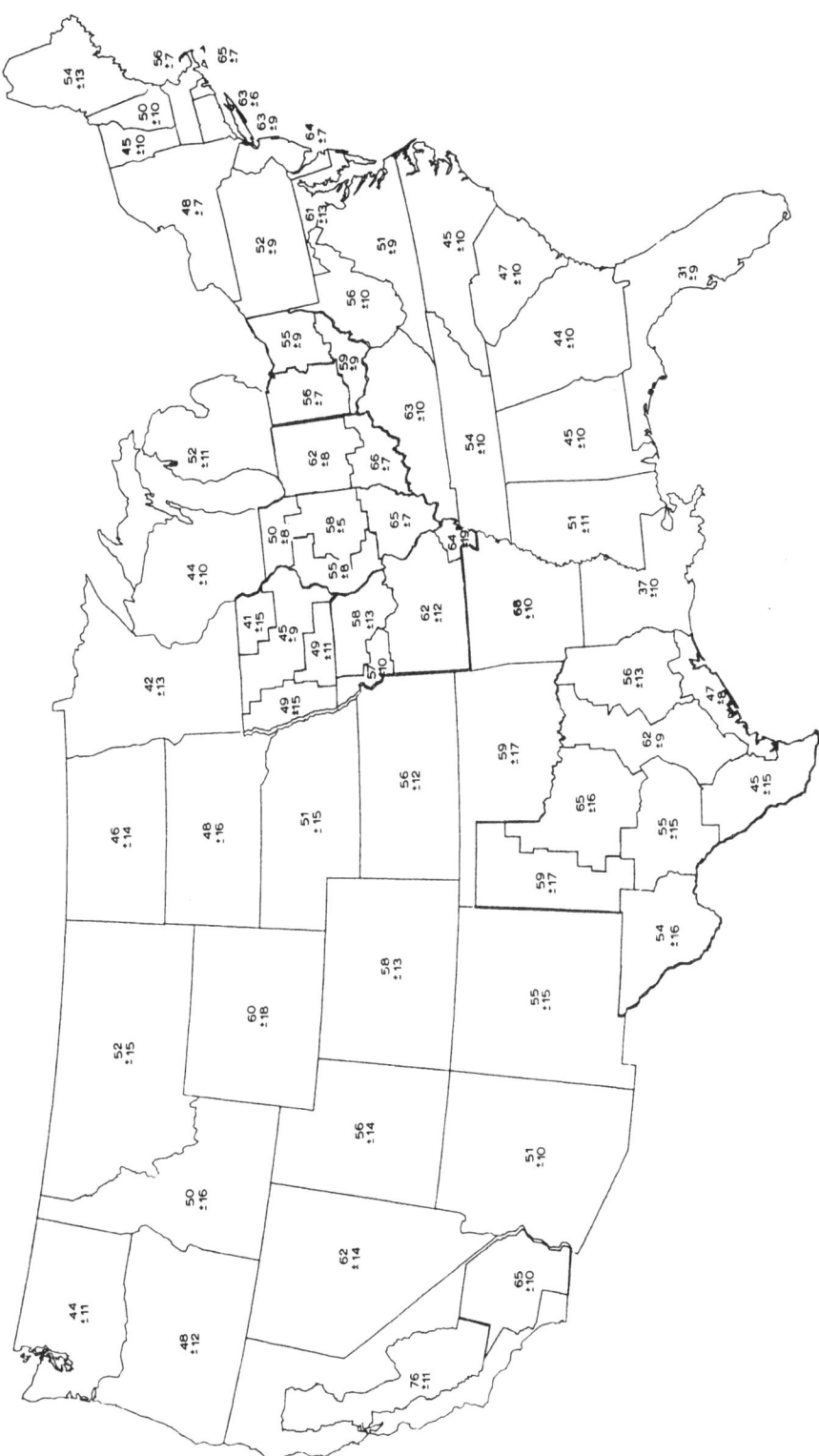

Fig. 5.7. Map of kriged estimates (concentrations in ppb) of max. 7-h O₃ for July 1984.

effects will have to await further input from researchers who desire to identify meaningful exposure parameters. Lefohn and Runeckles (1987) have discussed the application of alternative O_3 exposure surrogates to the 7-h seasonal mean for establishing meaningful standards to protect vegetation. In a related work, Lefohn *et al.* (1988*b*) explored the use of several exposure parameters and found that they appeared to perform as well as or better than the 7-h experimental-period mean for predicting exposure–response relationships for agricultural crops. The authors concluded that exposure indices do exist that offer an alternative to the long-term means (i.e., 7-h and 12-h means) that are presently used by research programs, such as the National Crop Loss Assessment Network.

REFERENCES

Adams, R. M., S. A. Hamilton, and B. A. McCarl. (1984). The economic effects of ozone on agriculture. EPA-600/3-84-090. Environmental Protection Agency, Corvallis, Oregon.

Akima, H. (1978). A method of bivariate interpolation and smooth surface fitting based on local procedures. Algorithm 474, Communications of Association of Computing Machinery.

Baafi, E. Y. and Y. C. Kim. (1983). Comparison of different ore reserve estimation methods using conditional simulation. *Mining Exp.*, **35,** 107–15.

Barry, C. R. (1964). Differences in concentration of surface oxidant between valley and mountaintop conditions in the southern Appalachians. *J. Air Pollut. Control Assoc.*, **14,** 238–39.

Dubrule, O. (1983). Two methods with different objectives: splines and kriging. *Math. Geol.*, **15,** 245–57.

Gandin, L. S. (1963). *Objective analysis of meteorological fields.* Jerusalem, Israel Program for Scientific Translations.

Grivet, C. D. (1980). Modeling and analysis of Air Quality Data. SIAM Institute for Mathematics and Society, Technical Report No. 43, Stanford University.

Heck, W. W., O. C. Taylor, R. M. Adams, G. E. Bingham, J. E. Miller, E. M. Preston, and L. H. Weinstein. (1984). *National Crop Loss Assessment Network (NCLAN) 1982 Annual Report.* EPA-600/3-84-049, 198–219. Corvallis, Oregon, US Environmental Protection Agency.

Issacs, E. H. (1984). *Risk qualified mappings for hazardous waste sites: A case study in distribution-free geostatistics.* M.S. thesis, Stanford University.

Kane, V. E., C. L. Begovich, T. L. Butz, and D. E. Myers. (1982). Interpolation of regional geochemistry using optimal interpolation parameters. *Comp. Geosci.*, **8,** 117–25.

Knudsen, H. P. and Y. C. Kim. (1978). A comparative study of the geostatistical ore reserve estimation method over the conventional methods. *Mining Eng.*, **30,** 54–58.

Knudsen, H. P. and Y. C. Kim. (1979). Development and verification of variogram models in roll front type uranium deposits. *Mining Eng.*, **31,** 1215–19.

Lefohn, A. S. and C. K. Jones. (1986). The characterization of ozone and sulfur dioxide air quality data for assessing possible vegetation effects. *J. Air Pollut. Control Assoc.*, **36,** 1123–29.

Lefohn, A. S. and V. A. Mohnen. (1986). The characterization of ozone, sulfur dioxide, and nitrogen dioxide for selected monitoring sites in the Federal Republic of Germany. *J. Air Pollut. Control Assoc.*, **36,** 1329–37.

Lefohn, A. S. and H. P. Knudsen. (1987). Section 6. Using kriging to estimate the monthly mean of the daily maximum 7- and 12-h ozone concentrations for 68 geographic regions in the United States. *National Crop Loss Assessment Network (NCLAN) 1985 Annual Report*. Corvallis, Oregon, United States Environmental Protection Agency.

Lefohn, A. S. and V. C. Runeckles. (1987). Establishing standards to protect vegetation—ozone exposure/dose considerations. *Atmos. Environ.* **21,** 561–8.

Lefohn, A. S., H. P. Knudsen, J. L. Logan, J. Simpson, and C. Bhumralkar. (1987). An evaluation of the kriging method to predict 7-hr seasonal mean ozone concentrations for estimating crop losses. *J. Air Pollut. Control Assoc.,* **37,** 595–602.

Lefohn, A. S., H. P. Knudsen, and L. R. McEvoy, Jr. (1988a) The use of kriging to estimate monthly ozone exposure parameters for the southeastern United States. *Environ. Pollut.,* **53,** 27–42.

Lefohn, A. S., J. A. Laurence, and R. J. Kohut. (1988b). A comparison of indices that describe the relationship between exposure to ozone and reduction in the yield of agricultural crops. *Atmos. Environ.,* **22,** 1229–40.

Matheron, G. (1963). Principles of geostatistics. *Econ. Geol.,* **63**(58), 1246–66.

Matheron, G. (1981). Splines and kriging, Syracuse University. *Geol. Contrib.,* **81**(8), 77–95.

Miller, P. R., O. C. Taylor, and M. Poe. (1986). Spatial variation of summer ozone concentrations in the San Bernardino Mountains. *Proceedings of the 79th APCA Annual Meeting,* Pittsburgh, PA, APCA.

Mohnen, V. A., A. Hogan, and P. Coffey. (1977). Ozone measurements in rural areas. *J. Air Pollut. Control Assoc.,* **82,** 5889–95.

Ripley, B. D. (1981). *Spatial statistics.* New York, John Wiley.

Stasiuk, W. N. and P. E. Coffey. (1974). Rural and urban ozone relationships in New York State. *J. Air Pollut. Control Assoc.,* **24,** 564–68.

Wahba, G. (1979). How to smooth curves and surfaces with splines and cross-validation. *Proceeding, 24th Design of experiments conference,* ed. by F. Dressel. New York, Academic Press.

Watson, G. S. (1984). Smoothing and interpolation by kriging and with splines. *Math. Geol.,* **84,** 601–15.

6

OZONE EXPOSURE INDICES: CONCEPTS FOR DEVELOPMENT AND EVALUATION OF THEIR USE

W. E. HOGSETT and DAVID T. TINGEY

US Environmental Protection Agency, Corvallis, Oregon, USA

and

E. HENRY LEE

Northrop Services Inc., Corvallis, Oregon, USA

6.1 INTRODUCTION

Most air pollutant research is concerned with quantifying the relationship between pollutant exposure of the organism and the ensuing biological response. The quantifying function for this relationship has been frequently referred to as "dose–response". For this discussion of exposure indices, or characterizations, we elect to refer to this process as "exposure–response;" a more generic term. Thus, this expression is not limited to the classical definition of dose as the product of the toxicant concentration absorbed by the organism and the duration of exposure. The exposure–response function requires a measure of the biological response and a measure of the pollutant exposure. The biological response is the culmination of a series of events, physical, biochemical, and physiological, beginning with pollutant uptake by the plant and ending with a measurable biological effect on the plant. In the case of crops, what is usually measured is the final agronomic yield. The measure of pollutant exposure has been some expression of the concentration of the pollutant and the time of the exposure period. At present, there is no agreement on a measure of exposure that best relates pollutant exposure over time (e.g. growing season) and the measurable response of crops or trees. A number of indices have, however, been developed, and their inherent assumptions, advantages, and disadvantages will be evaluated and discussed in this review. Defining the appropriate index that best relates exposure to plant response requires consideration of the underlying biological basis for the response, and this perspective will be used to evaluate currently applied indices and formulate new ones.

The development of an exposure index can, in turn, form the framework for development of secondary ambient air quality standards for O_3. However, before selecting an exposure index for the standard-setting process, it is necessary to explore and develop exposure indices that explain or account for as much variability in the exposure–response relationship as possible. This approach will provide the necessary understanding of environmental, chemical, and biological factors that influence plant response, as well as the most influential components of the exposure. Only with a thorough understanding of the key processes controlling the biological response is it possible to devise a "simple statistic" that can ultimately be used in the standard setting process, and that incorporates the biological basis of response.

The fundamental concepts of toxicology, which provide an understanding of the relationship between a toxic agent and the organism's response, offer a basis for understanding pollutant exposure and the concomitant biological response. A toxic effect is the outcome of a series of events that can be assigned into three phases: (1) exposure; (2) absorption, distribution, and metabolism (toxokinetic); and (3) target interaction (toxodynamic) (Ariens *et al.*, 1976).

- Exposure phase: when exposed to a toxicant, a toxic effect can occur only after absorption of the substance. The uptake of the gaseous substance is highly dependent upon: (1) the rate at which the substance, in an absorbable form, comes into contact with a surface capable of absorption, and (2) the concentration of the gaseous substance in the surrounding air. In plants or trees, the stomata are key factors controlling the rate of absorption, or the exposure phase. The barriers, in turn, are under genetic, metabolic, and climatic influence. Alteration of the toxicant to a more or less toxic substance can also occur during this phase. A number of biological, environmental, and exposure structure factors can influence this phase of the toxic response.
- Toxokinetic phase: this phase is denoted by the absorption, distribution, and metabolism of the pollutant, including all the processes affecting the relationship between the available pollutant and the pollutant concentration attained in the target tissue. Actions of the biological receptor in this phase include: (1) distribution of the toxicant, which involves absorption and distribution over the tissues, transport and binding to tissue components (transport via membranes), and excretion, if possible; and (2) metabolic alteration of the toxicant involving changes in its chemical properties (e.g., hydrophilicity or lipophilicity), leading to bioactivation or bioinactivation. Obviously, a wide variety of biological factors can influence this phase, including genetic and developmental stages of growth.
- Toxodynamic phase: this phase of the toxic response represents the interaction of the activated toxicant and the target tissue or receptor. The concentration of the toxicant attained in the target tissue determines the degree of the resultant biological response. This response may be irreversible enzyme inhibition, uncoupling of biochemical reactions,

synthesis of a lethal compound, removal of an essential metal, or interference with the general function of a cellular membrane.

These toxicological principles are conceptually similar to the phases outlined by Tingey and Taylor (1982) for plant uptake of O_3 and resulting effects. Injury is regulated by three processes: (1) gas-phase conductance controlling entry of the pollutant into the leaf (stomatal resistance), (2) liquid-phase conductance controlling the pollutant distribution and concentration at the target site (biochemical resistance), and (3) repair and compensation processes (homeostasis).

Quantifying the toxic response requires an expression of the toxicant, which incorporates some measure of concentration, time the toxicant is present, and temporal distribution of concentration. Before discussing this expression, defining certain frequently used terms may be helpful.

- Dose: a product of the concentration of the toxicant in the organism and the duration of exposure to the concentration. In many air pollution applications, "dose" has often been mistakenly used in reference to the product of ambient concentration and time of exposure. In the ambient environment, pollutant concentration is determined, in part, by meteorological or climatic events; the concentration measured is only potentially available for plant uptake and eventual interaction with target issues and is subject to many biological, environmental, and genetic influences.
- Dose–effect/dose–response: a primary symptom of a toxicant is defined as an effect. The response of an individual to a particular dose is called the dose–effect. Dose–response, on the other hand, is the determination of the percentage of a population that exhibits the defined effect with a certain dose of the substance (Ariens *et al.*, 1976).
- Dose–response curves: a functional (mathematical) description of the response of a population.
- Effective dose: a term in air pollutant studies introduced by Runeckles (1974), as the concentration of pollutant that was absorbed by vegetation over a time period, in contrast to that concentration in the ambient air.
- Exposure: a term in air pollutant toxicology referring to concentrations in the external surroundings of the plant, including a time component. We propose that the term "exposure", rather than "dose", be used in conjunction with air pollution studies since the exposure statistics, or indices, currently in use have only rarely reported actual tissue concentrations.

Typically in air pollution studies, a surrogate measure of exposure has been used in lieu of "dose". An exposure consists of the elements of concentration, duration, and dynamics (temporal distribution of concentrations). The most frequently used surrogate for the concentration element in the exposure is a measure of the pollutant concentration at a point in time in the air surrounding the plant (i.e., ambient concentration). This measure is an indicator of pollutant potentially available for plant uptake, and not the actual amount absorbed by the plant or the concentration of the bioactive component of the pollutant at the target site in the organism. This use of the ambient concentration as a surrogate is widely employed, primarily because of convenience, and it typically does not require additions to experimental design and expensive determination of stomatal activity (e.g., pollutant absorbed dose, Fowler and Cape, 1982; Runeckles, 1974). The relationship of this ambient concentration and the actual target tissue concentration of the pollutant is controlled by several factors including biological, genetic and environmental. Any surrogate measure of concentration or exposure must consider this relationship, as well as how and to what degree these factors alter the pollutant concentration to the target site.

Along with surrogate measures of tissue concentrations, there has been the use of surrogates in describing the exposure itself, attempting to account for both the elements of concentration and duration (time). Some measures have been used to characterize a single exposure (e.g., concentration), while others have attempted to represent cumulative effects from repeated exposures (e.g., "total impact" or seasonal mean) and include the temporal aspects of exposure (duration and frequency of occurrence of concentrations). The application of various surrogate measures of exposure is frequently clouded by the use of terms such as "acute" and "chronic" exposure. These terms typically refer to the length of exposure, but also have connotations of actual concentration level. An acute exposure is usually, but not always, a short-term exposure with high concentrations, whereas chronic implies long-term exposure, usually with low concentrations, but not necessarily. The concentrations, however, may be either constant or intermittent. Mean concentrations with various averaging times are frequently used to characterize these exposures, but averaging removes the temporal component of exposure. For this reason and others, the mean is not typically used in toxicology.

The toxic effect also requires an expression of the response itself, in addition to characterization of the exposure. Any number of variables may be measured for the response, but all usually relate ultimately to the growth and/or economic yield of crops or trees. It is the cumulative impact of "n" effects of "n" exposures over time that is being measured and reported as growth response. The underlying physiological processes affected by the exposure and culminating in the growth alterations are the factors to examine for their role in the susceptibility of the plant to the pollutant exposure. These biological processes and their interaction with the components of pollutant exposure should be incorporated into any indices characterizing exposure.

Evaluation and formulation of various exposure indices first requires a review of what components of the exposure, ambient environmental factors, and/or physiological processes are important in controlling or influencing the magnitude of the ensuing plant response. These factors should, in turn, be considered in developing indices that adequately characterize the distributions of hourly concentrations and temporal variations of the exposure. This approach of including those influential factors in a measure of exposure is based on the underlying assumption that expression of plant sensitivity is dependent on the appropriate concentration of pollutant under the necessary environmental conditions and at the proper time in the development stage of the plant.

Several factors influence the magnitude of the biological response in the pollutant exposure–response relationship. These factors include biological, environmental, and the three components of exposure; concentration, duration, number. All of these factors will aid in the subsequent evaluation of existing exposure indices and development of new approaches. In these discussions, emphasis will be on information from the published literature of O_3 exposure studies.

6.2 THE BIOLOGICAL BASIS OF SUSCEPTIBILITY

Plant response to O_3 is the result of a series of physical, biochemical, and physiological events. The toxic effect of O_3 occurs only if a sufficient amount reaches the target cellular sites within the plant leaf or needle. Three biological factors control the amount reaching the target site: (1) stomata, through which the O_3 diffuses into the leaf, by exercising some control on uptake; (2) the presence and/or activity level of detoxification systems for O_3 and its metabolites; and (3) the cell's/plant's ability to repair and/or compensate for the O_3 injury (Tingey and Taylor, 1982). Ozone uptake into the leaf and its movement through the cell wall to the membrane is discussed in two other papers presented in this symposium (Heath, 1988; Taylor *et al.*, 1988).

The interaction of O_3 with cellular processes causes a number of specific alterations in biochemical and physiological processes, which are measurable as biological responses, including reduced photosynthetic rate, altered stomatal conductance, premature senescence, foliar injury, altered assimilate allocation, and reduced growth and yield (US EPA, 1986). Plant growth and yield is one of the most commonly measured responses in a variety of crop and tree species following long-term exposure (US EPA, 1978, 1986; Heagle *et al.*, 1985; Kress *et al.*, 1985; Endress and Grunwald, 1985; Hogsett *et al.*, 1986; Temple *et al.*, 1986). Growth is the summation of a series of these processes related to uptake, assimilation, biosynthesis, and translocation, and an interruption in any one of these can lead to reduced growth and yield if the process is rate limiting.

The magnitude of this response is often dependent upon (1) the physical environment of the plant, including micro- and macro-climate both before and during the exposure (e.g., temperature, relative humidities, and light intensities), chemical environment (e.g., nutrition, soil water availability, interaction with other pollutants), and biological features of the environment (e.g., interaction with pests and pathogens); (2) the structure or dynamic aspects of the O_3 exposure; and (3) the biological factors controlling toxic entry/effects, including gas exchange, genetic potential, phenology, growth rates, and biotic interactions. A brief review of these factors and the degree to which each influences the magnitude of the exposure–response will aid in the subsequent review and evaluation of appropriate exposure indices.

6.2.1 Gas exchange

Stomata control the rate of O_3 uptake into the leaf and are influenced by various plant and environmental stimuli. Several studies with O_3 have demonstrated a correlation between stomatal behavior and exposure–response. Increased injury has been observed in trees with increased stomatal conductance patterns (Coyne and Bingham, 1982). Similarly, foliar injury of pea from O_3 or SO_3 was greatest when stomatal conductance was at a maximum (Kobriger et al., 1984). Ozone has been shown to induce stomatal closure, and a number of studies have directly correlated O_3 concentration and stomatal closure (US EPA, 1978). Likewise, transpirational water loss and O_3 absorption rates were positively correlated in a number of tree and herbaceous species (Thorne and Hanson, 1972, 1976). Manipulation of stomatal behavior, such as closing with water stress or abscisic acid, decreases O_3 sensitivity (Tingey and Hogsett, 1985). Ozone flux and foliar injury were positively correlated under field conditions (Mukammal, 1965). Using mean O_3 flux density and total O_3 uptake (mean flux density × time), Amiro et al. (1984) incorporated stomatal control of O_3 uptake and found these measures to be well correlated with observed onset of visible injury.

Short-term exposure to O_3 decreased stomatal conductance and decreased the integrated absorbed O_3 (Reich et al., 1985; Olszyk and Tingey, 1986). Amiro et al. (1984) showed that this reduction in stomatal conductance varied with cultivar and leaf age (phenology) and could result in decreased O_3 flux into the leaf.

Both woody and herbaceous plants display distinct diurnal patterns of stomatal conductance that would influence pollutant flux into a leaf. Unfortunately, very little data exist of continuous measurements over time. Water use strategy, however, indicates two types of behavior: isohydric and anisohydric (Larcher, 1975). The isohydric behavior, observed in many woody plants, is characterized by midday depression in transpiration rates (decreased conductance values), which corresponds to the time of increased O_3 concentrations in many geographical settings, and would thus suggest a decreased O_3 flux into the leaf at this time. Anisohydric behavior is characterized by a lack of stomatal regulation of water loss. Herbaceous plants exhibit both isohydric and anisohydric behaviors.

Although plant response to O_3 is the consequence of O_3 diffusion into the leaf interior, knowledge of its uptake rate is not sufficient for predicting subsequent responses for all species (US EPA, 1978, 1986). Other mechanisms also appear to be important in controlling plant response to O_3. For example, several studies found an inconsistent relationship between conductance, or O_3 uptake, and susceptibility (e.g., Tingey and Taylor, 1982; Taylor et al., 1982). The inconsistency implies that there is an internal, or biochemical, mechanism responsible, in part, for resistance.

6.2.2 Genetics

The influence of genetic potential and diversity in the response of plants to O_3 may be at several levels, but an obvious one to mention is the role of genetic determination in stomatal behavior. Sensitivity of plants to O_3 has been associated with increased stomatal conductance, and it has been suggested that this mechanism is genetically determined (Thorne and Hanson, 1976). Butler and Tibbitts (1979) suggested that resistance to O_3 in various cultivars may be attributed to the rate of O_3-induced closure of stomates, rather than stomatal frequency.

6.2.3 Developmental stages (phenology)

Plants do not appear to be equally sensitive to O_3 exposures at all stages in their life cycle. Such differential response would obviously influence any attempt to summarize or measure exposure and the ensuing response over a long period of time. Plant foliage is most sensitive to O_3 exposure just prior to or at maximum leaf expansion (US EPA, 1978). At this stage, stomata are functional and other barriers, such as internal cutin and secondary thickening of the cell walls, are minimal. Craker and Starbuck (1973) suggested that increased sensitivity of the intermediate aged leaf of tobacco was due to the increased uptake of O_3 when compared to the very young and very old leaves. A number of studies with both woody and herbaceous plants have shown plants to be most sensitive early in development, or just before senescence (US EPA, 1978; Blum and Heck, 1980). Similarly, current year needles in various conifer species exhibited maximum sensitivity to acute exposures during the early spring and early summer (Davis and Wood, 1972). Changes in sensitivity with leaf age could also be associated with changes in the levels of potential detoxification enzymes, such as superoxide dismutase (Tanaka and Sugahara, 1980). Flower buds mark another time of differential sensitivity. Petunias were most resistant to O_3 just before appearance of the flower bud (Hanson *et al.*, 1975). Crop maturity or stage of development also regulates the time of symptom expression (Haas, 1970).

6.2.4 Environmental factors

The environmental factors that appear to play a major role in exposure response are light, relative humidity, and soil moisture (US EPA, 1986). Exposure temperatures, however, do not exhibit any consistent pattern of influence on response. All of these environmental factors influence stomatal opening (leaf conductance), and thereby, O_3 uptake. Light is required to induce stomatal opening, and light conditions that are conducive to stomatal opening enhance O_3 injury. Increasing relative humidity increases stomatal aperture and foliar injury. Plants absorb more O_3 at high humidity compared to low humidity (McLaughlin and Taylor, 1981). Decreases in soil moisture are associated with reductions in plant sensitivity to O_3. The O_3 sensitivity is apparently related to reduced leaf conductance, which lessens O_3 uptake (Olszyk and Tibbitts, 1981; Tingey *et al.*, 1982; Tingey and Hogsett, 1985). Water stress does not confer a permanent tolerance to O_3; with relief of water stress, plants regain their sensitivity to O_3 (Tingey *et al.*, 1982).

Environmental conditions before and during the exposure are more influential than post-exposure conditions in determining the magnitude of the response (US EPA, 1986). Most studies evaluating the influence of environment have studied only a single environmental factor, not interactions, and have relied on visible foliar injury as the measured response, rather than growth (e.g., Dunning and Heck, 1977).

6.3 COMPONENTS OF OZONE EXPOSURE CAN MODIFY THE MAGNITUDE OF THE PLANT RESPONSE

A third factor influencing the magnitude of the plant's response to O_3 is the structure or manner of exposure. Changes in the exposure structure alters the magnitude of the response, but not the type of response (e.g., Musselman *et al.*, 1983; Hogsett *et al.*, 1985). The exposure structure is characterized by the concentration, duration of the concentration and exposure period and the elements of dynamics; that is temporal distribution of concentration occurrences in a diurnal and seasonal reference, peak or episode frequency, time between episodes or recovery time, etc. These components of exposure will be reviewed, with particular attention paid to the relative importance of each in contributing to the variability of the exposure–response relationship for subsequent evaluation of appropriate exposure indices.

6.3.1 Concentration

Several lines of evidence suggest that higher concentrations, or peaks, have a greater influence on plant response. The O_3 effect increases with increasing concentration, but approaches an asymptotic value at high concentrations. That is, continued increases in concentration cause only slightly more injury. White bean yields decreased with increasing O_3 exposure, but at a decelerating rate (Adomait *et al.*, 1987). Using either O_3 concentration or flux density, Amiro *et al.* (1984) demonstrated that higher O_3 concentrations were more important in determining foliar injury than lower ones. In addition, these authors showed that O_3 flux densities are better predictors of the occurrence of visible injury than only concentration and duration of the exposure. The use of flux density removes the variable of leaf diffusive conductance (Taylor and Tingey, 1983) and reflects the amount of pollutant absorbed. The genetic and environmental factors that influence stomatal conductance, and thus O_3 uptake, are implicit in the expression.

Peak (or high) concentrations, from the point of view of toxicology, are expected to cause greater influence. The greater the amount of absorbed pollutant, the more likely it is the toxicant will exceed the organism's ability to detoxify, repair, and/or compensate for the effect. Inherent in a peak are the two exposure components of concentration and duration, both of which are important in describing an exposure. However, based on foliar injury studies of tobacco (Tonneijck, 1984), concentration would appear to be more important than duration. Heagle and Heck (1980) also suggested that high concentrations over short time periods appeared to be more important in causing plant response than low concentrations over long time periods (Heck *et al.*, 1966; Heck and Tingey, 1971; Bennett, 1979; Amiro *et al.*, 1984; Ashmore, 1984).

6.3.2 Exposure duration

The O_3 impact increases with increasing exposure duration, but like the response to increasing concentrations, an asymptotic response is observed. Heagle and co-workers (1987) demonstrated a greater yield reduction in tobacco with the 12-h (0900–2059) daily exposure period than with a 7-h (0900–1559) period. Maximum concentrations and elevated O_3 concentrations frequently occur after 1559; consequently, the increased exposure period is more likely to reflect the increased number of occurrences of higher hourly O_3 concentrations which may impair plant yield.

6.3.3 Concentration and duration—cumulative effects

When plant yield is considered, the effect of O_3 on yield depends on the cumulated impact of the exposures that occurred during the growth of the plant (US EPA, 1986). In air pollution studies, an approach for incorporating cumulative effects has been the use of "dose", defined as the product of ambient concentration and time of exposure. However, equal doses of O_3 do not produce equal effects. The effect is influenced more by concentration than exposure duration (Guderian *et al.*, 1985). For example, using O_3 exposures of varying concentrations and duration, Nouchi and Aoki (1979) found that the degree of foliar injury was not a simple linear relationship with dose $(C \times t)$, but rather a logarithmic function. The degree of injury increased with both concentration and exposure duration, but was saturated at very high concentrations and long exposure durations. When the doses were equal, the most severe injury was associated with the high concentrations. Using their functional model for leaf injury, the predicted thresholds for injury were: $0 \cdot 196$ ppm $(384 \, \mu g \, m^{-3})$ for 1 h; $0 \cdot 119$ ppm for 3 h; and $0 \cdot 077$ ppm for 8 h. Similarly, Guderian *et al.* (1985) demonstrated that growth reductions in several crops were greater with increased concentration, rather than increased exposure time, for equivalent doses of SO_2. Concurring with these observations are the recent studies of Lee *et al.* (1988) and Lefohn *et al.* (1988), which compare the adequacy of several commonly used exposure indices, as well as newly proposed indices. These authors found that exposure indices that accumulated concentrations over the exposure period, as well as preferentially weighted elevated concentrations, best related plant response to O_3 exposure.

6.3.4 Concentration and duration—threshold

Several studies have shown that plants can tolerate some combinations of O_3 concentration and exposure duration without exhibiting foliar injury or growth reductions, illustrating that not all concentrations are equally effective in causing a response (e.g., Bennett, 1979; Amiro *et al.*, 1984; US EPA, 1986). These observations suggest the occurrence of a threshold concentration for a particular response. This is consistent with the toxicological perspective that peaks above some level are the most likely to cause a toxic response, or effects occur when the amount of pollutant absorbed exceeds the ability to repair or compensate for the impact. A threshold, however, could include both duration of exposure and peak concentration. Heck *et al.* (1966) demonstrated a sigmoid-type response in tobacco and pinto bean exposed to variable O_3 concentrations for varying time periods, presenting both a threshold concentration and distinct exposure time required before injury was observed. Nouchi and Aoki (1979) also demonstrated a nonlinear relationship of exposure concentration and duration with visible foliar injury in morning glory. With exposure reported as O_3 flux density and response reported as foliar injury, Amiro *et al.* (1984) found a nonlinear relationship in response of bean (foliar injury) to increasing O_3 flux densities with exposure duration and suggested a biochemical threshold. This threshold would consist of flux densities that the plant could experience without displaying visible foliar injury.

A problem in defining a threshold value is that evidence suggests the threshold is different for different plant responses. In studies in open-top chambers, Heagle *et al.* (1979) found that concentrations (expressed as daily 7-h means) required to cause visible foliar injury to corn were much lower than the concentrations required to reduce kernel yield.

6.3.5 Dynamics (temporal distribution of ozone concentrations)

The dynamics of the exposure can also influence the magnitude of the response of the plant to O_3. Dynamics refers to the temporal distribution of the hourly concentrations in either a diurnal or seasonal context, the frequency of occurrence of hourly concentrations, or the frequency of occurrence of episodes or events over the exposure period (a growing season).

Exposures with different distributions of hourly concentrations over the exposure period cause different magnitudes of response, but the particular response (i.e. reduced growth) is the same with each exposure regime (Musselman *et al.*, 1983; Hogsett *et al.*, 1985). In long-term exposure studies of alfalfa, timothy, and orchard grass in open-top chambers, an episodic exposure had greater influence on reducing growth than did a daily exposure regime with both exposure distributions having equal cumulative exposure or dose (Hogsett *et al.*, 1985; and unpublished data). In addition, the seasonal 7-h mean of the exposure did not adequately predict the growth results. The greater growth reductions occurred with the episodic regimes, which had the lower mean exposure value (Hogsett *et al.*, 1985). Uniform distributions of concentrations over a long exposure period appear to have less influence on growth than a variable distribution. For example, a single continuous exposure caused less ethylene production and foliar necrosis in bean plants than did repetitive O_3 peaks of the same concentration and accumulated time (Stan and Schicker, 1982). Continuous concentrations of O_3 with no breaks in the exposure resulted in what appeared to be an accommodation response in radish plants, whereby the final biomass of the treated plants was not different from the control plants (Walmsley *et al.*, 1980). In fact, the newly formed leaves on the exposed plants were expanding with less area and fewer stomata, thus reducing the potential for O_3 uptake.

Studies with SO_2 also demonstrate a similar response of greater sensitivity with variable concentration exposure regimes compared to constant concentrations of SO_2 (McLaughlin *et al.*, 1979; Male *et al.*, 1983).

Experimentally, several researchers have shown peak concentrations to be more important than the mean concentration over the exposure period in determining plant response (Heck and Tingey, 1971; Bennett 1979; Musselman *et al.*, 1983; Hogsett *et al.*, 1985). Nevertheless, studies have also shown that plant growth can be reduced with long-term exposures to low concentrations (e.g., US EPA, 1978; Musselman *et al.*, 1986), indicating that plants are also impaired by exposure to essentially constant concentrations.

6.3.6 Dynamics—recovery time

The recovery time, or respite time, in an exposure refers to the time period or the number of occurrences of low concentrations within an exposure structure. This component of exposure dynamics is best considered in the relationship of the distribution of hourly concentrations and the intensity of those concentrations. This relationship is another determinant of the magnitude of the response. For example, plant response to repeated peaks depends on the concentration of the peaks. If the peak O_3 concentration is relatively high (>0·10 ppm) the sensitivity, in terms of visible leaf injury, will increase as the time between peaks increases (Mukammal, 1965). A high O_3 concentration on the first day caused substantial injury, but an equal or higher concentration on the second day caused only slight injury. Conversely, when the peak concentrations were at or near the threshold concentration required for visible injury, injury to morning glory leaves decreased with increasing time between episodes (Nouchi and Aoki, 1979). In tobacco, a 1-h continuous exposure produced more foliar injury than two 0·5-h exposures separated in time by 1 to 3 h (Heck and Dunning, 1967). These various experiments would suggest that plants can recover from O_3 exposures only during pollutant-free times. With repeated daily peaks, there is no recovery period; consequently, the time between exposures is too short to allow the plant to recover. However, preliminary data with beans suggest that the concentrations of the peaks are probably more important than the frequency of peaks in determining the magnitude of growth reductions (Hogsett, unpublished data). In long-term exposure studies of alfalfa, timothy, and orchardgrass comparing episodic and daily peak regimes having the same exposure value, Hogsett *et al.* (1985) have shown that growth reductions were greatest with episodic peak occurrence rather than daily peak occurrence. The episodic regimes, however, have the greatest number of hourly occurrences of low concentrations (0·05 ppm or less), and the greatest number of days when these low levels were the daily maxima; all characteristics potentially considered respite time. These results would suggest that rather than simply pollutant-free days being necessary for recovery (respite time), that peak concentrations and the timing of those occurrences over the lifetime of the plants, as well as the timing and occurrence of the low concentrations, are crucial in consideration of respite time. Also, it may be necessary that the concentrations be very low for recovery to occur.

6.3.7 Dynamics—predisposition

Predisposition, like respite time, is also dependent on the temporal distribution of low concentrations over the exposure period. In particular, research has demonstrated the role of exposure to low concentrations of O_3 in predisposing the plants to subsequent higher concentrations of O_3, and the concentration and duration of these low concentrations determining the intensity of the biological response (Heagle and Heck, 1974; Runeckles and Rosen, 1977; Johnston and Heagle, 1982; Steinberger and Naveh, 1982).

Runeckles and Rosen (1974) reported that daily pretreatment with 0·02 ppm O_3 decreased susceptibility of bean leaves to subsequent acute exposures, but pretreatment with 0·04 ppm or above increased susceptibility. Bel W3 Tobacco responded in a similar fashion upon exposure to ambient oxidant for 7 days prior to an acute exposure (Heagle and Heck, 1974), or to 0·03 ppm for a 12-h period, prior to an exposure to higher concentrations (Steinberger and Naveh, 1982). Soybeans and wax beans exposed to O_3 below a level and duration that causes visible injury exhibited greater injury when exposed to higher concentrations of O_3 (Runeckles and Rosen, 1974; Johnston and Heagle, 1982). If the pretreatment concentration is high, the subsequent response to an acute, or high concentration, exposure is quite different from pretreatment with a low concentration. Mukammal (1965) reported that if a high concentration caused injury in tobacco on one day, an equal or higher concentration on the following day caused only a slight response in injury. However, Mukammal did not follow the phenomenon beyond two days.

The time period of increased susceptibility following pre-exposure is variable. Runeckles and Rosen (1977) demonstrated that with pretreatment at a low concentration, the period of increased sensitivity begins after 2 to 3 days and lasts for about 8 days. If pretreatment is with a higher concentration (0·05 ppm), there is a decrease in susceptibility beginning 8 days after sowing, followed by a period of marked increase in susceptibility to acute injury. The authors suggested that the later stages of decreased sensitivity (pretreatment with low concentrations) resulted from a dampening effect on stomatal action. This was deduced from the transpiration rate. Nouchi and Aoki (1979) demonstrated the cumulative effects of pretreatment with O_3 upon exposure to levels equal to the threshold level for visible injury on the sensitivity of morning glory leaves. The effects of the pretreatment decreased linearly with the lapse of time from the first exposure, approaching zero after 5 days; that is, if 5 days lapsed between the first and second exposure at threshold levels, then the observed injury is not increased above that observed with the first exposure.

Predisposition with other pollutants, such as SO_2, can also increase plant sensitivity to subsequent acute O_3 exposure (Ashmore and Onal, 1984; Hofstra et al., 1985).

The mechanism for predisposition caused by low concentrations is not known. Predisposition may increase susceptibility to subsequent exposures by a slow accumulation of a toxicant; a gradual toxodynamic phase in which there is a slow reaction with membranes or an interruption of enzymatic systems, such as CO_2 fixation, or a combination of events. The reversibility of the predisposition is poorly understood. Similarly, the role of respite, or recovery times, is not fully understood, nor are the changes in pollutant flux density and stomatal behavior on concurrent peak days or peak times following lower concentrations. The distinction between recovery time and predisposition is not very clear and is perhaps based on length of exposure period. Most observations of predisposition have been with foliar injury and short-term exposures, whereas the concept of recovery time, although supported with short-term foliar injury studies, is primarily deduced from long-term exposures involving variable concentration distributions over the exposure period. It is noteworthy that so little research has been done in this area since these earlier studies, yet the mechanistic basis of this phenomenon is important in understanding the response of plants to the component of exposure.

6.4 CURRENTLY EMPLOYED EXPOSURE INDICES

The preceding brief review has shown the experimentally deduced influence of the biological and environmental factors, and the components of the exposure (concentration, duration, number) on the magnitude of the response in the exposure–response relationship. We now can ask which, if any, of these factors have been included in any surrogate measures of exposure and if they account for the observed variability in the exposure–response relationships. The difficulty in selecting an appropriate exposure index was pointed out by Heagle and Heck (1980). They found ambient and experimental exposures had been summarized as (1) seasonal, monthly, weekly, or daily means; (2) peak hourly means; (3) number of hours above a selected concentration; or (4) number of hours above a selected concentration interval. However, none of these indices adequately addresses the relationships among ambient concentration, tissue concentrations, hourly concentration distributions over the exposure period, exposure duration, interval between episodes, and plant growth stages. In fact, most currently used exposure indices were not selected from a biological basis, replete with experimental data to support the selection. Although not an explicit oversight, the choice of an appropriate exposure index should result from an understanding of those factors controlling entry of O_3 to the target tissue and the subsequent biological response.

In addition to the lack of biological considerations, development of an appropriate index is also made difficult by lack of a clear demonstration of those exposure components that are most influential in the plant response (i.e., concentration, duration, frequency, time between events). The influence of temporal fluctuations in concentrations, which is characteristic of ambient air, needs resolution to aid in the selection of an exposure index. Until these issues are addressed, the selection of the appropriate index will continue to be discretionary.

Most of the currently employed indices used to characterize O_3 exposure, or other pollutants for that matter, consider the exposure (i.e., concentration and duration) to be the most important feature controlling the biological response. Two examples of biological considerations that have not been included are (1) the sensitivity of the plant at the time of exposure, and (2) the amount of pollutant absorbed by the plant. This issue of sensitive developmental stages is difficult to resolve because it has not clearly been shown whether the various stages of plant growth are differentially sensitive to exposures relative to ultimate yield, although data exist to demonstrate sensitive growth stages (e.g., Tingey et al., 1971; Blum and Heck, 1980; Adomait et al., 1987). Even though the previous indices have considered exposure the most important feature in the exposure–response relationship, primary emphasis has been on the ambient concentrations with little attention paid to the elements of the exposure dynamics (i.e., temporal and seasonal variations in concentration).

To understand what an appropriate exposure index is, it is worthwhile to consider the strengths and limitations of the currently used exposure indices, as well as their underlying assumptions. A thorough assessment of these indices should help in the selection or even in the development of new indices that attempt to incorporate a biological basis into their specification. These current indices can be grouped according to their approach to characterization: (1) averaging, (2) cumulative, (3) event only, (4) concentration weighting, (5) multi-component weighting, or (6) dose. It was not possible to evaluate all previously used exposure indices; however, many of the most commonly used indices are evaluated.

6.4.1 Averaging approach (mean)

This exposure index is a mathematical mean of the hourly O_3 concentrations over various time periods, including the daily 7- and 12-h means over the growing season, as well as seasonal averages of the 1-h maximum concentrations. At least recently, because of its application in the National Crop Loss Assessment Network (NCLAN), the 7-h seasonal mean has been the most frequently used index to describe yield response to O_3 (Heck et al., 1982, 1984a, b; Kress and Miller, 1983; Kress et al., 1985; Heagle et al., 1987). The long-term, or seasonal mean, concentration is a summation of all hourly concentrations for the selected time period divided by the number of observations (Cure et al., 1986). For the 7-h seasonal mean, the usual time period has been the 7 daylight hours 9 am–4 pm over the growing season, because this was thought to correspond to the period of great plant sensitivity and the highest level of O_3 (Heck et al., 1983). More recently, averaging time has been extended to 12 h (Heagle et al., 1987). The longer averaging time was implemented to include more of the elevated O_3 concentrations. Various studies have used a range of averaging times from 1 to 24 h (e.g., US EPA, 1986).

The seasonal mean of daily maximum 1-h O_3 concentrations (M1) is another average characterization of exposure (Heck et al., 1984a). In a comparison using three crop studies conducted in Rayleigh, NC, the M1 index predicted yield losses essentially as well as the 7-h seasonal mean (M7) index (Cure et al., 1986). But, as with the 7-h seasonal mean, the 1-h mean does not correlate well with short-term indicators of O_3 exposure (Heck et al., 1984a).

Mathematically, the mean contains all hourly concentrations that make up the exposure period, but treats all concentrations equally, therefore implying that all concentrations of O_3 are equally effective in causing a response. A mean also minimizes the contributions of higher, or peak, concentrations over the long-term exposure period to the response. The mean treats low-level, long-term exposures the same as high concentration, short-term exposures; a scenario that the literature does not support (see Section 6.3). An infinite number of hourly distributions, ranging from those containing many peaks to those containing none, can yield the same 7-h seasonal mean. Cure *et al.* (1986) reported that mean characterizations of O_3 exposure were much less sensitive to variations in yearly O_3 patterns than were the 1-h max.

In addition to being relatively insensitive to variations in yearly O_3 concentrations (Cure *et al.*, 1986), the 7-h seasonal mean has other limitations. For example, there is poor association between the peak hourly O_3 concentrations and the 7-h and 12-h seasonal averages for selected sites in the US (Lefohn and Benedict, 1985). Also, Lefohn and Jones (1986) presented data showing that not all monitoring sites in the USA experience their highest O_3 concentration within the 7-h time period of 9:00 am to 4:00 pm.

The use of a mean exposure index for characterizing an exposure(s) implies certain assumptions, as follows.

- A seasonal mean assumes that crop yield reductions result from the accumulation of daily O_3 effects over the growing season (Cure *et al.*, 1986).
- A mean assumes that the distribution of hourly O_3 concentrations (over the averaging time) is not highly skewed and that the distribution is unimodal. Ambient O_3 concentration distributions are frequently skewed toward the higher concentrations.
- A mean assumes that all concentrations, within the selected averaging time, are equally effective in eliciting the measured response.
- A mean assumes that peak events do not need to be given special consideration. This is not consistent with results that indicate short-term peak concentrations are important in determining response (Section 6.3).
- A mean does not specifically include an exposure duration component. It cannot distinguish between two exposures of the same concentration but of different duration.
- A mean assumes that the selected time interval, over which concentrations are averaged, is the period of highest hourly occurrences of the pollutant.
- A mean assumes that the selected time interval, over which concentrations are averaged, corresponds to the period when the plant is most active with respect to gas exchange or stomatal conductance and toxicological responses.

6.4.2 Cumulative approach

The cumulative approach to characterizing an exposure involves summing all hourly concentrations over the exposure period without a censoring, or threshold, concentration. This exposure index is called "total exposure" and has units of ppm (some studies have used ppm-h). The total exposure is not an actual measure of dose in a toxicological context, or that concentration within the organism. As with a mean index, the peak, or high, concentration values are not given additional weight, and temporal distributions of concentrations are not considered. However, a time component (time exposure period) is specifically included because exposures to the same concentration (mean) but to different durations will have different total exposures. In some short-term (<24-h) studies, total exposure has been calculated as the product of concentration × time (ppm-h) (Guderian et al., 1985).

The cumulative index, like the mean index, assumes that crop yield reductions result from the accumulation of daily O_3 effects over the growing season. This index, as a nondiscriminating summation of all hourly concentrations over the exposure period, treats all concentrations as equally effective in eliciting a plant response, similar to the mean index. However, if a concentration, threshold, or censor value, is used to discriminate which concentrations are summed, a cumulative exposure index that weights the higher concentrations is the result, and this type of index will be discussed as part of the weighted indices below.

6.4.3 One time event approach

These measures of exposure are peak measures representing the seasonal maximum, 1-h (P1), and maximum, 7-h (P7), concentrations (Heck et al., 1984a). This index neither averages, nor cumulates concentrations. Peak indices (that is, extreme values) display greater variability than do seasonal mean indices (M7, M1). Although they have been used to develop exposure–response models, they place too much emphasis on the single day having the maximum O_3 concentration and ignore other important factors influencing response (Heck et al., 1984a; Cure et al., 1986). This characterization assumes that plant response is not cumulative, but rather is the result of a single event, such as a maximum hourly concentration over the entire growing season.

6.4.4 Concentration weighting approach

This approach focuses on the importance of concentration and in particular, the higher concentrations, in eliciting the biological response. Other potentially influential exposure aspects, such as the developmental changes in sensitivity of the plant during exposure or the absorbed dose are not considered in this type weighting scheme. The currently employed weighting indices typically use either a discontinuous or a continuous weighting function to emphasize the higher concentrations.

The discontinuous weighting approach is used by most of the cumulative exposure indices, including (1) the number of hours (or days) with peak concentrations above a threshold (e.g., Nosal, 1983; Ashmore, 1984; Adomait et al., 1987; Lefohn et al., 1988); (2) the integrated sum of absolute hourly concentrations above a particular threshold (Lefohn and Benedict, 1982); (3) the summation of the fraction of the concentration above a selected threshold (Oshima et al., 1976); and (4) the number of episodes, or events with consecutive hourly concentrations above a threshold (Lefohn and Benedict, 1982). A cumulative exposure index with a threshold concentration was used successfully by Oshima and co-workers (1976) in predicting the impact of O_3 on alfalfa yield in a California study. Other types of cumulative indices have been used to predict yield losses in other crops; for example, the monthly cumulative concentrations above 0·08 ppm (Adomait *et al.*, 1987) or the number of days above a threshold (Legassicke and Ormrod, 1981). An example of this discontinuous weighting scheme for a cumulative index is illustrated in Fig. 6.1, with a 0·08 ppm censor, or threshold, value. Only those hourly concentrations equal to 0·08 ppm or greater are cumulated for the exposure index (the shaded area in the 7-day profile), thus weighting those higher concentrations in an index describing the exposure and response.

A discontinuous weighting approach gives only those concentrations above the censor value credence in eliciting the biological response and no import to those concentrations below the censor value (Fig. 6.1). However, recent studies (Lee *et al.*, 1988; Lefohn *et al.*, 1988) have shown that concentrations below the threshold also contribute to the plant response. The applicability of the threshold-based cumulative index is limited since it depends on defining a concentration level (a distinct threshold) at which an effect is observable. The experimental data to define this concentration value are lacking, and there is no evidence that the threshold is constant for a species, among species, or through the growing season. The cumulative indices also fail to consider the temporal variation in distribution of concentrations.

The second approach to weighting concentrations is the use of a continuous function whereby all concentrations are raised in value by a predetermined exponent, or a mathematical function (for example, sigmoid or allometric) to add increased value to the increasing concentrations (Fig. 6.2). This approach does not censor any hourly concentrations from the exposure characterization,

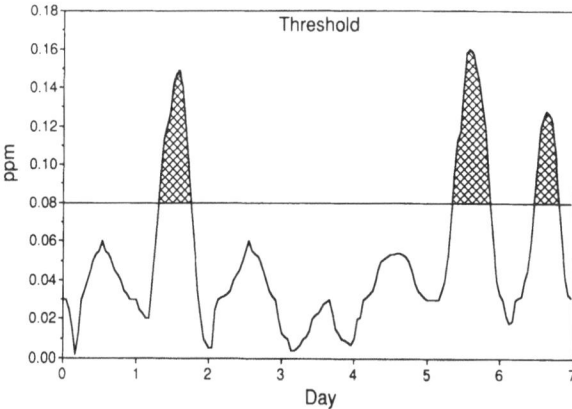

Fig. 6.1. A 7-day O_3 exposure regime illustrating the hourly concentrations summed with a discontinuous weighting function or threshold value (the shaded area). In this case 0·08 ppm is the threshold value, and all concentrations equal to or greater than 0·08 ppm are summed to give a weighted, cumulative index of exposure. The total exposure = 8·45 ppm; the sum 08 (threshold of 0·08 ppm) = 3·72 ppm.

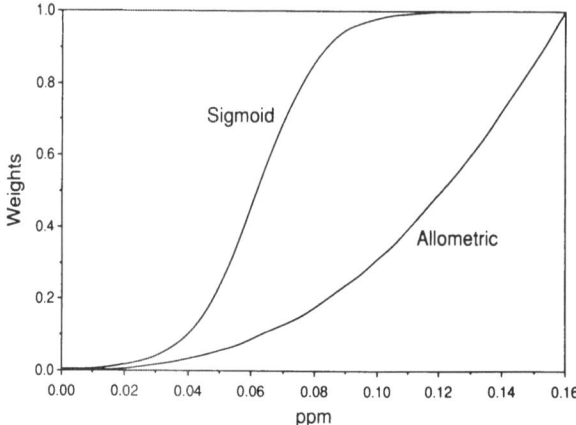

Fig. 6.2. Two continuous weighting functions for giving weight or import to hourly O_3 concentrations over an exposure period for use with cumulative exposure indices.

but rather gives credence (although not equal) to all concentrations in eliciting a biological response. With a continuous weighting scheme, either allometric (the concentration raised to a power) or functional, the higher the concentrations, the more import in the toxic response. The lower concentrations are still included, but are weighted to a lesser degree, reflecting their anticipated lesser import in the toxic response.

Allometric weighting functions have been used to preferentially weight the higher concentrations in exposure characterizations (Nouchi and Aoki, 1979; Larsen *et al.*, 1983). Larsen *et al.* (1983) described an index which they termed "impact (I)", in which the hourly concentrations are raised to an exponent (2·618) and summed over the exposure period. In essence, the index gives greater weight to the higher concentrations and cumulates hourly concentrations for the exposure value. The "impact" index assumes a cumulative effect of O_3 and the importance of peak events without using the threshold concept. Larsen and Heck (1984) also proposed the "effective mean" as an exposure index to give greater weight to peak concentrations. However, this index has all the other limitations of a long-term mean. The "effective mean", calculated for the usual 7-h period, has been used in analysis of some of the NCLAN data (Larsen and Heck, 1984). An example of an allometric weighting of a 7-day exposure is shown in Fig. 6.3. The shaded area of the profile represents the weighted concentration value compared to the actual hourly concentration values.

Sigmoid functions have also been used to weight hourly pollutant concentrations in calculating a cumulative exposure index (Lefohn and Runeckles, 1987; Lee *et al.*, 1988). When concentration is weighted with a sigmoid function, a multiplicative weighting factor, which depends on the concentration, is used for each of the hourly O_3 concentrations (Fig. 6.2). There are two main differences between the allometric and sigmoid weighting functions. The value of the sigmoid weighting function approaches a limiting value of one, while the weighting value of the allometric function continues to increase with con-

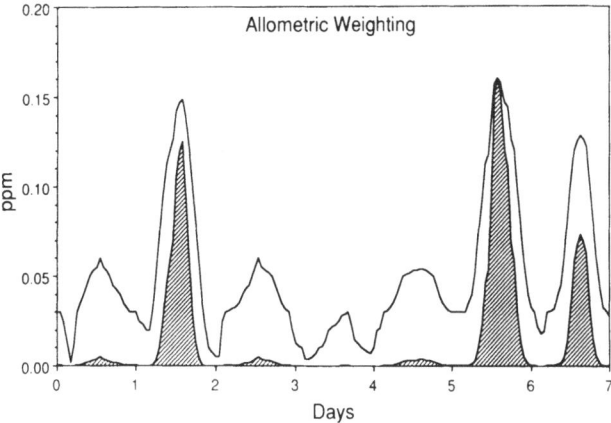

Fig. 6.3. A 7-day O_3 exposure regime illustrating the concentration value given the hourly concentrations using an allometric function (shaded areas) to weight concentrations over the exposure period. Total exposure value = 8·46 ppm; weighted exposure sum = 2·43 ppm. Note the lack of weight given lower and mid-range concentrations on days 1, 3, 5.

centration (Fig. 6.2). A sigmoid function depending on the inflection point, places more weight on mid-range concentrations (Fig. 6.2). If the sigmoid function has an inflection point below 0·07 ppm, it will place greater weight on mid-range (0·05–0·09 ppm) O_3 concentrations. The sigmoid weighting of a 7-day exposure is illustrated in Fig. 6.4. The shaded areas represent the weighted concentration values. Note the increased weight given to the low to mid-range concentrations with the sigmoid weighting scheme, as well as the higher concentrations compared to the allometric weighting shown in Fig. 6.3.

The following assumptions are made with indices that weight concentrations, depending whether a continuous or discontinuous function is employed.

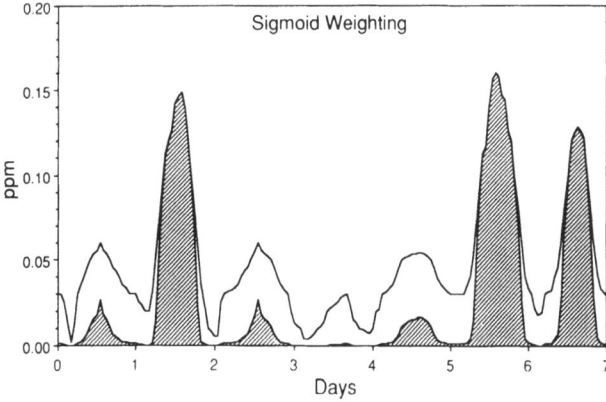

Fig. 6.4. A 7-day O_3 exposure regime illustrating the concentration value given hourly concentrations using a sigmoid weighting function (shaded areas) to weight concentrations over the exposure period. Same regime as in Fig. 6.1 and 6.3. Total exposure sum = 8·46 ppm; sigmoid weighted exposure sum = 4·67 ppm. Note that concentrations exceeding 0·1 ppm are all given full value and the lower and mid-range concentrations are given more weight than in the allometric weighting scheme.

- Weighting concentrations and cumulating values assume that higher concentrations play a greater role in plant response, and that the observed response is the result of cumulative exposure over a period of time.
- Weighting concentrations with a discontinuous function assume that only those concentrations equal to or greater than the threshold, or censor value, have import in causing the observed plant response.
- Weighting concentrations with a continuous function (for example, exponential or sigmoid function) assume that not any one particular concentration is of paramount significance, but rather, that as concentrations increase, there are increasing impacts on the biological system.
- Weighting concentrations assume that other aspects of the exposure, particularly temporal considerations such as duration of episode, time between episodes, or recovery time, and differential plant sensitivity are not important factors influencing the biological response.

6.4.5 Multicomponent indices

These indices are referred to as "multicomponent", since they attempt to address more than one characteristic of exposure in the relationship of ambient pollutant exposure and biological response. These indices include the weighting of concentration as well as weighting temporal features of the exposure (Lee *et al.*, 1988; Lefohn and Runeckles, 1987).

Temporal features of exposure include both the temporal distributions of concentrations over the exposure period and the time component of differential plant sensitivity over the exposure period, that is, phenological stages. These temporal components, like concentration, are influential in the biological response. The multicomponent indices explicitly incorporate differential weighting of various components of the exposure. This reflects the variations in these components over the time period of concern and their perceived importance in the relationship of exposure to biological response.

More exposure indices have not considered the recovery time between episodes, or peak O_3 exposure events. Ozone episodes occur intermittently over the growing season and can be viewed as temporal variations in hourly O_3 concentrations. The length of an episode is the consecutive occurrence (in hours) of elevated hourly concentrations (above a given threshold), and the time (hours or days) between episodes is a measure of the recovery time. Both of these measures may have a bearing on the plant's detoxification rate, or repair abilities. Only a few studies have attempted to include episode length or respite time in an exposure index. Lefohn and Runeckles (1987) proposed an O_3 exposure index that incorporates the length of time between episodes as well as the duration of the episode. The formulation of the index was derived from the original work of Mancini (1983), describing the effects of exposure to time-variable concentrations of toxicants for aquatic organisms. This index provides expression, or characterization, of the separate contributions of episodes (peak events) and the respite time between events (a possible period of recovery or repair) to the toxic effect observed. The index also includes the cumulative aspects of the exposure by summing weighted concentrations using a sigmoid weighting scheme to assign import to the higher concentrations, yet still including the lower concentrations (see Fig. 6.2). The efficacy of this index, however, has not been tested with any crop data sets.

The importance of recovery time in an exposure index is not clearly resolved. For example, a study with alfalfa and two grasses exposed to two different exposure regimes with equal "total exposure" found that the episodic regimes caused greater growth reductions than did daily peak exposures (Hogsett *et al.*, 1985; personal communication from Hogsett). The episodic exposure regimes, however, had a greater portion of the exposure period occupied by low concentrations and days between peak events (possible recovery periods) compared to the daily peak regimes. This observation would suggest that recovery time is not defined by concentration alone, but may perhaps have a temporal distribution element that is important, as well. The definition of recovery time and the role this component of exposure plays in the biological response remains to be discovered.

Another temporal aspect influencing plant response is the change in plant sensitivity with different stages of plant development. Most exposure indices, such as the "impact" (Larsen *et al.*, 1983) or functionally-weighted concentration indices (Lefohn *et al.*, 1988) that differentially weights concentrations, also implicitly weights the temporal component of differential plant sensitivity over a growing season or life cycle of the plant. This implicit weighting is one, thus implying that the plant is uniformly sensitive throughout the exposure period. A uniform weighting of the time component does not address the observation of differential sensitivity of the plant over a growing season (e.g., Adomait *et al.*, 1987).

An exposure index that differentially weights the time component to address the changing sensitivity of the plant over the season was proposed by Lee *et al.* (1988). The authors developed and evaluated an index that incorporates a temporal weighting scheme in conjunction with weighting hourly concentrations and cumulates the values over the exposure period. The temporal weighting scheme relates the timing of the exposure to the proposed time of plant sensitivity. Performance of this multicomponent exposure index, termed the "phenologically weighted cumulative impact" index (PWCI), was evaluated with several sets of NCLAN data, and various weighting functions for the temporal and concentration components of the exposure were compared. Allometric and sigmoid functions were compared for efficacy in weighting concentrations and it was concluded that a sigmoid function is preferable. Two features probably contribute to the enhanced performance of the sigmoid functions in weighting concentrations. First, the sigmoid functions give more weight to the mid-range O_3 concentrations than do the allometric functions (see Fig. 6.2), which may be quite important biologically. Second, with the sigmoid function, the approach to a plateau in weighting is a surrogate for the concept that higher O_3 concentrations can induce stomatal closure restricting O_3 uptake and, consequently, the impact.

Weighting functions were compared to discover which best described phenological aspects of plant sensitivity, including the exponential and gamma

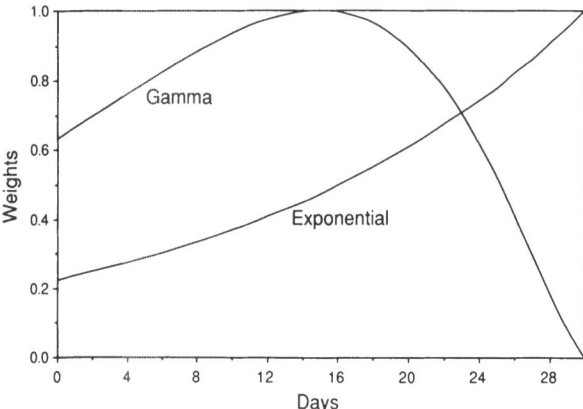

Fig. 6.5. Two continuous weighting functions for weighting time period of exposure or the phenological stage of the plant exposed. Concentrations are given the assigned weight on the days shown.

(Lee *et al.*, 1988). These functions are illustrated in Fig. 6.5. The application of exponential decay in sensitivity from the time component assumed an exponential decay in sensitivity from the time of final harvest. Various rates of decline were tested using a range of exponentials to weight the exposure over the growth period. The other continuous function tested was the gamma function, which de-emphasized early exposures and exposures just before harvest and was found to be better than the exponential function as a surrogate for differential plant sensitivity. The selection of the gamma function as the best descriptor for weighting exposures relative to developmental or sensitivity stages during the growth and exposure period is not surprising and seems to reflect a biological basis by (1) giving increasing weight to the exposures as growth progresses, (2) reaching a maximum at some time sufficiently ahead of harvest, and (3) declining in value as final harvest is approached (Fig. 6.5). Plants may be able to compensate or repair impacts of exposure early in development, and exposures that occur just prior to harvest simply do not have sufficient time to be expressed as injury. Consequently, exposures at the early and later stages in development and/or the exposure period are given less weight than those concentrations occurring during the middle of the exposure season. The weighting of the concentrations over a hypothetical 30-day exposure period and plant life cycle with an exponential and a gamma function is illustrated in Figs 6.6 and 6.7, respectively. The determination of a weighting scheme to account for temporal differential sensitivity in plants has only been briefly examined by these authors, and future work should examine other species, functional descriptors, and exposure scenarios for the general applicability of this approach. However, the PWCI, using a sigmoid concentration weighting and gamma weighting of the exposure period, was rated "overall best" along with two other cumulative, peak-weighting measures when compared to a number of currently used and recently developed exposure indices in predicting yield response in several crop species from NCLAN data (Lee *et al.*, 1988).

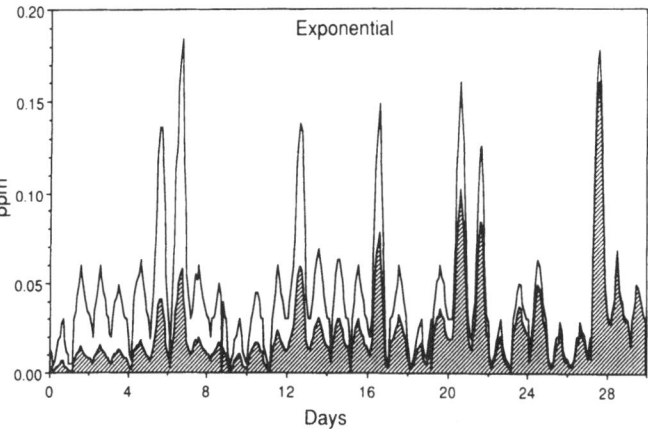

Fig. 6.6. A 30-day O_3 exposure regime illustrating the exponential weighting of the exposure period of a hypothetical plant with a 30-day life cycle. The weighted concentrations are shown in shaded areas, and reflect the days and concentrations presumed to have the most effect on the final yield of the plant (the days nearest the end of the life cycle and final harvest). Total exposure sum = 31·36 ppm; exponential weighting sum = 16·48 ppm.

The use of multicomponent exposure indices is based on the following assumptions.

- The impact of O_3 is cumulative on plant response.
- All concentrations have some impact on plant response, but the higher concentrations exhibit greater influence. The use of an allometric weighting scheme for all concentrations assumes that there is an exponential increase in sensitivity as concentration increases, whereas the use of a sigmoid weighting scheme assumes that mid-range concentrations are important, and that there is a plateau in the biological response curve.

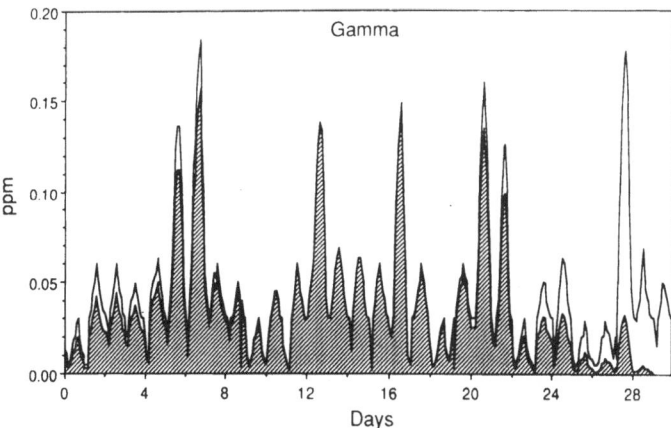

Fig. 6.7. A 30-day O_3 exposure regime illustrating the gamma weighting of the exposure period of a hypothetical plant with a 30-day life cycle. The weighted concentrations are shown as shaded areas, and reflect the presumed import of concentrations in the early to mid-life cycle and decreasing import of exposure in later life stages. Total exposure sum = 31·36 ppm; gamma weighting sum = 23·62 ppm.

- More than one factor, or exposure component, is active concurrently in influencing the exposure response.
- Plant sensitivity to O_3 changes with the course of plant development over the exposure period, or growing season. Phenological stages and timing are similar for all species and environmental conditions.
- Recovery time is important in the response, and at least in formulations used thus far, recovery time is defined by concentrations below some threshold value.

6.4.6 Dose approach

The previous exposure indices use the ambient O_3 concentration to describe a biological response and not the concentration of pollutant absorbed by the plant. This approach is one of convenience. Dose, by the classical toxicological definition, is (1) the product of concentration of toxicant in the organism and the duration of exposure, (2) a measure of pollutant exposure that is directly related to the biological response, and (3) reflects the influence of biological, environmental, and chemical factors controlling absorption of the pollutant. Several researchers have attempted to bridge this gap between ambient concentrations, actual dose, and the factors controlling uptake of pollutant.

Runeckles (1974) proposed an index of exposure called the "effective dose", which was a direct measure of the concentration of pollutant absorbed by the plant. Fowler and Cape (1982) suggested an exposure index called "pollutant absorbed dose" (PAD), which calculates the flux of pollutant to the plant as a product of concentration, time, and stomatal, or canopy, conductance. Similarly, Taylor et al. (1982) defined the dose to which a plant responds as the "internal flux" ($mg\ m^{-2}\ h^{-1}$). These values provide an estimate of the pollutant dose taken up by the plant and, thus, available to elicit a toxic response. A problem with this formulation, however, is that the same dose may, in different conditions, lead to different effects because of the time period over which the dose is received. Obtaining this measure of exposure requires an investment in experimental design to take stomatal measurements over time with the crop of trees under study and/or atmospheric conductance over time. The "effective dose" has been employed to describe foliar injury response in plants, but not in yield. The PAD, however, has not actually been used in describing any exposure response data.

Other researchers have attempted to include environmental measurements as indirect measures of O_3 uptake into an index of exposure. Mukammal and co-workers (1982) described a site-specific exposure index that related variability in the injury to a combination of O_3 exposure, pan evaporation, and plant maturity, incorporating these factors into a cumulative, weighted measure relating O_3 exposure to leaf injury. In a similar approach, Adomait et al. (1987) incorporated rainfall and temperature into a cumulative exposure index to account for O_3 flux in an exposure–response function describing yield in white beans. Further investigations are needed to generalize the environmental considerations and alleviate the site-specific nature of these indices.

6.4.7 Evaluation of approaches

Most of these exposure indices were not developed or selected from a biological basis, nor were they developed from an experimental approach designed to arrive at the appropriate indices that account for the variability in biological response. Recently, researchers have used a retrospective analysis of existing data (largely from the NCLAN program) in evaluating the relative efficacy of existing and proposed exposure indices with a large number of crop data sets (Lee *et al.*, 1988; Lefohn *et al.*, 1988). Lee *et al.* (1988), using eight crop data sets (2 years of data for each of four species), compared 25 exposure indices based on (1) averaging over various time-periods, (2) censored or integrated summations, (3) peak-weighted cumulative indices with various weighting functions, (4) multicomponent weighting, or peak and phenology, and (5) other indices that measure average episode length, number of episodes, days between episodes, and indices that combine many of the above components. From the comparison, the authors concluded that the best indices of exposure were cumulative in nature and weighted the higher concentrations. Lefohn and co-workers (1988) formed a similar conclusion from a two year study of soybean and wheat, and by comparing the following indices: (1) fixed or maximum 7-h seasonal means, (2) number of hours above certain threshold concentrations, (3) integrated summations of concentrations above three different minimum values, and (4) two sigmoid functions weighting concentrations over the exposure period.

6.5 FACTORS THAT SHOULD BE CONSIDERED IN SELECTING AN EXPOSURE STATISTIC

This review has outlined biological, environmental, and exposure dynamics features that influence the magnitude of the biological response, contributing to the observed variability, and, thus, are necessary to consider incorporating into a measure of the exposure that best describes the exposure–response relationship. Many of the currently used indices to not consider (1) the evidence showing varying response with different exposure structures or the diurnal distribution of pollutants over time; (2) the degree of temporal resolution required; (3) the response of plants to apparent threshold concentrations; (4) the diurnal stomatal behavior; (5) the pollutant uptake or effective concentration or dose; and (6) plant growth stage, maturity, and plant vigor. An understanding of these processes in the pollutant–plant interaction can lead to development of exposure indices that correctly account for all or most of all the variability in the response relationship.

However, the ultimate goal of this endeavor to develop the appropriate exposure indices may indeed be two goals. One is developing the "best" exposure index that incorporates all contributing features and accounts for all of the variation in the exposure response, and the second, more practical goal is that of specifying indices useful to the standard setting process. An index for standard should be simple, not site-specific, and as generic as possible. This objective may represent a compromise of the features included in the formulation of the first goal. However, to best determine what components should be included in the minimal, or simplistic, formulation for standards, it is necessary to fully understand the degree of influence of all factors contributing to the variation. Factors that should be considered in the development of exposure indices are given below.

- Peak concentrations are important and should be given appropriate weight. Functional, or continuous, weighting of all concentrations may be a better approach than threshold, or discontinuous, weighting.
- Both the intensity and distribution of concentrations over the exposure period should be considered.
- Duration of concentration and duration of exposure period need to be included.
- Elements of concentration and duration are inseparable; both are required to reflect the episodicity and time-varying nature of the exposure.
- Cumulative nature of the exposure needs to be represented as cumulated concentration over time. Environmental factors may also be important as a means to modify the cumulation of concentration in an attempt to create a partial surrogate for uptake.
- Uptake rate should be included, perhaps as a surrogate(s) reflecting the two factors controlling the rate: stomatal and biochemical resistance. This would require either direct measures of gas exchange properties (for example, stomatal conductance, boundary layer resistance) or surrogates (such as environmental measurements) that reflect or account for gas exchange and pollutant uptake.
- It is important to define temporal distribution of pollutant concentrations versus the gas exchange characteristics of the plant.
- Temporal distribution of pollutant concentrations versus temporal aspects of plant sensitivity (developmental stages of plant growth) needs to be included. The phenological stages of plant growth and sensitivity should be addressed at several levels, including the relationship of the differential stages of sensitivity and the final growth response, as well as the variable response between species and families, within species, and within a growth period.
- Exposure history, including, for example, the predisposition and recovery periods within the exposure, need to be considered.
- Exposure period should be accounted for; that is, short-term versus long-term exposure. Should a different index be used depending on length of exposure period?

6.6 ELEMENTS OF EXPERIMENTAL DESIGN FOR EXPOSURE INDICES STUDIES

This review has indicated that there are other, perhaps better, exposure indices than averages or means, but those indices and their characteristics still need to be studied explicitly to better understand them and their inherent variation. New fumigation, or exposure studies, are required for much of this examination. For example, the NCLAN studies, by having O_3 concentration data for only a 7- or 12-h period each day in the modified ambient treatment chambers over one season do not allow further analysis of the role of exposure duration or temporal variation in concentration distribution and the ensuing growth response. The variations in response to pollutant concentrations point to the need for further studies addressing appropriate weighting schemes for concentration. Weighting schemes that may account for the observed variation between species, among species, and the role of environment in moderating the response are needed. Also, studies are needed to examine the factors of uptake or molecular sensitivity that may be under genetic or environmental control in variability of the exposure–response.

In addition to the examples of suggested experiments, the following generic characteristics of experimental design should be included in all studies to examine exposure indices.

- Collect 24 h of pollutant concentration data for all treatment levels each day over the exposure period to relate temporal distributions of concentrations and intermediate and final plant responses to the exposure. Without complete exposure data, it is not possible to relate components of the exposure with their influence over time in chronic exposure studies. The exposures should be related to a diurnal pattern of occurrence of pollutant.
- Include temporal variation in plant growth, sensitivity, and exposure characteristics.
- Include periodic measurements of plant response; for example, sequential biomass harvests, stomatal resistance, transpiration, and pollutant uptake.
- Include a range of exposure treatments in the design.
- Design studies that vary pollutant exposures with plant phenological stages. Studies should also consider exposure history, or predisposition and respite time.
- Include adequate replication in design, since the responses measured usually exhibit a high degree of variability.
- Include some measure of pollutant uptake or a surrogate that reflects uptake rate.
- Utilize both simple and complex designs; that is, the inclusion of variation in all components of concentration and time within the same experiment is not the only approach for these studies. A simple design approach would be better to define the characteristics that have importance (e.g. predisposition or phenological stages), thus developing a mechanistic knowledge base before defining more complex studies to validate hypotheses.

The suggestions put forth in this review and in the brief proposals of experimental work reveal a number of approaches to develop appropriate exposure indices. Knowing the sources of variability and their contribution to the biological response provides information as to which should be included to develop the "best" index. However, the formulation may not have the elements of simplicity or other characteristics that best serve the standard setting process. In that case, a compromise may be required whereby an index is chosen that accounts for most of the variation, but not all, and is easily adapted to the predictive models.

REFERENCES

Adomait, E. J., J. Ensing, and G. Hofstra. (1987). A dose-response function for the impact of ozone on Ontario-grown white bean and an estimate of economic loss. *Can. J. Plant Sci.*, **67**, 131–6.

Amiro, B. D., T. J. Gillespie, and G. W. Thurtell. (1984). Injury response of *Phaseolus vulgaris* to ozone flux density. *Atmos. Environ.*, **18**, 1207–15.

Ariens, E. J., A. M. Simonis, and J. Offermerier. (1976). *Introduction to general toxicology.* New York, Academic Press.

Ashmore, M. R. (1984). Effects of ozone on vegetation in the United Kingdom. In *Evaluation of the effects of photochemical oxidants on human health, agricultural crops, forestry, materials and visibility.* Proceedings of the international workshop, Goteburg, Sweden, ed. by P. Greenfelt, 92–104. Goteburg, Sweden, Swedish Environmental Research Institute.

Ashmore, M. R. and M. Onal. (1984). Modification by sulphur dioxide of the responses of *Hordeum vulgare* to ozone. *Environ. Pollut., Ser. A,* **36,** 31–43.

Bennett, J. H. (1979). Foliar exchange of gases. In: *Methodology for the assessment of air pollution effects on vegetation: a handbook,* ed. by W. W. Heck, S. V. Krupa, and S. N. Linzon. Chapter 10. Pittsburg, PA, Air Pollution Control Assoc.

Blum, U. and W. W. Heck. (1980). Effects of acute ozone exposures on snap bean at various stages of its life cycle. *Environ. Exp. Bot.,* **20,** 73–85.

Butler, L. K. and T. W. Tibbitts. (1979). Stomatal mechanisms determining genetic resistance to ozone in *Phaseolus vulgaris* L. *J. Am. Soc. Hortic. Sci.,* **104,** 211–13.

Coyne, P. I. and G. E. Bingham. (1982). Variation in photosynthesis and stomatal conductance in an ozone-stressed ponderosa pine stand: Light response. *For. Sci.,* **28,** 257–73.

Craker, L. E. and J. S. Starbuck. (1973). Leaf age and air pollutant susceptibility: Uptake of ozone and sulfur dioxide. *Environ. Res.,* **6**(1), 91–4.

Cure, W. W., J. S. Sanders, and A. S. Heagle. (1986). Crop yield response predicted with different characterizations of the same ozone treatments. *J. Environ. Qual.,* **15,** 251–4.

Davis, D. D. and F. A. Wood. (1972). The relative susceptibility of eighteen coniferous species to ozone. *Phytopathology,* **62,** 14–19.

Dunning, J. A. and W. W. Heck. (1977). Response of bean and tobacco to ozone: Effect of light intensity, temperature, and relative humidity. *J. Air Pollut. Control. Assoc.,* **27,** 882–6.

Endress, A. G. and C. Grunwald. (1985). Impact of chronic ozone on soybean growth and biomass partitioning. *Agric. Ecosystem Environ.,* **13,** 9–23.

Fowler, D. and J. N. Cape. (1982). Air pollutants in agriculture and horticulture. In *Effects of gaseous air pollution in agriculture and horticulture,* ed. by M. H. Unsworth and D. P. Ormrod, 3–26. London, Butterworth Scientific.

Guderian, R., D. T. Tingey, and R. Rabe. (1985). Effects of photochemical oxidants on plants. In *Air pollution by photochemical oxidants: formation, transport control, and effects on plants,* ed. by R. Guderian, 129–333. Berlin, Springer-Verlag.

Haas, J. H. (1970). Relation of crop maturity and physiology to air pollution incited bronzing of *Phaseolus vulgaris*. *Phytopathology,* **60,** 407–10.

Hanson, G. P., L. Thorne, and D. H. Addis. (1975). The ozone sensitivity of *Petunia hybrida* Vilm. as related to physiological age. *J. Am. Soc. Hortic. Sci.*, **100,** 188–90.

Heagle, A. S. and W. W. Heck. (1974). Predisposition of tobacco to oxidant air pollutant injury by a previous exposure to oxidants. *Environ. Pollut.*, **7,** 247–51.

Heagle, A. S. and W. W. Heck. (1980). Field methods to assess crop losses due to oxidant air pollutants. In *Crop loss assessment.* Proceedings of E. C. Stakman Commemorative Symposium, ed. by P. S. Teng, and S. V. Krupa, 296–305. Misc. Publication No. 7. St. Paul, University of Minnesota.

Heagle, A. S., R. B. Philbeck, and W. M. Knott. (1979). Thresholds for injury, growth, and yield loss caused by ozone on field corn hybrids. *Phytopathology,* **69,** 21–6.

Heagle, A. S., W. W. Cure, and J. O. Rawlings. (1985). Response of turnips to chronic doses of ozone in open-top field chambers. *Environ. Pollut., Ser. A,* **38,** 305–19.

Heagle, A. S., W. W. Heck, V. M. Lesser, and J. O. Rawlings. (1987). Effects of daily ozone exposure duration and concentration fluctuation on yield of tobacco. *Phytopathology,* **77,** 856–62.

Heath, R. L. (1988). Biochemical mechanisms of pollutant stress. In *Assessment of crop loss from air pollutants.* Proceedings of the international conference, Raleigh, N.C., USA, ed. by W. W. Heck, O. C. Taylor, and D. T. Tingey, 259–86. London, Elsevier Applied Science.

Heck, W. W. and J. A. Dunning. (1967). The effects of ozone on tobacco and pinto beans as conditioned by several ecological factors. *J. Air Pollut. Control Assoc.,* **17,** 112–4.

Heck, W. W. and D. T. Tingey. (1971). Ozone time-concentration model to predict acute foliar injury. In *Proceedings of the Second International Clean Air Congress,* December, 1970, ed. by H. M. Englund and W. T. Beery, 249–255. New York, Academic Press.

Heck, W. W., J. A. Dunning, and I. J. Hindawi. (1966). Ozone: Nonlinear relation of dose and injury to plants. *Science* (*Washington*), **151,** 577–8.

Heck, W. W., O. C. Taylor, R. M. Adams, G. Bingham, J. Miller, E. Preston, and L. Weinstein. (1982). Assessment of crop loss from ozone. *J. Air Pollut. Control Assoc.,* **32,** 353–61.

Heck, W. W., R. M. Adams, W. W. Cure, A. S. Heagle, H. E. Heggestad, R. J. Kohut, L. W. Kress, J. O. Rawlings, and O. C. Taylor. (1983). A reassessment of crop loss from ozone. *Environ. Sci. Technol.,* **17,** 573a–81a.

Heck, W. W., W. Cure, J. O. Rawlings, L. J. Zaragoza, A. S. Heagle, H. E. Heggestad, R. J. Kohut, L. W. Kress, and P. J. Temple. (1984*a*). Assessing impacts of ozone on agricultural crops. I. Overview. *J. Air Pollut. Control Assoc.,* **34,** 729–35.

Heck, W. W., W. Cure, J. O. Rawlings, L. .J. Zaragoza, A. S. Heagle, H. E. Heggestad, R. J. Kohut, L. W. Kress, and P. J. Temple. (1984*b*). Assessing impacts of ozone on agricultural crops. II. Crop yield functions and alternative exposure statistics. *J. Air Pollut. Control Assoc.,* **34,** 810–7.

Hofstra, G., A. E. G. Tonneijck, and O. B. Allen. (1985). Cumulative effects of low levels of SO_2 on O_3 sensitivity in bean and cucumber. *Atmos. Environ.,* **19,** 195–8.

Hogsett, W. E., D. T. Tingey, and S. R. Holman. (1985). A programmable exposure control system for determination of the effects of pollutant exposure regimes on plant growth. *Atmos. Environ.,* **19,** 1135–45.

Hogsett, W. E., M. Plocher, V. Wildman, D. T. Tingey, and J. P. Bennett. (1986). Growth response of two varieties of slash pine seedlings to chronic ozone exposures. *Can. J. Bot.,* **63,** 2369–76.

Johnston, J. W., Jr and A. S. Heagle. (1982). Response of chronically ozonated soybean plants to an acute ozone exposure. *Phytopathology,* **72,** 387–9.

Kobriger, J. M., T. W. Tibbitts, and M. L. Brenner. (1984). Injury, stomatal conductance, and abscisic acid levels on pea plants following ozone plus sulfur dioxide exposures at different times of the day. *Plant Physiol.*, **76**, 823–6.

Kress, L. W. and J. E. Miller. (1983). Impact of ozone on soybean yield. *J. Environ. Qual.*, **12**, 276–81.

Kress, L. W., J. E. Miller, and H. J. Smith. (1985). Impact of ozone on winter wheat yield. *Environ. Exp. Bot.*, **25**, 211–28.

Larcher, W. (1975). *Physiological plant ecology*. Berlin, Springer-Verlag.

Larsen, R. I. and W. W. Heck. (1984). An air quality data analysis system for interrelating effects, standards, and needed source reduction: Part 8. An effective mean O_3 crop reduction mathematical model. *J. Air Pollut. Control Assoc.*, **34**, 1023–34.

Larsen, R. I., A. S. Heagle, and W. W. Heck. (1983). An air quality data analysis system for interrelating effects, standards, and needed source reductions: Part 7. An O_3–SO_2 leaf injury mathematical model. *J. Air Pollut. Control Assoc.*, **33**, 198–207.

Lee, E. H., D. T. Tingey, and W. E. Hogsett. (1988). Evaluation of ozone exposure indices in exposure–response modeling. *Environ. Pollut.*, **53**, 43–62.

Lefohn, A. S. and H. M. Benedict. (1982). Development of a mathematical index that describes ozone concentration, frequency, and duration. *Atmos. Environ.*, **16**, 2529–32.

Lefohn, A. S. and H. M. Benedict. (1985). Exposure consideration associated with characterizing ozone ambient air quality monitoring data. In *Evaluation of the scientific basis for ozone/oxidants standards*, 17–31. Pittsburg, Air Pollution Control Association.

Lefohn, A. S. and C. K. Jones. (1986). The characterization of ozone and sulfur dioxide air quality data for assessing possible vegetation effects. *J. Air Pollut. Control Assoc.*, **36**, 1123–9.

Lefohn, A. S. and V. C. Runeckles. (1987). Establishing standards to protect vegetation—ozone exposure/dose considerations. *Atmos. Environ.*, **21**, 561–8.

Lefohn, A. S., J. A. Laurence, and R. J. Kohut. (1988). A comparison of indices that describe the relationship between exposure to ozone and reduction in the yield of agricultural crops. *Atmos. Environ.* **22**, 1229–40.

Legassicke, B. C. and D. P. Ormrod. (1981). Suppression of ozone-injury on tomatoes by ethylene diurea in controlled environments and in the field. *HortScience*, **16**, 183–4.

McLaughlin, S. B. and G. E. Taylor, Jr. (1981). Relative humidity: Important modifier of pollutant uptake by plants. *Science (Washington)*, **221**, 167–9.

McLaughlin, S. B., D. S. Shriner, R. K. McConathy, and L. K. Mann. (1979). The effects of SO_2 dosage kinetics and exposure frequency on photosynthesis and transpiration of kidney beans (*Phaseolus vulgaris* L.) *Environ. Exp. Bot.*, **19**, 179–91.

Male, L., E. Preston, and G. Neely. (1983). Yield response curves of crops exposed to SO_2 time series. *Atmos. Environ.*, **17**, 1589–93.

Mancini, J. L. (1983). A method for calculating effects, on aquatic organisms, of time varying concentrations. *Water Res.*, **17**, 1355–62.

Mukammal, E. I. (1965). Ozone as a cause of tobacco injury. *Agric. Meteorol.*, **2**, 145–65.

Mukammal, E. I., H. H. Neumann, and G. Hofstra. (1982). Ozone injury to white bean (*Phaseolus vulgaris* L.) in southwestern Ontario, Canada: correlation with ozone dose, pan evaporation, plant maturity and rainfall. In *Effects of gaseous air pollution in agriculture and horticulture*, ed. by M. H. Unsworth, and D. P. Ormrod, 470–1. London, Butterworth Scientific.

Musselman, R. C., R. J. Oshima, and R. E. Gallavan. (1983). Significance of pollutant concentration distribution in the response of 'red kidney' beans to ozone. *J. Am. Soc. Hortic. Sci.*, **108**, 347–51.

Musselman, R. C., A. J. Huerta, P. M. McCool, and R. J. Oshima. (1986). Response of beans to simulated ambient and uniform ozone distributions with equal peak concentrations. *J. Am. Soc. Hortic. Sci.*, **111**, 470–3.

Nosal, M. (1983). Atmosphere–biosphere interface: probability analysis and an experimental design for studies of air pollutant-induced plant response. 98 pp. RMD Report 83/25. Edmonton, Alberta, Research Management Division, Alberta Environment.

Nouchi, I. and K. Aoki. (1979). Morning glory as a photochemical oxidant indicator. *Environ. Pollut.*, **18**, 289–303.

Olszyk, D. M. and T. W. Tibbitts. (1981). Stomatal response and leaf injury of *Pisum sativum* L. with SO_2 and O_3 exposures. *Plant Physiol.*, **67**, 539–44.

Olszyk, D. M. and D. T. Tingey. (1986). Joint action of O_3 and SO_2 in modifying plant gas exchange. *Plant Physiol.*, **82**, 401–5.

Oshima, R. J., M. P. Poe, P. K. Braegelmann, D. W. Baldwin, and V. Van Way. (1976). Ozone dosage-crop loss function for alfalfa: A standardized method for assessing crop losses from air pollutants. *J. Air Pollut. Control Assoc.*, **26**, 861–5.

Reich, P. B., A. W. Schoettle, and R. G. Amundson. (1985). Effects of low concentration of O_3, leaf age and water stress on leaf diffusive conductance and water use efficiency in soybean. *Physiol. Plant.*, **63**, 58–64.

Runeckles, V. C. (1974). Dosage of air pollutants and damage to vegetation. *Environ. Conserv.*, **1**, 305–8.

Runeckles, V. C. and P. M. Rosen. (1974). Effects of pretreatment with low ozone concentrations on ozone injury to bean and mint. *Can. J. Bot.*, **52**, 2607–10.

Runeckles, V. C. and P. M. Rosen. (1977). Effects of ambient pretreatment on transpiration and susceptibility to ozone injury. *Can. J. Bot.*, **55**, 193–7.

Stan, H.-J. and S. Schicker. (1982). Effect of repetitive ozone treatment on bean plants—stress ethylene production and leaf necrosis. *Atmos. Environ.*, **16**, 2267–70.

Steinberger, E. H. and Z. Naveh. (1982). Effects of recurring exposures to small ozone concentrations on Bel W3 tobacco plants. *Agric. Environ.*, **7**, 255–63.

Tanaka, K. and K. Sugahara. (1980). Role of superoxide dismutase in the defense against SO_2 toxicity and induction of superoxide dismutase with SO_2 fumigation. Research Report, National Institute of Environmental Studies, **11**, 155–64.

Taylor, G. E., Jr and D. T. Tingey. (1983). Sulfur dioxide flux into leaves of *Geranium carolinianum* L. Evidence for a nonstomatal or residual resistance. *Plant Physiol.*, **72**, 237–44.

Taylor, G. E., Jr, S. B. McLaughlin, and D. S. Shriner. (1982). Effective pollutant dose. In *Effects of gaseous air pollution in agriculture and horticulture*, ed. by M. H. Unsworth and D. P. Ormrod, 458–60. London, Butterworth Scientific.

Taylor, G. E., Jr, P. J. Hanson, and D. D. Baldocchi. (1988). Pollutant deposition to individual leaves and plant canopies: Sites of regulation and relationship to injury. In *Assessment of crop loss from air pollutants*. Proceedings of the international conference, Raleigh, N.C., USA, ed. by W. W. Heck, O. C. Taylor, and D. T. Tingey, 227–57. London, Elsevier Applied Science.

Temple, P. J., O. C. Taylor, and L. F. Benoit. (1986). Yield response of head lettuce (*Latuca sativa* L.) to ozone. *Environ. Exp. Bot.*, **26**, 53–8.

Thorne, L. and G. P. Hanson. (1972). Species differences in rates of vegetal ozone absorption. *Environ. Pollut.*, **3**, 303–12.

Thorne, L. and G. P. Hanson. (1976). Relationship between genetically controlled ozone sensitivity and gas exchange rate in *Petunia hybrida* Vilm. *J. Am. Soc. Hortic. Sci.*, **101**, 60–3.

Tingey, D. T. and George E. Taylor, Jr. (1982). Variation in plant response to ozone: A conceptual model of physiological events. In *Effects of gaseous air pollution in agriculture and horticulture*, ed. by M. H. Unsworth and D. P. Ormrod, 113–38. London, Butterworth Scientific.

Tingey, D. T. and W. E. Hogsett. (1985). Water stress reduces ozone injury via stomatal mechanism. *Plant Physiol.*, **77**, 944–7.

Tingey, D. T., W. W. Heck, and R. A. Reinert. (1971). Effect of low concentrations of ozone and sulfur dioxide on foliage, growth and yield of radish. *J. Am. Soc. Horic. Sci.*, **96**, 369–71.

Tingey, D. T., G. L. Thutt, M. L. Gumpertz, and W. E. Hogsett. (1982). Plant water status influences ozone sensitivity of bean plants. *Agric. Environ.*, **7**, 243–54.

Tonneijck, A. E. G. (1984). Effects of peroxyacetyl nitrate (PAN) and ozone on some plant species. In *Proceedings of the OECD workshop on ozone,* Goteborg, Sweden. Goteborg, Swedish Environmental Research Institute.

US Environmental Protection Agency. (1978). Air quality criteria for ozone and other photochemical oxidants. EPA-600/8-78-004. US Environmental Protection Agency, Environmental Criteria and Assessment Office. North Carolina, Research Triangle Park.

US Environmental Protection Agency. (1986). Air quality criteria for ozone and other photochemical oxidants. Vol. III. EPA-600/8-84/020cF. US Environmental Protection Agency, Environmental Criteria and Assessment Office. North Carolina, Research Triangle Park.

Walmsley, L., M. R. Ashmore, and J. N. B. Bell. (1980). Adaptation of radish *Raphanus sativus* L. in response to continuous exposure to ozone. *Environ. Pollut., Ser. A,* **23**, 165–77.

Session III

YIELD ASSESSMENT USING
FIELD APPROACHES FOR
MEASURING CROP LOSS

(Michael H. Unsworth, *Chairman*; Howard E. Heggestad,
Co-Chairman)

7

FACTORS INFLUENCING OZONE DOSE–YIELD RESPONSE RELATIONSHIPS IN OPEN-TOP FIELD CHAMBER STUDIES

A. S. HEAGLE

US Department of Agriculture, Raleigh, North Carolina, USA

L. W. KRESS

US Department of Agriculture, Research Triangle Park, North Carolina, USA

P. J. TEMPLE

University of California, Riverside, California, USA

R. J. KOHUT

Boyce Thompson Institute, Ithaca, New York, USA

J. E. MILLER

US Department of Agriculture, Raleigh, North Carolina, USA

and

H. E. HEGGESTAD

US Department of Agriculture, Beltsville, Maryland, USA

7.1 INTRODUCTION

The National Crop Loss Assessment Network (NCLAN) consisted of a group of government and non-government organizations cooperating to determine immediate and potential effects of air pollution on crop production (Heck *et al.*, 1982*b*, 1983*b*, 1984*b*). The major NCLAN objective was to obtain valid relationships between seasonal exposure to different levels of ozone (O_3) and yields of important agricultural crops. Prior to NCLAN, results for only five field experiments of this type were available. The NCLAN program produced significant O_3 dose–yield relationships from 38 field experiments. These dose–response results were needed for estimates of the annual impact on the US economy caused by a range of ambient O_3 levels. These estimates are needed by the US EPA in setting national ambient air quality standards for O_3.

Methods to perform NCLAN research were first discussed by NCLAN participants in a series of meetings in 1980. The participants chose open-top field chambers as the major research tool because the chambers allowed adequate control of a range of O_3 concentrations regardless of wind direction, allowed for a control treatment (less than ambient O_3 level), allowed treatment replication at a moderate cost, and did not cause a major change of microclimate within them. We will discuss how the open-top chambers affect microclimate, plant gas exchange, and yield as well as potential effects of NCLAN methods on crop response to O_3.

The NCLAN program has made significant strides to understand the possible interactions between O_3 and soil moisture, O_3 and sulfur dioxide (SO_2), and O_3 and cultivar. We will summarize results of NCLAN O_3 dose–yield response experiments, with and without added variables. Strengths and weaknesses of the NCLAN program will be identified and research needs will be discussed.

7.2   NCLAN METHODS

7.2.1   Chamber design and function

The open-top chambers used in NCLAN studies were designed in 1969 with several criteria. In order of importance, these criteria were: (1) an adequate size allowing crop growth to maturity; (2) uniform distribution of added gaseous pollutants; (3) minimal alteration of microclimate; (4) durability; (5) portability; and (6) reasonable cost. The chambers (Fig. 7.1) are cylinders (3.05 m diameter $\times 2.44$ m tall) with an aluminum channel frame. The chamber perimeter is covered with clear polyvinyl chloride film that is UV resistant, and flexible. The lower film is double-layered with the inner layer perforated by six rows of 2.5-cm holes spaced at 17.8 cm. The lower film serves as an air duct with an inlet tube attached to a fan box that houses an axial-blade fan. The

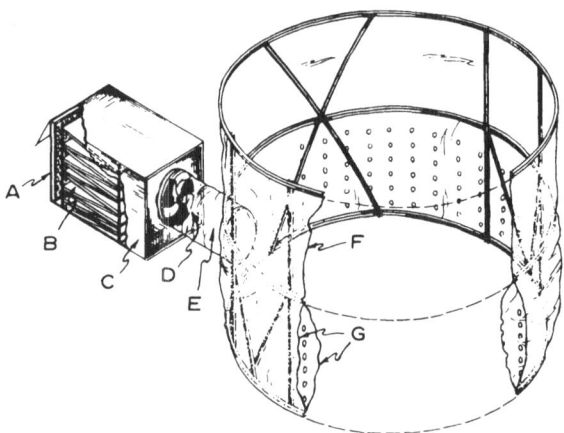

Fig. 7.1. Diagram of an open-top field chamber used in NCLAN dose–response studies. A, particle filter; B, activated-charcoal filter; C, housing for filters and fan; D, axial blade fan; E, air duct from fan to chamber; F, upper plastic panel; G, lower double-wall plastic panel with perforated inner wall.

moving air (approximately 2500 cfm ($1 \cdot 18 \, \text{m}^3 \, \text{s}^{-1}$)) inflates the duct and passes through the holes and into the chamber plot area, then out the open top, with approximately three air changes per minute. The fan box holds a particle filter and can hold an activated charcoal filter comprised of 12 separate elements.

7.2.2   Cultural methods

Standard agronomic procedures were employed for planting dates, fertilizer rates, and herbicide application. At most sites, plots were selected after plant emergence based on plant appearance, size and stand density (enough to allow thinning to a uniform stand in all plots) and soil uniformity. In most studies, soil nutrient analyses were performed for all plots before exposures began, after crop maturity, or both. Once plots were selected, weeds were controlled by hand although in a few cases, recommended herbicides were used. Insects and diseases were controlled before levels became high enough to affect yield. If required, recommended chemical applications were made to all experimental plots.

One NCLAN protocol was to irrigate as needed to minimize moisture stress. Although attempts were made to standardize irrigation protocols, this was not possible because of variables associated with soil type, moisture measurement methods, crop moisture requirements, and irrigation methods, all of which can interact with the stage of crop growth and weather. Thus, different methods were used to measure soil moisture (tensiometer, psychrometer, neutron probe, gravimetric), to irrigate (fixed spray nozzle, hand-held nozzle, drip-tubing, furrow), and different irrigation rates were used. For various reasons, slight to moderate moisture stress probably occurred for short periods in some experiments even though the protocol called for well-watered conditions.

7.2.3   Crop and cultivar selection

NCLAN O_3 dose–yield response studies focused on the most economically important crops (i.e., field corn, soybean, wheat, cotton) with a few notable exceptions (e.g., turnip and kidney bean) (Table 7.1). Within each crop, the most widely grown cultivars within each region were chosen, although exceptions exist. For soybean, field corn, turnip, winter wheat, cotton, and alfalfa, NCLAN produced dose–response data for nine, six, four, three, three, and two cultivars, respectively. For 11 other species, only one cultivar has been tested.

7.2.4   Exposure methods

Ozone concentrations in ambient air (AA) are affected by regional and local weather patterns. Daily 7-h (1000 to 1700 hours EDT) and 24-h mean O_3 concentrations at Raleigh, NC (Fig. 7.2) are similar to the fluctuations seen at all NCLAN sites. All NCLAN O_3 exposures used levels of O_3 in AA as baseline concentrations (Heagle *et al.*, 1979*b*). Consequently, treatment levels of O_3 also showed variation due to weather conditions. Ozone concentrations in treatment plots were monitored with chemiluminescence- or UV photometric-type instruments on a time-shared basis (Heagle *et al.*, 1979*b*).

TABLE 7.1

A summary of ozone dose–yield response experiments in open-top field chambers performed in the NCLAN program

Year	Crop species	Cultivars	Site location	Daily duration (h) and method of O_3 addition[a]	Seasonal exposure duration[b]	Chamber O_3 levels	Reps per O_3 level	Number of chambers	Factors other than O_3 studied	Reference	Significant interactions between O_3 and other factors tested	Comments
1980	Phaseolus vulgaris (dry bean)	Red kidney	Ithaca, NY	7C	26	5	4	20	—	Kohut and Laurence, 1983	—	
	Glycine max (soybean)	Corsoy 79	Argonne, IL	7C	59	5	4	20	—	Kress and Miller, 1983	—	Virus infection caused need for covariate analyses
	Brassica rapa (turnip)	Purple top, Tokyo cross, Shogoin Just right	Raleigh, NC	7C	91	5	4	20	4 cultivars	Heagle et al., 1985	—	Acute injury response to low O_3 levels
	Arachis hypogea (peanut)	NC-6	Raleigh, NC	7C	86	6	4	24	—	Heagle et al., 1983a	—	
	Lactuca sativa (lettuce)	Empire	Riverside, CA	7C	<50	6	4	24	—	Heck et al., 1983b	—	Winds (80 to 90 mph) damaged chambers and plants one month before scheduled harvest
1981	Zea mays (field corn)	Pioneer 3780 PAG 397	Argonne, IL	7C	84	6	3	18	2 hybrids	Kress and Miller, 1985a	—	
	Soybean	Hodgson	Ithaca, NY	7C	62	5	4[d]	20	effects of thinning	Kohut et al., 1986	—	2 reps for sequential harvests; 2 reps untouched
	Soybean	Davis	Raleigh, NC	7C	86	6	8[c]	48	4 SO_2 levels	Heagle et al., 1983b	—	
	Soybean	Essex, Williams	Beltsville, MD	7C	73	4	4[c]	16	6 SO_2 levels	Heck et al., 1984c	—	No weekend exposure for first 14 days
	Gossipium hirsutum (cotton)	Acala SJ-2	Shafter, CA	7C	33	6	4[c]	24	2 soil moisture levels	Temple et al., 1985b	soil moisture	Less O_3 response at low soil moisture
	Lycopersicon esculentum (tomato)	Murrieta	Tracy, CA	7C	66	5	6[c]	30	6 SO_2 levels	Heck et al., 1984c	—	
1982	Triticum aestivum (winter wheat)	Vona	Ithaca, NY	7C	69	5	4	20	—	Kohut et al., 1987	—	Powdery mildew
	Winter wheat	Abe, Arthur, Roland	Argonne, IL	7C	87	5	4	20	3 cultivars	Kress et al., 1985c	cultivars	Numbers of heads m^{-1} used as covariate

Year	Crop	Cultivar	Location	Exposure					Treatment	Reference	Covariate	Comments
	Cotton	Stoneville-213	Raleigh, NC	7C	88	5	4c	20	4 SO₂ levels	Heagle et al., 1986a	soil moisture	Verticillium wilt and salt accumulation (loss of two plots)
	Soybean	Williams, Forrest	Beltsville, MD	7C	70	5	6c	30	3 SO₂ levels 2 soil moisture levels	Heggestad et al., 1985	—	
	Tomato	Murrieta	Tracy, CA	7C	65	5	6c	30	6 SO₂ levels	Heck et al., 1984c	—	
	Dry bean	Red kidney	Ithaca, NY	7C	66	5	4c	20	Effects of physiology measurements	Heck et al., 1984c	—	
	Sorghum vulgare (grain sorghum)	DeKalb A28⁺	Argonne, IL	7C	84	6	3	18		Kress and Miller, 1985b	—	
	Hordeum vulgare (barley)	Poco	Shafter, CA	7P	50	6	4	24	—	Temple et al., 1985a	—	Barley stripe mosaic virus—23 days of added O₃—no significant yield response to O₃
	Cotton	Acala-SJ-2	Shafter, CA	7P	59	6	4c	24	2 soil moisture levels	Temple et al., 1985b	—	
1983	Soybean	Davis	Raleigh, NC	7C, 7P	87	NF, CF, 3C, 3Pe	2e	16	C vs P O₃ addition	Heagle et al., 1986b	—	
	Lettuce	Empire	Shafter, CA	7P	72	6	4	24		Temple et al., 1986	—	Number of heads used as covariate
	Winter wheat	Abe, Arthur	Argonne, Il.	7C	87	5	3	15		Kress et al., 1985c	—	
	Soybean	Amsoy 71, Corsoy 79, Pella, Williams	Argonne, IL	7C	92	4	3	12	4 cultivars	Kress, pers. comm.	cultivar	
	Nicotiana tabacum (flue-cured tobacco)	McNair 944	Raleigh, NC	7C, P 12P	87	NF, CF, 3 7C 3 7P, 3 12Pf	2f	22	C vs P for 7h 12h vs 7h for P	Heagle et al., 1987a	—	Plants at 12 h yielded less than those at 7 h
	Soybean	Davis	Raleigh, NC	7C	88	4	6c	24	2 soil moisture	Heagle et al., 1987b	soil moisture	Severe moisture stress
	Soybean	Williams, Corsoy 79	Beltsville, MD	7C	69	5	6c	30	2 soil moisture 2 cultivators	Heggestad et al., 1985	soil moisture	More O₃ effect at low moisture at NF level for one CV
	Barley	CM-72	Shafter, CA	7P	78	6	4c	24	2 soil moisture	Temple et al., 1985a	—	Low ambient O₃ levels—no significant yield response to O₃
	Winter wheat	Vona	Ithaca, NY	7P	38	4	4c	16	4 SO₂	Kohut et al., 1987	—	No effects of SO₂ on yield
	Soybean	Amsoy 71, Corsoy	Argonne, IL	7C	92	5	4c	20	4 SO₂ 2 cultivars	Kress et al., 1986	cultivar	

(Continued)

TABLE 7.1 (continued)

Year	Crop species	Cultivars	Site location	Daily duration (h) and method of O3 addition[a]	Seasonal exposure duration[b]	Chamber O3 levels	Reps per O3 level	Number of chambers	Factors other than O3 studied	Reference	Significant interactions between O3 and other factors tested	Comments
1984	Soybean	Davis	Raleigh, NC.	7C	84	6	4[c]	24	2 soil moisture	Heagle et al., 1987b	—	Moderate moisture stress
	Festuca arundinacea (tall fescue) with Trifolium repens (ladino clover)	Kentucky 31 with Regal	Raleigh, NC	12P	94	6	4[c]	24	2 soil moisture	Heagle et al., 1989	—	Moderate moisture stress
	Medicago sativa (alfalfa)	WL-514	Shafter, CA	12P	95	5	4[c]	20	2 soil moisture	Temple et al., 1988a	—	Moderate moisture stress
	Phleum pratense (timothy) mixed with Trifolium pratense (red clover)	Champlain with Arlington	Ithaca, NY	12P	55	4	8[c]	32	4 SO2	Heck et al., 1984a	—	No effects of SO2 on total yield
	Alfalfa	WL-312	Argonne, IL	12P	95	5	8[c]	40	4 SO2	Heck et al., 1984a	—	Frost damage caused need for spring planting—Total "season" was only 55 days, leaf rust present. No significant yield response to O3 or SO2
1985	Alfalfa	WL-514	Shafter, CA	12P	100 (year 2)	5	4[c]	20	2 soil moisture	Temple et al., 1988a	—	Trends indicate less O3 effects if plants were water stressed
	Cotton	Acala-SJ3	Riverside, CA	12P	62	4	6[c]	24	3 soil moisture	Temple, pers. comm.		Crop did not mature due to late planting date caused by procurement delays
	Cotton	McNair 235	Raleigh, NC	12P	84	5	4[c]	20	2 soil moisture, frusta	Heagle et al., 1988	soil moisture	No O3 effect at low soil moisture. No frusta effect
	Tall-fescue with ladino clover	Kentucky 31 with Regal	Raleigh, NC	12P	100 (year 2)	6	4[c]	24	2 soil moisture	Heagle et al., 1989		

Crop	Cultivar	Location	Exposure[a]	Duration %[b]				Treatments	Reference	Notes
Field corn	Pioneer 3780 FR20A × FR634 FR20A × FR35 FR23 × LH74	Argonne, IL	12P	82	7	6[c]	42	3 SO$_2$ levels, hybrids	Kress et al., 1988	hybrids
Red clover with timothy	Champlain with Arlington	Ithaca, NY	12P	66 (year 2)	4	8[c]	32	4 SO$_2$ levels	Heck et al., 1985	Reduced population of clover in second year, high variability. No significant yield response to O$_3$ or SO$_2$
Soybean	Corsoy 79	Argonne, IL	12P	82	5	4[c]	20	2 soil moisture	Heck et al., 1985	Spider mites, low O$_3$ levels in August. Water stress affected CF more than other WS plots. No significant yield response to O$_3$
1986 Soybean	Young	Raleigh, NC	12P	86	5	6[c]	30	2 soil moisture, rain exclusion lids	Heagle. pers. comm.	
Soybean	Corsoy 79	Argonne. IL	12P	86	5	6[c]	30	2 soil moisture, rain exclusion lids	Kress, pers. comm.	soil moisture

[a] C = constant O$_3$ addition to ambient air in nonfiltered-air (NF) chambers; P = O$_3$ addition in proportion to ambient O$_3$ concentrations in NF chambers.

[b] Seasonal exposure duration is defined as the percentage of the period from crop emergence (estimated at seven days after planting seed for annual crops or on 15 April for fall planted perennial crops) to when plants were physiologically mature (or were harvested) for which daily O$_3$ additions were made.

[c] Replicates include different levels of SO$_2$, soil moisture, and/or exposure dynamics.

[d] Replicates include two undisturbed and two disturbed (physiological measurement) plots per O$_3$ level.

[e] There were two replicates for the CF and NF treatments with no O$_3$ added, and two replicates at each of three O$_3$ levels added, and two replicates at each of three O$_3$ levels for 7-h P or 7-h C O$_3$ additions.

[f] Same as for e except two replicates of three levels of 12-h P O$_3$ additions were included.

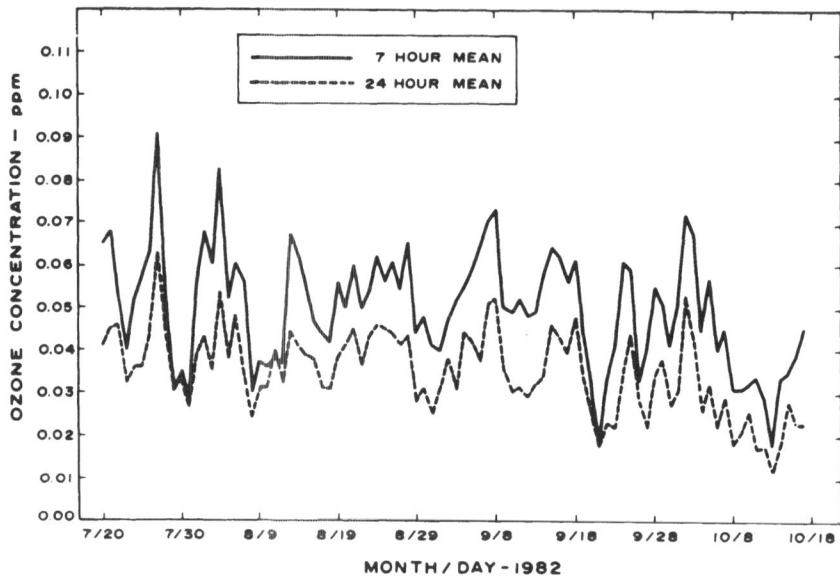

Fig. 7.2. Daily 7 h (1000 to 1700 hours EDT) and 24 h mean O_3 concentrations in ambient air at Raleigh, NC, during part of 1982.

This method provided an O_3 value for each plot every 12 to 45 min depending on the number of plots shared and sample time per plot. This method is adequate to characterize daily and seasonal O_3 concentration statistics (Heck *et al.*, 1984a). For the first 2 years, O_3 was added in three or four different but constant amounts for 7 h (1000 to 1700 h EDT) per day to chambers receiving nonfiltered air (NF). We will refer to this method of addition as 7C. Charcoal-filtered-air (CF) chambers, NF chambers without O_3 added, and plots without chambers (AA plots) were included in most experimental designs. Thus, a series of O_3 treatments was established, each with a constant relationship to the diurnal change in the ambient O_3 concentration (Fig. 7.3). Ozone was not added during periods of rainfall; instead, all chambered plots received either NF or CF air.

The 7C method had several strong points and several weak points. From 1000 to 1700 h EDT, more photosynthesis and transpiration occurs for most plants than for other periods, and O_3 concentrations in ambient air are usually higher than for other 7 h periods. However, most plants are also active at other times and O_3 levels usually remain elevated until near sunset in most areas. With 7C addition, treatment O_3 levels as much as five times greater than that in AA sometimes occurred when AA O_3 was low (e.g., when exposures began each day and during cloudy conditions). Improved technology allowed the NCLAN protocol gradually to shift from 7C to proportional-to-ambient (P) O_3 dispensing, starting in 1982 (Table 7.1). With P O_3 addition, the various treatment concentrations become more divergent (absolute differences become greater) with increased O_3 in AA (Fig. 7.3), and the range in concentrations for a given treatment mean is greater than with constant addition (Fig. 7.4).

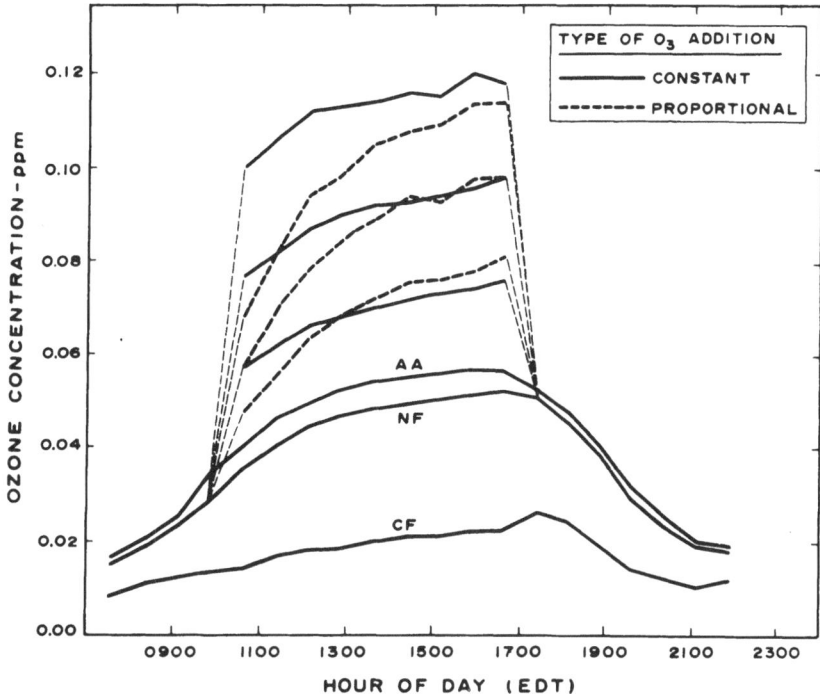

Fig. 7.3. Seasonal mean diurnal curves at Raleigh, NC, for the period shown in Fig. 7.2. AA = ambient air (no chamber); NF = nonfiltered-air chamber; CF = charcoal-filtered-air chamber; upper curves were in NF chambers with constant or proportional amounts of O_3 added for $7 \, h \, d^{-1}$.

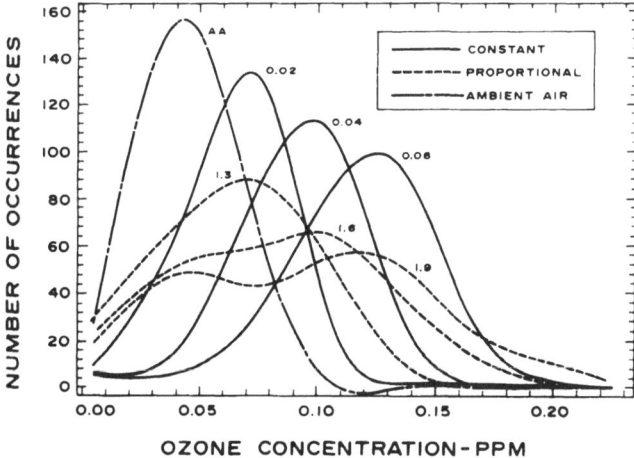

Fig. 7.4. Frequency distribution of O_3 concentrations resulting from constant or proportional-to-ambient additions of O_3 for the period in 1982 at Raleigh shown in Fig. 7.2.

The P O_3 additions eliminated the occurrence of unrealistically high levels of O_3 during periods of low ambient O_3 concentrations.

A study in 1983 measured the effect of 7-h proportional (7P) and 12-h (1000 to 2200 h) proportional (12P) O_3 additions on yield of flue-cured tobacco (Heagle *et al.*, 1987*a*). More yield loss of tobacco occurred in the 12P than in the 7P treatments. After 1983, all NCLAN studies but one used 12P O_3 additions (Table 7.1).

7.2.5 Experimental designs

NCLAN designs initially were simple, with five or six O_3 levels randomized within four replicates or blocks (Table 7.1). Gradually, the designs became more complex and experiments were designed to determine whether certain environmental or methodological factors affected plant response to O_3. These factors included SO_2, genotype (cultivars), soil moisture, frusta (truncated cones), rain exclusion lids, exposure dynamics (P versus C addition) and daily exposure duration (7 h versus 12 h exposures). Because chamber numbers were limiting, some of these factorial experiments used unreplicated designs (Table 7.1). Companion plots (an AA plot near each treatment plot) as possible covariates were included in several designs, but these rarely improved the results statistically.

7.3 POTENTIAL EFFECTS OF NCLAN METHODS ON O_3 DOSE–RESPONSE RESULTS

7.3.1 Effects of chambers on microclimate and plant gas exchange

The standard open-top configuration (without frustum, rain exclusion lid or other additions) and air flow pattern affect the microclimate in several ways (Table 7.2). Inside the chambers, air temperature and dew point temperature are often slightly higher, whereas light intensity and wind velocity are usually lower, than outside. Chamber panels (walls) cause "rain shadows" and thus cause a need for irrigation during most experiments to reduce lateral gradients in soil moisture levels within the plots.

In commercial fields, canopy closure usually occurs during vegetative growth. Then, lower leaves become shaded, gradually senesce, and cease contributing to growth and yield. In most NCLAN research, chambers were placed over plots after crop emergence. With this procedure, portions of "border rows" directly under the chamber frame and those too near the plastic walls (either inside or outside) were often stunted and sometimes even killed. This effect is greatest when row spacing is 1 m because more of the chamber frame falls directly on border rows than with narrower row spacing. With 1 m row spacing, canopy closure occurred for interior portions of plot rows in the center of chambers, but because of border-row stunting, often did not occur adjacent to the chamber wall. Thus, while the chamber walls caused an overall decrease in light within the chamber, the chamber caused increased light in some experiments for lower portions of plant canopies adjacent to chamber

TABLE 7.2
Open-top chamber effects on microclimate

Variable	Effects in chamber plant-growth area	References
Air temperature	Increases of up to 3·7°C, but usually less than 2·0°C; greatest increase on calm, sunny, and hot days at midday; mean seasonal increase probably less than 1·0°C	Heagle et al., 1973 Heagle et al., 1979b Olszyk et al., 1980 Weinstock et al., 1982
Leaf temperature	Slight increase by amounts caused by chamber effects on air temperature	Weinstock et al., 1982
Light (PAR)	Decreased by as much as 20%; greatest decrease with low sun angle on sunny days; mean seasonal decrease approximately 12%; in northern hemisphere, north chamber positions receive more light than southern positions: Normal vertical gradient (decreased light with decreased height) in plant canopy at chamber periphery does not exist if border row plants are not adequate	Heagle et al., 1979b Olszyk et al., 1980
Windspeed	Seasonal mean velocity decreased; but velocity never less than 2–3 km h^{-1}, in chambers; for plants taller than 120 cm, more air movement near base of plant canopy than near top during calm periods	Heagle et al., 1979b Weinstock et al., 1982
Relative humidity	Up to 10% increase or decrease depending on ambient conditions, soil moisture, and type of plant canopy; usually less than 5% difference	Heagle et al., 1973 Weinstock et al., 1982
Dew point	Up to 2°C higher when maximum differences in air temperatures occur; less than 0·5°C when cloudy or at night	Weinstock et al., 1982
Rainfall	Less direct rainfall at some chamber positions when rain is accompanied by wind; rain intercepted by panels is concentrated at base of chamber walls	Heagle et al., 1979b
Soil moisture	No reported effects for grapes when plots irrigated; probably more rapid drying of surface due to constant aspiration near ground.	Weinstock et al., 1982

walls. The potential impact of this chamber effect on dose–response results has not been studied.

There are some data showing the effects of chambers on rates of gas exchange by plants (Table 7.3). Stomatal conductance was less inside than outside the chambers for kidney bean and cotton (Heck *et al.*, 1983*a*), as well as lettuce (Heck *et al.*, 1984*a*), but no chamber effects on conductance were found for soybean (Heck *et al.*, 1984*a*) or grapes (Weinstock *et al.*, 1982). Carbon dioxide uptake rates were similar inside and outside of chambers for soybean at Ithaca (Heck *et al.*, 1982*a*) but were slightly greater in chambers at Beltsville (Heck *et al.*, 1984*a*). Various measures of microclimate, evaporation rates, and comparative depletion rates of sulfur hexafluoride and O_3 were used to show that while leaf boundary layer resistance was decreased by chambers (Unsworth *et al.*, 1984*a*) the estimated rate of O_3 uptake by a soybean canopy inside a chamber was similar to that previously reported for soybeans growing outside (Unsworth *et al.*, 1984*b*).

7.3.2 Effects of chambers on growth and yield

Most NCLAN experiments included AA plots to measure chamber effects which are defined here as the difference in plant growth or yield in AA plots from that in the NF chamber treatment with no O_3 added. For most such comparisons, O_3 levels in the NF chambers were similar but slightly lower than those in AA. Direct comparisons to determine the statistical significance of observed results have not been made for most studies.

Plants in NF chambers are commonly taller than those in AA (Kress and Miller, 1985*a*; Olszyk *et al.*, 1980; Heagle *et al.*, 1979*c*). This effect may be a response to decreased light or because plants in chambers are moved less by wind than plants in the open. The chamber effect on height is probably the only consistently observed response. The amount of foliar injury (chlorosis and necrosis) was greater in AA than in NF for sorghum (Kress and Miller, 1985*b*) and soybean (Kress and Miller, 1983), but was less in AA than in NF for tobacco (Heagle *et al.*, 1973), soybean (personal communication from Lance Kress), and peanuts (Heagle *et al.*, 1983*a*), while injury of winter wheat was apparently not affected by chambers (Heagle *et al.*, 1979*a*; Kress *et al.*, 1985*c*). Whether these differences in injury were caused by chamber effects on plant response to O_3 or by chamber effects on the amount of chlorosis and necrosis caused by insects, fungi, or other factors is not known.

Plant yield in AA can be equal to, greater than, or less than that in NF chambers; variable results have occurred with the same species (Table 7.4). For example, yield of winter wheat was less in AA than in chambers at Argonne (Kress *et al.*, 1985) but the reverse was true at Raleigh (Heagle *et al.*, 1979*a*). The chamber effect on soybean yield at Argonne was negligible (Kress and Miller, 1983) but soybean yield at Raleigh was less in AA than in NF chambers (Heagle *et al.*, 1987*b*). With soybean at Beltsville, AA and NF yields were similar with moderate soil moisture deficit but yield was less in AA than NF when plants were well-watered (Heggestad *et al.*, 1988). Peanut at Raleigh yielded more in NF chambers than in AA in 1979, but the reverse was

TABLE 7.3

Open-top chamber effects on plant gas exchange

Crop	Response measured	Measurement method	Chamber effect	References
Grape	Stomatal conductance	LICOR 65-auto porometer	None detectable	Weinstock et al., 1982
Soybean	CO_2 uptake (net photosynthesis)	Cuvette, CO_2 depletion	No consistent difference	Heck et al., 1982a
Kidney bean	Stomatal conductance (abaxial)	LICOR 1600-steady state porometer	Decreased by approximately 22%	Heck et al., 1983a
Cotton	Stomatal conductance (abaxial)	LICOR 1600-steady state porometer	Decreased by approximately 22% in the afternoon	Temple et al., 1988b
Head luttuce	Stomatal conductance	LICOR 1600-steady state porometer	Slightly less	Heck et al., 1984a
Soybean	Stomatal conductance	Cuvette, CO_2 depletion[a]	Slightly greater	Heck et al., 1984a
Soybean	Leaf boundary layer resistance		Less in chambers (with frusta)	Unsworth et al., 1984a
Soybean	Canopy resistance to O_3 uptake	[b]	Similar to open field (with frusta)	Unsworth et al., 1984b

[a] Evaporation rate of water from blotting paper leaf replicas, with measures of leaf temperature, air relative and absolute humidity.
[b] Measure of relative decay rate of SF_6 (inert) and O_3.

TABLE 7.4

Effects of open-top chambers on plant yield[a]

Crop	Year	Response measured	Units	Ozone Concentration[b] AA	NF	Plant response AA	NF	Ratio NF/AA
Winter wheat								
3 cv	1982	Seed yield	kg ha^{-1}	0·042	0·041	5443	4673	0·86
2 cv	1983	Seed yield	kg ha^{-1}	0·045	0·044	6538	5804	0·89
4 cv	1976	Seed yield	g plant^{-1}	0·060	0·060	4·51	5·12	1·14
Soybean								
Corsoy-79	1980	Seed yield	kg ha^{-1}	0·050	0·050	2507	2591	1·03
Corsoy-79	1986	Seed yield	kg ha^{-1}	0·038	0·037	4489	4700	1·05
Davis	1983	Seed yield	g m^{-1}	0·057	0·052	359	451	1·26
Davis	1984	Seed yield	g m^{-1}	0·051	0·045	296	434	1·47
Hodgson	1981	Seed yield	g plant^{-1}	0·035	0·035	11·5	11·1	0·97
Grain sorghum DeKalb A28 +	1982	Seed yield	kg ha^{-1}	0·039	0·040	8560	7643	0·89
Field corn								
Coker 16	1976	Seed yield	g plant^{-1}	0·060	0·070	228	240	1·05
Pioneer 3780	1981	Seed yield	kg ha^{-1}	0·044	0·044	10675	10743	1·01
PAG 397	1981	Seed yield	kg ha^{-1}	0·044	0·044	12190	12911	1·06
Peanuts	1979	Marketable pod wt	g plant^{-1}	0·052	0·049	82	94	1·15
NC-6	1980	Marketable pod wt	g plant^{-1}	0·056	0·056	158	122	0·77
Cotton								
McNair 235	1985	Lint & seed wt	g m^{-1}	0·043	0·040	307	424	1·38
Acala SJ-2	1981	Lint wt	g m^{-1}	0·077	0·071	670	611	0·91
Acala SJ-2	1982	Lint wt	g m^{-1}	0·047	0·044	828	715	0·86
Tobacco								
Bel W3	1972	Shoot fresh wt	g plant^{-1}	[c]	[c]	1409	1738	1·23
McNair 944	1983	Shoot dry wt	g plant^{-1}	0·068	0·057	550	609	1·11
Alfalfa								
WL-514	1984	Seasonal yield (10% moisture)	g m^{-2}	0·049	0·045	2882	2792	0·97
	1985	Seasonal yield (10% moisture)	g m^{-2}	0·042	0·038	2991	2971	0·99
Ladino clover & Tall fescue	1984	Shoot dry wt	g m^{-2}	0·045	0·043	994	1117	1·12
Regal & KY31	1985	Shoot dry wt	g m^{-2}	0·048	0·047	897	913	1·02
Lettuce								
Empire	1983	Mean head fw	g m^{-1}	0·054	0·052	829	815	0·98

[a] For experiments with well-watered plants where AA plots were included in treatment plot randomization. Some data predate NCLAN studies.

[b] Ozone levels are seasonal 7 h d^{-1} means except for McNair 235 cotton and ladino clover–tall fescue which show 12 h d^{-1} means.

[c] The seasonal mean was not characterized.

true in 1980 (Heagle *et al.*, 1983a). Because of the numerous interacting factors (e.g., seasonal climate, species and cultivars, pests or biotic disease, soil factors, cultural practices) the cause for these different responses is unknown. The differences in results preclude prediction of chamber effects on yield for any of the crops used in NCLAN studies. Although chamber effects on yield are common, there are no results showing that this will result in a changed yield response to O_3. There is a need to perform experiments specifically designed to measure open-top chamber effects on crop yield and on crop response to O_3.

7.3.3 Effects of chamber quadrant on yield

Differences in microclimate at various chamber positions result from differences in sun and rain shadows caused by the chamber panels. Ingress of ambient air causes greater air turbulence in the downwind half of chambers than in the upwind half. In the Northern Hemisphere, less total seasonal irradiation occurs in the southern than in the northern half of chambers. During one growing season at Raleigh, NC, seasonal irradiation (PAR) was 15% less than ambient in the southern half of a chamber but was 10% less than ambient in the northern half (Heagle *et al.*, 1979*b*). Differences in sunlight cause differences in other climatic and edaphic factors but the extent of these effects in open-top chambers has not been adequately characterized.

Differences in yield at different chamber positions (Table 7.5) are probably related to seasonal differences in sunlight. Plants grown in the northern half of chambers generally yielded better than those grown in the southern half. While these position effects on yield were significant for most NCLAN experiments, there were usually no significant interactions between the position effect and plant response to O_3. Exceptions to this generality occurred with tobacco (Heagle *et al.*, 1987*a*) and in the second year of a 2-year study with fescue grown with ladino clover (Heagle *et al.*, 1989). Since no biologically meaningful interpretations could account for these interactions, plot means were used in regression analyses. Because differences in yield can occur at different positions within chambers without significantly affecting plant response to O_3, there is reason to believe that some level of microclimate change in chambers is acceptable for purposes of measuring effects of O_3 on yield.

7.3.4 Effects of chambers on O_3 fluctuation and O_3 gradients

Ozone concentrations in ambient air are usually similar over large areas. In open-top chambers, the regular diurnal change in O_3 concentrations is similar to that in ambient air except with 7C exposures which caused a sudden increase in O_3 concentration when exposures began each morning and a sudden decrease when exposures ended each day. The sudden changes are greatly diminished with 12P additions. Abrupt, brief (momentary) concentration changes greater than 0·01 ppm (1 ppm v/v = 1960 μg m^{-3}) are rare in ambient air except with sudden changes in sunlight or rain. However, momentary changes in O_3 concentration are greater in open-top chambers than in ambient air because of the chamber structure which allows some ambient air to enter through the open top (ingress). If wind speed were constant, the effects of ingress could be corrected by adjusting the amount of added O_3. However, wind speed is usually extremely variable causing variable rates of ingress and, thus, variable degrees of dilution of chamber air with ambient air. The degree to which ingress affects O_3 concentration is directly related to wind velocity, wind velocity variability, and the difference between ambient O_3 concentrations and chamber O_3 concentrations. The greatest fluctuations occur when

TABLE 7.5

Chamber quadrant effects on crop yield

Crop	Year	Location	Response measured	Units	Yield per Quadrant (% of mean)				References
					NE	NW	SE	SW	
Soybean									
Davis	1981	Raleigh	Seed wt	g m^{-1}	280(103)	290(107)	260(96)	256(94)	Heagle et al., 1983b
Davis	1982	Raleigh	Seed wt	g m^{-1}	367(105)	387(110)	334(95)	313(89)	Heagle et al., 1986
Young	1986	Raleigh	Seed wt	g m^{-1}	351(117)	343(114)	244(81)	267(89)	Personal communication (A. S. Heagle)
Cotton									
Stoneville	1982	Raleigh	Unginned cotton	g m^{-1}	285(107)	266(100)	272(102)	240(90)	Heagle et al., 1986
Acala SJ-2	1982	Shafter	Lint wt	g m^{-1}	152(101)	176(118)	132(88)	143(95)	Personal communication (P. J. Temple)
Peanuts									
NC-2	1980	Raleigh	Pod wt	g plant^{-1}	101(97)	100(96)	110(106)	104(100)	Heagle et al., 1983a
Clover-fescue well-watered	1984	Raleigh	Shoot wt	g m^{-2}	1013(103)	1002(102)	937(95)	980(100)	Heagle et al., 1988
Clover-fescue well-watered	1985	Raleigh	Shoot wt	g m^{-2}	643(106)	616(102)	561(92)	615(101)	Heagle et al., 1988
Barley									
Poco	1982	Shafter	Seed wt	g m^{-2}	261(93)	301(108)	296(106)	260(93)	Personal communication (P. J. Temple)
				Means	(104)	(106)	(96)	(95)	

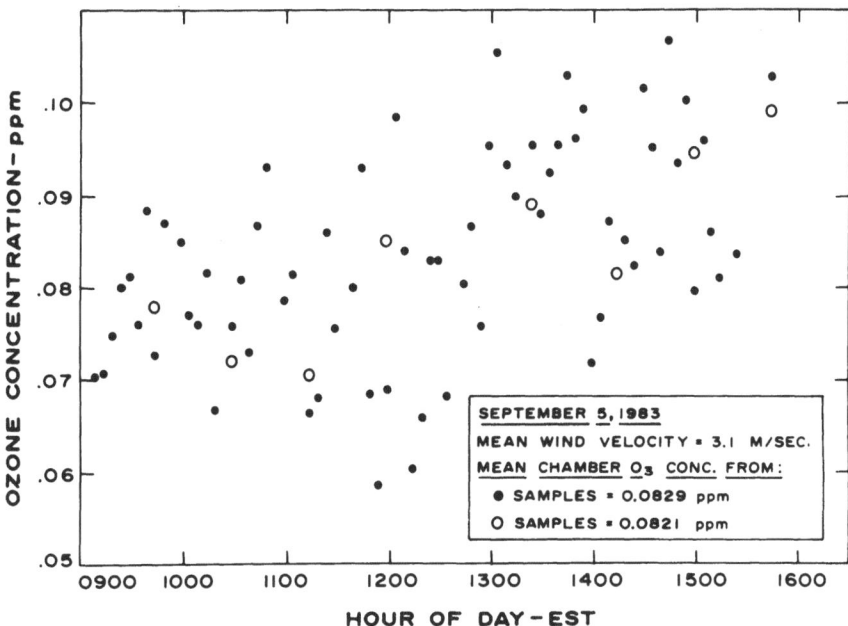

Fig. 7.5. Fluctuation of O_3 concentration in an open-top field chamber with O_3 added (a mean of 0·047 ppm added for 7 h) to the ambient O_3 concentration (7 h mean = 0·036 ppm). The chamber contained full-grown soybeans with a closed canopy. Ozone was sampled at the top of the soybean canopy (1·2 m) each 5 min (●) or each 45 min (○).

wind is strong and highly variable and when the difference in O_3 concentration between the chamber and ambient air is greatest (e.g., in CF chambers when ambient O_3 is high and in NF chambers with the highest added O_3 level). In such situations, momentary fluctuations in the 0·04 ppm range can occur (Fig. 7.5). It is not known whether these fluctuations affect plant response to O_3. A truncated cone (frustum) which decreases the amount of ingress by about 50% was used in the Shafter and Riverside experiments. Frusta were shown not to affect plant response to O_3 for one cultivar of cotton (Heagle *et al.*, 1986*a*).

Lateral gradients in O_3 concentration within open-top chambers are small and reflect wind direction and velocity (Heagle *et al.*, 1979*b*). However, there are significant vertical gradients in chambers that can differ from those in AA because of the chamber configuration and method of air introduction into chambers. For a crop growing in an open field, there is decreased air movement, higher humidity and lower O_3 levels within the canopy than above it (Heck *et al.*, 1985). In open-top chambers, there is little or no canopy-induced vertical O_3 gradient because air enters the chambers uniformly from a height of 15 to 120 cm. For chambers with O_3 added, ambient air ingress causes O_3 concentrations to decrease with increased height above 120 cm, which is the reverse of the open-field situation. However, in CF chambers there is an increase in O_3 concentration with increase in height that resembles the vertical gradient in ambient air.

7.3.5 Seasonal exposure duration

The main NCLAN objective was to provide estimates of crop loss in commercial US agriculture caused by seasonal exposure to ambient levels of O_3. Because crops in commercial production are exposed to O_3 in ambient air from emergence to harvest, the most appropriate experimental protocol would be to duplicate this experimentally. However, this was not usually possible because of limited resources which limited the number of chambers for most studies to 24. Thus, because of a relatively small plot size, it was critical to take measures to decrease plot-to-plot variability. This was usually achieved by planting an entire field, or in some cases individual plots, waiting for plant emergence and early crop growth (2 to 3 weeks) and selecting plots based on crop and soil appearance. Once plots were selected, electrical service was installed, chambers were placed on the plots, and dispensing and monitoring lines and irrigation systems were installed before exposures began. The time needed for each of these operations was dependent on the weather, available personnel, and sometimes if equipment and instruments were available and functional. All of these factors decreased the percentage of the seasonal exposure duration (SED). We define SED as the percentage of the period from crop emergence (estimated at 7 days after planting seed for annual crops or on 15 April for fall-planted perennial crops) to when plants were physiologically mature (or were harvested) for which daily O_3 additions were made. The SED for each NCLAN study is shown in Table 7.1.

Presumably, the effects of daily exposures to O_3 are accumulative, gradually causing premature leaf senescence and thereby decreasing growth and decreasing yield. Therefore, exposures for a whole season could cause greater effects than would exposures for part of a season. There are no data to determine whether this is true, however, and experiments with SED as a variable are needed to answer this question.

7.4 RESULTS OF DOSE–RESPONSE STUDIES

7.4.1 The Weibull model

The use of the Weibull model in O_3 dose–yield response studies has been described previously (Rawlings and Cure, 1985) and will be used in this chapter to discuss NCLAN results. The Weibull model relates yield (y) to O_3 concentration by: $y = \alpha \exp[-(x/\sigma)^c]$, where $\alpha =$ maximum yield at 0 ppm O_3; σ is the seasonal mean O_3 concentration at which α is reduced by 63%; c is a dimensionless shape parameter; and x is the seasonal (period of O_3 addition) mean O_3 concentration in ppm. The Weibull function fits most responses observed in NCLAN studies (except those showing greater yield at any O_3 treatment above the lowest O_3 treatment). This function allows direct comparisons of O_3 affects on yield across all experiments in terms of a proportional yield response when the α term is set at 100 for all separate models. We will discuss results in terms of proportional yield response when the α term is set at 100 for all separate models. We will discuss results in terms

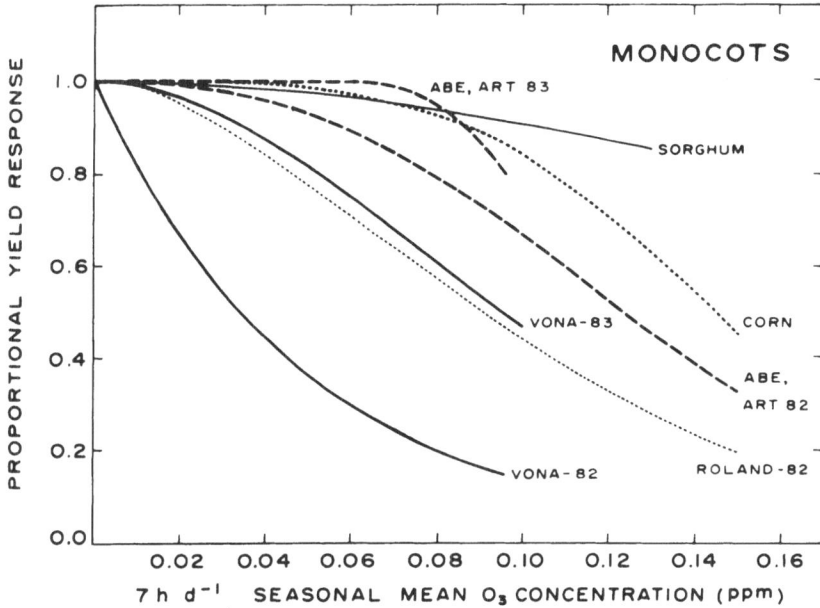

Fig. 7.6. Proportional yield response curves for monocots in NCLAN experiments with O_3 as the only independent variable. Curves were derived from experiments outlined in Table 7.1 using Weibull models with the α value set at 100.

of proportional yield response or as the estimated yield loss at five seasonal mean O_3 concentrations representative of ambient O_3 levels in the United States. The seasonal 7 h or 12 h mean was used as the independent variable for experiments with 7 h or 12 h O_3 additions, respectively. For loss estimates, the 'control' concentration was set at 0·025 ppm for 7 h-addition experiments and at 0·020 ppm for 12 h-addition experiments. Values for the Weibull parameters used here are shown in Chapter 16 of this volume (Rawlings *et al.*, 1988).

The first NCLAN designs were relatively simple, with 7C O_3 addition with plants generally grown under well-watered conditions. Results from 7C studies for monocots (Fig. 7.6) show a wide range in estimated proportional response curves. For dicots (Fig. 7.7), most curves fall within a relatively narrow range (compared to monocots) except for lettuce, which was more resistant to low O_3 levels than any other crop tested. The apparent sensitivity of turnip is misleading because low levels of O_3 caused extensive 'acute' injury after a cool rainy period (Heagle *et al.*, 1985).

Even though the NCLAN studies attempted to minimize edaphic, biotic or climatic interactions, there was a wide range in estimated responses to O_3 as Figs 7.6 and 7.7 illustrate. That much of this variation was due to factors other than genotype (species or cultivar) can be seen in the wheat results (Fig. 7.6). For 'Vona' wheat, estimated yield loss at 0·050 ppm (compared to 0·025 ppm) was 39% in 1982 but only 14% in 1983 (Table 7.6). For 'Abe' and 'Arthur' at Argonne, IL, estimated yield loss at 0·050 ppm was 6% in 1982 but was 0% in 1983. With estimates that vary this much from year to year, conclusions about

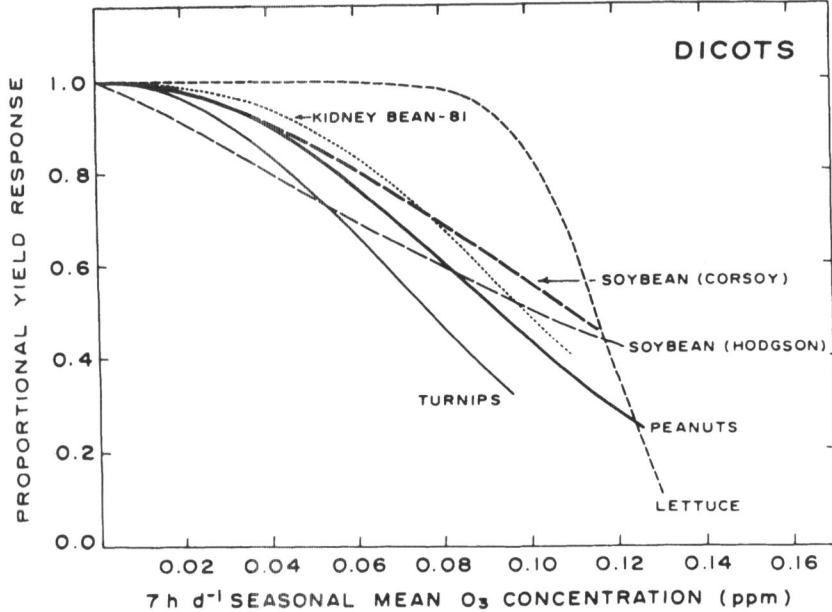

Fig. 7.7. Proportional yield response curves for dicots in NCLAN experiments with O_3 as the only independent variable. Curves were derived from experiments outlined in Table 7.1 using Weibull models with the α value set at 100.

the relative sensitivity of crop species based on Figs 7.6 and 7.7 are not warranted.

7.4.2 Season effects on dose–response results

Variable conditions that occur over seasons at one site and in the same season at different sites can be expected to affect plant growth and yield and may affect plant response to O_3. For the best measure of site or season effects on loss estimates derived from dose–response studies, experiments should be performed in which all variables except weather and ambient O_3 levels are relatively constant. Daily exposure duration and method of O_3 addition should be the same and the SED should be similar. Soil moisture or SO_2 as planned experimental variables should also be the same across experiments. The only acceptable variables when measuring season or site effects are those caused by weather and the seasonal sequence, frequency, and severity of O_3 episodes. All of these factors affect the seasonal mean O_3 concentration.

NCLAN results provide some indication of the degree of variability in O_3 dose–yield estimates caused by seasonal factors, but true repeats were rare because of program needs to test the effects of other variables. For example, kidney bean was tested for 2 years but the SED was 26% in 1980 and 66% in 1982. For a given seasonal (period of O_3 addition) mean O_3 concentration, estimated kidney bean loss was about six times greater in 1982 than in 1980 (Table 7.6). Cotton ('Acala SJ-2') was exposed for two seasons with two soil moisture levels as a variable (Temple *et al.*, 1985*b*). In 1981, O_3 additions were 7C with the SED at 33%; in 1982, O_3 additions were 7P with the SED at 59%. The 1981 season was normal, but the 1982 season was cloudier and cooler than normal. Estimated cotton yield decrease at given seasonal O_3 concentrations was about twice as great in 1982 as in 1981 (Table 7.6). Winter wheat ('Vona') was exposed with the SED at 69% in 1982 and at 38% in 1983. Ozone additions were 7C in 1982 and 7P in 1983 when four SO_2 levels were also included. Estimated losses for 'Vona' were approximately twice as great in 1982 as in 1983 (Table 7.6). Field corn ('Pioneer 3780') was tested with 7C exposures in 1981 (SED = 84%) and with 12P exposures in 1985 (SED = 82%). The loss estimates over years were different when the seasonal 12-h d^{-1} mean O_3 concentration was used as the independent variable, but were similar when the 7 h mean was used (personal communication from Lance Kress). Soybean ('Davis') was tested for four seasons with 7C additions and SEDs ranging from 84 to 86% (Table 7.1). However, added experimental variables included SO_2 in 1981, an exposure-addition method in 1982, and soil moisture in 1983 and 1984. In 1981, the response of 'Davis' to O_3 was similar whether plants grew in sandy loam soil (Block I) or in clay-loam soil (Block II). For 1982, 1983, and 1984, the yield loss response of 'Davis' to O_3 was similar (considering only well-watered plots) but was less than that in 1981 (Table 7.7, Fig. 7.8(A)). Soybean ('Corsoy 79') was tested over 3 years at Argonne but with variable SEDs, variable daily O_3-addition durations and with SO_2, cultivars, and soil moisture as added factors (Table 7.1). In 1980, with 7C addition and the SED at 61%, the estimated loss at 0.05 ppm was 10%. In 1983, with 7C additions and the SED at 92% the estimated loss at 0·05 ppm according to the Weibull model was 0% (Table 7.7, Fig. 7.8(B)). This was an extreme example of different estimates resulting from different dose–response models since linear and quadratic polynomial models gave loss estimates of 14 and 8%, respectively, for the 1983 Argonne 'Corsoy' data (personal communication from Lance Kress). In 1986, the O_3 additions were 12P with SED at 86%. Estimated loss for 'Corsoy' at 0·04 ppm (seasonal 12 h day mean compared to 0·02 ppm) for well watered plots was 2% (Table 7.7). 'Corsoy' and 'Amsoy' were included in two studies in 1983 (Tables 7.1, 7.7). One included four cultivars and an SED of 92%, the other included four SO_2 levels and an SED of 92%. For each cultivar, estimated losses were greater in the experiment with SO_2 (Table 7.7). Soybean ('Williams') was tested for 3 years with SEDs ranging from 69 to 73%. Sulfur dioxide and/or soil moisture were included as factors in all experiments (Table 7.1). In spite of the different designs and seasonal weather, the loss estimates were similar at low O_3 levels (Table 7.7). The exposure protocol and SED for winter wheat ('Abe' and 'Arthur') were nearly identical for two seasons at Argonne (Table 7.1) making this as close to a true repeat over years as any NCLAN study (Kress *et al.*, 1985). However, estimated yield decrease at 0·05 ppm was 6% in 1982 and 0% in 1983 (Table 7.6). Possible cause for the difference was weather, which favored slow growth, high O_3 levels, and foliar injury during May 1982 but not during May 1983.

TABLE 7.6

Weibull model estimates of yield loss for thirteen crop species at different O₃ concentrations and results of tests to determine if other factors affect the O₃ response

Crop	Cultivars	Year	Daily O₃, exposure duration (h) and additional methods	Seasonal exposure duration (%)	Factors other than O₃ combined in Weibull[b]	Factors showing significant interaction with O₃	Coefficient of variation (CV)	Estimated percentage yield loss per seasonal mean ozone concentration—ppm[f]				
								0·04	0·05	0·06	0·07	0·08
Dry bean	Red kidney	1980	7C	26	—	—	6·4	1	1	2	4	5
	Red kidney	1982	7C	66	—	—	15·5	4	9	15	23	31
Turnip	Purple top, Shogoin Just right Tokyo cross	1980	7C	91	4 cultivars	—	33·6[a]	10	19	24	40	50
Peanut	NC-6	1980	7C	86	—	—	7·3	7	13	20	28	37
Lettuce	Empire	1983	7P	72	—	—	28·2	0	0	0	1	2
Tomato	Murrieta	1981	7C	66	6 SO₂	—	11·8[e]	1	3	5	8	12
	Murrieta	1982	7C	65	6 SO₂	—	12·3[e]	8	16	30	44	58
Sorghum	DeKalb A28	1982	7C	84	—	—	5·1	1	2	3	4	5
Tobacco	McNair 944	1983	7C, 7P,	87	2 durations, 2 dynamics	—	5·3	3	6	9	13	18
			12P		2 durations 2 dynamics	—		7	12	17	23	28
Field corn	Pioneer 3780, CAPPAG 397	1981	7C	84	2 cultivars	—	9·9[d]	0	1	3	4	7
Field corn	Pioneer 3780, LH74 × FR23	1985	12P	82	2 cultivars, 3 SO₂		15·4[d]	2	3	6	10	15
	FR20A × FR35, FR20A × FR634	1985	12P	82	2 cultivars, 3 SO₂		15·4[d]	3	7	11	17	25

Crop	Cultivar	Year	Exposure		Factors	Soil moisture	Seasonal exposure					
Winter wheat	Vona	1982	7C	69	—		15·6	26	39	50	59	67
	Vona	1983	7P	38	—		19·3	7	14	21	28	36
	Abe, Arthur	1983	7C	87	4 SO_2		4·3[d]	0	0	1	2	5
	Abe, Arthur	1982	7C	87	2 cultivars		10·9[d]	3	6	9	14	20
	Roland	1982	7C	87	2 cultivars		10·9[d]	9	16	23	31	38
Cotton	Acala SJ-2	1981	7C	33	—	Soil moisture (WW)	5·6	3	7	13	21	30
						Soil moisture (WS)	12·8	1	2	3	7	12
	Acala SJ-2	1982	7P	59	2 H_2O		17·8	6	15	26	40	55
	Stoneville 213	1982	7C	88	4 SO_2, Tops	—	6·7[e]	4	9	16	24	33
Tall fescue-Ladino clover	McNair 235	1985	12P	84	2 H_2O	—	14·5	7	13	21	30	40
	Kentucky 31	1984	12P	94	2 H_2O	—	5·6	5	8	12	17	22
	with Regal	1985	12P	100	2 H_2O	—	12·1	6	11	17	24	32
Alfalfa	WL-514	1984	12P	95	2 H_2O	—	7·6	6	9	13	17	20
	WL-514	1985	12P	100	2 H_2O[g]	—	8·3	4	7	10	14	18
Red clover-timothy	Champlain with Arlington	1984	12P	55	4 SO_2	—	18·8	9	19	31	44	59

[a] Seasonal exposure duration is defined as the percentage of the period from crop emergence (estimated at 7 days after planting seed for annual crops or on 15 April for fall planted perennial crops) to when plants were physiologically mature (or were harvested) for which daily O_3 additions were made.

[b] Within each O_3 level, results for factors with no interaction with O_3 were combined for Weibull analyses.

[c] Analysis of variance coefficients of variation unless otherwise indicated.

[d] Split-plot AOV, CV.

[e] Over-parameterized model CV.

[f] Estimated loss based on Weibull analyses with yield at 0·025 ppm considered as the control yield for 7-h exposures and yield at 0·020 ppm as the control for 12-h exposures. Seasonal means are for 7- or 12-h exposures, as indicated in column four.

[g] Regression analysis (polynomial) showed a slightly greater response to O_3 in well watered (WW) plots than in water-stressed (WS) plots, but Weibull analyses did not.

TABLE 7.7

Weibull model estimates of soybean yield loss at different O₃ concentrations and results of tests to determine if other factors affect O₃ response

Soybean cultivars	Year	Daily O_3 exposure duration (h)	Seasonal exposure duration (%)[a]	Factors other than combined in Weibull[b]	Factors showing significant interaction with O_3	Coefficient of variation (CV)[c]	Estimated percentage yield loss per seasonal mean Ozone concentration-ppm[f]				
							0·04	0·05	0·06	0·07	0·08
Corsoy 79	1980	7C	61	—	—	9·4	6	10	16	22	28
Hodgson	1981	7C	62	2 plant densities	—	18·0	9	15	21	27	32
Essex	1981	7C	73	6 SO_2	—	12·7[d]	8	15	22	30	38
Williams	1981	7C	73	6 SO_2	—	9·2[d]	8	16	24	32	42
Williams	1982	7C	70	3 SO_2, $2H_2O$	—	12·1[d]	7	13	19	24	30
Forrest	1982	7C	70	3 SO_2	Soil moisture (WW)	15·7[d]	3	9	21	39	60
Forrest	1982	7C	70	3 SO_2	Soil moisture (WS)	15·7[a]	13	21	28	35	41
Williams, Corsoy 79	1983	7C	69	2 cultivars	Soil moisture (WW)	4·1[e]	7	13	18	24	30
Williams, Corsoy 79	1983	7C		2 cultivars	Soil moisture (WS)	4·1[e]	6	11	15	19	23
Corsoy 79	1983	7C	92	—	cultivar	6·6[e]	0	0	2	5	13
Pella	1983	7C	92	—	cultivar	6·6[e]	7	11	15	19	22
Williams	1983	7C	92	—	cultivar	14·9[e]	8	13	17	21	25

Amsoy 71	1983	7C	92	—	cultivar	14.9[e]	2	4	9	15	24
Amsoy 71	1983	7C	92	4 SO₂	cultivar	19.8[a]	4	8	12	18	25
Corsoy 79	1983	7C	92	4 SO₂	cultivar	19.1[a]	3	6	11	17	24
Davis	1981	7C	86	4 SO₂	—	12.1	12	19	25	30	35
Davis	1982	7C,P	87	2 O₃ dynamics	—	7.3	5	10	16	23	31
Davis	1983	7P	88	—	Soil moisture (WW)	11.4	4	7	12	16	21
Davis	1984	7P	84	2 H₂O	—	12.9	4	7	12	18	24
Corsoy 79	1986	12P	86	—	Soil moisture (WW)	5.6	2	4	8	13	21
Corsoy 79	1986	12P	86	—	Soil moisture (WW)	5.6	0	0	0	0	1
Young	1986	12P	86	2 H₂O	moisture (WS) —	3.6	6	11	17	25	34

[a] Seasonal exposure duration is defined as the percentage of the period from crop emergence (estimated at 7 days after planting seed for annual crops or on 15 April for fall planted perennial crops) to when plants were physiologically mature (or were harvested) for which daily O₃ additions were made.

[b] Within each O₃ level, results for factors with no interaction with O₃ were combined for Weibull analyses. (WW) = well-watered; (WS) = water-stressed.

[c] Analysis of variance coefficients of variation unless otherwise indicated.

[d] Over-parameterized model CV.

[e] Split-plot AOV, CV.

[f] Estimated loss based on Weibull analyses with yield at 0·025 ppm considered as the control yield for 7 h exposures and yield at 0·020 ppm as the control for 12-h exposures. Seasonal means are for 7 or 12 h exposures, as indicated in column three.

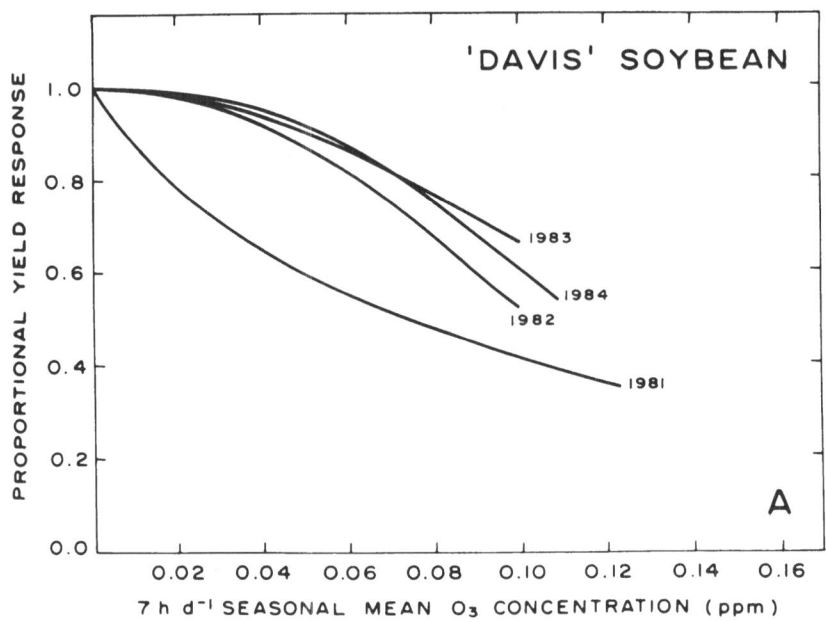

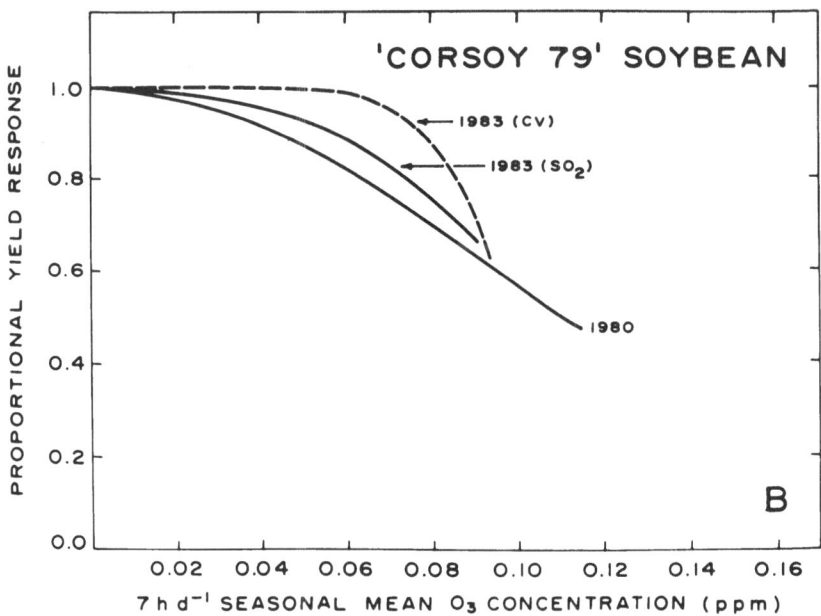

Fig. 7.8. Proportional yield response of soybean to O_3 (considering only well-watered plants): (A) for 'Davis' in four seasons at Raleigh, NC: (B) for Corsoy in two studies at Argonne, IL, in 1983 and one study at Argonne, IL, in 1980. Curves were derived from experiments outlined in Table 7.1 using Weibull models with the α value set at 100.

NCLAN data provide strong evidence that O_3 dose–yield response relationships can vary greatly from year to year. A better understanding of how national loss estimates are affected by seasonal weather patterns and seasonal progression of O_3 episodes is needed. Experimental validations of NCLAN loss estimates have not been attempted.

7.4.3 Cultivar selection

Ideally, for each crop species, cultivars chosen for NCLAN studies should have been of a sensitivity level representative of the mean sensitivity level (weighted for acreage planted) of cultivars in commercial production. However, data showing relative cultivar sensitivity to O_3-induced yield decrease were not available. Thus, NCLAN chose for the most part to use the most widely grown cultivars in commercial production. Ozone dose–yield response relationships have been developed for one cultivar each of kidney bean, peanut, lettuce, tomato, sorghum, tobacco, tall fescue-ladino clover, alfalfa, and red clover-timothy. More work has been done with soybean (nine cultivars) than for any other crop because of its relative sensitivity and economic importance, while six field corn hybrids, and three cotton and winter wheat cultivars have been tested. However, direct comparisons among cultivars in the same experiment were not common in NCLAN. For soybean, one experiment included four cultivars (personal communication from Lance Kress), and four experiments included two cultivars (Heck *et al.*, 1984*c*; Heggestad *et al.*, 1985; Kress *et al.*, 1986). Significant cultivar effects occurred in all but one (Heggestad *et al.*, 1985) study. Loss estimates at low O_3 levels (Table 7.7) were sometimes similar for cultivars showing statistically significant differences (Heck *et al.*, 1984*c*; Kress *et al.*, 1986), probably because of greater differences in estimated effects at higher O_3 levels. 'Williams' soybean was more sensitive than other cultivars in two studies where cultivar effects were significant (Heck *et al.*, 1984*c*; personal communication from Lance Kress). This was the only consistent result concerning relative soybean cultivar sensitivity (Table 7.7).

Two experiments directly measured relative corn cultivar response. In one study (Kress and Miller, 1985*a*) no differences occurred. In the other study (Kress *et al.*, submitted) two crosses containing the same parent were more sensitive than two crosses that did not (Table 7.6). For winter wheat, 'Roland' was more sensitive than 'Abe' or 'Arthur' in 1982. 'Abe' and 'Arthur' did not significantly differ in sensitivity in each of two experiments (Kress *et al.*, 1985*c*) (Table 7.6).

7.4.4 Ozone addition method and daily exposure duration

Addition of O_3 in constant amounts results in a series of diurnal O_3 exposure curves with a relatively constant relationship with the diurnal O_3 curve in AA (Figure 7.3). With P addition, the absolute deviation from the AA O_3 level changes as the AA O_3 level changes. With increasing AA O_3 concentration, the curves for the P treatments show an increased divergence in concentration (absolute values) from the AA curve and from each other (Fig. 7.3). With P addition, the range of O_3 concentrations (peaks and valleys) is greater than for

C additions (Fig. 7.4). Daily proportional addition of 2.0 times ambient O_3 provides a seasonal regime that probably encompasses the range of O_3 levels likely to occur in ambient air.

Effects of 7C and 7P O_3 additions have been directly compared only for 'Davis' soybean (Heagle *et al.*, 1986*b*) and 'McNair 235' tobacco (Heagle *et al.*, 1987*a*). For both tests, the seasonal mean O_3 levels at each increment of O_3 addition were similar whether C or P additions were made. Regression analyses showed no significant difference in dose–response relationships caused by the type of O_3 addition (Fig. 7.9(A)). No evidence of acute foliar injury caused by a single 'peak' exposure occurred for either study. Possibly, if ambient O_3 levels were high enough that P addition occasionally caused moderate or severe injury, one 'peak' exposure could affect yield and might affect dose–response results. Further comparisons appear warranted since approximately half of the NCLAN studies used constant O_3 addition and half used proportional addition (Table 7.1).

Our understanding of the importance of daily exposure duration is also extremely limited. We would expect exposure for 12 h or longer each day to cause more yield loss over a season than $7\,h\,d^{-1}$ exposures. Our only direct test of the daily exposure duration effect was with tobacco. While not significantly different, tobacco exposed for $12\,h\,d^{-1}$ yielded less than tobacco exposed for $7\,h\,d^{-1}$ (Heagle *et al.*, 1987*a*) (Fig. 7.9(B)). 'Corsoy' soybean was exposed to 7C or 12P additions at Argonne, but in different years (Table 7.1). In this experiment, possible confounding caused by seasonal weather and other variables probably invalidated a meaningful comparison of C and P addition. Since about 70% of NCLAN data was with $7\,h\,d^{-1}$ exposures, research to compare the effects of $7\,h\,d^{-1}$ and $12\,h\,d^{-1}$ (or longer) exposures is needed.

7.4.5 Effects of soil moisture on response to ozone

The potential effect of soil moisture deficit (SMD) on crop yield response to O_3 raises more questions as to the applicability of NCLAN dose–response results than does any other environmental variable. This is because SMD is common in commercial crop production, and it is widely assumed that plant response to SMD (decreased gas exchange between leaf and air) decreases O_3 uptake and therefore decreases the impact of a given O_3 concentration in the air. For this reason, NCLAN protocol was to irrigate to minimize the effects of SMD which, if great enough, could have confounded experimental results. Recognizing the need to address this issue, NCLAN made a major effort to measure the effects of SMD on yield response to O_3.

The addition of SMD as a variable greatly increased the complexity of NCLAN experiments. There were problems associated with controlling SMD in the field, problems with measuring soil moisture, and problems with measuring SMD effects on leaf water potential and gas exchange rates between leaves and air. Controlling SMD was difficult because there generally was no practical way to exclude rain from the plots that would not compromise the experiments. Thus, soil moisture control was possible only during dry periods by applying differential irrigation, although manual 'rain caps' were used in the

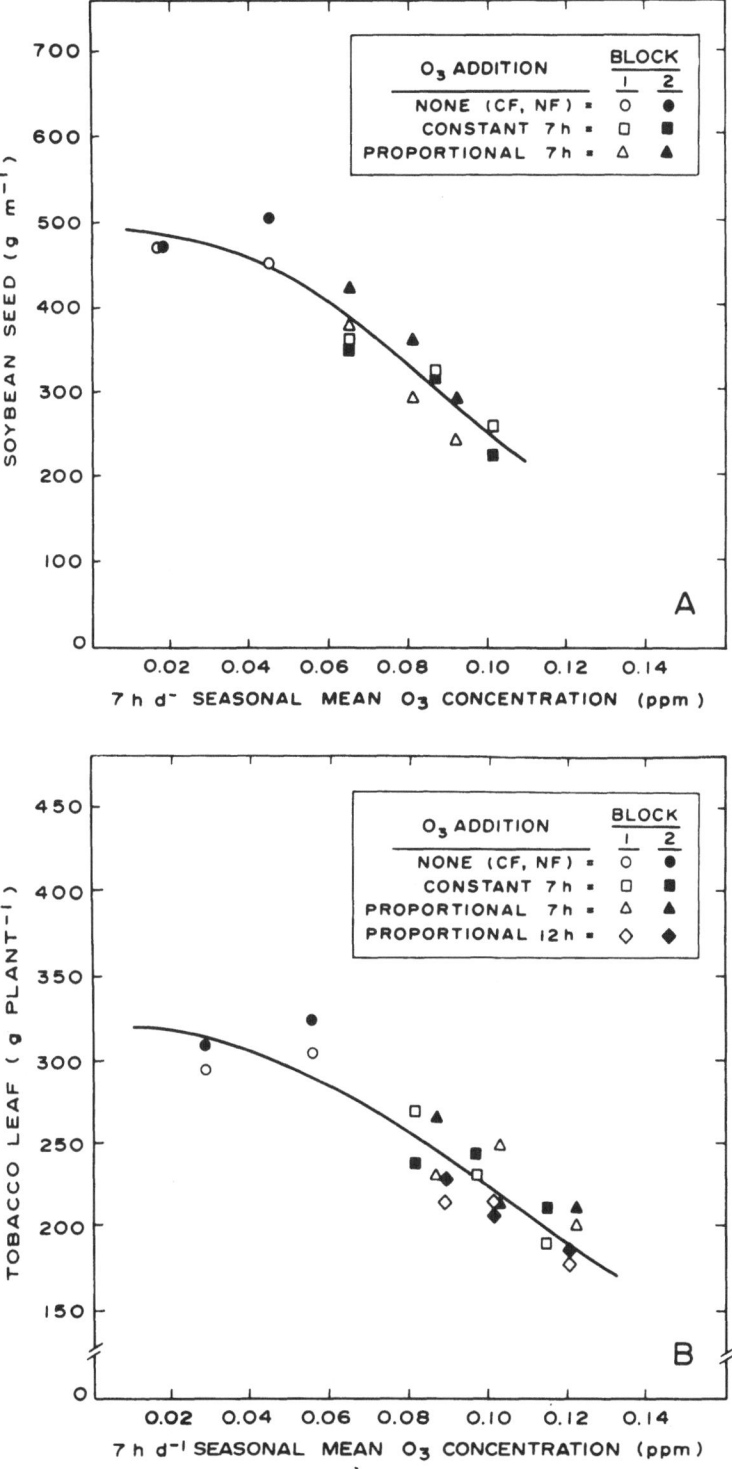

Fig. 7.9. Effects of O₃ addition method and daily exposure duration on crop yield: (A) response of 'Davis' soybean to 7 h constant (7C) and 7 h proportional-to-ambient (7P) O₃ addition at Raleigh, NC, in 1982; (B) response of 'McNair 235' tobacco to 7C, 7P and 12P O₃ addition at Raleigh, NC, in 1983. The estimated curves were derived from experiments outlined in Table 7.1 using polynomial regression analysis with all data combined.

1985 and 1986 Argonne soybean studies, and plastic film was placed over the soil between rows in the 1983 Beltsville soybean study. It was possible to obtain differential SMD throughout an entire season only at Riverside and Shafter, CA, where summer rainfall is rare. For most NCLAN studies involving SMD, two levels of soil moisture were used. We will refer to these as well-watered (WW) and as water-stressed (WS) with variable degree, frequency, and duration of stress. Measurement of SMD requires elaborate equipment, it is time consuming, and it is often inconclusive. Measurement of plant response to SMD (leaf moisture potential and gas exchange rates) is also expensive and time consuming. For all these reasons, plant response to SMD during NCLAN experiments has not been measured to a degree that allows an in-depth interpretation of cause–effect relationships.

The effects of soil moisture on crop response to O_3 were studied in four cotton experiments, six soybean experiments, and in one experiment each for a clover–fescue mixture, alfalfa, and barley (Table 7.1). One cotton study (Riverside 1985) and the barley experiment did not show a significant O_3 response at either moisture level and will not be discussed further. Results of the SMD studies were varied. Significant interactions between SMD and O_3 yield–response relationships occurred only for three soybean studies (Table 7.7), two cotton studies and one alfalfa study (Table 7.6). With 'Acala SJ-2' cotton at Shafter, CA, the moisture by O_3-effect interaction was significant in 1981 (Fig. 7.10(A)) but not in 1982 (Temple *et al.*, 1985*b*). The weather at Shafter in 1981 was typically hot and dry. Plants in the WS plots showed frequent wilting and yielded less than those in the WW plots. Estimated yield loss at a seasonal $7\,h\,d^{-1}$ mean of $0·05\,ppm\,O_3$ (compared to $0·025\,ppm$) was 2% in the WS plots and 7% in the WW plots (Table 7.6). In 1982, the weather at Shafter was atypically cool and cloudy with no wilting in either soil moisture treatment. Estimated yield loss at $0·05\,ppm\,O_3$ was 15% (moisture treatments combined). Results with 'McNair 235' cotton at Raleigh (Heagle *et al.*, 1988) were similar to those at Shafter in 1981 (Fig. 7.10(B)). Periods of low rain caused SMD, low leaf moisture potential, wilting and decreased yield in the WS plots. The analysis of variance showed a significant O_3 effect on yield in the WW but not in the WS plots. Polynomial regression analyses showed a significant O_3-effect by soil moisture interaction but Weibull analyses did not (Heagle *et al.*, 1988). Weibull estimates for yield loss at a seasonal $12\,h\,d^{-1}$ mean of $0·05\,ppm$ (compared to $0·02\,ppm$) were 4% in the WS plots and 18% in the WW plots (Fig. 7.10(B)). With alfalfa at Riverside, regression analyses (but not analyses of variance) showed a significant O_3–soil moisture interaction for 1985 and for the combined 1984 and 1985 data. The effect was for a slightly greater O_3 response in the WW than in the WS plots (Temple *et al.*, 1988*a*). Weibull analyses showed a homogeneous response over moisture levels (personal communication from Matt Somerville).

For soybeans, a significant O_3-effect by soil moisture interaction occurred in three of the six studies (Table 7.7). 'Davis' soybeans were tested in 1983 and in 1984 at Raleigh (Heagle *et al.*, 1987*b*). In 1983, the weather was much hotter and dryer than normal with plants in the WS plots exhibiting frequent severe

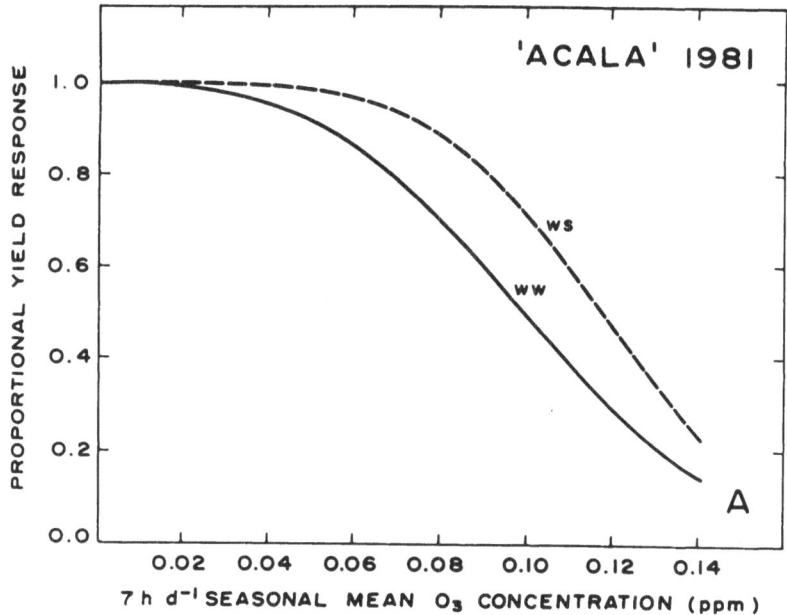

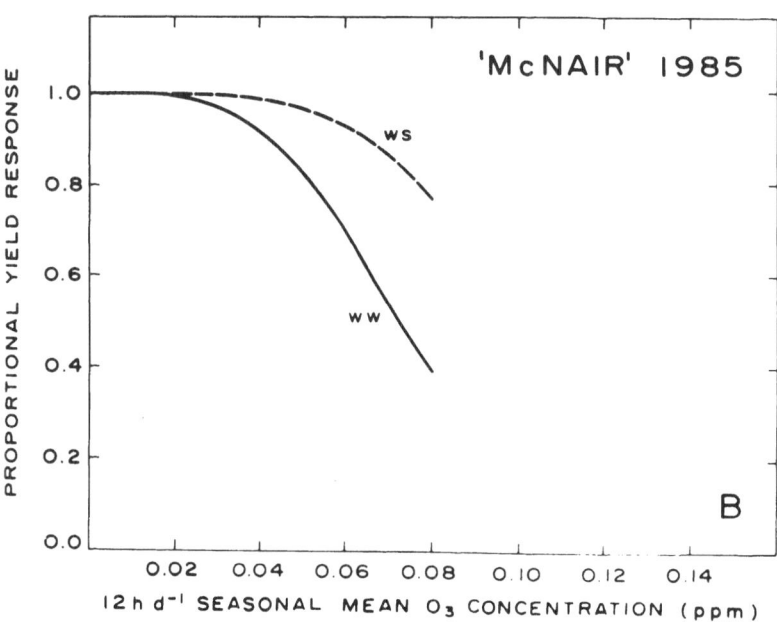

Fig. 7.10. Proportional yield response to O_3 of cotton grown with intermittent periods of soil moisture stress (WS) or grown with well-watered conditions (WW): (A) for 'Acala SJ-2' at Shafter, CA, in 1981; (B) for 'McNair 235' at Raleigh, NC, in 1985. Curves were derived from experiments outlined in Table 7.1 using Weibull analyses with the α value set at 100.

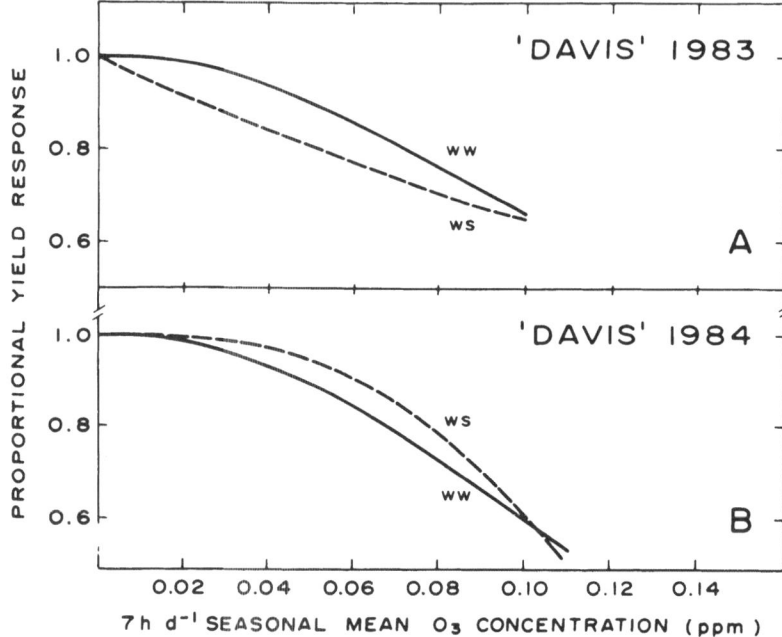

Fig. 7.11. Proportional yield response of Davis soybean to O_3 for plants grown with intermittent periods of water-stress (WS) or for well-watered (WW) plants: (A) 1983; (B) 1984. Curves were derived from experiments outlined in Table 7.1 using Weibull analyses with the α value set at 100. For the 1983 WS plots, the O_3 effect was not significant in the analysis of variance and the Weibull model was not significant.

wilting and depressed yield. The analysis of variance for the WS plots showed no significant yield response to O_3. Because there was a significant yield decrease with increased O_3 only in the WW plots, we assume an O_3–soil moisture interaction occurred (Fig. 7.11). In 1984, plants in the WS plots were only moderately stressed by SMD and there were no significant O_3–soil moisture interactions. Weibull estimates of proportional yield response to O_3 for the WW and WS plots separately for 1983 and 1984 are shown in Fig. 7.11. At Argonne in 1986, the interaction between O_3 and soil moisture was significant for 'Corsoy 79' with greater yield loss response in the WW than in the WS plots (Table 7.7). At Beltsville in 1982, there was no significant O_3-effect by soil moisture interaction with 'Williams' soybean, but for 'Forest', the interaction was significant with the SMD effect on yield response to O_3 dependent on the O_3 concentration (Table 7.7). In 1983 the interaction was significant for 'Williams' and 'Corsoy' (combined cultivar analyses) (Fig. 7.12; Table 7.7).

The NCLAN results show that SMD can decrease the response of crops to O_3 under some conditions but not under other conditions. Probably the occurrence of O_3 by SMD interactions was dependent on the degree of SMD-induced plant moisture stress. At present, there are not enough data to allow estimates of the effects of a range of SMD levels on O_3 dose–yield

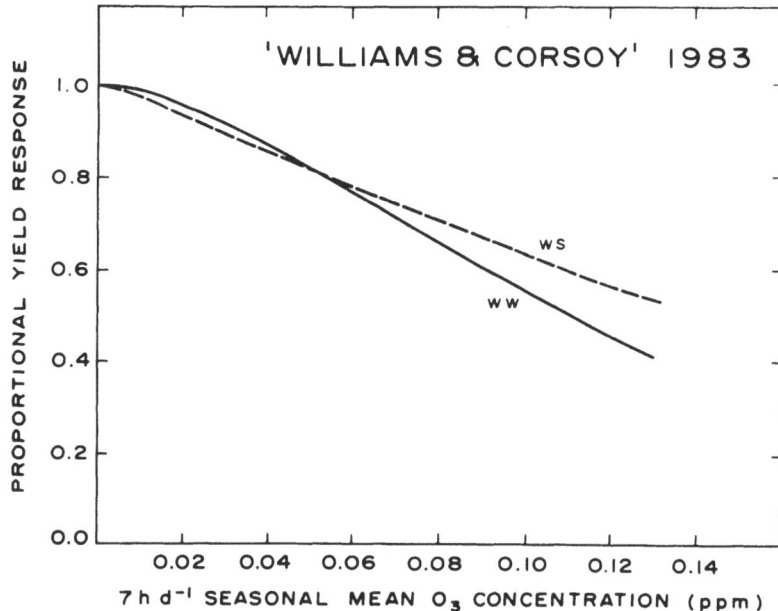

Fig. 7.12. Proportional yield response of Williams and Corsoy soybean (data combined) for plants with intermittent periods of water-stress (WS) or for well-watered (WW) plants. Curves were estimated by Weibull analyses with the α value set at 100.

response models. A high level of long-range research is needed which should include: (1) the ability to control and measure soil moisture at more than two levels for desired periods; (2) measurements that relate SMD to leaf gas exchange rates and that relate SMD to rates of physiological response to O_3; and (3) repetition of tests for a given crop genotype over several years.

7.4.6 Effects of sulfur dioxide on response to ozone

With increased fossil-fuel combustion there is a potential for increased levels of SO_2 to occur with phytotoxic levels of O_3 on a regional basis. Mixtures of O_3 and SO_2 can cause effects that are additive, greater than additive, or less than the additive effects of each gas singly (Reinert *et al.*, 1975). Prior to the NCLAN program, there were only a few studies on crop yield response to seasonal exposures to mixtures of O_3 and SO_2. Twelve NCLAN experiments addressed crop response to O_3–SO_2 mixtures (Table 7.1). Included were four soybean experiments, two experiments each with tomato and a clover–timothy mixture, and one experiment each with cotton, winter wheat, field corn, and alfalfa. Sulfur dioxide was usually added in constant amounts for $4\,h\,d^{-1}$ for $7\,d\,wk^{-1}$. Even when additions were constant however, wide momentary and daily SO_2 fluctuations occurred because of different rates of ambient-air ingress. For the red clover–timothy and field corn study, SO_2 regimes were programmed to simulate the dynamics of exposures that occur near point sources. In all studies, the mean SO_2 concentration for two or more of the treatments was higher than that which occurs on a regional basis in the US.

There was a significant SO_2 effect on yield in all but the winter wheat, cotton, clover–timothy and field corn studies. Except for the alfalfa (Heck *et al.,* 1984*a*) and 1985 clover–timothy study, the O_3 effect on yield was significant. However, there were no cases where O_3 by SO_2 interactions significantly affected yield. Results for one cotton and one soybean study are shown in Figs 7.13(A) and 7.13(B), respectively. Thus, for all O_3 and SO_2 mixture studies, final O_3 dose-yield models were fit using data combined across SO_2 levels.

7.5 RESEARCH NEEDS

The NCLAN program is an outstanding example of close research coopera-tion among federal, state, and private agencies. NCLAN success was directly related to a high level of coordination of research objectives and methods that for the most part were uniform at each of the regional sites. Overall, NCLAN was highly successful in accomplishing its broadly stated objectives related to developing dose–response functions for economically important crops and evaluating the economic impact. The studies have also been instrumental in exploring the complexity of the impact of O_3 in the field and the potential significance of other environmental variables. In spite of this success, major questions concerning crop response to O_3 remain. These questions involve the use of open-top chambers, the experimental exposure regimes, cultivars, and the role of experimental procedures and environmental variability in the reproducibility of results (Table 7.8).

Open-top chambers cause major changes in wind velocity, light profile, air velocity profiles, and O_3 concentration-fluctuation compared to ambient. Chambers can affect plant growth and yield. Thus, there is a lingering question as to whether plant response to O_3 in open-top chambers is similar to that in commercial fields.

After NCLAN began, new methods allowed proportional instead of constant O_3 addition and 12 h instead of 7 h daily exposure duration. Thus, NCLAN loss estimates are derived from data obtained in experiments using different types of daily exposure regimes. Research is needed to show how yield response using one set of regimes relates to that using others. A major variable in NCLAN studies was the seasonal exposure duration (SED) which ranged from 26% to 100%. More yield loss would probably occur with increased SED, but research is needed to verify this.

Relationships between the sensitivity of cultivars used in dose–response studies and the mean sensitivity of cultivars in production have not been established. Results in NCLAN studies indicate a fairly narrow response range for soybean cultivars, but more work is needed for other crops.

The degree to which results are repeatable from site to site or from year to year is not known. True repeats in NCLAN studies were rare but they show that major season or site effects can occur. Thus, there is a need to measure the degree of predictability in field O_3 dose–yield response studies. In such studies, the effects of factors such as climate (rainfall, humidity, soil moisture,

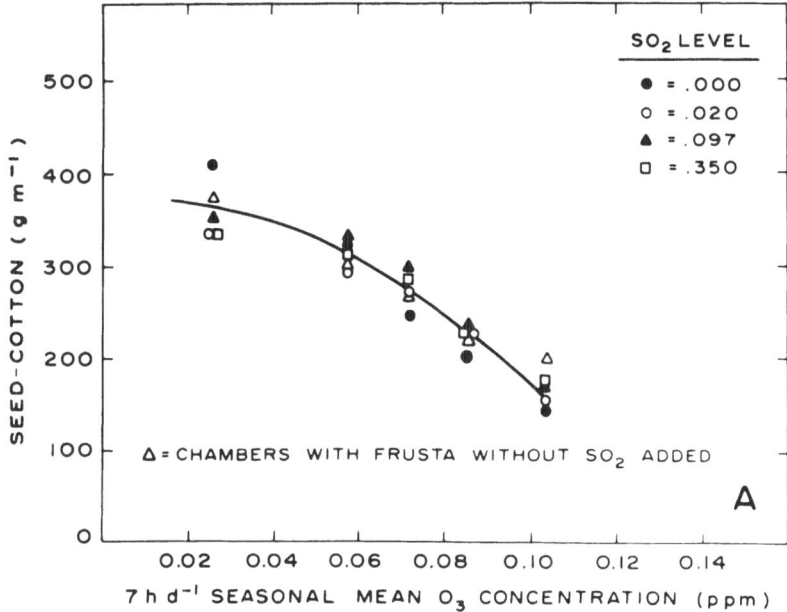

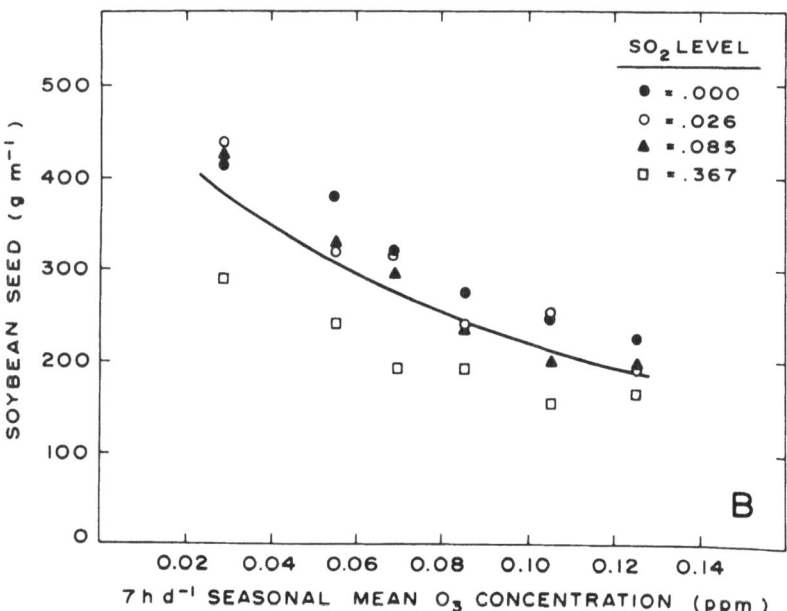

Fig. 7.13. Yield response of (A) Stoneville 213 cotton and (B) 'Davis' soybean to combinations of O_3 and SO_2 at Raleigh, NC. Curves were estimated by polynomial analyses using combined SO_2 data.

TABLE 7.8

Research needs relating to dose–response studies with gaseous pollutants in open-top
field chambers

Research needs	Approaches	Comments
I. Direct or indirect measures of how chambers affect plant response to gaseous pollutants	Compare dose–response results for plants in open-top chambers with results for plants in the open	This approach would be the most definitive for this research need but also the most difficult; it requires development of an open-air exposure system to duplicate exposures in open-top chambers; allows direct comparison of results within and outside of open-top chambers over a range of doses
	Measure gas exchange rates, morphology, growth, and yield for plants grown in nonfiltered-air chambers and in the open with an equivalent pollutant dose in both exposures	This approach would show whether chamber effects on exposure dynamics, climate, or soil affect plants in ways that could change plant response to pollutants; it is limited to one dose at a given location and time
II. Studies on the potential interactions among pollutant dose–response results and genetic, cultural, climatic, and edaphic factors	Conduct dose–response studies to determine potential interactive effects of: • Crop genotype (cultivars) • Seasonal climate (repeats over years) • Plant moisture stress (for a range of moisture levels) • Biotic factors (diseases or insects) • Agricultural chemicals	These studies would expand the relevance of NCLAN data that are limited in terms of understanding the effects of these variables; data would increase the feasibility of extrapolating NCLAN and other dose–response results to different years and different locations
III. Studies on how the timing of exposures and exposure dynamics affect dose–response results	Conduct dose–response studies that include the following factors as variables: • Type of pollutant addition (constant vs. proportional) • Plant growth stages for which exposures occur • Magnitude and timing of peak pollutant episodes	Results of these studies would allow adjustment of dose–response data to a common base since these factors were variable in NCLAN and in previous studies

light, and temperature) and the seasonal sequence of ambient O_3 episodes (peaks) on plant response should be measured. Knowledge of the effects of climate and O_3-related variables on O_3 response would aid in the search for the best way to characterize the O_3 dose in chronic dose–response studies.

A major objective of NCLAN was to include measures of physiological response to better understand cause–effect relationships for O_3-induced yield loss. Funding was sufficient to initiate such work but was insufficient to sustain it. We recommend a high level of attention to physiological response in all future research.

ACKNOWLEDGEMENTS

We thank Virginia M. Lesser, Susan E. Spruill, Matthew C. Somerville, and John R. Rawlings for analyses leading to many of the Weibull model dose–response estimates used in this report.

REFERENCES

Heagle, A. S., D. E. Body, and W. W. Heck. (1973). An open-top field chamber to assess the impact of air pollution on plants. *J. Environ. Qual.*, **2**, 365–8.

Heagle, A. S., S. Spencer, and M. B. Letchworth. (1979a). Yield response of winter wheat to chronic doses of ozone. *Can. J. Bot.*, **57**, 1999–2005.

Heagle, A. S., R. B. Philbeck, H. H. Rogers, and M. B. Letchworth. (1979b). Dispensing and monitoring ozone in open-top field chambers for plant effects studies. *Phytopathology*, **69**, 15–20.

Heagle, A. S., R. B. Philbeck, and W. M. Knott. (1979c). Thresholds for injury, growth, and yield loss caused by ozone on field corn hybrids. *Phytopathology*, **69**, 21–6.

Heagle, A. S., M. B. Letchworth, and C. Mitchell. (1983a). Injury and yield responses of peanuts to chronic doses of ozone in open-top field chambers. *Phytopathology*, **73**, 551–5.

Heagle, A. S., W. W. Heck, J. O. Rawlings, and R. B. Philbeck. (1983b). Effects of chronic doses of ozone and sulfur dioxide on injury and yield of soybeans in open-top field chambers. *Crop. Sci.*, **23**, 1184–91.

Heagle, A. S., W. W. Cure, and J. O. Rawlings. (1985). Response of turnips to chronic doses of ozone in open-top field chambers. *Environ. Pollut., Ser. A* **38**, 305–19.

Heagle, A. S., W. W. Heck, V. M. Lesser, J. O. Rawlings and F. Mowry. (1986a). Injury and yield response of cotton to chronic doses of ozone and sulfur dioxide. *J. Environ. Qual.*, **15**, 375–82.

Heagle, A. S., V. M. Lesser, J. O. Rawlings, W. W. Heck, and R. B. Philbeck. (1986b). Response of soybeans to chronic doses of ozone applied as constant or proportional additions to ambient air. *Phytopathology*, **76**, 51–6.

Heagle, A. S., W. W. Heck, V. M. Lesser, and J. O. Rawlings. (1987a). Effects of daily ozone exposure duration and concentration fluctuation on yield of tobacco exposed to chronic doses of ozone. *Phytopathology*, **77**, 856–62.

Heagle, A. S., R. B. Flagler, R. P. Patterson, V. M. Lesser, S. R. Shafer and W. W. Heck. (1987b). Injury and yield response of soybean to chronic doses of ozone and soil moisture deficit. *Crop Sci.*, **27**, 1016–24.

Heagle, A. S., J. E. Miller, R. P. Patterson and W. W. Heck. (1988). Injury and yield response of cotton to chronic doses of ozone and soil moisture deficit. *J. Environ. Qual.* (in press).

Heagle, A. S., J. Rebbeck, S. R. Shafer, V. M. Lesser, U. Blum, and W. W. Heck. (1989). Effects of long-term ozone exposure and soil moisture deficit on yield of ladino clover and tall fescue. *Phytopathology* (in press).

Heck, W. W., O. C. Taylor, R. M. Adams, G. Bingham, J. E. Miller, E. M. Preston, and L. H. Weinstein. (1982a). *National Crop Loss Assessment Network (NCLAN) 1981 Annual Report*, 190 pp., EPA 600/3-83-049. Corvallis Environmental Research Laboratory, Office of Research and Development, US EPA, Corvallis, OR.

Heck, W. W., O. C. Taylor, R. Adams, G. Bingham, J. Miller, E. Preston, and L. Weinstein. (1982b). Assessment of crop loss from ozone. *J. Air Pollut. Control Assoc.*, **32**, 353–61.

Heck, W. W., O. C. Taylor, R. M. Adams, G. Bingham, J. E. Miller, E. M. Preston, L. H. Weinstein, R. G. Amundson, R. J. Kohut, J. A. Laurence, W. C. Cure, A. S. Heagle, T. J. Gish, H. E. Heggestad, L. W. Kress, G. E. Neely, J. O. Rawlings, and P. Temple. (1983a). *National Crop Loss Assessment Network (NCLAN) 1982 Annual Report*, 249 pp., EPA 600/3-84-049. Corvallis Environmental Research Laboratory, Office of Research and Development, US EPA, Corvallis, OR.

Heck, W. W., R. M. Adams, W. W. Cure, A. S. Heagle, H. E. Heggestad, R. J. Kohut, L. W. Kress, J. O. Rawlings, and O. C. Taylor. (1983b). A reassessment of crop loss from ozone. *Environ. Sci. Technol.*, **17**, 573A-80A.

Heck, W. W., O. C. Taylor, R. M. Adams, J. E. Miller, E. M. Preston, L. H. Weinstein, R. G. Amundson, W. C. Cure, A. S. Heagle, T. J. Gish, H. E. Heggestad, D. A. King, L. W. Kress, R. J. Kohut, J. A. Laurence, J. Miller, G. E. Neely, J. O. Rawlings, and P. Temple. (1984a). *National Crop Loss Assessment Network (NCLAN) 1983 Annual Report*, 227 pp., EPA 600/3-85-061. Corvallis Environmental Research Laboratory, Office of Research and Development, US EPA, Corvallis, OR.

Heck, W. W., W. W. Cure, J. O. Rawlings, L. J. Zaragoza, A. S. Heagle, H. E. Heggestad, R. J. Kohut, L. W. Kress, and P. J. Temple. (1984b). Assessing impacts of ozone on agricultural crops: I. Overview. *J. Air Pollut. Control Assoc.*, **34**, 729–35.

Heck, W. W., W. W. Cure, J. O. Rawlings, L. J. Zaragoza, A. S. Heagle, H. E. Heggestad, R. J. Kohut, L. W. Kress, and P. J. Temple. (1984c). Assessing impacts of ozone on agricultural crops: II. Crop yield functions and alternative exposure statistics. *J. Air Pollut. Control Assoc.*, **34**, 810–17.

Heck, W. W., O. C. Taylor, R. M. Adams, J. E. Miller, D. T. Tingey, L. H. Weinstein, R. G. Amundson, A. S. Heagle, D. A. King, R. G. Kohut, L. W. Kress, J. A. Laurence, A. S. Lefohn, V. M. Lesser, J. R. Miller, G. E. Neely, P. J. Temple, and J. O. Rawlings. (1985). *National Crop Loss Assessment Network (NCLAN) 1984 Annual Report*, 228 pp., EPA/600/3-86/041. Environmental Research Laboratory, Office of Research and Development, U.S. EPA, Corvallis, OR.

Heggestad, H. E., T. J. Gish, E. H. Lee, J. H. Bennett, and L. W. Douglas. (1985). Interactions of soil moisture stress and ambient ozone on growth and yield of soybeans. *Phytopathology*, **75**, 472–7.

Heggestad, H. F., E. L. Anderson, T. J. Gish, and E. H. Lee. (1988). Effects of O_3 and soil water deficit on roots and shoots of field-grown soybeans. *Environ. Pollut.* (in press).

Kohut, R. J. and J. A. Laurence. (1983). Yield response of Red Kidney Bean *Phaseolus vulgaris* to incremental ozone concentrations in the field. *Environ. Pollut., Ser. A*, **32**, 233–40.

Kohut, R. J., R. G. Amundson and J. A. Laurence. (1986). Evaluation of growth and yield of soybean exposed to ozone in the field. *Environ. Pollut.*, **41**, 219–34.

Kohut, R. J., R. G. Amundson, J. A. Laurence, L. Colavito, P. van Leuken, and P. King. (1987). Effects of ozone and sulfur dioxide on yield of winter wheat. *Phytopathology*, **77**, 71–4.

Kress, L. W. and J. E. Miller. (1983). Impact of ozone on soybean yield. *J. Environ. Qual.*, **12**, 276–81.

Kress, L. W. and J. E. Miller. (1985a). Impact of ozone on field corn yield. *Can. J. Bot.*, **63**, 2408–15.

Kress, L. W. and J. E. Miller. (1985b). Impact of ozone on grain sorghum yield. *Water, Air, Soil Pollut.*, **25**, 377–90.

Kress, L. W., J. E. Miller and H. J. Smith. (1985c). Impact of ozone on winter wheat yield. *Environ. Expt. Bot.*, **25**, 211–28.

Kress, L. W., J. E. Miller, H. J. Smith and J. O. Rawlings. (1986). Impact of ozone and sulfur dioxide on soybean yield. *Environ. Pollut.*, **41**, 105–23.

Kress, L. W., P. M. Irving, W. Prepejchal and H. J. Smith. (1988). Impact of O_3 and SO_2 on four field corn hybrids. I. Growth and yield. *Phytopathology* (in press).

Olszyk, D. M., T. W. Tibbitts, and W. M. Hertzberg. (1980). Environment in open-top field chambers utilized for air pollution studies. *J. Environ. Qual.*, **9**, 610–15.

Rawlings, J. O. and W. W. Cure. (1985). The Weibull function as a dose–response model to describe ozone effects on crop yields. *Crop Sci.*, **25**, 807–14.

Rawlings, J. O., V. M. Lesser, and K. A. Dassel. (1988). Statistical approaches to assessing crop losses. In *Assessment of crop loss from air pollutants*. Proceedings of the international conference, Raleigh, N.C., USA, ed. by W. W. Heck, O. C. Taylor and D. T. Tingey, 389–416. London, Elsevier Applied Science.

Reinert, R. A., A. S. Heagle, and W. W. Heck. (1975). Plant response to pollutant combinations. In *Response of plants to air pollution*, ed. by J. B. Mudd and T. T. Kozlowski, 159–77, New York, Academic Press.

Temple, P. J., O. C. Taylor, and L. F. Benoit. (1985a). Effects of ozone on yield of two field-grown barley cultivars. *Environ. Pollut., Ser. A*, **39**, 217–25.

Temple, P. J., O. C. Taylor, and L. F. Benoit. (1985b). Cotton yield responses to ozone as mediated by soil moisture and evapotranspiration. *J. Environ. Qual.*, **14**, 55–60.

Temple, P. J., O. C. Taylor, and L. F. Benoit. (1986). Yield response of head lettuce (*Lactuca sativa* L.) to ozone. *Environ. Expt. Bot.*, **26**, 53–58.

Temple, P. J., L. F. Benoit, R. W. Lennox, C. A. Reagan, and O. C. Taylor. (1988a). Combined effects of ozone and water stress on alfalfa growth and yield. *J. Environ. Qual.* **17**, 108–13.

Temple, P. J., R. S. Kupper, R. L. Lennox, and N. Rahr. (1988b). Physiological and growth responses of differentially irrigated cotton to ozone. *Environ. Poll.*, **53**, 255–263.

Unsworth, M. H., A. S. Heagle, and W. W. Heck. (1984a). Gas exchange in open-top field chambers I. Measurement and analysis of atmospheric resistances to gas exchange. *Atmos. Environ.*, **18**, 373–80.

Unsworth, M. H., A. S. Heagle, and W. W. Heck. (1984b). Gas exchange in open-top field chambers II. Resistance to ozone uptake by soybeans. *Atmos. Environ.*, **18**, 381–5.

Weinstock, L., W. J. Kender, and R. C. Musselman. (1982). Microclimate within open-top air pollution chambers and its relation to grapevine physiology. *J. Am. Soc. Hortic. Sci.*, **107**, 923–9.

8

THE USE OF OPEN FIELD SYSTEMS TO ASSESS YIELD RESPONSE TO GASEOUS POLLUTANTS

A. R. McLEOD

Central Electricity Research Laboratories, Leatherhead, Surrey, UK

and

C. K. BAKER

University of Nottingham, UK

8.1 OPEN-FIELD EXPOSURE METHODS

8.1.1 Introduction

Early assessments of air pollution damage to crops were often based on surveys of plants growing near industrial and urban sources that produced decreases in pollutant concentration and foliar damage with distance from the source. Such assessments led to a consensus in the 1930s that polluted plants were affected only if they were visibly damaged. Attempts were made to determine concentration limits below which plants were protected from damage, but as early as 1923 Stoklasa suggested that growth and yield might be reduced even in the absence of visible signs of injury. Other workers (e.g., Bleasdale, 1952) later obtained results that supported the notion of 'invisible damage,' and much work since then has amply confirmed it. A problem in early studies of gaseous pollutant effects on plants was the difficulty of defining which statistical exposure parameters could be related to plant response. In recent years, more attention has been paid to the range of temporal and spatial distributions of pollutant concentrations and their importance for vegetation response (Roberts, 1984; Hogsett *et al.*, 1987a).

In order to characterise the pollution levels that produced effects, experimental chambers were used in which plants could be exposed to closely regulated concentrations of a specific pollutant. In closed chambers it is not easy to prevent changes in microclimate that may affect growth, and unless air-change rates are large enough, leaf boundary layer resistance limits pollutant uptake (Black, 1982). Open-top chambers (OTC) were also developed to minimise plant enclosure without sacrificing the desired control over pollutants. Plants inside an OTC experience a good approximation to outside conditions, but there may still be chamber effects that may influence plant response to pollutants. These effects include reductions in light intensity and rainfall, increases in air temperature and humidity (Heagle *et al.*, 1988), reduced dew formation and persistence (Olszyk *et al.*, 1986*a*), reduced frost formation (Weigel *et al.*, 1987), and modified profiles of wind speed and pollutant concentration in the canopy (Colls and Baker, 1988). However, an important advantage of open-top and closed chambers is that ambient pollutants may be excluded by filtration, and pollutant addition can provide a range of treatment concentrations in several chambers.

In the last decade, closed and open-top chambers have been developed to optimise pollutant flux using high air flows and to minimise changes in environmental conditions. However, the disadvantages of chamber enclosures encouraged the further development of methods for studying the effects of pollution on plants growing in the open. These have ranged from surveys of ambient exposure effects (Colvill *et al.*, 1985) and the use of chemical treatments to protect against pollutant injury (Temple and Bisessar, 1979) to a variety of experimental systems for dispensing a protective flow of charcoal-filtered air over the crop (Olszyk *et al.*, 1986*a*) or fumigating with a pollutant/air mixture without chambers (Mooi and van der Zalm, 1986).

8.1.2 Field surveys of ambient pollution effects

Since the early days of air pollution research, field surveys of damage to plants have been undertaken to assess the effects of emissions. Frequently, the aim of the survey has been to determine the extent of foliar injury caused by high concentrations of a predominating pollutant emitted from a point source. The method is particularly suitable for this purpose, as damage severity decreases with distance from the source. Early this century, Crowther and Ruston (1914) grew lettuce (*Lactuca sativa* L.), radish (*Raphanus sativus* L.) and wallflowers (*Cheiranthus* sp. L.) in standard soils in a transect across Leeds, England, and showed large decreases in growth with increasing urban pollution. In the most polluted areas, some plants died, probably from the combination of attenuated sunlight and gaseous pollution. Urban air in most Western cities is much cleaner today, but in some industrial and population centres in China air pollution is still severe and has been extensively monitored through field surveys and experimental plantings (Yu, 1984).

Jones *et al.* (1977) cite field survey reports that soybeans and alfalfa growing at sites near coal-fired power stations showed no yield decreases unless at least 5% of the leaf area was destroyed. However, field surveys around industrial installations in Eastern Europe have revealed decreased growth and yield in the absence of visible injury. Various crop species were grown in containers of standardized soil at increasing distances from pollution sources. Guderian and Strattman (1962*a*, *b*) grew 12 crops at 6 sites located 325 to 6000 m from an iron-ore smelter and related yield reductions to continuous SO_2 records at each site. A 3-year study at 11 sites by Warteresiewicz (1979) showed a subtle effect of an industrial source of SO_2 and arsenic on barley: grain yield decreased with pollution even though total shoot dry matter was not affected. In California, peroxyacetyl nitrate (PAN) and O_3 are the principal pollutants, and field surveys have been used to assess visible injury symptoms and yield depression in trial plantings and to relate this to O_3 concentration data to model economic losses in alfalfa (*Medicago sativa* L. cv. 'Moapa 69') (Oshima *et al.*, 1976). Field surveys become more difficult to interpret when there is no foliar injury attributable to a specific pollutant, and growth responses may be affected by multiple pollutant exposure (e.g., heavy metals and SO_2; Dreisinger and McGovern, 1970; Warteresiewicz, 1979) and other environmental factors along pollutant gradients (e.g., air temperature and drought; Colvill *et al.*, 1985). It is also difficult to determine which characteristics of pollutant exposure (peaks or long-term means) have produced any growth effects.

Field surveys of plant damage caused by area sources include the use of indicator plants sensitive to a particular pollutant to assess pollution zones on a national or regional scale. In Japan, the air pollution by PAN around the Tokyo conurbation has been assessed by exposing the sensitive species morning glory (*Pharbitis nil* Choisy cv. 'Scarlet O'Hara') and peanut (*Arachis hypogea* L.) (Furukawa, 1984). In Holland, a nationwide network for pollution monitoring has operated at 40 locations across the country since 1976, exposing identical sets of plants sensitive to different pollutants (e.g., Gladiolus *Gladiolus gandavensis* L. cv. 'Sneeuprinses' to hydrogen fluoride (HF) and tobacco *Nicotiana tabacum* L. cv. 'Bel-W_3' to O_3). The results have clearly shown that damage to tobacco coincides with sunny weather and is greater in the western half of Holland and that damage to *Gladiolus* is greatest in the southwest (Floor and Posthumus, 1977).

The pattern of air pollution has changed considerably in Europe and the United States over the past several decades, and it might be expected that field surveys and pollutant monitoring data would reveal changes in crop yields that reflect the changes in pollutant levels. However, agricultural practices and crop varieties have also changed dramatically. There has been a large post-war increase in crop yields in industrialised countries through the use of nitrogenous fertilisers. Many of these contain no sulphur, and atmospheric sulphur deposition may now prevent soil sulphur deficiency in some regions where SO_2 pollution may once have reduced yield (Cowling and Jones, 1978). Better yields have also resulted from plant breeding for increased yield capacity (Austin *et al.*, 1980) and the use of a widening spectrum of agrochemicals that suppress competitive organisms and regulate growth. It is therefore unlikely that any useful assessment can be made of the historical trends of yield response to air pollutants.

8.1.3 Chemical protection against ambient pollution

A technique that maintains the ambient growing conditions of crops whilst permitting assessment of air pollution effects is the application of chemicals that protect plants against pollutants. Various insecticides, fungicides, nematicides, and antioxidants confer a variable degree of protection against O_3 and have been used in the field to reduce visible leaf damage during episodes of high ambient O_3 concentration (e.g., Temple and Bisessar, 1979) and to assess resultant yield loss (e.g., Bisessar, 1982). The substances may reduce damage by causing breakdown of the O_3 on the leaf surface, by inducing stomatal closure, or by modifying plant metabolism. A comprehensive review of the types and use of chemicals that protect against oxidants has been compiled by Mooi (1982). For use on field-grown crops, the mode of action and administration of the protectant is important. Some substances are non-systemic, protecting only the plant parts contacted, and are therefore administered as foliar sprays; others are systemic and may be absorbed by the plant from foliar sprays or from solutions applied to the soil.

The technique is suitable for field experiments when periods of elevated O_3 concentrations are predicted during the summer, and protectants can be applied every 7 to 10 days. Beneficial effects are then believed to demonstrate the influence of O_3 on visible damage and yield (Bisessar, 1982; Littlejohns et al., 1976). However, one disadvantage with the method is that the protectant itself may affect growth, particularly if there is a direct effect on pests and pathogens. This factor may sometimes be assessed by comparing different fungicides with and without antioxidant properties during the same trial but should strictly be determined by applying the protectant to plants grown in an O_3-free environment. Certain protectants may be effective against O_3 but not against other oxidants such as PAN (Pell, 1976), whilst other compounds (e.g., the nematicide fensulphothion) are known to increase the effect of oxidants on vegetation (Miller et al., 1976).

8.1.4 Air exclusion systems

An alternative protection against ambient pollution is to use methods developed for blowing pollutant-free air over the crops, which have been described as air exclusion systems (AES). Perforated ducts (diameter 25 to 35 cm) are laid between rows of a crop and used to blow charcoal-filtered air over the plants, displacing polluted air upwards. Studies using this technique are listed in Table 8.1. Jones et al. (1977) first used an AES with a soybean crop (Glycine max L. Merr. cv. 'Essex') to reduce peak SO_2 exposures from a local power station. Four perforated ducts between adjacent crop rows directed filtered air up and through the canopy whenever ambient SO_2 exceeded 0·1 ppm (266 μg m^{-3}) between 0700 and 1900 during the period June to September. The efficiency of pollutant exclusion was approximately 80% in a test during an ambient exposure of >0·375 ppm and wind velocity of approximately 0·6 m s^{-1}. However, in subsequent trials during ambient exposures with and without a full canopy, the exclusion efficiency was only 33 to 69% (Noggle, 1980). The system tended to eliminate peak SO_2 concentrations but did not completely eliminate SO_2 from the canopy. Subsequent experiments at several sites used eight ducts per exclusion plot oriented perpendicular to a line from the power plant source. Each plot had a 1·3-m tall fibreglass barrier along the side nearest the pollutant source, inclined toward the plot to deflect polluted air above the plants. The systems were used in soybean and wheat crops to study the effect of minimising exposure to peak SO_2 concentrations (Noggle, 1980).

Shinn *et al.* (1977) developed an AES in which charcoal-filtered air was blown from three perforated ducts, in a plot surrounded by a transparent barrier 0·6 m high to improve exclusion efficiency. The system was used for fumigation with SO_2, H_2S, or O_3 during ambient air exclusion. The flows of SO_2 and H_2S or O_3 production were at predetermined rates but were controlled by timer clocks to permit variations in the daily exposure duration. A gradient of concentration was obtained by increasing the size of the perforations along the ducts, and so the system was termed a linear-gradient chamber (LGC). Similar LGC systems were used for fumigation studies by Laurence *et al.* (1982) and Reich (1982), achieving a pollution gradient by a variation in the number of holes rather than their size. However, the surrounding fence was omitted, and, because unfiltered air was used, ambient pollutants were not excluded from the canopy.

Thompson and Olszyk (1985) investigated the efficiency of AES design using a variety of duct and outlet patterns. Holes oriented 45° upward, horizontally, or 45° downward were used to avoid trapping ambient pollutants in the canopy and the numbers of ducts, perforations, and air flows were optimised to increase exclusion efficiency. In the final design used with alfalfa in 1983, 50 to 70% of the ambient O_3 was excluded from the canopy between the ducts. The equipment was also modified by varying the hole sizes in sections of the ducts to give a range of O_3 exclusion efficiency and exposure and to provide ambient air exclusion with simultaneous SO_2 fumigation at a range of concentrations (Olszyk *et al.*, 1986*b*). A computer-control system was developed to permit regulation of the exposure concentration.

Air exclusion systems permit the determination of ambient air pollution effects and also allow the addition of pollutants to the canopy. They have the advantage of producing fewer changes to environmental conditions than may occur when using chambers. However, the duct location between rows makes them suited only to row crops, with physical limitations on crop height and row width. Air flow and pollutant distribution within the plant canopy may also differ from ambient conditions.

8.1.5 Open-field fumigation systems

Gas released from a single source adjacent to the target vegetation has been used as the simplest form of open-air fumigation. Dowding (1987) reported burning 1·3 kg of sulphur close to some trees to obtain a qualitative measure of SO_2 effects upon the leaf-surface microorganisms, and Spierings (1967) developed a 'gun' system to direct a controlled flow of pollutants onto sections of tree branches. Skye (1968), Jürging (1975), and Beetham (1980) released pure gases at measured rates from gas cylinders to expose vegetation but made only limited attempts to measure and control the exposures, which were characterised by large fluctuations in concentration. The release of pollutant gases from a single source close to the vegetation can, however, be improved by attention to design. A system has been constructed at Ruohonieni, Finland, (personal communication from L. Karenlampi), with which 2·6 ppm SO_2 in air is released at $60 \, m^3 \, h^{-1}$ from an outlet at the centre of a 5-m diameter plot

TABLE 8.1

Experimental methods of open-field exposure of vegetation to air pollutants

Exposure system	Vegetation	Pollutant	Reference
Single-point gas release			
Single-point sources from 100% gas cylinders	Lichens	SO_2	Skye (1968)
		SO_2, HCl, HF NH_3, CO	Jürging (1975) Beetham (1980)
Short-line source releasing 100% SO_2	Lichens	SO_2	Moser et al. (1980)
Single-point source from burning 1·3 kg sulphur	Trees	SO_2	Dowding (1987)
Single-gas outlet releasing an air/SO_2 mixture inside 5-m diameter circle of 1-m fencing. Duration of gas release automatically controlled	Coniferous trees	SO_2	Personal communication from L. Karenlampi
Multiple-source gas releases			
Vertical network of pipes dispensing an air/SO_2 mixture continuously. Fumigation stopped if concentration limit exceeded	Coniferous, deciduous and fruit trees	SO_2	DeCormis et al. (1975) Bonte et al. (1981)
Horizontal pipes dispensing an air/SO_2 mixture or an air/SO_2 + NO_2 mixture when wind was perpendicular to pipes. Fumigation stopped if concentration limit exceeded	Soybean	SO_2	Miller et al. (1980b)
		$SO_2 + NO_2$	Irving and Miller (1984)
Horizontal network of pipes dispensing an air/SO_2 mixture continuously	Prairie grassland	SO_2	Lee and Lewis (1977)
Horizontal network of pipes dispensing an air/SO_2 mixture continuously	Young trees and grasses	SO_2	Runeckles et al. (1981)
Single horizontal pipe dispensing an air/gas mixture when the wind was perpendicular to the pipe to produce a concentration gradient	Soybean	SO_2	Miller et al. (1980a)
	Field corn	O_3, SO_2	Miller et al. (1981a,b)
Two horizontal pipes dispensing an air/SO_2 mixture when the wind was perpendicular to the pipe to produce a concentration gradient	Desert plants	SO_2	Thompson and Olszyk (1985)
Vertical pipes hanging from horizontal supports between rows dispensing an air/HF mixture	Vine	HF	Cantuel (1980)

Description	Crop	Pollutant	Reference
Air-exclusion linear gradient chamber. Three perforated ducts dispensing charcoal-filtered air continuously between crop rows. Pollutants added for set durations, and concentration gradient produced by a gradient of hole size	Snap beans	$H_2S + O_3$, $SO_2 + O_3$	Shinn et al. (1977)
Linear gradient chamber as above dispensing unfiltered air/pollutant mixtures between crop rows during set periods. Concentration gradient produced by a gradient of hole spacing	Soybean	SO_2, HF	Laurence et al. (1982)
	Soybean	$SO_2 + O_3$	Reich et al. (1982)
Computer control of gas-release rate			
20-m square formed by four line sources dispensing an air/SO_2 mixture on upwind sides. Exposures a fixed amount above ambient with control of gas release rate	Winter cereals	SO_2	Greenwood et al. (1982), Colls et al. (1987)
Network of vertical pipes within the crop designed to give horizontally uniform exposure to an air/SO_2 mixture with control to follow a predefined target data set	Winter cereals	SO_2	McLeod et al. (1985), McLeod and Roberts (1987)
20-m circle of perforated vertical pipes controlled against both measured concentration and wind velocity to maintain a constant target concentration	Field beans	SO_2	Mooi and van der Zalm (1986)
Two horizontal pipes dispensing an air/SO_2 mixture when the wind was perpendicular to the pipe to produce a concentration gradient downwind. Control of release rate against measured concentration	Desert plants	SO_2	Olszyk et al. (1987)
Air-exclusion system operated continuously to dispense charcoal-filtered air between crop rows. A gradient of O_3 exclusion and added SO_2 generated with control of SO_2 release rate against measured concentration	Lettuce, wheat, alfalfa	SO_2	Olszyk et al. (1986b), Thompson and Olszyk (1985)
60-m circle of vertical pipes releasing an air/SO_2 mixture at the top of a forest canopy. Control of gas-release rate against measured concentration	Mature pine forest	SO_2	Personal communication from J. E. Hallgren
50-m circle of vertical pipes releasing an air/SO_2/O_3 mixture at two heights. Control of gas-release rate against measured concentration and wind velocity	Young coniferous trees	SO_2, O_3	McLeod et al. (1987)
15-m circle of vertical pipes releasing an air/SO_2 mixture for fumigation trials. Control of gas-release rate against measured concentration	Grassland	$SO_2 + O_3$, CO_2	Personal communication from K. Lewin
Four 5-cm perforated ducts between rows releasing an air/O_3 mixture. Computer control under development	Grasses and young conifers	O_3	Personal communication from A. Davison

surrounded by a 1-m plastic wall. The wall reduces dispersion and dilution of the SO_2 by wind and results in exposure of potted trees within the plot to SO_2 concentrations of 0·04 to 0·1 ppm. An electronic timer and valve is incorporated to permit control of the exposure duration. Gas release from a single point produces a gradient in exposure with distance from the source, but the technique may be suitable for simulation of the intermittent acute exposure typical of real point-sources, which produce pollutant episodes in areas with a low background concentration.

Many experiments have exposed larger areas of crops and trees to pollutants released from networks of pipes and gas sources (Table 8.1). The earliest attempts to expose large areas of vegetation were undertaken in the United States and France by releasing a dilute mixture of SO_2 from a number of horizontal pipes suspended above the canopy (DeCormis *et al.*, 1975; Lee and Lewis, 1977; Miller *et al.*, 1980*b*; Runeckles *et al.*, 1981). These became known as Zonal Air Pollution Systems (ZAPS), and at Corvallis, Oregon, the method was used to expose 0·5-ha areas of prairie grassland to SO_2 (Preston and Lee, 1984). The design of the release pipework was based upon early trials, and two sets of four exposure plots were established at nearby sites. At three plots at each site, dilute SO_2 was released in air at fixed rates throughout four or five growing seasons. Exposure concentrations depended upon the effect of meteorological conditions on pollutant dispersion. Effects on a range of components of the Northern Mixed Prairie ecosystem were assessed in a comprehensive study (Lauenroth and Preston, 1984).

Miller *et al.* (1980*b*) used a ZAPS to expose 783-m^2 areas of soybean crop to SO_2. This system operated only when the wind was perpendicular to the release pipes so that adjacent control plots were not fumigated. The intermittent fumigation produced a frequency distribution that was neither normal nor log-normal but was likened to a point-source exposure. The gas release was manually adjustable but could also be shut off automatically whenever a set SO_2 concentration was exceeded. This was the first step in achieving some automatic control of the exposure concentration in an open-field plot. The method was simplified by using only one or two release pipes upwind of a plot to produce concentration gradients downwind during intermittent fumigation. Different sample locations received a range of measured exposures, and the technique was used for both SO_2 and NO_2 fumigation of soybeans (Irving and Miller, 1984) and SO_2 fumigation of desert plants (Olszyk *et al.*, 1987).

The first European open-air fumigation system using multiple sources exposed young trees to SO_2 in France (DeCormis *et al.*, 1975). An SO_2/air mixture was dispensed from 128 vertical release pipes over a 28- × 60-m plot sited 100 m away from a control plot. The gas was distributed through an underground pipe network with each 2-m-high outlet pipe at an equal distance from the air turbine. The SO_2 was released at a constant rate from six sets of holes in each pipe to distribute the gas throughout the 2-m height. Some control of exposure was achieved by switching off the gas supply whenever measured concentration exceeded 0·07 ppm and a mean exposure of 0·05 ppm was achieved over 3 years. A multiple-source system for fumigating vines with HF was also constructed in France (Cantuel, 1980). An HF/air mixture was dispensed from a network of perforated flexible pipes hanging from supports between the vine rows in a 9- × 32-m plot. Hydrogen fluoride was dispensed continuously by vaporising a controlled flow of dilute hydrofluoric acid at 150°C into the air supply pipes. However, only limited control was possible because exposure was measured retrospectively by deposition to NaOH-impregnated filter paper.

Attempts were made to overcome the limitations of constant rates of gas-release by using a computer to control the exposure in open-field fumigation systems (Table 8.1). Greenwood *et al.* (1982) described a system that released dilute SO_2 in air from one or two upwind sides of a 20- × 20-m square of perforated pipe. The pipes were positioned 0·2 m above the crop canopy and acted as line sources of SO_2. The gradient of concentration with distance from the pipe provided an acceptably uniform exposure for 5 to 15 m from the source (Colls *et al.*, 1987), and the central 10-m square of the plot was used for crop sampling. The computer switched the line sources according to wind direction so that only one or two pipes on the upwind edge were in operation. The SO_2 concentration was monitored at the centre of a fumigated plot, and the ambient concentration was determined upwind. The target concentration for each treatment plot was a chosen amount above ambient, and the microcomputer controlled the rate of gas-release using a motorised needle valve to maintain the elevation desired. Fumigation was turned off when averaged wind velocity fell below 1 m s^{-1} and poor dispersion would have produced concentrations above target values. The system has been used in winter wheat (*Triticum aestivum* L. cv. 'Bounty') using a single exposure plot (Baker *et al.*, 1982) and in winter barley (*Hordeum vulgare* L. cv. 'Igri') using three plots at different concentrations (Baker *et al.*, 1986).

McLeod *et al.* (1985) also developed a crop fumigation system with computer control of gas-release. A simulation model of short-range dispersion from patterns of sources was used to determine a design that achieved a horizontally uniform exposure across a 9-m diameter central crop sampling area in a 27-m diameter plot (McLeod and Fackrell, 1983). A gas-source pattern was constructed from a network of polypropylene pipe resting on the soil surface, and dilute SO_2 was released from outlet assemblies at the top of vertical pipes within the crop. The sources formed a circle 0·5 m above the canopy around the perimeter and a grid of 3-m spacing throughout the plot at 1·5 m above the canopy. With a source-strength ratio of 10 to 1 between the low- and high-level sources, a horizontally uniform exposure was achieved across the central 9-m diameter area during neutral conditions of atmospheric stability. The control computer was used to prevent fumigation at low-wind velocities ($<1 \text{ m s}^{-1}$), to switch the four sectors of low-level gas sources so that only two sectors on the upwind side released gas, and to restrict increases in gas-release rate at high wind velocity ($>6 \text{ m s}^{-1}$) to prevent excessive use of gas. The system differed from the system described in the above paragraph (Greenwood *et al.*, 1982) in that a fixed sequence of hourly mean SO_2

concentrations measured at a rural site in central England over an entire growing season was used as the target values. This permitted fumigation at a site where ambient SO_2 was too low to form the basis of treatment concentrations, by addition or multiplication, but has the disadvantage that exposure was not linked to meteorological conditions.

Common features of the systems of Greenwood et al. (1982) and McLeod et al. (1985) were that (1) gas release rate depended on measured concentration in the fumigated plot, (2) changes in wind direction were used to alter the sections of sources in operation, (3) low wind velocities caused the system to turn off, and (4) a single SO_2 analyser was used to monitor each plot sequentially every 20 min. Between sampling occasions, the concentration varied with wind velocity fluctuations.

Mooi and van der Zalm (1986) developed an open-field system in the Netherlands that included controlled SO_2 release rate against measured concentration with additional control adjustments determined by wind velocity changes. The system consisted of a single 30-m diameter circle of 96 vertical emission pipes, each with 3 emission holes. The circle was divided into 16 segments controlled by valves connected to a computer, so that only 3 segments on the upwind side released gas. Air was pumped into the pipe network at $20\,l\,min^{-1}$, and SO_2 was added at a rate determined by mass flow valves controlled by the computer. From 15 March to 29 July 1985, potted bean plants (Vicia faba L. cv. 'Minica') were fumigated using a constant target concentration of 0·075 ppm SO_2. The computer adjusted the SO_2 supply rate every 16 min according to the difference between the observed concentration and the target value. However, the SO_2 supply rate was also modified every minute in response to changes in wind velocity and an adaptive amplification factor calculated from the trends in wind velocity and concentration. The system maintained the set-point, with deviations of about 15% and rare peaks outside this range, and it achieved an arithmetic mean of 0·05 ppm SO_2 (Fig. 8.1). It also produced a horizontal gradient of 20% across the experimental zone of plants. The successful development of a feedback control system responding to wind velocity changes presents an advantage for open-field systems in which a single pollutant monitor is used to monitor sequentially on several plots.

Open-field exposure techniques have also been developed to permit open-air exposures of trees. McLeod et al. (1987) described a system in which 50-m diameter circles of gas sources are used to expose 25-m diameter areas planted with 1-m high coniferous trees to controlled concentrations of SO_2 and O_3. The gas sources are connected to form four sectors of a circle, and electronic valves are used to release gas from two sectors on the upwind side. The rate of gas release is controlled in relation to both measured concentration and wind velocity. A system for fumigation of a single 60-m diameter area of a mature Scots Pine forest has been constructed near Umea, Sweden (personal communication from J.-E. Hallgren). Dilute SO_2 in air is released from two outlets near the top of 9-m-high vertical pipes, which are connected to form eight switchable sectors of a circle. Only one or two sectors release gas on the

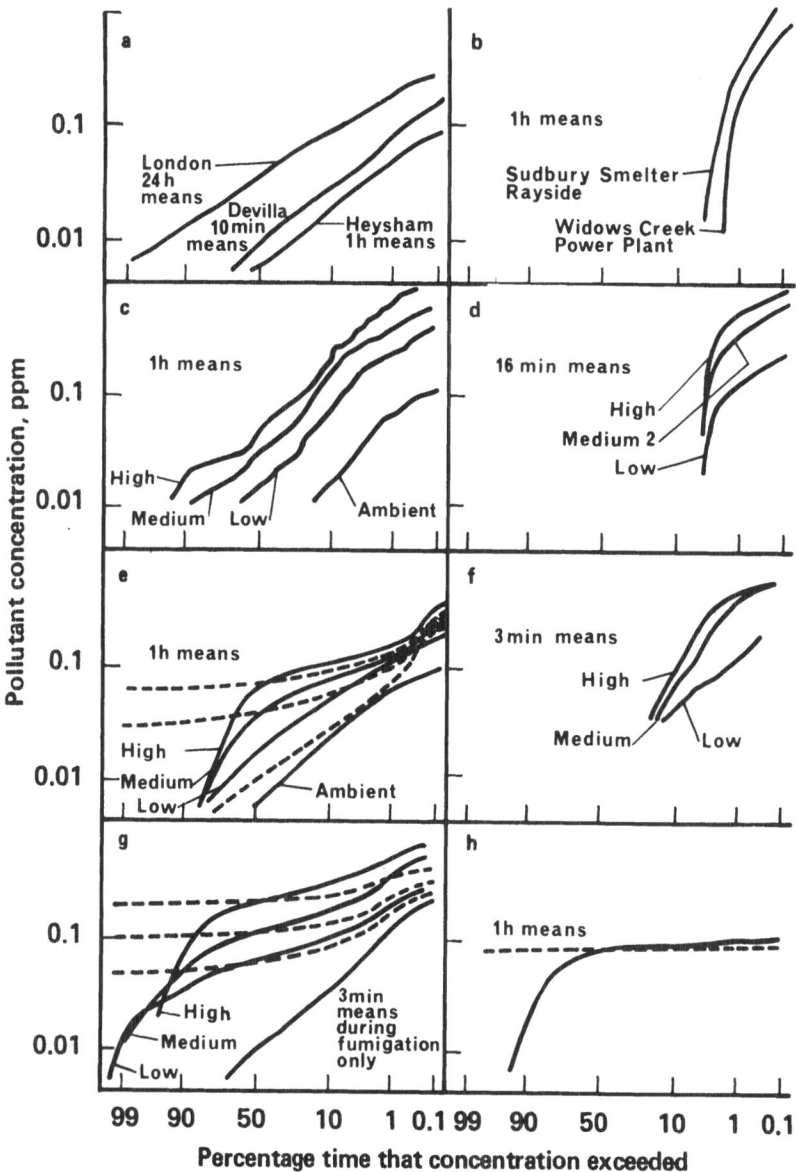

Fig. 8.1. Cumulative frequency distributions of SO_2 concentration at ambient sites and during experimental open-field exposures. ——, observed values; – – –, target values (see text for details). (a) Ambient sites—area sources; (b) ambient sites—point sources; (c) ZAPS—Corvallis; (d) ZAPS—Argonne; (e) open-air fumigation—Littlehampton; (f) LGC—Ithaca; (g) open-air fumigation—Sutton Bonington; (h) open-air fumigation—Wageningen.

upwind side, and the gas release rate is controlled against measured concentration. A system has also been constructed to be used for CO_2 enrichment of a 15-m diameter open-field plot (personal communication from K. Lewin). The system, which is termed a free air controlled enrichment (FACE) facility, releases gas from perforations in 32 vertical pipes surrounding the plot, and each pipe is controlled according to wind direction with a valve. Although developed for CO_2, it has been evaluated using SO_2 and would be suitable for use with O_3. A mass flow controller is used to regulate the delivery of gas, and a computer system has been developed to permit control of gas-release rate against measured concentration in the plot.

Other computer-controlled systems recently described and under construction include a portable pipework system for SO_2 and O_3 exposure of native plant communities, developed for the U.S. National Park Service by Northrop Services, Inc., and described by Hogsett et al. (1987b); a small ZAPS system for fumigation with O_3 (personal communication from V. Runeckles); and a system of perforated ducts for fumigation with O_3 (personal communication from A. Davison).

8.2 FEATURES OF OPEN-FIELD SYSTEMS THAT AFFECT CROP RESPONSE

8.2.1 Pollutant exposure patterns

The definition of pollutant exposure parameters on which to base dose–response studies has received much discussion (e.g., Heck and Heagle, 1985). Whatever the parameters chosen to examine plant response, an experimental exposure should simulate the concentration frequency distribution and the diurnal, seasonal, and autocorrelative characteristics of the pollutant at sites of interest. On the basis of concentration frequency distributions, pollutant exposures may be divided into single point-source simulations, which produce short episodes of high concentration in an area with a low background, and area or multiple-source simulations, which have a more uniform distribution of concentration values. The standard geometric deviation of pollutant concentrations from point sources may be higher than that from area sources (Knox and Lange, 1974). The concentration frequency distributions of multiple or area-source exposure to SO_2 (Fig. 8.1(a)) are generally log-normal (e.g., London (Warren Spring Laboratory, 1982), Heysham (Harrison and McCartney, 1980), and Devilla (Nicholson et al., 1980)). Systems in which SO_2 is released continuously at a constant rate (e.g., Lee and Lewis, 1977) have produced the closest approximation to such log-normal frequency distributions (Fig. 8.1(c)). However, the diurnal and seasonal patterns are dependent upon wind velocity and stability changes, and this ZAPS system produced higher concentrations at night, when lower wind velocities reduced dispersion, than during the day. Nocturnal peaks may be characteristic of certain pollutants and locations, but midday peaks are also common (Heck and Heagle, 1985). Point sources may also produce frequency distributions that are approximately log-normal (e.g., Sudbury smelter site (Dreisinger and McGovern, 1970), and some (e.g., Widows Creek Power plant (Noggle, 1980)) that are not log-normal (Fig. 8.1(b)). Open-field exposure systems that are restricted to intermittent gas release (e.g., according to wind direction constraints) or that produce exposures of high concentration but controlled duration (e.g., single-point gas release) are poorly suited for dose–response studies of area-source exposure but may provide a good simulation of a point-source exposure (Fig. 8.1(d) and (f)).

Systems with computer control of gas-release rate introduce the possibility of fumigating at fixed concentrations above measured ambient values (e.g., Colls *et al.*, 1987; Fig. 8.1(g)) or at fixed concentrations above a sequence of ambient monitoring data (McLeod *et al.*, 1985; Fig. 8.1(e)). With both systems, the frequency distributions reveal similar periods when the concentrations were higher or lower than the target values. The former case results from the release of pollutant at concentrations higher than the desired exposure value, poor dispersion, poor control-system response, or equipment failure. The latter case results from control system response (e.g., Fig. 8.1(g), which shows data during fumigation only) and periods without fumigation either through equipment failure, or because wind velocity fell below 1 m s^{-1} and the fumigation was suspended (e.g., Fig. 8.1(e)). Similar effects were also observed in the study of Mooi and van der Zalm (1986) who used a constant target concentration throughout their study (Fig. 8.1(h)).

Computer control permits greater adherence to targets of diurnal concentration patterns during fumigation, but the long-term pattern is influenced by periods of nonfumigation and diurnal patterns of dispersion conditions. Equipment failure may be random, but periods of low wind velocity and stability changes may have site-specific patterns that modify the long-term diurnal pattern of concentration. Overcoming these deficiencies should be a major objective of future designs.

Few polluted atmospheres contain a single phytotoxic gas, and Mansfield (1988) has described the importance of pollutant interactions with respect to crop yield. Open-field exposure systems have rarely used more than one pollutant (Tables 8.1 and 8.2), perhaps because of technical difficulties and cost. Shinn *et al.* (1977) used O_3 with H_2S or SO_2, whilst Irving and Miller (1984), using SO_2 with NO_2, found that the mixture was corrosive to aluminium pipes and each gas required separate distribution pipes. In Europe, the cost of cylinders of NO_2 is approximately 10 times that of SO_2, and large electrical O_3 generators are very expensive. However, the use of open-field systems for specific multiple pollutant exposures presents no serious technical problems, and a system has been developed for simultaneous fumigation with SO_2 and O_3 (McLeod *et al.*, 1987).

8.2.2 Spatial gradients of pollutant

Experimental release of pollutant gases at greater than ambient concentrations inevitably leads to differences in exposure with distance from a source. In the earliest open-air fumigation experiments (DeCormis *et al.*, 1975; Miller *et al.*, 1980*b*), although field plots were considered to have received a uniform exposure, gradients in concentration were often identified. The system of DeCormis *et al.* (1975) produced an SO_2 concentration on the downwind edge of the treated area that was 2·4 times that on the upwind edge at a wind velocity of $1·5 \text{ m s}^{-1}$. Miller *et al.* (1980*b*) observed differences in concentration of 18% for sample locations within the same treatment plot of a ZAPS

TABLE 8.2

Studies of pollutant effects on crop growth and yield using experimental open-field exposures

Crop/cultivar	Site	Pollutant and exposure method	Comments	References
Medicago sativa L. Northrup King 512	Riverside, CA	Air exclusion of ambient oxidants	Comparison with NCLAN open-top chambers and closed chambers. Photosynthesis and stomatal conductance measured	Thompson and Olszyk (1985) Olszyk *et al.* (1986c)
Triticum aestivum L. Yecora Rojo	Riverside, CA	Air exclusion of ambient oxidants with four levels of added SO_2	Comparison with NCLAN open-top chambers	Olszyk *et al.* (1986b)
Lactuca sativa L. Empire	Riverside, CA	Air exclusion of ambient oxidants with four levels of added SO_2	Comparison with NCLAN open-top chambers	Olszyk *et al.* (1986b)
Hordeum vulgare L. Sonja	Littlehampton, UK	Unreplicated fumigation plots at 3 SO_2 levels and ambient plots	Disease effects on growth reported	McLeod *et al.* (1985, 1988)
Hordeum vulgare L. Igri	Sutton Bonington, UK	Unreplicated fumigation plots at 3 SO_2 levels and ambient plots	Growth and yield effects assessed	Baker *et al.* (1986) Baker *et al.* (1987)
Triticum aestivum L. Bounty	Sutton Bonington, UK	Single SO_2 fumigation plot and ambient plot	Report of winter cold stress effects	Baker *et al.* (1982)
Triticum aestivum L. Rapier	Littlehampton, UK	Unreplicated fumigation plots at 3 SO_2 levels and ambient plots	Disease, pest and winter cold stress effects reported	McLeod and Roberts (1987)

Species / cultivar	Location	Method	Assessment	Reference
Phaseolus vulgaris L. California Light Red Kidney	Ithaca, NY	Replicate linear gradient chambers to give a range of SO_2 exposure	Study of SO_2 effect on disease development	Reynolds *et al.* (1987)
Vicia faba L. Minica	Wageningen, Netherlands	Single SO_2 fumigation plot and ambient plot	Yield assessed	Mooi and van der Zalm (1986)
Hordeum vulgare L. Igri	Sutton Bonington, UK	Unreplicated fumigation plots at 3 SO_2 levels and ambient plots	Report of effects of agrochemical spray	Baker and Fullwood (1986)
Phaseolus vulgaris L. GV50	Davis, CA	Air exclusion linear gradient chamber with fumigation of H_2S and H_2S plus O_3	Pod yield assessed	Bennett *et al.* (1980)
Glycine max (L.) Merr. Northrup King 1492	Argonne, IL	Unreplicated fumigation plots using SO_2, NO_2, and $SO_2 + NO_2$	Seed yield and leaf chlorophyll assessed	Irving and Miller (1984)
Zea mays L. Pioneer 3780 PAG 397	Argonne, IL	SO_2	Grain yield assessed	Miller *et al.* (1981*b*)
			14 hybrids compared	Miller *et al.* (1981*a*)
Lolium perenne L. S23	Vancouver, Canada	O_3	Growth and root/shoot ratio assessed	Runeckles *et al.* (1981)
Glycine max (L.) Merr. Wells	Argonne, IL	Unreplicated fumigation plots at 3 SO_2 levels and ambient plot	Seed yield and quality assessed	Sprugel *et al.* (1980)
Glycine max (L.) Merr. Corsoy Williams	Argonne, IL	Unreplicated fumigation plots at 3 and 5 SO_2 levels and ambient plots	Photosynthesis measured	Muller *et al.* (1979)
		Single plot fumigated from a single SO_2 line source with upwind control plots	Seed yield assessed	Miller *et al.* (1980*a*)
Vitis viniflora L. Petit Manseng	France	Single plot fumigated with HF and ambient control plot		Cantuel (1980)
Glycine max (L.) Merr. Hark	Ithaca, NY	Linear gradient chamber with fumigation of SO_2 with O_3	Seed yield assessed	Reich *et al.* (1982)

TABLE 8.3

The grain yield of winter barley (*Hordeum vulgare* cv. 'Sonja') subject to open-air fumigation with SO_2 at Littlehampton, UK, in 1982–1983

SO_2 concentration[a] (ppm)	Grain yield[b] $(t\,ha^{-1})$		
	Fumigation plot	Companion control plots	
0·010 (ambient)	4·6(0·3)	4·4(0·5)	4·3(0·3)
0·023	6·0(0·5)	5·1(0·3)	4·4(0·4)
0·038	5·1(0·1)	4·2(0·3)	5·1(0·3)
0·058	4·2(0·3)	5·0(0·4)	5·7(0·4)

[a] Arithmetic mean throughout the period of fumigation from 1 November 1982 to 5 July 1983.
[b] Values are arithmetic means of 5 replicate samples per plot. Each fumigated plot had 2 companion control plots. Standard errors are shown in parentheses.

system. These effects both result from the cumulative drift of pollutant from multiple pipe sources across the plot. Lee and Lewis (1977) found areas of high concentration 'hot spots' near the pipe sources of their ZAPS system, which were excluded from vegetation sampling. However, a detailed analysis of concentration distribution using sulphation plates revealed a good uniformity in horizontal exposure averaged over 3 months (Preston and Lee, 1984). The large plot size (80 × 74 m), with 10 m between adjacent gas sources, may have reduced the effect of cumulative drift of pollutant from several pipe outlets compared to smaller ZAPS systems.

The problem of achieving horizontal uniformity of exposure has been avoided by using a LGC or by adapting the ZAPS design to produce a gradient in concentration downwind of one or two horizontal pipes (Irving and Miller, 1984; Thompson and Olszyk, 1985). The range of exposure concentrations was included in the analysis of plant response. McLeod and Fackrell (1983) investigated the design of open-air fumigation systems using a computer simulation model of gas dispersion and produced a gas-source pattern that achieved an improved uniformity of horizontal exposure (McLeod *et al.*, 1985). Many open-air studies, however, have accepted a known variability in exposure concentration across a treatment plot (e.g., 20%, Mooi and van der Zalm, 1986). Assessment of exposure differences across sampled areas is necessary in an open-air study, and sulphation plates (Preston and Lee, 1984) or diffusion tubes (Colls, 1986) provide a suitable mean value of exposure. However, mean values over long periods with variable wind direction may hide the short-term gradients present with a constant wind direction. The frequency distribution of exposure at a treatment plot perimeter may therefore be different from that in the centre, and the horizontal variation of exposure should also be examined by sampling across the plot.

The production of vertical gradients of pollutant in the canopy by fumigation systems should also be considered. Inside a dense crop, the concentration of

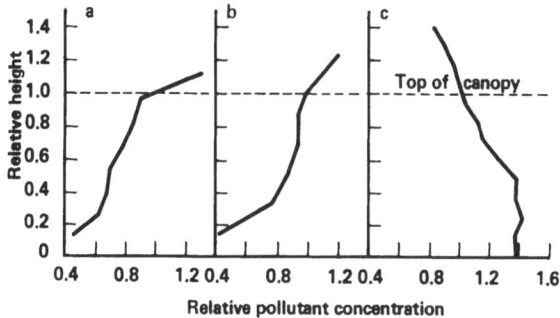

Fig. 8.2. Vertical gradients of pollutants within a crop canopy (a) NO_2 at an ambient site and (b) and (c) SO_2 during experimental open-field exposures.

ambient pollutants generally decreases towards the ground because of deposition to the canopy and the soil (Colls, 1986; Fig. 8.2(a)). Open-air exposure systems may produce different profiles; with an LGC, the profile was reversed, the highest concentration being near the canopy base where the pollutant was introduced (Shinn *et al.*, 1977; Fig. 8.2(c)). Open-field fumigation systems may have too little fetch to establish an equilibrium of pollutant exchange between the atmosphere and the canopy, but vertical profiles that show a concentration decreasing with depth in the canopy can be achieved (McLeod *et al.* 1985; Fig. 8.2(b)). Such gradients may be destroyed by the ventilation fans used in some modern chambers to produce high air flow and reduce aerodynamic boundary layer resistance to pollutant flux. Vertical gradients in the field may reflect variation in turbulent mixing with height and may be indicative of different levels of pollutant deposition to different plant organs. This may influence plant response and the occurrence of insects and fungal pathogens, which have a localised distribution in the canopy (Preston and Lee, 1984), and which may respond to pollutant exposure to indirectly influence crop growth (McLeod and Roberts, 1987).

8.2.3 Meteorological conditions

A major advantage of many designs of open-field exposure systems is minimal interference with weather and microclimate, particularly snow, frost, dew, and rain. LGC and AES experiments produce changes in wind velocity and dew formation resulting from the enhanced air flow in the canopy (Olszyk *et al.*, 1986*b*). Computer-controlled fumigation using fixed target concentrations (e.g., McLeod *et al.*, 1985; Mooi and van der Zalm, 1986) produces exposures that are unrelated to current weather conditions, but this is avoided by generating targets from the measured ambient concentration (e.g., Greenwood *et al.*, 1982). Some fumigation systems may be restricted to certain wind directions, and others have limited gas release to wind velocities greater than $1 \, m \, s^{-1}$. Except for periods of low wind velocity, it has proved possible to operate some exposure systems throughout the winter and in all weather conditions in Great Britain.

8.2.4 Pollutant uptake

When crops are exposed to air pollutants, the severity of phytotoxic effect depends on how much of the available dose is taken up by the plants. Pollutant uptake by a canopy has been represented by resistance analogue diagrams (Unsworth, 1981; Fowler, 1985). The pollutant flux (F) under ambient conditions is limited by an atmospheric resistance (r_a) plus a canopy resistance (r_c), which under dry conditions is the resultant of three parallel resistances representing pathways to the soil, cuticle, and stomata, respectively. The flux F may be written as

$$F = C/r_{tot} = C/(r_a + r_c)$$

where r_{tot} is the sum of the resistances r_a and r_c, and C is the pollutant concentration in the atmosphere. When dew was present on a cereal canopy, Fowler (1985) reported that r_c for SO_2 became negligible because the resistance of deposition to the dew tended to zero, and under such conditions the SO_2 flux was limited only by C and atmospheric mixing. Experimental monitoring locations are usually near the crop, within the logarithmic concentration gradient that exists for several metres above an extensive crop canopy. With a dew-wetted canopy, flux will increase, and measured concentration will decrease to an extent that depends upon the proximity of the monitoring point to the crop.

In an open-field fumigation experiment, the exposure concentration is achieved by the release of pollutant. When the canopy is wet, the SO_2 flux will increase, and concentration at the plot centre will decrease because the pollutant plumes are more rapidly depleted as they pass over the vegetation. DeCormis et al. (1975) observed such a decrease in measured concentration when operating an open-air fumigation system with a constant rate of SO_2 release during rainfall. Horizontal gradients in concentration across the plot will increase, but with a constant rate of gas release the pollutant flux is limited by the fixed source strengths. However, a computer-controlled system will attempt to maintain its target concentration by increasing the amount of pollutant released from the pipework, and the flux near the sources may be higher than expected relative to the concentration at a central monitoring location. If plume depletion reduces monitored concentration to zero, the control system may increase pollutant release subject only to software restrictions. McLeod et al. (1985) prevented such uncontrolled increases in pollutant release by restricting increases in gas release to 10% every 20 min, but other studies have not reported such restraints. An increased flux may be of greater significance when the target concentration is a constant value or part of a sequence that is unrelated to ambient conditions of rain or dew. Fowler (1985) reported an increase of canopy resistance from zero after 2 to 3 h of dew formation, which was interpreted to result from a decrease in dew pH until liquid-phase resistance limited uptake. He also pointed out that SO_2 uptake by dew on leaves may be different from that on rain-wetted leaves because the former is initially pure water, but the latter may be in equilibrium with ambient pollutant concentrations. However, rain is unlikely to be in equilibrium with the pollutant concentration in a fumigation experiment, and increased pollutant deposition may be important when rainfall persists for several days. This deposition may influence crop growth via foliar and soil effects and may alter the germination of fungal pathogens in moisture films. Elkiey and Ormrod (1981) suggested that rainfall outdoors may be a major factor affecting plant injury response to gaseous pollutants. Enhanced deposition during rainfall is clearly a shortcoming of computer-controlled fumigation systems.

8.2.5 Concentration fluctuation

The fluctuations in pollutant concentration within open-top chambers and air-exclusion systems may result from ambient air ingress into the chamber or canopy (Heagle *et al.*, 1988). However, open-field fumigation produces short-term ($<10\,s$) fluctuations that are caused by turbulent mixing processes dispersing the pollutant plumes from pipework sources. A source of small size (e.g., 3-mm hole) compared with the local turbulence scales over vegetation produces an instantaneous plume that meanders within the boundaries of the mean plume. The instantaneous plume generates most fluctuations near the source but later increases in size until it may become equal in dimension to the mean plume at a downwind distance approximately 100 times source height.

Fackrell and Robins (1982) have investigated the effect of source size on concentration fluctuation and found that the highest peak-to-mean ratio was produced from the smallest source orifice. These fluctuations may be important with respect to plant responses, but they are often undetectable using standard air pollutant monitors that have a response time of about 1 min. Methods to reduce these fluctuations could include static mixing vanes, passive rotating vanes driven by the wind, or active motor-driven vanes located at each source to disperse and dilute the instantaneous plume. Source buoyancy and momentum could also influence the concentration field; Preston and Lee (1984) checked this aspect of their ZAPS system and determined that their plume dispersion was influenced only by wind and turbulence. Investigation of the concentration field from a pair of sources by Sawford *et al.* (1985) has revealed that high concentrations relative to the mean occur less frequently than for isolated sources because the stochastic nature of the turbulence processes effecting dispersion results in only partial correlation between plume occurrence at a fixed point. There may therefore be an advantage in the use of many gas sources (e.g., ZAPS system) rather than a few to produce a steady concentration.

8.3 RESPONSES OF CROP GROWTH AND YIELD USING EXPERIMENTAL OPEN-FIELD EXPOSURES

8.3.1 Experimental design and objectives

For the determination of yield effects that result from fertiliser or agrochemical trials, replicate treatment plots are used, often laid out to special designs (e.g., Latin squares) to overcome the confounding variability of growth within a single field (Dyke, 1974). Such designs are difficult to use in open-air fumigation studies not only because the released gas may drift, necessitating plot separation, but also because large plot areas have been required to achieve a realistic pollutant exposure. Open-field fumigations also use expensive equipment. Consequently, few open-air exposure trials have included replication of treatment plots, and this has resulted in considerable debate about the validity of results from such studies. Replications using current designs would occupy a great deal of space and be very expensive. Primarily due to the expense, small open-air exposure systems are required for the determination of crop-yield response. The air-exclusion system of Olszyk et al. (1986a) and the FACE facility (personal communication from K. Lewin) approach this requirement. For annual crops, repetition of experiments over time may corroborate the functional form of any yield response generated from an unreplicated experiment. Baker et al. (1986) performed an unreplicated SO_2 exposure of barley during two seasons and found close agreement between the trends in yield response to pollutant dose in each year. However, certain ecosystem responses such as parasite–predator–pest interactions may require larger areas of exposed vegetation, and, whether replicated or not, they can improve our understanding of such processes. The results of many open-air exposures are unreplicated (Table 8.1) and should be examined with attention to the statistical limitations of the design.

8.3.2 Crop responses

A comparison of the crop responses from open-field exposures includes a variety of methods and suffers from a paucity of data for a range of crops for which the same technique is used (Table 8.2). Unlike the NCLAN study, which achieved considerable uniformity of treatments and equipment, the dose–responses from open-field exposures incorporate the variable effects of all the factors described in the previous section. Using a single ZAPS, it was possible to observe the different response patterns of ryegrass and wheat grass to SO_2 (Runeckles et al., 1981), and AES systems were used by Olszyk et al. (1986b) to compare the dose–response of lettuce and wheat to SO_2 and O_3, but comparisons between studies are questionable. Site factors including climatic features and variation in ambient pollutants between widely separated locations could influence crop response. Heggestad and Bennett (1984) suggested that the yield reduction at low SO_2 concentrations in a ZAPS exposure of soybean (Sprugel et al., 1980) could have resulted from interaction with ambient O_3. McLaughlin and Taylor (1985) collected data from studies exposing soybean to SO_2 and found a reasonable comparison between the response of two cultivars 'Wells' and 'Hark', exposed using a ZAPS (Sprugel et al., 1980) and an LGC (Reich et al., 1982) at sites where the ambient O_3 was similar.

Olszyk et al. (1986c) compared the response of alfalfa to O_3 in an AES and in OTCs and found that under the cooler conditions of fall and winter plant growth in an unfiltered AES was more representative of the field environment than growth in the conditions of an unfiltered OTC. Indeed, a strength of open-air pollution experiments for crop studies has been the information provided about responses under ambient field conditions. Cold, pests, diseases, and agrochemicals may all modify the yield response to pollutants. Effects of some pollutants (e.g., SO_2) are known to be most severe in winter because low light and growth may limit plant capacity for detoxification (e.g., Davies, 1980) and because the plant resistance to freezing is reduced (Davison and Bailey, 1982). Using an open-air system, Baker *et al.* (1982) reported the first field observations of winter damage attributable to SO_2 exposure following a period when the mean minimum air temperature was $-7.8°C$. Leaf injury was greater on plants in the treated plot (0·1 ppm) than in the remainder of the crop (approximately 0·015 ppm). It was concluded that damage was caused by ambient cold stress after SO_2 exposure had rendered the plants cold sensitive.

In addition to cold stress, the interaction of fungal pathogens and insect pests with air pollutants is widely known from laboratory studies (Laurence *et al.*, 1983). The importance of fungal pathogens under field conditions was assessed by McLeod (1988), who measured their natural occurrence in open-air fumigation plots and concluded that effects of SO_2 on fungi had influenced yield response of the crop (McLeod *et al.*, 1988). Effects of SO_2 on insect pests were also examined and, together with fungi, had considerable importance for crop yield, economic cost of crop production, and other ecosystem interactions under field conditions (McLeod and Roberts, 1987).

It is well established that certain agrochemicals may afford protection against oxidant damage (see Section 8.1.3). However, an open-field exposure showed an increase in leaf damage of SO_2-fumigated barley after agrochemical application (Baker and Fullwood, 1986). A mixture of two contact herbicides, a fungicide, and a growth regulator was applied to barley and, after two days the severity of leaf scorch was proportional to SO_2 treatments of 0·04, 0·08, and 0·12 ppm above an ambient of 0·016 ppm. A spell of cold nights with a minimum temperature of $-3.2°C$ may have contributed to this effect.

In Great Britain, two different computer-controlled fumigation systems have been used to expose winter barley to SO_2 for 7 months (Baker *et al.*, 1986; McLeod *et al.*, 1988). The yield responses to mean SO_2 concentration throughout the exposure periods are shown in Fig. 8.3 and Tables 8.3 and 8.4. Baker *et al.* (1986) found a decrease in yield of *H. vulgare* cv. 'Igri' with increasing SO_2 above the 0·015-ppm ambient level whilst McLeod *et al.* (1988) found an increase in yield of *H. vulgare* cv. 'Sonja' at 0·02 ppm with a decreasing trend in yield at higher concentrations. The different response shape is of great importance because 95% of rural areas in Western Europe and North America are exposed to annual mean SO_2 concentrations <0·019 ppm (Fowler and Cape, 1982). The difference in these responses may result from many of the features discussed in the previous section, particularly the following: cultivar; exposure system design; exposure concentration regime; crop husbandry; and site factors including ambient pollution, meteorology, and soil.

In addition to the differences in SO_2 exposures (Fig. 8.1(e) and (f)), the

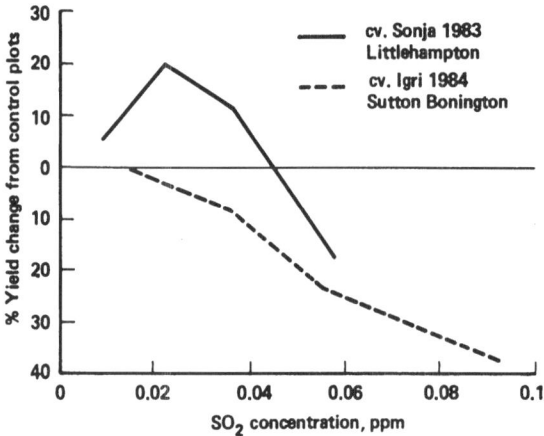

Fig. 8.3. Yield response · of winter barley to SO_2 determined using two different open-field exposure systems.

number of degree days below zero during growth of the Sutton Bonington crop was 14·8 with a daily minimum temperature of −6·7°C in November. At Littlehampton, there were no degree days below zero and a daily minimum temperature of −4·8°C in February, a difference which may have influenced any effect of SO_2 on growth responses to winter cold stress (Baker *et al.*, 1982; McLeod and Roberts, 1987). The crop at Sutton Bonington received applications of fungicide, herbicide, and growth regulator as determined by local farming practice, whilst at Littlehampton herbicides and growth regulators were not required and fungicides were used only when disease thresholds were exceeded. Both crops were managed according to accepted farming methods, but the possibility for effects of SO_2 on insect pests, fungal pathogens, and saprophytic microorganisms (McLeod and Roberts, 1987; McLeod, 1988) was

TABLE 8.4

The grain yield of winter barley (*Hordeum vulgare* cv. 'Igri') subject to open-air fumigation with SO_2 at Sutton Bonington, UK in 1983–1984

SO_2 concentration[a] (ppm)	Grain yield[b] $(t\,ha^{-1})$
0·015 (ambient)	Control 1 8·3(0·4)
	Control 2 7·6(0·3)
0·036	7·2(0·3)
0·055	6·0(0·2)
0·093	4·9(0·4)

[a] Arithmetic mean throughout the period of fumigation from 24 January to 8 July 1984.
[b] Values are arithmetic means of 6 replicate samples per plot. Standard errors are shown in parentheses.

greater in the Littlehampton experiment. In contrast, SO_2 interacted with agrochemicals to cause damage at Sutton Bonington (Baker and Fullwood, 1986). Further research is needed to assess the importance of different crop management systems on crop response to pollutants.

8.4 CONCLUSIONS AND RESEARCH NEEDS

Open-field exposure systems have evolved rapidly over the past decade and now offer an alternative approach to investigate effects of air pollutants on crops with minimal alteration of environmental conditions. The choice of an open-field exposure design is currently dependent upon four principal factors: (1) pollutant exposure regime, (2) experimental objective, (3) crop species, and (4) financial resources. The equipment for large open-field exposure systems is expensive, and available funds may be a primary factor in the choice. Crop species will also influence the design, number, and location of pollutant sources, but the pollutant exposure regime and the experimental objective are most important. The simulation of a point-source exposure lends itself to more simple designs utilising a single pollutant source with intermittent fumigation. However, realistic simulation of an area-source exposure or secondary pollutant, such as O_3, may require more equipment and a sophisticated control system. To attain the most suitable open-field system in the future, the following research needs are suggested.

8.4.1 Exposure system hardware

Improvements in the dispersion of gas from pipework sources to achieve a realistic crop exposure to pollutants are still possible. Vertical concentration gradients inside crop canopies should be representative of ambient measurements, and horizontal gradients across treatment plots minimised or the concentration gradient incorporated into the assessment of vegetation response. The importance to crop response of concentration fluctuations caused by meandering plumes requires investigation, and, if appropriate, new features such as static or active mixing vanes should be introduced at gas outlets to break up and dilute the plumes. Continuation of fumigation at wind velocities $<1\,\mathrm{m\,s^{-1}}$ should be possible by using greater precision of flow control and by dispensing pollutant mixtures close to the target concentration. The importance of deposition to surface moisture on the crop must be evaluated, and, if necessary, surface moisture sensors must be critically compared to establish specifications for their use in limiting feedback control. Some of these matters may require assessment under controlled conditions in wind tunnels before undertaking trials under the variable conditions of the field.

8.4.2 Computer control software

Computer control systems offer great potential for improving experimental simulation of ambient exposure to air pollutants. Systems engineers could optimise control parameters in relation to lag times and equipment performance. The most suitable algorithms to control gas release should be determined by theoretical analysis and trials. Deficiencies in attaining the desired diurnal and seasonal patterns may be corrected by using appropriate gain factors.

Existing open-air studies have used exposure concentrations taken from a fixed sequence of ambient data or by using a fixed elevation above the measured ambient value. The former results in exposure that may be unrelated to the prevailing weather conditions, and the latter may produce no episodes of high concentration during the entire study. Computer control systems may permit the generation of target values in relation to statistical descriptions of pollutant concentration and real-time measurement of weather conditions (e.g., wind velocity, solar radiation), using theoretical relationships or probability functions based on ambient measurements. All these possibilities may produce improvements in the realism of pollutant exposures.

8.4.3 Definitive designs

There may be advantages in producing definitive designs for open-field exposure systems to reduce any variability in crop response resulting from differences in equipment. A small exposure system, $<200\,m^2$ in plot size, would permit replication of treatments within a field for the accurate assessment of dose–response functions in the same way that OTC systems have been used in the NCLAN study. However, studies of ecological processes (e.g., Preston and Lee, 1984) would benefit from a larger plot size ($>500\,m^2$) so that spatial limitations on plant, invertebrate, and microbial populations are reduced. Two requirements for open-field exposure systems could therefore be established to fulfill slightly different roles. The use of standard designs would then enhance the value of studies of deposition, exposure dynamics, and control systems that would be required.

8.4.4 Crop response studies

The importance of insect pests and fungal pathogens in modification of crop responses to pollutants is well-established (McLeod and Roberts, 1987; Manning, 1988). Open-air exposure systems have an advantage in permitting assessment of natural pest and pathogen occurrence (McLeod, 1988) and may be used for artificial introduction of disease at different pollutant levels (Reynolds et al., 1987). Major crops that are regularly affected by pests and pathogens in the field should be identified and receive a high priority for evaluation of response to pollutants using open-field exposure methods. Parallel investigations using conventional OTCs would permit important comparisons with the existing dose–response data of the NCLAN study.

Open-field exposure systems are an important method in the variety of technical approaches for assessing gaseous pollutant effects on crop yield. The next generation of open-field systems is unlikely to overcome all the technical problems described in this paper, but our experience during the next few years with the systems now in operation or under construction will provide us with much of the information to meet this objective. Open-field systems should be used to complement the established methods using closed and open-top chambers, to provide additional dose–response data with minimal alteration of environmental conditions. The correct understanding and integration of deposition, micrometeorology, and ecology may then lead to a more complete knowledge of vegetation response to air pollutants in the environment.

ACKNOWLEDGEMENTS

The work of A. R. McLeod was undertaken at the Central Electricity Research Laboratories and is published with the permission of the Central Electricity Generating Board. The work of C. K. Baker was funded by the Department of the Environment, the Ministry of Agriculture, Fisheries and Food, and the Natural Environment Research Council. The authors thank J. J. Colls, B. E. A. Fisher, T. M. Roberts, A. G. Robins and M. H. Unsworth for helpful discussions.

REFERENCES

Austin, R. B., J. Bingham, R. D. Blackwell, L. T. Evans, M. A. Ford, C. L. Morgan, and M. Taylor. (1980). Genetic improvements in wheat yields since 1900 and associated physiological changes. *J. Agric. Sci., Camb.*, **94**, 675–89.

Baker, C. K. and A. E. Fullwood. (1986). Leaf damage following crop spraying in winter barley exposed to sulphur dioxide. *Crop Prot.*, **5**, 365–7.

Baker, C. K., M. H. Unsworth, and P. Greenwood. (1982). Leaf injury on wheat plants exposed in the field in winter to SO_2. *Nature*, **299**, 149–51.

Baker, C. K., J. J. Colls, A. E. Fullwood, and G. G. R. Seaton. (1986). Depression of growth and yield in winter barley exposed to sulphur dioxide in the field. *New Phytologist*, **104**, 233–41.

Baker, C. K., A. E. Fullwood and J. J. Colls. (1987). Tillering and leaf area of winter barley exposed to sulphur dioxide in the field. *New Phytologist*, **107**, 373–85.

Beetham, P. A. (1980). *The effect of sulphur dioxide in vivo on epiphytic lichens.* PhD Thesis, University College of Wales, Aberystwyth.

Bennett, J. P., K. Barnes, and J. H. Shinn. (1980). Interactive effects of H_2S and O_3 on the yield of snap beans. *Environ. Exptl Bot.*, **20**, 107–14.

Bisessar, S. (1982). Effect of ozone, antioxidant protection, and early blight on potato in the field. *J. Am. Soc. Hortic. Sci.*, **107**, 597–9.

Black, V. J. (1982). Effects of sulphur dioxide on physiological processes in plants. In *Effects of gaseous air pollution in agriculture and horiculture*, ed. by M. H. Unsworth and D. P. Ormrod, 67–91. London, Butterworths Scientific.

Bleasdale, J. K. A. (1952). Atmospheric pollution and plant growth. *Nature*, **169**, 376–7.

Bonte, J., L. Decormis and A. Tisne. (1981). Étude des effets à long terme d'une pollution chronique par SO_2 à concentration subnécrotique (50 μg m^{-3}). In *Rapport Final Contrat* No. 79-03 PA 77-156. Montardon, Morlaas, France, Institut National de la Recherche Agronomique.

Cantuel, J. (1980). Influence de la pollution par les composes fluorés sur la Vigne. In *Rapport D'Activite 1980*, ed. by J. Bonte, 24–29, Montardon, Morlaas, France Préfecture des Pyrénées—Atlantique, Section Agriculture.

Colls, J. (1986). Measurement of nitrogen dioxide profiles by diffusion tubes within a barley canopy. *Atmos. Environ.*, **20**, 239–42.

Colls, J. J. and C. K. Baker. (1988). The methodology of open-field fumigation. In *Air Pollution and Ecosystems*, ed. by P. Mathy. Proceedings of an international symposium, Grenoble, France, 361–71. D. Reidel Pub. Co., Dordrecht, Holland.

Colls, J. J., C. K. Baker, and G. G. R. Seaton. (1987). Computer control of SO_2 fumigation in a study of crop response to pollution. In *Computer applications in agricultural environments*, ed. by J. A. Clark, K. Gregson and R. A. Saffell, 143–58. London, Butterworths Scientific.

Colvill, K. E., D. C. Horsman, M. L. Roose, T. M. Roberts, and A. D. Bradshaw. (1985). Field trials on the influence of air pollutants, and sulphur dioxide in particular, on the growth of ryegrass *Lolium perenne* L., *Envir. Pollut.*, **39**, 235–66.

Cowling, D. W. and L. H. P. Jones. (1978). Sulphur and amino acid fractions in perennial ryegrass as influenced by soil and atmospheric supplies of sulphur. In *Sulphur in forages*, ed. by J. C. Brogan, 15–31. Dublin, Anas Foras Taluntais.

Crowther, C. and A. G. Ruston. (1914). Town smoke and plant growth. *J. Agric. Sci., Camb.*, **6**, 387–94.

Davies, T. (1980). Grasses more sensitive to SO_2 pollution in conditions of low irradiance and short days. *Nature*, **284**, 483–5.

Davison, A. W. and I. F. Bailey. (1982). SO_2 pollution reduces the freezing resistance of ryegrass. *Nature*, **297**, 400–2.

DeCormis, L., J. Bonte, and A. Tisne. (1975). Experimental technique for determining the effect on vegetation of sulfur dioxide pollutants applied continuously in subnecrotic doses. *Pollut. Atmos.*, **17**, 103–7.

Dowding, P. (1987). Leaf yeasts as indicators of air pollution. In *Microbiology of the phyllosphere*, ed. by N. J. Fokkema and J. Van den Heuvel, 121–36. Cambridge, Cambridge University Press.

Dreisinger, R. B. and P. C. McGovern. (1970). Monitoring atmospheric sulfur dioxide and correlating its effects on crops in the Sudbury area. In *Impact of air pollution on vegetation conference*, ed. by S. N. Linzon. Toronto, Ontario Department of Energy and Resource Management.

Dyke, G. V. (1974). *Comparative experiments with field crops*, London, Butterworths Scientific.

Elkiey, T. and D. P. Ormrod. (1981). Sulphur and nitrogen nutrition and misting effects on the response of bluegrass to ozone, sulphur dioxide, nitrogen dioxide or their mixture. *Water Air Soil Pollut.*, **16**, 177–86.

Fackrell, J. E. and A. G. Robins. (1982). The effects of source size on concentration fluctuations in plumes. *Boundary-Layer Meteorol.*, **22**, 335–50.

Floor, H. and A. C. Posthumus. (1977). Biologische Erfassing von ozon- und PAN-immissionen in den Niederlanden 1973, 1974 und 1975. *VDI-Ber.* **270**, 183–90.

Fowler, D. (1985). Deposition of SO_2 on to Plant Canopies. In *Sulfur dioxide and vegetation*, ed. by W. E. Winner, H. A. Mooney and R. A. Goldstein, 389–402. Stanford, CA, Stanford University Press.

Fowler, D. and J. N. Cape. (1982). Air pollutants in agriculture and horticulture. In *Effects of gaseous air pollution in agriculture and horticulture*, ed. by M. H. Unsworth and D. P. Ormrod, 3-26. London, Butterworths Scientific.

Furukawa, A. (1984). Defining pollution problems in the Far East—a case study of Japanese air pollution problems. In *Gaseous air pollutants and plant metabolism*, ed. by M. J. Koziol and F. R. Whatley, 59–74. London, Butterworths Scientific.

Greenwood, P., A. Greenhalgh, C. K. Baker, and M. H. Unsworth. (1982). A computer-controlled system for exposing field crops to gaseous air pollutants. *Atmos. Environ.*, **16**, 2261–6.

Guderian, R. and H. Stratmann. (1962a). Freilandversuche zur Ermittlung von

Schwefeldioxidwirkungen auf die Vegetation. I. Teil: Übersicht zur Versuchsmethodik und Versuchsauswertung. *Forsch. Ber. d. Landes Nordrhein-Westfalen*, Köln und Opladen, Westdeutscher Verlag, Nr. 1118.

Guderian, R. and H. Stratmann. (1962*b*). Freilandversuche zur Ermittlung von Schwefeldioxidwirkungen auf die Vegetation. III. Teil: Grenzwerte schädlicher SO_2—Immissionen für Obst—und Forstkulturen sowie landwirtschaftliche und gärtnerische Pflanzenarten, *Forsch. Ber. d. Landes Nordrhein-Westfalen*, Köln und Opladen, Westdeutscher Verlag, Nr. 1920.

Harrison, R. M. and H. A. McCartney. (1980). Ambient air quality at a coastal site in rural North West England. *Atmos. Environ.*, **14**, 233–44.

Heagle, A. S., L. W. Kress, P. J. Temple, R. J. Kohut, J. E. Miller and H. E. Heggestad. (1988). Factors influencing ozone dose-yield response relationships in open-top field chambers. In *Assessment of crop loss from air pollutants*. Proceedings of the international conference, Raleigh, North Carolina, USA, ed. by W. W. Heck, O. C. Taylor and D. T. Tingey, 141–179, London, Elsevier Applied Science.

Heck, W. W. and A. S. Heagle. (1985). SO_2 effects on agricultural systems: a regional outlook. In *Sulfur dioxide and vegetation*, ed. by W. E. Winner, H. A. Mooney and R. A. Goldstein, 418–430. Stanford, CA, Stanford University Press.

Heggestad, H. E. and J. H. Bennett. (1984). Impact of atmospheric pollution on agriculture. In *Air pollution and plant life*, ed. by M. Treshow, 357–96. Chichester, UK, John Wiley.

Hogsett, W. E., D. Olszyk, D. P. Ormrod, G. E. Taylor, and D. T. Tingey. (1987*a*). *Air pollution exposure systems and experimental protocols: Volume 1: A review and evaluation of performance*. Corvallis, OR, U.S. Environmental Protection Agency, 600/3-87/037a.

Hogsett, W. E., D. Olszyk, D. P. Ormrod, G. E. Taylor, and D. T. Tingey. (1987*b*). *Air pollution exposure systems and experimental protocols: Volume 2: Description of facilities*. Corvallis, OR, U.S. Environmental Protection Agency, 600/3-87/37b.

Irving, P. M. and J. E. Miller. (1984). Synergistic effect on field-grown soybeans from combinations of sulfur dioxide and nitrogen dioxide. *Can. J. Bot.*, **62**, 840–6.

Jones, H. C., N. L. Lacasse, W. S. Liggett, and F. Weatherford. (1977). *Experimental air exclusion system for field studies of SO_2 effects on crop productivity*. Muscle Shoals, AL, Tennessee Valley Authority, No. E-EP-77-5.

Jürging, P. (1975). Epiphytische Flechten als Bioindikatoren de Luftverunreinigung. *Biblioteca Lichenologica*, **4**, Vaduz, Cramer.

Knox, J. B. and R. Lange. (1974). Surface air pollutant concentration frequency distributions: implication for urban modelling. *J. Air Pollut. Control Assoc.*, **24**, 48–53.

Lauenroth, W. K. and E. M. Preston. (1984). *The effects of SO_2 on a grassland*, ed. by W. K. Lauenroth and E. M. Preston. New York, Springer-Verlag.

Laurence, J. A., D. C. Maclean, R. H. Mandl, R. E. Schneider, and K. S. Hansen. (1982). Field tests of a linear gradient system for exposure of row crops to SO_2 and HF. *Water Air Soil Pollut.*, **17**, 399–407.

Laurence, J. A., P. R. Hughes, L. H. Weinstein, G. T. Geballe, and W. H. Smith, (1983). *Impact of air pollution on plant–pest interactions: Implications of current research and strategy for future studies*. Ithaca, NY, Cornell University, Ecosystems Research Centre Report No. 20.

Lee, J. J. and R. A. Lewis. (1977). Zonal air pollution system: design and performance. In *The bioenvironmental impact of a coal-fired power plant, 3rd interim report*, ed. by J. J. Lee and R. A. Lewis. 322–44. Corvallis, OR, US Environmental Protection Agency, EPA-600/3-78-021.

Littlejohns, D. A., A. D. McLaren, and J. W. Aylesworth. (1976). The effect of foliar sprays in controlling ozone damage in white beans, *Can. J. Plant Sci.*, **56**, 430.

Manning, W. J. (1988). Biotic factors affecting crop loss assessment. In *Assessment of crop loss from air pollutants.* Proceedings of the international conference, Raleigh, N.C., ed. by W. W. Heck, O. C. Taylor, and D. T. Tingey, 365–386. London, Elsevier Applied Science.

Mansfield, T. A. (1988). Problems of crop loss assessment when there is exposure to two or more gaseous pollutants. In *Assessment of crop loss from air pollutants.* Proceedings of the international conference, Raleigh, N.C., USA, ed. by W. W. Heck, O. C. Taylor, and D. T. Tingey, 317–344. London, Elsevier Applied Science.

McLaughlin, S. B. and G. E. Taylor. (1985). SO_2 effects on dicot crops: some issues, mechanisms and indicators. In *Sulfur dioxide and vegetation,* ed. by W. E. Winner, H. A. Mooney and R. A. Goldstein, 227–63. Stanford, CA, Stanford University Press.

McLeod, A. R. (1988). Effects of open-air fumigation with sulphur dioxide on the occurrence of fungal pathogens in winter cereals. *Phytopathology,* **78,** 88–94.

McLeod, A. R. and J. E. Fackrell. (1983). A prototype system for open-air fumigation of agricultural crops. 1. Theoretical design. Leatherhead, UK, CEGB No. TPRD/L/2474/N83.

McLeod, A. R. and T. M. Roberts. (1987). The importance of the total ecosystem response to air pollution. Preprint No. 87-36.1. *Proceedings of the 80th Annual Meeting of the Air Pollution Control Association,* New York, June 21–26. Pittsburgh, Air Pollution Control Association.

McLeod, A. R., J. E. Fackrell, and K. Alexander. (1985). Open-air fumigation of field crops: Criteria and design for a new experimental system. *Atmos. Environ.,* **19,** 1639–49.

McLeod, A. R., R. A. Skeffington, and K. A. Brown. (1987). Effects of acid mist, ozone and environmental factors on conifers: chamber and open-air studies. In *Direct effects of dry and wet deposition on forest ecosystems—in particular canopy interactions,* Air pollution research report 4. Brussels, Commission of the European Communities.

McLeod, A. R., T. M. Roberts, K. Alexander, and D. M. Cribb. (1988). Effects of open-air fumigation with sulphur dioxide on the growth and yield of winter barley. *New Phytologist.,* **109,** 67–78.

Miller, P. M., H. Tomlinson, and G. S. Taylor. (1976). Reducing severity of O_3 damage to tobacco and beans by combining benomyl or carboxin with contact nematicides. *Plant Dis. Reptr.,* **60,** 433–36.

Miller, J. E., H. J. Smith and P. B. Xerikos. (1980a). Variability of the yield response of field-grown soybean cultivars to SO_2. In *Argonne National Laboratory Report No. ANL-80-115, Part III,* 15–18. Argonne, IL, Argonne National Laboratory.

Miller, J. E., D. G. Sprugel, R. N. Muller, H. J. Smith, and P. B. Xerikos. (1980b). Open-air fumigation system for investigating sulphur dioxide effects on crops. *Phytopathology,* **70,** 1124–28.

Miller, J. E., W. Prepejchal, and H. J. Smith. (1981a). Relative sensitivity of field corn hybrids to ozone: a field study. In *Argonne National Laboratory Report No. ANL-81-85, Part III,* 30–6, Argonne, IL, Argonne National Laboratory.

Miller, J. E., H. J. Smith, and W. Prepejchal. (1981b). Evidence for extreme resistance of field corn to intermittent sulfur dioxide stress. In *Argonne National Laboratory Report No. ANL-81-85, Part III,* 27–29. Argonne, IL, Argonne National Laboratory.

Mooi, J. (1982). *Gebruiksmogelijkheden van anti-oxidentia: een literatuurstudie.* Wageningen, The Netherlands, Research Institute for Plant Protection, No. R273.

Mooi, J. and A. J. A. van der Zalm. (1986). *Research on the effects of higher than ambient concentrations of SO_2 and NO_2 on vegetation under semi-natural conditions. The developing and testing of a field fumigation system: Final report.* Wageningen, The Netherlands, Research Institute for Plant Protection, No. R317.

Moser, T. J., T. H. Nash, and W. D. Clark. (1980). Effects of a long-term field sulfur dioxide fumigation on Arctic caribou forage lichens, *Can. J. Bot.,* **58,** 2235–40.

Muller, R. N., J. E. Miller, and D. G. Sprugel. (1979). Photosynthetic response of field-grown soybeans to fumigations with sulphur dioxide. *J. Appl. Ecol.*, **16**, 567–76.

Nicholson, I. A., D. Fowler, I. S. Paterson, J. N. Cape, and J. W. Kinnaird. (1980). Continuous monitoring of airborne pollutants. In *Ecological impacts of acid precipitation*. Proceedings of an international conference, Sandefjord, Norway, ed. by D. Drablos and A. Tollan, 144–5. Ås, Norway, SNSF project.

Noggle, J. C. (1980). *The effect of atmospheric emissions from the Widows Creek Coal-Fired Power Plant on yield of soybeans and wheat during 1977 and 1978.* Muscle Shoals, AL, Tennessee Valley Authority, Report No. TVA/AQB-I80-6.

Olszyk, D. M., G. Kats, P. J. Dawson, A. Bytnerowicz, J. Wolf, and C. R. Thompson. (1986a). Characteristics of air exclusion systems vs. chambers for field air pollution studies. *J. Environ. Qual.*, **15**, 326–34.

Olszyk, D. M., A. Bytnerowicz, G. Kats, P. J. Dawson, J. Wolf, and C. R. Thompson. (1986b). Effects of sulfur dioxide and ambient ozone in winter wheat and lettuce. *J. Environ. Qual.*, **15**, 363–69.

Olszyk, D. M., A. Bytnerowicz, G. Kats, P. J. Dawson, J. Wolf, and C. R. Thompson. (1986c). Crop effects from air pollutants in air exclusion systems vs. field chambers. *J. Environ. Qual.*, **15**, 417–22.

Olszyk, D. M., A. Bytnerowicz, and C. A. Fox. (1987). Sulfur dioxide effects on plants exhibiting Crassulacean Acid Metabolism. *Environ. Pollut.*, **43**, 47–62.

Oshima, R. J., M. P. Poe, P. K. Braegelmann, D. W. Baldwin, and V. Van Way. (1976). Ozone dosage-crop loss function for alfalfa: a standardised model for assessing crop losses from air pollutants. *J. Air Pollut. Control Assoc.*, **26**, 861–65.

Pell, E. J. (1976). Influence of benomyl soil treatment on pinto bean plants exposed to PAN and O_3. *Phytopathology*, **66**, 731–33.

Preston, E. M. and J. J. Lee. (1984). The field exposure system. In *The effects of SO_2 on a grassland,* ed. by W. K. Lauenroth and E. M. Preston, 45–60. New York, Springer Verlag.

Reich, P. B., R. G. Amundson, and J. P. Lassoie. (1982). Reduction in soybean yield after exposure to ozone and sulfur dioxide using a linear gradient exposure technique, *Water Air Soil Pollut.*, **17**, 29–36.

Reynolds, K. L., M. Zanelli, and J. A. Laurence. (1987). Effects of sulphur dioxide exposure on the development of common blight in field-grown red kidney beans. *Phytopathology*, **77**, 331–34.

Roberts, T. M. (1984). Effects of gaseous air pollution in agriculture and forestry. *Atmos. Environ.*, **18**, 629–52.

Runeckles, V. C., K. T. Palmer, and H. Trabelski. (1981). Effects of field exposures to SO_2 on Douglas Fir, *Agropyron spicatum* and *Lolium perenne. Silva Fennica*, **15**, 405–15.

Sawford, B. L., C. C. Frost, and T. C. Allan. (1985). Atmospheric boundary layer measurements of concentration statistics from isolated and multiple sources. *Boundary-Layer Meteorol.*, **31**, 249–68.

Shinn, J. H., B. R. Clegg, and M. L. Stuart. (1977). *A linear-gradient chamber for exposing field plants to controlled levels of air pollutants.* Livermore, CA, Lawrence Livermore Laboratory, UCRL Reprint No. 80411.

Skye, E. (1968). Lichens and air pollution. *Acta Phytogeogr. Swec.*, **52**, 1–123.

Spierings, F. (1967). Method for determining the susceptibility of trees to air pollution by artificial fumigation. *Atmos. Environ.*, **1**, 205–10.

Sprugel, D. G., J. Miller, R. N. Muller, H. J. Smith, and P. B. Xerikos. (1980). Sulfur dioxide effects on yield and seed quality in field-grown soybeans. *Phytopathology*, **70**, 1129–33.

Stoklasa, J. (1923). *Die Beschädigungen der vegetation durch Rauchgase und Babriks-exhalationen.* Berlin, Urban and Schwarzenburg.

Temple, P. J. and S. Bisessar. (1979). Response of white bean to bacterial blight, ozone, and antioxidant protection in the field. *Phytopathology*, **69**, 101–3.

Thompson, C. R. and D. M. Olszyk. (1985). A field air-exclusion system for measuring the effects of air pollutants on crops. Palo Alto, CA, Electric Power Research Institute, EPRI EA-4203 Final Report Project 1908-3.

Unsworth, M. H. (1981). The exchange of air pollutants and carbon dioxide between vegetation and the atmosphere. In *Plants and their atmospheric environment*, ed. by J. Grace, E. D. Ford and P. G. Jarvis, 111–38. Oxford, Blackwell.

Warren Spring Laboratory. (1982). The investigation of air pollution: National survey of smoke and sulphur dioxide, 1980/81. Stevenage, UK, Warren Spring Laboratory.

Warteresiewicz, M. (1979). Effect of air pollution by sulphur dioxide on some plant species in Upper Silesian Industrial Region, *Achiwum Ochrony Srodowiska*, **1**, 95–166.

Weigel, H. J., G. Adaros, and H.-J. Jäger. (1987). An open-top chamber study with filtered and non-filtered air to evaluate the effects of air pollutants on crops. *Environ. Pollut.*, **47**, 231–34.

Yu, Shu-Wen. (1984). Air pollution problems and the research conducted on the effects of gaseous pollutants on plants in China. In *Gaseous air pollutants and plant metabolism*, ed. by M. J. Koziol and F. R. Whatley, 49–55. London, Butterworths Scientific.

9

ESTIMATING CHANGES IN PLANT GROWTH AND YIELD DUE TO STRESS

J. A. LAURENCE

Boyce Thompson Institute, Ithaca, New York, USA

and

D. S. LANG

Dataright Inc., Milaca, Minnesota, USA

9.1 INTRODUCTION

Assessment of reduced productivity of agricultural and forested ecosystems due to insects, diseases, weather, and other elements of the environment has been the subject of research for over 100 years, but only recently have quantitative methods been applied to the problem (Teng, 1987a). Part of the problem in recognizing and quantifying losses in agriculture and forestry is that plant improvement and the advent and application of inorganic fertilizers have contributed such gains to productivity that many previously severe constraints to production have become of minor importance. Could anyone demonstrate a deleterious effect of degraded air quality on gross production of maize or soybeans over the last 40 years?

In fact, it is the obstacle of outstanding productivity in North American agriculture that we who are interested in assessment of losses are faced with today. Our ability to detect a loss is dependent upon the facility to account sequentially for the contribution of a distinct set of variables, of declining importance, to the level where the measured loss is described by the variable under consideration. In many cases in the past, a single variable has been considered, a situation we know to be unrealistic. In the future, it is clear that we must consider the interaction of production factors to understand and assess accurately changes in performance and productivity of systems in agriculture.

We do not intend for this paper to serve as a comprehensive review of previous crop loss assessments, nor is it a primer for model fitting; those topics have been treated in depth elsewhere (James and Teng, 1979; France and Thornley, 1984; Krupa and Kickert, 1987). Rather, it is our intention to review briefly statistical and epidemiological methods used to assess changes in productivity of plants due to biotic as well as abiotic agents, discuss the coupling of such methods with more mechanistically (or process) based models, and make recommendations for future efforts to relate the results of assessment to regulatory activities.

9.1.1 General introduction to crop loss assessment

Zadoks (1987) has presented a concise review of the history of efforts in the assessment of crop losses due to disease and insects, dividing the last 150 years into three periods: the exploratory, the emergency, and the implementation. Similar periods may be applied to efforts to assess losses due to man-made environmental stress, although they might more appropriately be called descriptive, directed, and holistic. Interestingly, the time periods even coincide roughly, even though most consider concerns over environmental quality to be a relatively recent development.

In terms of research the descriptive period for air quality effects on plant productivity began with a report by Schröter (1908) on the assessment of damage to forest trees caused by fumes from smelters in Saxony in the mid to late 1800s. During the next 50 years, reports of losses were primarily anecdotal, with the exception of the pioneering studies stimulated by the problems around the smelter at Trail, BC, and other studies related to the local occurrence of air pollutants (National Research Council, 1939). As with our colleagues assessing losses due to more traditional pests, early attempts at assessing losses due to environmental stresses were limited by lack of adequate methods, suitable equipment, and adequate understanding of pollution dispersion and atmospheric chemistry. When considered as a "pathogen" that causes a "disease," air pollutants leave a lot to be desired. They rarely occur alone. They are regulated by law, but do not observe jurisdictional boundaries; they are often assigned singular thresholds of action, but then act in concert at lesser levels over long periods and with deleterious, but largely unknown consequences. Often, the methods necessary to produce, apply, and measure known concentrations of the pollutants are complex and expensive, making standardization even more difficult.

Zadoks describes the emergency period occurring generally from World War I to 1967 when the Food and Agricultural Organization (FAO) of the United Nations published a manual of procedures (FAO, 1971). Food supplies and transportation systems were often disrupted by war and economic crisis, and the need to protect against additional loss became essential. The descriptive period continued for assessing losses due to environmental stress until the 1950s and early 1960s, when it was recognized that ozone injured plants (Middleton, 1961). Serious efforts to assess losses associated with air pollution on a regional and national basis began at that time. It was not until Benedict and his colleagues (1971) related published reports of crop productivity to county-level emissions of air pollutants that a systematic assessment (although not based on direct experimentation) of the importance of pollutants to agriculture was available.

In essence, Benedict's study initiated a period of directed research (which in our opinion continues today), driven in many cases by needs of regulatory agencies. The period is characterized by experimentation on the direct effects of air pollutants on plants and factors that modify the productivity of plants. The research conducted involves primarily empirical studies, has used many different methodologies for exposing plants to pollutants, has been conducted under both controlled and uncontrolled conditions in the field as well as the laboratory, and has ranged greatly in complexity. The research has responded to the needs of the various interested parties in their search for methods and meaning. As our methods become honed, we are moving towards a more holistic period when we can apply our knowledge to address the questions of concern—the regulation of man-made environmental stresses (Laurence and McCune, 1980). In the end, in the same way that results of our colleagues who work with pests and pathogens are transferred to producers and used in a control strategy, our results must be transferred to legislators, the controllers of our 'pests'.

It would seem appropriate to apply what has been learned during the directed period for assessing losses caused by biotic agents to facilitate our transition to the holistic period.

9.2 METHODS OF ASSESSMENT

There are four major methods for assessing the effects of a stress on a plant. First, the responses of a plant given a known stress may be described statistically, and a measure of the change in productivity may be produced. Such models are frequently used and vary in complexity based on the length of the experiment, the number of variables measured, and the number of locations and time periods over which the experiment is conducted. Critical point models and those that use, for instance, the area under a disease progress curve (a season or epidemic-long integration of stress severity) to describe plant growth or yield are generally of this type (James and Teng, 1979).

9.2.1 Critical point models

There are several good examples in the literature of critical point models to assess the effects of disease or environmental stress (Oshima *et al.*, 1975, 1976, 1977, 1979; James and Teng, 1979). These models have in common the characteristic that yield can be predicted from a measurement of disease, insect populations, or stress at a single point in time.

9.2.1.1 *Examples of critical point models*

Romig and Calpouzos (1970) conducted a field experiment to describe the relationship between stem rust and the yield of spring wheat in Minnesota. Four different epidemics were initiated in field plots by varying the time of inoculation and the date and rate of fungicide application. Nondiseased plots were maintained through the use of protective fungicides. Estimates of disease severity were made at eight times during the growing season at documented stages of plant development. At maturity, the plots were harvested and the yield of grain was measured.

Average percent yield loss was plotted versus average percentage disease severity for the four epidemics and the control. Observations of disease severity made when the caryopsis was three-fourths of its final size correlated well with final yield. The loss function derived was

$$\text{Yield} = -25 \cdot 53 + 27 \cdot 17 \ln(\text{average percent disease severity})$$

with a correlation coefficient of $0 \cdot 993$. The authors point out that they could not be sure that the fungicide treatments did not affect yield other than by control of the pathogen, but, based on extensive experience, they believed the method to be valid. In addition, they point out that the critical point model fits well with physiological studies that demonstrate the importance of photosynthesis at the time of anthesis in determining yield. They also point out that their equation may not work with all epidemics or cultivars.

The models developed as part of National Crop Loss Assessment Network (NCLAN) experiments (Heck *et al.*, 1984*a*, *b*) estimate yield of several crops based on an experiment-long average of O_3 concentrations, in essence a critical point measurement. Other critical point models have used an assessment at a specific time in the growing season, before harvest (James and Teng, 1979; Kim and MacKenzie, 1987).

Critical point models are generally based on a regression of yield on some measure of the stress applied. These dose–response models may use either linear or nonlinear functions as a mathematical form (Madden *et al.*, 1981; Teng, 1987*a*), and are, in most cases, quite easy to apply. They are, of course, limited by all the constraints that must be applied to regression models and are usually restricted in their applicability from one geographic region to another, and perhaps among different populations (e.g., cultivars) of plants. Some researchers have attempted to generalize the models to wider areas by comparing the results of similar experiments and developing common models to describe the response of the plants over wide geographic areas (Rawlings and Cure, 1985).

Critical point models may not be applicable to multiple stress situations, for instance, in Europe where mixtures of several pollutants occur frequently and in ratios that vary with location. The experiments and analyses conducted to develop these models are based on experiments that vary only one stress. Consequently, when the models are applied, stresses that were not present in the experiments on which the model is based may significantly affect the ability of the model to predict yield accurately.

While models that use area under the disease progress curve (AUDPC) may be considered a separate type of assessment model (James and Teng, 1979) because they represent an integration of disease severity over time, they are similar to critical point models in that yield loss is related to a single, cumulative measure of disease. They can be used, however, to chart the progress of an epidemic and, therefore, provide interim assessments of the effect of a stress on plant productivity. Models developed by Oshima and co-workers (1976) that used an accumulation of O_3 concentrations as a predictor of yield are similar to AUDPC models. A similar analysis has been applied to selected NCLAN experiments (Lefohn *et al.*, 1988).

The critical point and AUDPC models represent a transition from emergency to implementation in terms of assessment of crop loss due to biotic agents. They provide tools that can be used to make a rational decision about the value of a control or the importance of a stress. Although the predictions may not be responsive to all the possible combinations of stress that may occur, they do provide a guideline for a decision by a producer. The models of Oshima and co-workers are good examples of research in the directed period; they have been used by regulatory agencies in setting of standards. In the case of the NCLAN experiments, the models developed have provided a first assessment of the importance of a single air quality factor (Adams *et al.*, 1985; Heck *et al.*, 1984*a*, *b*). If, in fact, they approach the true value, they have provided a valuable tool for the assessment process.

A subset of the dose–response experiments used to develop critical point models that involves only two treatments is the use of antioxidant chemicals to control the effects of O_3 on plants (Smith *et al.*, 1987; Toivonen *et al.*, 1982). This methodology is an extension of that used in plant pathology, entomology, and weed science that involves protecting one set of plants from a stress and not protecting another. It is fraught with similar difficulties and uncertainties: what are the interactions of the chemical with the rest of the plant's environment (including other pollutants that may be present) and perhaps the plant itself? Is the chemical equally effective at all levels of stress? What is the shape of the response curve? Although the model is simplistic and intuitively easy to understand (loss in productivity = protected − control), the results are confounded with possible interactions. In the case of insects and pathogens, at least some measure of severity or intensity can be obtained from visual assessment of the plant and this may be used to help validate the results. In addition, by varying the concentration of the chemical or the frequency of treatment, and relating the dose of chemical to disease severity, a number of levels of partial control may be applied (i.e., as with the four epidemics of Romig and Calpouzos). In the case where an antioxidant is used, only O_3 dose can be used as a surrogate: no measure of plant uptake or response to the treatment is available, and there is no method to judge the efficacy of partial control. The major drawback of such an experiment is the lack of information on the shape of the function describing the response of plants to various levels of a treatment. While such studies might be used to validate other loss assessment projects, they must be carefully repeated over several years to obtain necessary ranges of treatments levels and estimates of precision. In addition, experiments using protectants are probably suited for local but not regional use, an important consideration depending on the goal of the assessment (Toivonen *et al.*, 1982). As assessment of crop losses due to

environmental stresses moves towards the implementation period, local
assessments will become more important.

9.2.2 Multiple point models

In the second method of assessment (actually a variation on the first), a
plant may be exposed to a known series of stress or stresses, and a number of
variables may be measured and related to the observed response. For instance,
the dose of the stress may be divided temporally within a growing season,
components of plant growth may be measured, and the experiment may be
repeated at several locations or over several growing seasons. Models such as
these are often called multiple point models since they rely on a time or event
course or attempt to relate levels of multiple stresses to plant growth or yield.

Such models are exemplified by those of Burleigh *et al.* (1972), James *et al.*
(1972) and Benson *et al.* (1982). In these cases, assessments of stress,
measured as disease severity or O_3 dose, are made at several time intervals or
at several growth stages. These variables are then used in multiple regression
equations to predict final yield of the crop. At the same time, the analysis
helps to identify the most critical stage of development in terms of the effect of
stress on yield.

9.2.2.1 Example of multiple point models

Benson *et al.* (1982) used experimental data collected by investigators in
various parts of the United States to develop yield loss–O_3 models. When
modeling alfalfa, they produced a daily biomass loss value, while with other
major crops, a single yield loss value was used. Daily O_3 dose statistics were
calculated from the experimental data; however, the daily dose statistics were
generally combined for periods of up to 2 weeks, creating O_3 doses for a
number of growth periods. Regression analyses were conducted relating yield
to O_3 dose in the various summary periods (i.e., daily for alfalfa, 14 days for
wheat). The following examples are models developed from the analysis:

Alfalfa: $Y = ax + bx^2 + cx^3$

  where Y = actual loss in biomass for a given day and

        x = the sum of the hourly ozone concentrations for the day.

Corn : $Y = ax_1 + bx_2 + \cdots + lx_{12}$

  where Y = season yield loss in weight per 100 kernels and

        x_1 through x_{12} are 7-day summed ozone dose statistics for

        periods 1 through 12 in the growing season.

Air quality data were predicted for Minnesota based on interpolation of
actual monitoring data from 1979 and 1980. Appropriate measures of O_3 dose
were calculated for each county and used in conjunction with the yield loss
models to predict losses to the crops under consideration. Predicted losses

were applied to agricultural production statistics and subsequently used in a detailed economic analysis.

These models are multiple point models because they use assessments of O_3 at a number of times in the growing season (daily for alfalfa, weekly for corn) to predict final yield loss. Such models should improve precision over critical point models because many more assessments of the stress are made.

As Benson *et al.* (1982) point out, models are simplifications of the real world and depend, in large part, on the data sets from which they are built. Because the data used to build the models came from different geographical areas, the results must be interpreted carefully. More important than the actual loss estimate is the method used to derive the estimate. The models developed in this study are an improvement on critical point models that do not account for pollutant episodes or variation in plant sensitivity based on stage of development. More recently, Lee (personal communication), in an attempt to account for variation in the sensitivity of plants to O_3, has included plant phenology in loss models derived from experiments conducted at Corvallis, OR.

9.2.3 Correlative studies

The third method of assessing crop loss utilizes correlative studies relating reported or measured productivity to predicted or measured levels of a stress (similar to Benedict *et al.*, 1971; Benson *et al.*, 1982; or Oshima *et al.*, 1976). These studies do not depend on the experimental imposition of a known level of stress on a plant, but rely on naturally occurring variation to provide the experimental range. In general, it is desirable to measure as many variables as possible in such a study, and the results often provide valuable, underlying information on the interrelationship between production factors (e.g., insects and pathogens, pathogens and weeds, etc.).

9.2.3.1 *Example of correlative studies*

Oshima *et al.* (1976) took advantage of a natural gradient of O_3 concentration that occurs in the South Coast Air Basin of California. Field plots were established at 13 locations that were known to differ in seasonal O_3 dose (concentration × time). Alfalfa was planted in a standard soil mix in containers. Irrigation and insect control were applied uniformly to all plots. Because of the uniformity in environment in southern California, the only major variable that differed from plot to plot was O_3 concentration. Plants were harvested at five times during the season, and measurements of mass were made. At the end of the season, yield of alfalfa was related to a summation of the hourly O_3 concentrations greater than 0·10 ppm (a threshold introduced to weight the higher concentration exposures) with the following equation:

$$\text{Yield (grams)} = 162 \cdot 4 - 0 \cdot 01503 \times \text{Dose}$$

where dose is in ppm-h and

$$r = 0 \cdot 83$$

This experiment resulted in a direct assessment of the effects of O_3 on yield without the usual experimental manipulation to control O_3 concentration, but represents a correlative approach because the exposures were uncontrolled. Unfortunately (or perhaps fortunately), few areas have the well-established gradient in pollutant concentration that permits this direct experimentation.

9.2.4 Survey techniques

An additional approach to assessing the effects of stress on plant productivity is available through disease survey (Stynes, 1980; Teng and Oshima, 1983; Weise, 1980). In this case, detailed assessments of a sample of plants are made at several locations in the region of interest. Using methods such as multiple regression or principal components analysis, groups of important variables can be identified and used to relate levels of various stresses to yield at the end of the season. The assessments may be conducted at several times, thus making stage of development an additional variable. Principal components analysis may be used to produce sets of new variables when there is a high degree of correlation between the various measures of stress (as there is likely to be when repeated measures of the same plants are made) (Shane, 1987).

9.2.4.1 Example of survey techniques

Stynes (1980) used an extensive disease survey to assess the importance of several production factors on yield of wheat in South Australia. Forty-two farms, with two sites at each farm, were surveyed at four times during the growing season. Measures of soil properties, plant growth, pathogen development, and climate were made in addition to detailed descriptions of cultural practices and cropping history. Soils were characterized physically and chemically.

To avoid intercorrelation, variables were grouped and analysed to produce simplified variables, if possible. For instance, soil salinity measurements taken during the study were described using orthogonal polynomials that characterized the dynamics of salinity through the season. After initial transformation and grouping, principal components analysis was used to reduce the number of independent variables needed. For example, the original 14 soil variables could be reduced to 8 principal components that contained 98% of the original variation. Canonical correlation was used when groups of variables were likely to be related as would be expected with the distribution of root pathogens and soil properties.

Regression analyses were used to develop models of wheat yield based on the important variables, groups of variables, or principal components that were identified in the additional analyses. Variables were entered into the equations in an order thought to be biologically significant, and intercorrelation was avoided by using transformed variables or avoiding the use of variables that were closely related. An example of a model produced using this method predicted yield of wheat from information collected at anthesis. Twelve variables were used, resulting in a model that explained 84% of the variation.

Seven subsets of variables accounted for the variation in yield: plant morphological variables, 31%; plant element concentrations, 9%, soil properties, 14%; climate, 7%; pathogens, 11%; and cultural and cropping practices, 4%.

The models were verified in a subsequent sampling of a smaller number of farms. In general, the models performed well except that an epidemic of stem rust could not be accounted for because the disease was not included in the original survey (a major limitation of survey and regression methods). The study does illustrate the use of natural variation in stress to model the relationships between factors that affect plant growth and yield and represents a method that could be used more widely in air pollution research.

9.3   DISCUSSION OF ASSESSMENT TECHNIQUES

With the exception of the Stynes study, we have addressed the problem of assessing stresses independently of one another, but now must face the fact that plants must grow and produce their yield in an environment that is rarely so forgiving as to provide single stresses. In fact, plants integrate, on a very short time step (perhaps hours or days), the effects of weather, soil, moisture, insects, disease, air quality, weeds, and many other factors that constitute the production system. Teng (1987*b*) provides an excellent review of this holistic philosophy and points out that in most cases, a biological system "cannot be properly understood or managed by ad hoc knowledge on its components alone". If we are now to move from the directed period to the holistic period, it is necessary to understand how abiotic stress interacts in the system and determines the potential importance of degraded air quality in a production system.

One method of achieving this goal is to use simulation modeling to describe a production system (de Wit and Goudriaan, 1978). The model might begin with a crop growth model that then incorporates modules to describe the response of the plant to the various stresses that affect productivity. Information from experiments that address the interactions of stresses on plant productivity (Johnson *et al.*, 1986; Kohut *et al.*, 1987; Reynolds *et al.*, 1987) are valuable, but may not be available. In that case, researchers may build the model using a best guess of how stresses may interact and then compare the results of the theoretical model to what is observed under field conditions. Validation and verification of simulation models can be a complex task that is extremely important to assure the accuracy of the prediction and modeling processes.

A major problem in almost all assessments is arriving at an estimate of loss in the production system, not just an estimate of loss due to a single cause (Teng and Oshima, 1983). For instance, if assessments are conducted for several individual stresses without consideration of other production factors (e.g., assessment of O_3 without consideration of disease or insect and weed pests), an estimate of the importance of the single component is actually an estimate of the end result of the plant's integration of multiple stresses. Such an integration will include both additive and multiplicative effects that may result in a synergism or an antagonism. Some experiments (Johnson et al., 1986; Kohut et al., 1987; Reynolds et al., 1987) have been conducted to assess the effects of multiple stresses on yield, but they are difficult to perform. However, an integrated picture of how plants respond to all factors in the environment is essential to accurate assessments. Correlative models and process-based simulation models offer the best possibilities, at present, for providing assessments of plant productivity under multiple stress conditions.

To complicate the assessment process, experiments necessary to acquire the information needed for the models that address air pollutants, environmental stress, or multiple stresses often involve considerable manipulation of plants and their environment. Questions may always be asked about whether the experiments are adequate and are valid simulations of the real environment.

While all of the above methods have advantages and disadvantages, the solution to the puzzle probably lies in coupling several of the models and adopting a systems approach to assess losses. For instance, a mechanistic model of crop growth may contain a module relating the response of the plant to various levels of O_3, insect pests, pathogens, and weeds that might occur over the course of a growing season. Such a systems approach has rarely been used in studying environmental stress, perhaps due to the time, complexity, and expense of conducting the research necessary. The systems approach has been used successfully for managing pests in agriculture (Zadoks, 1987) and should be extended to account for factors such as air quality.

The complexity, intended use, and validity of various types of models must also be considered when selecting a method. For instance, critical point models relating plant response to a known O_3 level over the course of a growing season may provide an adequate tool for the economist who desires a rough estimate for a national model, but may be of little use to the farm manager who needs to know how O_3 interacts with all the other stresses present, or to the regulatory agency that may not be able to accept a retrospective view of the pollution situation. In these cases, multiple point, multiple stress models may be of more value, but such models often depend on how the groups of variables are interrelated, making interpretation of the effects of a single stress (which may be of interest to the regulator or economist) impossible. The process-based mechanistic model may be of most use to a researcher looking for pathways of action in the plant, but such a model may well operate on too detailed a level for the economist or regulator.

9.4 VALIDATION OF LOSS MODELS

Because a model is a simplified representation of a real system, it is particularly important to assure that the results generated are reasonable and valid. The complexity of this task is, of course, related to the complexity of the model under consideration. Teng (1981) has reviewed the philosophy and methods of model verification and validation and pointed out that it is important to evaluate the behavior and sensitivity of the model (how it functions under a variety of circumstances) as well as how closely it relates to field observation. In the case of simulation models, the process can be very complex and involve both subjective and objective evaluations of the performance of individual parameters as well as the model as a whole.

Validation of statistical models, such as those used in critical point or multiple point models, usually depends on testing to determine how closely observations collected independently from those used in model development fit the equation used in the model. It is important to abide by the assumptions implicit in regression when using the model to predict future events. In addition, it must be understood that those models are constructed under particular sets of environmental conditions, and if predictions are made in environments that are different from those present during model development, the results of the predictions may be suspect.

In the end, it is a combination of objective testing, subjective analysis, and practical experience with a model that will provide validation. Development of common models, based on multiple data sets such as those produced by NCLAN, that attempt to describe the response of a species over a wide region may work in some cases, but not in others. However, it is only through continued testing and use that the accuracy of any model can be verified.

9.5 CONCLUSIONS AND RECOMMENDATIONS—RELATING NCLAN TO CROP LOSS ASSESSMENT

One would hope that we have now reached the transition from directed to holistic research when addressing the issues of environmental quality. That degraded air quality is affecting the health and productivity of crops and forests on a large scale should be beyond question. However, the role of air quality in production systems and the importance of the effects, both ecologically and economically, remain largely unknown. NCLAN has served a valuable role in this transition for several reasons. First, it represented a coordinated expansion of the pioneering work of Oshima and others who first addressed experimental assessment of crop loss due to air pollution. Second, the program provided a new methodology of evaluation and assessment over that used by Benedict and his colleagues. Third, NCLAN raised the level of consciousness of the scientific community, the regulators, and the public to a level where the importance of air quality as a production factor is recognized. Finally, NCLAN made attempts (although abortive due to lack of sufficient funding) to approach the assessment from a systems level; experiments were planned and in some cases conducted to assess the interactions of other production factors with air quality. So, the program represents a step in the transition. But what of the future?

Based on current trends in crop loss assessment in other disciplines, it seems that we must turn our attention towards methods that will provide an accurate picture of the role of air quality in crop production. Pest management systems such as EPIPRE, currently in use in The Netherlands (Zadoks, 1987), or SIRITAC in Australia (Ives and Hearn, 1987) represent the most developed and tested technology. Other pest management systems are under development. It is incumbent upon us, the scientists researching the effects of air quality on plant productivity, to link our studies to those currently under development. Furthermore, there is a need to revise the way in which we think of losses in productivity. Typically, reduced productivity has been evaluated in monetary terms that describe the magnitude of a loss at a given time or as a percentage of production. This approach ignores the variation from year to year (MacKenzie and King, 1980), change in production, and the intensity of the necessary husbandry. Instead, losses should be expressed as an energy balance; kilocalories invested in production of a crop account for the use of fertilizers while reflecting the value of improved plants, disease resistance, or improved air quality that eliminates the need for an application of a chemical, the increased use of machinery, or increased acreage in production. At the same time, an energy balance provides a constant, unaffected by world economics, by which we can estimate the true cost of production.

Because of the political nature of air pollution problems, there will always be a need for estimates of the marginal cost of individual pollutants. Using a systems approach, we should be able to provide regulatory agencies with a sound estimate of the real importance of pollutants in crop production systems. As Teng and Oshima (1983) pointed out, crop loss assessment developed as a result of failures in crop production systems, and it is time for production methods to be influenced by loss assessment. Similarly, the need for assessment of losses due to air pollutants has evolved as a result of failures in the regulatory process that is designed to protect the welfare of our nation.

REFERENCES

Adams, R. M., S. A. Hamilton, and B. A. McCarl. (1985). An assessment of the economic effects of ozone on U. S. Agriculture. *J. Air Pollut. Control Assoc.*, **35**, 938–43.

Benedict, H. M., C. J. Miller, and J. S. Smith. (1971). *Assessment of economic impact of air pollutants on vegetation in the United States: 1969–71*. Menlo Park, CA, Stanford Research Institute, EPA-650/5-78-002.

Benson, F. J., S. V. Krupa, P. S. Teng, D. E. Welsch, C. Chen, and K. W. Kromroy. (1982). *Economic assessment of air pollution damage to agricultural and silvicultural crops in Minnesota*. Report to the Minnesota Pollution Control Agency, Minneapolis, MN.

Burleigh, J. R., M. G. Eversmeyer, and A. P. Roelfs. (1972). Estimating damage to wheat caused by *Puccinia recondita tritici*. *Phytopathology*, **62**, 944–6.

de Wit, C. T. and J. Goudriaan. (1978). *Simulation of ecological processes*. Wageningen, PUDOC.

Food and Agriculture Organization of the United Nations. (1971). *Crop loss assessment methods. FAO manual on the evaluation and prevention of losses by pests, diseases, and weeds*. Commonwealth Agricultural Bureau, Slough, UK.

France, J. and J. H. M. Thornley. (1984). *Mathematical models in agriculture*. London, Butterworths Scientific.

Heck, W. W., W. W. Cure, J. O. Rawlings, L. J. Zaragoza, A. S. Heagle, H. E. Heggestad, R. J. Kohut, L. W. Kress, and P. J. Temple. (1984*a*). Assessing impacts of ozone on agricultural crops: I. Overview. *J. Air Pollut. Control Assoc.*, **34**, 729–35.

Heck, W. W., W. W. Cure, J. O. Rawlings, L. J. Zaragoza, A. S. Heagle, H. E. Heggestad, R. J. Kohut, L. W. Kress, and P. J. Temple. (1984*b*). Assessing impacts of ozone on agricultural crops: II. Crop yield functions and alternative exposure statistics. *J. Air Pollut. Control Assoc.*, **34**, 810–17.

Ives, P. M. and A. B. Hearn. (1987). The SIRATAC system for cotton pest management in Australia. In *Crop loss assessment and pest management*, ed. by P. S. Teng, 251–68. St. Paul, MN, APS Press.

James, W. C. and P. S. Teng. (1979). The quantification of production constraints associated with plant disease. *Appl. Biol.*, **4**, 201–67.

James, W. C., C. S. Shih, W. A. Hodgson, and L. C. Callbeck. (1972). The quantitative relationship between late blight of potato and tuber yield. *Phytopathology*, **62**, 92–6.

Johnson, K. B., E. B. Radcliffe, and P. S. Teng. (1986). Effects of interacting populations of *Alternaria solani*, *Verticillium dahliae*, and the potato leafhopper (*Empoasca fabae*) on potato yield. *Phytopathology*, **76**, 1046–52.

Kim, C.-H. and D. R. MacKenzie. (1987). Empirical models for predicting yield loss caused by a single disease. In *Crop loss assessment and pest management*, ed. by P. S. Teng, 126–32. St Paul, MN, APS Press.

Kohut, R. J., R. G. Amundson, J. A. Laurence, L. J. Colavito, P. van Leuken, and P. King. (1987). Effects of ozone and sulfur dioxide on yield of winter wheat. *Phytopathology*, **77**, 71–4.

Krupa, S. V. and R. N. Kickert. (1987). An analysis of numerical models of air pollutant exposure and vegetation response. *Environ. Pollut.*, **44**, 127–58.

Laurence, J. A. and D. C. McCune. (1980). Assessing losses caused by abiotic diseases—an overview. In *Crop loss assessment. E. C. Stakman Commemorative Symp.*, 290–5. St. Paul, Minnesota Agricultural Experiment Station, Misc. Publ. 7.

Lefohn, A. S., J. A. Laurence, and R. J. Kohut. (1988). A comparison of indices that describe the relationship between ozone exposure and agricultural crop yield reduction. *Atmos. Environ.*, **22**, 1229–43.

MacKenzie, D. R. and E. King. (1980). Developing realistic crop loss models for plant diseases. In *Crop loss assessment. E. C. Stakman Commemorative Symp.*, 85–9, St. Paul, Minnesota Agricultural Experiment Station, Misc. Publ. 7.

Madden, L. V., S. P. Pennypacker, and C. H. Kingsolver. (1981). A comparison of crop loss models. *Phytopath. Z.*, **101**, 196–201.

Middleton, J. T. (1961). Photochemical air pollution damage to plants. *Ann. Rev. Plant Physiol.*, **12**, 431–48.

National Research Council of Canada. (1939). *Effect of Sulphur Dioxide on Vegetation*. Ottawa, Canada.

Oshima, R. J., O. C. Taylor, P. K. Braegelmann, and D. W. Baldwin. (1975). Effect of ozone on the yield and plant biomass of a commercial variety of tomato. *J. Environ. Qual.*, **4**, 463–4.

Oshima, R. J., M. P. Poe, P. K. Braegelmann, D. W. Baldwin, and V. Van Way. (1976). Ozone dosage–crop loss function for alfalfa: A standardized method for assessing crop losses from air pollutants. *J. Air Pollut. Control Assoc.*, **26**, 861–5.

Oshima, R. J., P. K. Braegelmann, D. W. Baldwin, V. Van Way, and O. C. Taylor. (1977). Reduction of tomato fruit size and yield by ozone. *J. Am. Soc. Hort. Sci.*, **102**, 289–93.

Oshima, R. J., P. K. Braegelmann, R. B. Flagler, and R. R. Teso. (1979). The effects of ozone on the growth, yield, and partitioning of dry matter in cotton. *J. Environ. Qual.*, **8**, 474–9.

Rawlings, J. O. and W. W. Cure. (1985). The Weibull function as a dose–response model to describe ozone effects on crop yields. *Crop Sci.*, **25**, 807–14.

Reynolds, K. L., M. Zanelli, and J. A. Laurence. (1987). Effects of sulfur dioxide on the development of common blight in field-grown red kidney beans. *Phytopathology*, **77**, 331–4.

Romig, R. W. and L. Calpouzos. (1970). The relationship between stem rust and loss in yield of spring wheat. *Phytopathology*, **60**, 1801–5.

Schröter, E. (1908). Die Rauchquellen ein konigreiche Sachsen und ihr Einfluss auf die Forstwirtschaft. Sammlung von Abhandlungen und Rauchschader, Heft 2. Berlin, Paul Parey.

Shane, W. W. (1987). The use of principal components analysis and cluster analysis in crop loss assessment. In *Crop loss assessment and pest management*, ed. by P. S. Teng, 139–49. St. Paul, MN, APS Press.

Smith, G., B. Greenhalgh, E. Brennan, and J. Justin. (1987). Soybean yield in New Jersey relative to ozone pollution and antioxidant application. *Plant Dis.*, **71**, 121–5.

Stynes, B. A. (1980). Synoptic methodologies for crop loss assessment. In *Crop loss assessment. E. C. Stakman Commemorative Symp.*, 166–75. St. Paul, Minnesota Agricultural Experiment Station, Misc. Publ. 7.

Teng, P. S. (1981). Validation of computer models of plant disease epidemics: a review of philosophy and methodology. *Z. Pflanzenkr. Pflanzenschutz*, **88**, 49–63.

Teng, P. S. (1987a). Quantifying the relationship between disease severity and yield loss. In *Crop loss assessment and pest management*, ed. by P. S. Teng, 103–113. St. Paul, MN, APS Press.

Teng, P. S. (1987b). The systems approach to pest management. In *Crop loss assessment and pest management*, ed. by P. S. Teng, 160–7. St. Paul, MN, APS Press.

Teng, P. S. and R. J. Oshima. (1983). Identification and assessment of losses. In *Challenging problems in plant health*, ed. by T. Kommedahl and P. H. Williams, 69–81. St. Paul, American Phytopathological Society.

Toivonen, P. M. A., G. Hofstra, and R. T. Wukasch. (1982). Assessment of yield losses in white bean due to ozone using the antioxidant EDU. *Can. J. Plant. Pathol.*, **4**, 381–6.

Weise, M. V. (1980). Comprehensive and systematic assessment of crop yield determinants. In *Crop loss assessment. E. C. Stakman Commemorative Symp.*, 262–9. St. Paul, Minnesota Agricultural Experiment Station, Misc. Publ. 7.

Zadoks, J. C. (1987). Rationale and concepts of crop loss assessment for improving pest management and crop protection. In *Crop loss assessment and pest management*, ed. by P. S. Teng, 1–5. St Paul, MN, APS Press.

Session IV

THE VALUE OF PHYSIOLOGICAL UNDERSTANDING IN CROP LOSS ASSESSMENT

(David T. Tingey, *Chairman*; Robert G. Amundson, *Co-Chairman*)

10

POLLUTANT DEPOSITION TO INDIVIDUAL LEAVES AND PLANT CANOPIES: SITES OF REGULATION AND RELATIONSHIP TO INJURY

GEORGE E. TAYLOR, Jr, PAUL J. HANSON

Oak Ridge National Laboratory, Oak Ridge, Tennessee, USA

and

DENNIS D. BALDOCCHI

National Oceanic and Atmospheric Administration, Oak Ridge, Tennessee, USA

10.1 INTRODUCTION

The exchange of gases across the atmosphere–leaf interface is a subject of fundamental importance in applied and basic research in the plant sciences. This exchange is a subset of the more general phenomenon of gas–liquid interactions (Danckwerts, 1970), which have been applied to a range of environmental issues (e.g., Liss and Slater, 1974). However, the exchange at the atmosphere–leaf interface is somewhat unique because the deposition surface is physiologically active, and is highly variable in its capacity to function as a sink. At the level of the individual leaf, an example of the uniqueness of foliar gas exchange is carbon dioxide (CO_2) assimilation, in which the focus has been the degree to which gas-phase (e.g., stomata) and liquid-phase (e.g., carboxylation) processes govern CO_2 assimilation (Farquhar and Sharkey, 1982).

The unique attributes of gas exchange at the atmosphere–leaf interface are (1) pronounced and viscous/stagnant layers in the gas phase, adjacent to the leaf surface (boundary layer) and in the leaf interior (substomatal chamber and intercellular space); (2) variable diffusive resistance in the gas phase due to changes in stomatal porosity; (3) array of extensive deposition sites both on the leaf surface and within the leaf interior; (4) biochemically variable capacity for mesophyll tissues to assimilate trace gases; and (5) mix of functionally varying leaf surfaces in the plant canopy, comprised of different age classes and species and each having its own distinct local microclimate such that the canopy's effective deposition/sink potential varies in space and time.

This concept of the atmosphere–leaf gas exchange provides an appropriate basis with which to address the deposition of pollutant gases to individual leaves and plant canopies. Because the physiological sites of action for most gaseous pollutants are cells of the leaf interior, any factor that influences pollutant deposition will uncouple the relationship between the pollutant's concentration in the atmosphere and the corresponding concentration at sensitive cellular sites of action. It is proposed that the pronounced variation in plant response as a function of genotype, environment, pollutant exposure dynamics, or presence of other pollutants is in part accountable to differences in pollutant deposition rates. Plant scientists should recognize the role of deposition processes because it may help explain patterns in pollutant effects on a plant's physiological status, growth, development, and productivity.

Pollutant deposition can be addressed at different scales of resolution (Fig. 10.1). The most common is that of the individual leaf with a focus on foliar sites of deposition, and the role of cuticular, stomatal, and mesophyll resistances in governing trace gas exchange rates (Bennett and Hill, 1973). At a finer scale of resolution, physicochemical processes are emphasized at the interface of the intercellular space and mesophyll cell surface in the leaf interior (Tingey and Taylor, 1982). A third scale of resolution is the plant canopy, focusing on atmospheric processes governing turbulence in terrestrial landscapes and the distribution of pollutant deposition sites within plant canopies (Fowler, 1980). These varied scales of resolution make it necessary to extend the analysis of pollutant deposition beyond conventional aspects associated with individual leaves, to include exchange processes operating at levels of the leaf interior (scaling down) and the plant canopy (scaling up). An analogous framework for investigating the control of transpiration from terrestrial vegetation has been developed (Jarvis and McNaughton, 1986).

The objective of this chapter is twofold: (1) to characterize the processes governing the deposition of pollutants to individual leaves and plant canopies, and (2) to relate deposition to the pollutant's effects on a plant's physiology and growth. The focus is principally dry-deposited, regionally distributed gases of importance because of either their phytotoxicity (e.g., ozone, O_3; sulfur dioxide, SO_2; nitrogen dioxide, NO_2; peroxyacetyl nitrate, PAN; hydrogen sulfide, H_2S; hydrogen peroxide, H_2O_2) or their influence on biogeochemical cycling in terrestrial ecosystems (e.g., carbonyl sulfide, COS; ammonia, NH_3; nitric acid vapor, HNO_3).

10.2 METHODOLOGIES FOR MEASURING DEPOSITION TO INDIVIDUAL LEAVES

Many methodologies exist to investigate pollutant deposition (Table 10.1), but only a few are in routine use. The most common approach is based on the principle of mass balance in which the vegetation surface (i.e., leaf, plant, whole canopy) is enclosed in an environmentally controlled air reservoir, and pollutant deposition is estimated by comparing the concentration of the gas at the inlet and outlet (Sestak *et al.*, 1971). The technique can be modified to investigate deposition to individual leaves (Legge *et al.*, 1977), whole

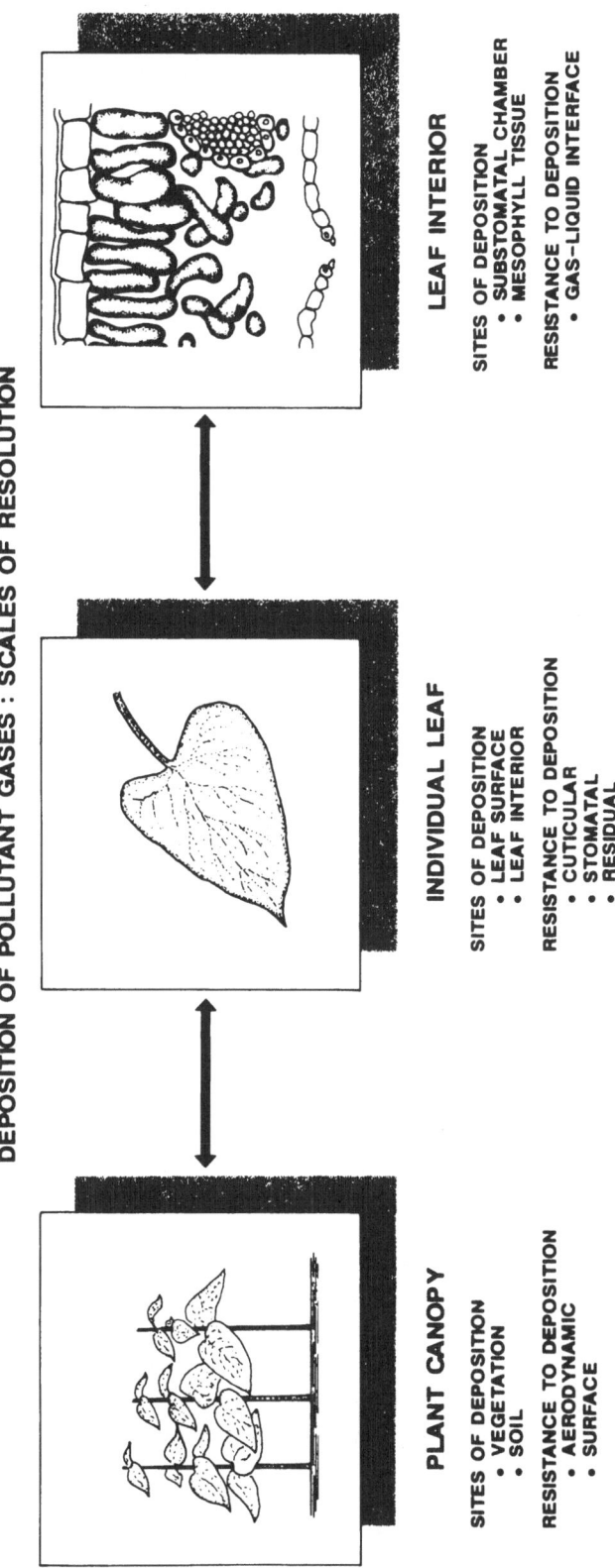

Fig. 10.1. Scales of resolution used to investigate the deposition of pollutant gases to plant canopies, individual leaves, and tissues of the leaf interior.

TABLE 10.1
Methodologies for characterizing the deposition of gases to individual leaves and
plant canopies

Methodological principle	Type	Reference
Mass balance	Split leaf	Petit et al. (1977)
	Individual leaf	Legge et al. (1977)
	Whole plant	Bennett and Hill (1973)
	Low stature crop	Unsworth et al. (1984)
Surface/tissue accumulation	—	Lindberg and Lovett (1985)
Isotope addition	Radioactive	de Cormis (1968)
	Stable	Hutchinson (1972)
Isotope dilution	Stable	Okano et al. (1986)
Process-level model	Leaf interior	Danckwerts (1970)
	Individual leaf	Mansfield (1973)
	Plant canopy	Wesely and Hicks (1977)
Micrometeorology	Eddy correlation	Wesely and Hicks (1977)
	Flux gradient	Galbally (1971)

plants/shoots (Bennett and Hill, 1973), stands/swards/low-stature crops (Unsworth *et al.*, 1984) and individual surfaces of hypostomatous and hyperstomatous leaves using split-leaf cuvettes (Petit *et al.*, 1977; Harris *et al.*, 1983). The mass balance approach is versatile, having been successfully applied to a range of pollutant gases and adapted for use in both the laboratory (Hill, 1971) and field (Hallgren *et al.*, 1982; Atkinson *et al.*, 1986). One of the principal assets of the mass balance approach is that it allows concurrent measures of multiple gases (e.g., CO_2, H_2O, pollutant gas) so that the physiological status of the plant can be continuously monitored during exposure.

Isotopic labeling is also a powerful methodology for investigating deposition rates of trace gases to vegetation surfaces, but it has been underutilized. Two methodological approaches are (1) the direct addition of a labeled gas to a leaf (Garsed, 1985) and the dilution of a label in previously spiked tissues (Okano *et al.*, 1986). Both radioactive (e.g., $^{35}SO_2$, $H_2^{35}S$) and stable (e.g., $^{34}SO_2$, $^{15}NO_2$, $^{15}NH_3$) isotopes have been used. The principal advantage of isotopic labeling is the opportunity to (1) characterize the distribution of deposition sites both on the leaf surface and leaf interior, and (2) to follow the labeled gas and its metabolites within the physiologically active tissues of the leaf interior (e.g., Garsed and Read, 1977).

Plant surface/tissue accumulation methods have been used but are limited to those pollutant gases exhibiting extremely high rates of deposition (e.g., HNO_3, hydrogen fluoride). The technique may significantly underestimate deposition of pollutant gases due to the translocation out of the leaf and re-emission and/or volatilization (Taylor *et al.*, 1985).

Process-level modeling of pollutant gas deposition is used most commonly to address the dynamic nature of the deposition. The most common technique operates at the level of individual leaves and is based on the resistance analogy (Unsworth *et al.*, 1976). Anticipated development and application of models with finer scale resolution (e.g., the two-layer stagnant film model) will help address processes operating within the gas phase and liquid phase of the leaf interior, thus bridging the gap between leaf physiology and cellular biochemistry.

Micrometeorological techniques can be used to measure or calculate the flux of a trace gas to vegetated surfaces in the field, given that the landscape surface is extended, flat, and horizontally homogeneous. The eddy correlation technique (Wesely and Hicks, 1977) measures vertical turbulent flux directly by calculating the mean covariance between fluctuations of the vertical velocity pollutant gas concentration. The flux gradient technique (Galbally, 1971) provides an inferential estimate of vertical fluxes to landscapes and is based on the proportionality between the product of the mean vertical concentration gradient and an eddy exchange coefficient. More detailed discussion of micrometeorological approaches are presented in Businger (1986).

10.3 PROCESSES GOVERNING DEPOSITION OF GASES AND PARTICLES

Deposition of gaseous pollutants to individual leaves and plant canopies occurs because of the chemical potential gradient between the atmosphere and sites of deposition, either on exterior foliar surfaces or cells of the leaf interior. Deposition (J) can be represented in a one-dimensional analysis as the ratio of the chemical potential gradient (C) to the sum of physical, chemical, and biological resistances ($\sum R$) to diffusion along the source-to-sink pathway. Mathematically, this is expressed in a form analogous to Ohm's law (for resistances in series):

$$J = C / \sum R$$

The C is commonly set equal to that of the concentration in the turbulent atmosphere, assuming a sink concentration equal to zero (Black and Unsworth, 1979). This assumption is not valid for those gases having a biogenic source or compensation point within the leaf interior (e.g., NH_3, NO), in which case the sink concentration is greater than zero (Farquhar *et al.*, 1980). The $\sum R$ is characterized as a catena of serial resistances in both the gas (atmosphere and leaf interior) and liquid phases.

The conductivity of the diffusive pathway for gaseous pollutants is a consequence of resistances operating in series and in parallel. The quantitative role of a single site of resistance operating in series can be estimated as the proportion it contributes to the total pathway resistance. The units of diffusive resistance at the individual leaf level (R_l) are s cm^{-1}, calculated as the ratio of concentration to deposition. At the whole canopy level, it is common to characterize pollutant deposition to a landscape with the deposition velocity (V_d), expressed in units of cm s^{-1}. Whereas the reciprocal of leaf resistance (i.e., leaf conductance, g_l) has the same units as V_d (cm s^{-1}), g_l and V_d characterize processes operating at different levels of resolution and are not quantitatively relatable without recognition of the canopy's leaf area index, nonfoliar sites of deposition, and the role of aerodynamic boundary layer resistance in governing deposition (Hosker and Lindberg, 1982).

10.4 PATHWAYS OF DRY DEPOSITION IN PLANT CANOPIES AND INDIVIDUAL LEAVES

10.4.1 Plant canopy

Surface exchange processes occurring between the atmosphere and plant canopies are characterized by an array of sources and sinks, varying with the plants' physiological state, species composition, canopy architecture, atmospheric turbulence, chemical characteristics of pollutant, and the environment. At the landscape level, pollutant molecules may be deposited to individual leaves (both exterior and interior surfaces,) stems/boles, and soil surface elements (e.g., soil, litter, water), and the effectiveness of these competing sinks for pollutants is not equivalent. For example, in controlled-environment gas-exchange studies with trace levels of NO_2, deposition to landscape surface elements characteristic of eastern forests ranged from 0 to 2·8 nmol cm^{-2} h^{-1} (Fig. 10.2) and were lowest for inanimate elements (e.g., teflon, water), intermediate for bark and individual leaves, and greatest for litter (P. J. Hanson and K. E. Rott, personal communication). For most pollutant gases, the significance of these various sites of deposition is poorly characterized. In agricultural ecosystems, O_3 deposition to the soil may constitute 30 to 40% of the total deposition to the terrestrial landscape (Leuning *et al.*, 1979a), whereas the value for more water-soluble pollutants (e.g., SO_2) may exceed 50%. The capacity of various sinks within terrestrial landscapes to remove pollutants from the atmosphere is important because the concentration profile within the canopy will vary according to the effectiveness of the sinks (Bennett and Hill, 1975).

10.4.2 Individual leaf

10.4.2.1 Leaf surface

At the level of the individual leaf, the analysis of deposition sites for pollutant gases is highly significant for physiological investigations but analytically difficult to address. Unlike that for CO_2 assimilation, cuticular resistance (R_c) to pollutant gases is not infinitely high and may be lower than that of the stomatal resistance (R_s). For example, the ratio of surface ($J_{SURFACE}$) to total (J_{TOTAL}) deposition can range from 0·1 to greater than 0·8 for gases of varying physicochemical properties (Fig. 10.3). This pattern is inversely correlated to the solubility of the gas in water (Taylor *et al.*, 1983), suggesting that in the

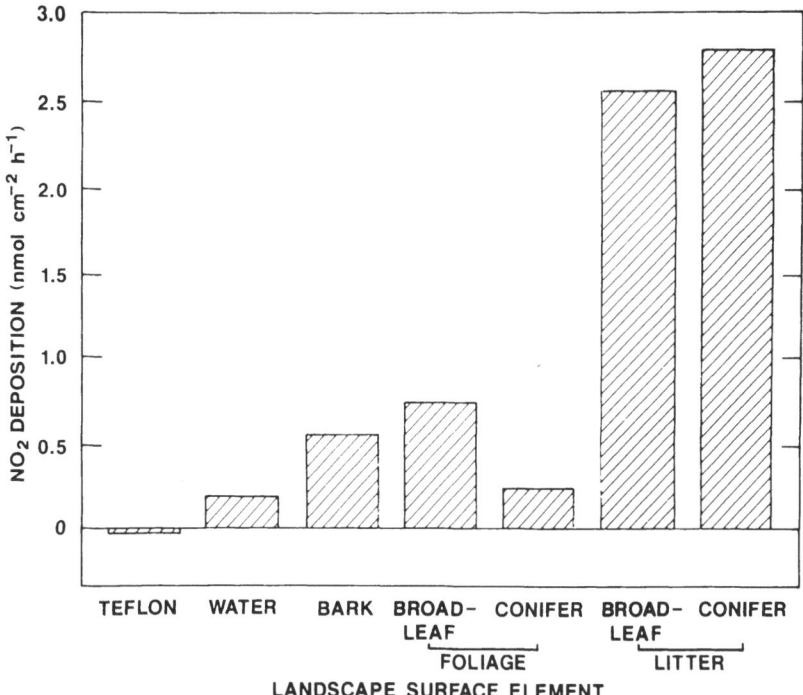

Fig. 10.2. Deposition of NO_2 to various elements of a terrestrial landscape. For comparison of deposition among elements, deposition is normalized to projected surface area. The NO_2 concentration in the atmosphere averaged 0·055 ppm or 142 μg m^{-3} (K. E. Rott and P. J. Hanson, personal communication). The deposition to teflon, an inanimate, inert surface, is shown as a control/reference surface.

absence of leaf surface wetness, $J_{SURFACE}$ is a function of lipid solubility (Gaffney *et al.*, 1987). Lendzian (1982) reported that the reactivity of cuticular components to many of the most common trace gases (e.g., O_3, SO_2) exceeds that for H_2O vapor by approximately an order of magnitude, suggesting that, unlike effluxing H_2O vapor molecules, the cuticular pathway may be quantitatively important for some trace gases. Thus, in contrast to that of CO_2 and H_2O, the diffusive path for pollutant gases may include a significant surface deposition component that is dependent on the chemical characteristics of the gas and reactivity of the cuticular surface. Experimentally, surface deposition includes adsorption and chemical oxidation/reaction reactions with the structural materials of the leaf cuticle.

10.4.2.2 *Leaf interior*

Within the leaf interior, the transport of pollutant gases from the substomatal chamber/intercellular space to cell surfaces is not well understood. The processes governing deposition at this scale of resolution can be addressed by the two-layer, stagnant film model (Fig. 10.4) of Danckwerts (1970). In this

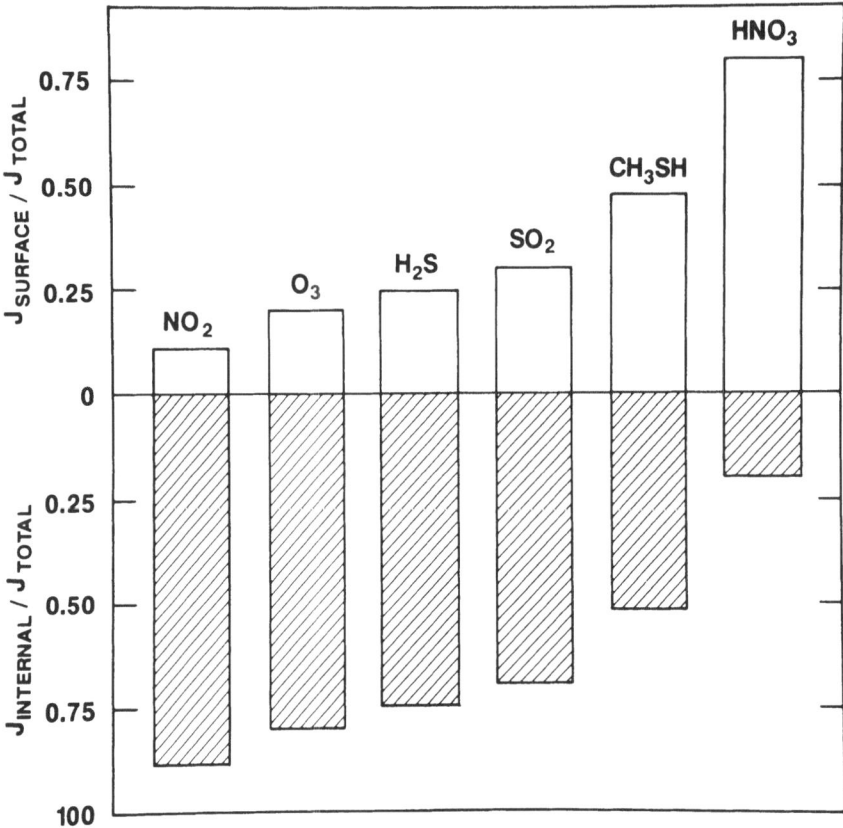

Fig. 10.3. Gaseous pollutant deposition to the leaf surface ($J_{SURFACE}$) versus that to the leaf interior ($J_{INTERIOR}$) for a range of physicochemically dissimilar pollutant gases (P. J. Hanson, personal communication). Abbreviation: CH_3SH, methyl mercaptan.

model, deposition is governed by molecular processes in the gas phase and liquid phase, and diffusion and chemical partitioning of molecules across the phase interface is taken into account (e.g., partition coefficient). In many applications in the environmental sciences, the concept of a stagnant layer is unrealistic because turbulence in both phases effectively eliminates a stagnant layer (Broecker and Peng, 1974). However, for atmosphere–leaf exchange, consecutive stagnant layers (i.e., transfer via molecular diffusion) exist because stomata create an environment in the gas phase of the leaf interior that is physically separated from the turbulence exterior to the leaf boundary layer, and the apoplasm is uncoupled from the symplast's physical/bulk transport processes. The pathway through the stagnant layer begins in the unsteady laminar boundary layer outside the leaf, fully develops in the stomata and substomatal chamber, and extends into the intercellular space of the leaf interior. Because the diffusive path within the intercellular space of the leaf interior may be highly convoluted, the potential thickness of the unstirred stagnant layer in the gas phase is not simply the distance between the abaxial

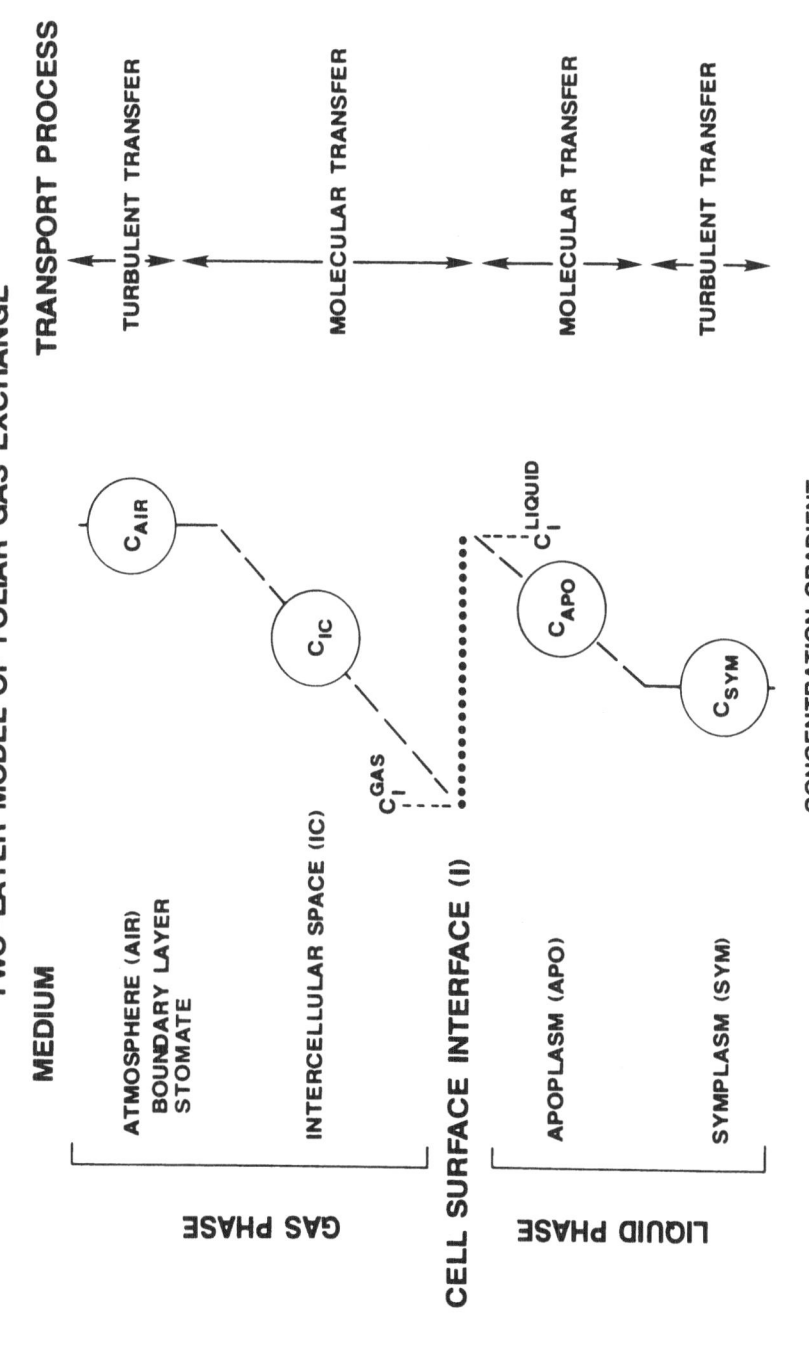

Fig. 10.4. Application of the two-layer, stagnant film model (Danckwerts, 1970) to characterize the processes governing the deposition of pollutant gases within the leaf interior. Abbreviations: concentration (C) in free atmosphere (C_{AIR}), intercellular space (C_{IC}), cell surface in the gas phase (C_I^{GAS}) and liquid phase (C_I^{LIQUID}), apoplasm (C_{APO}), and symplasm (C_{SYM}).

and adaxial surfaces of the leaf, which is commonly a few tenths of a millimeter. Many consequences of a pronounced stagnant layer for CO_2 exchange at the leaf–atmosphere interface have been explored by Leuning (1983) and Smith (1985).

As a site of gaseous pollutant deposition, the leaf interior is unique because it is a highly porous and physiologically complex organ. The surface area for deposition in the leaf interior (A_{mes}) is at least an order of magnitude greater than the projected leaf area (A), and typical values range from 15 to 40 for mesophytes and as high as 70 for xerophytes (Nobel and Walker, 1985). Assuming an analogy to CO_2 assimilation, deposition of gaseous pollutants should increase with A_{mes}/A because the leaf's mesophyll resistance would be reduced (Nobel and Walker, 1985; Parkhurst, 1986). Given the leaf's internal morphology, the ease of diffusion and site of deposition for gaseous pollutants are likely to vary appreciably among gases and in accordance with their physical and chemical properties. A notable example is the dissimilarity in the sites of CO_2 deposition and H_2O evaporation within the leaf interior. Whereas the substomatal chamber is the source of 80 to 90% of the transpired H_2O, the same region of the leaf assimilates only 10 to 20% of the total CO_2 (Cowan, 1977). The majority of the CO_2 molecules are deposited in the photosyntheti-cally active mesophyll tissue, which is consistent with its physiological function. In an analogous fashion, it is proposed that the leaf surface and substomatal cavity are deposition sites for highly reactive, water-soluble gases (e.g., HNO_3, SO_2) (Taylor and Tingey, 1982), whereas the mesophyll tissue is the primary site of deposition of less soluble gases (e.g., O_3) (Taylor *et al.*, 1982) (Fig. 10.5). Process-level models of diffusion that address chemical potential

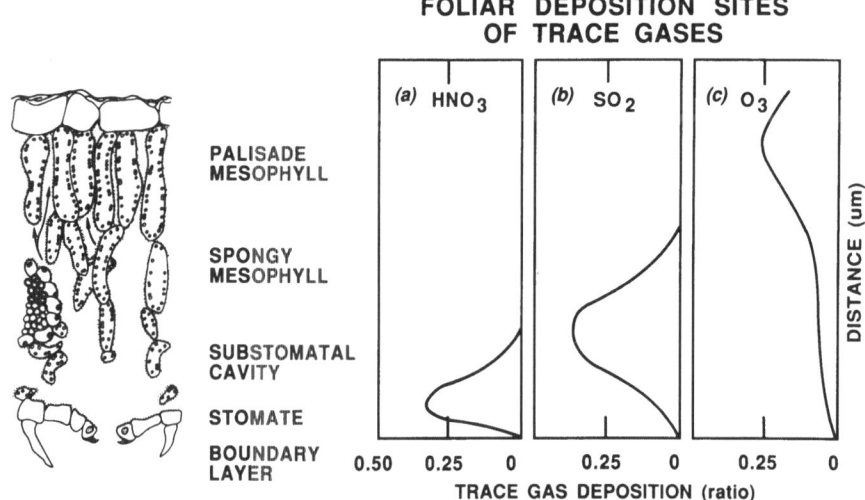

FOLIAR DEPOSITION SITES OF TRACE GASES

Fig. 10.5. Proposed deposition sites on the leaf surface, substomatal cavity, and tissues of the leaf interior for (a) HNO_3, (b) SO_2, and (c) O_3. The three pollutants differ significantly in their physicochemical properties, notably their chemical reactivity in solution and their H_2O solubility. The relationship is expressed as the frequency that pollutant molecules are deposited at a particular location along the diffusive pathway into the leaf.

gradients and diffusion processes in a finite element, three-dimensional framework (Parkhurst, 1986) present a more highly defined approach to characterizing deposition of pollutant gases.

The diffusion of a trace gas within the stagnant layer of the leaf's intercellular space and across surfaces of mesophyll cells is influenced by processes operating at the molecular level. Parkinson and Penman (1970) proposed that the bidirectional flux of gas molecules in the leaf interior influences deposition rates, as effluxing H_2O molecules physically collide and chemically react with influxing pollutant gases within the stomata and substomatal cavity. This aspect of CO_2 exchange is now routinely accounted for in estimating net CO_2 assimilation rates by individual leaves using gas-exchange systems (e.g., Ball, 1987). This phenomenon has a parallel in deposition research, being referred to as Stefan's flow or diffusiophoresis (Fowler, 1980). The importance of molecular processes vary among gases, being most important for gases that are large molecules (Slinn, 1987), highly reactive, and water soluble (Taylor *et al.*, 1983).

Molecular interactions within the gas phase of the leaf interior are not limited to H_2O vapor alone. Chemical reactions may affect pollutant gases within the leaf interior (Tingey and Taylor, 1982). A variety of hydrocarbons (i.e., terpenoids) are produced during secondary plant metabolism, volatilized into the gas phase of the leaf interior, and diffused through the stomata to the free atmosphere. Many of these are known to undergo ozonolysis and thus may consume O_3 molecules. Based on ozonolysis reaction kinetics and a molecule's residence time in the gas phase of the leaf interior, it is postulated that only a few of the terpenoid emissions are likely to undergo significant rates of ozonolysis (Tingey and Taylor, 1982). The possibility that these gas-phase interactions are involved directly in causing physiological dysfunction has been proposed for volatile organics in general (Gaffney *et al.*, 1987) and ethylene (Mehlhorn and Wellburn, 1987) specifically. However, based on measurements of hydrocarbons in the boundary layer of foliar surfaces, Russi (1986) concluded that the ozonolysis reaction kinetics are insufficient to be a significant source of phytotoxic agents such as peroxides.

Pollutant molecules in the intercellular space or substomatal cavity are partitioned across the gas-to-liquid interface at a rate determined by the solubility of the gas in the apoplasm (Nobel, 1974) and its chemical reactivity in the liquid phase (Liss and Slater, 1974). Hill (1971) reported a sevenfold difference in deposition rates among five gaseous pollutants and concluded that the variation was due to differences in H_2O solubility. In a comparison of deposition rates of five sulfur-containing gases of dissimilar physicochemical characteristics, Taylor *et al.* (1983) concluded that water solubility accounted for 40 to 50% of the variation in deposition. However, chemical reactivity in the liquid phase also plays a role because some gases (e.g., SO_2) exhibit the phenomenon of flux enhancement due to the ionization of the gas in solution (Liss, 1971) so that the partitioning of the gas across the gas–liquid interface is greater than that predicted by Henry's Law coefficient (Liss and Slater, 1974). Notable examples are SO_2 (Liss, 1971) and CO_2 (Bolin, 1974).

10.5 CONTROL OF DEPOSITION RATES TO PLANT CANOPIES AND INDIVIDUAL LEAVES

10.5.1 Plant canopy

The canopy's aerodynamic (R_a) and quasi-boundary layer (R_b) resistances vary substantially as a function of the canopy's architecture, and typical values are reported in Fig. 10.6(a) and (b) for a deciduous forest, tall herbaceous crop (*Zea mays*), and short-stature crop (*Hordeum vulgare*). Midday R_a values (Fig. 10.6(a)) vary inversely as a function of the aerodynamic roughness of the canopy, and midday values are of the order of 0·10, 0·17, and 0·25 s cm^{-1}, respectively, for the three canopy types. A typical diurnal course of R_b (Fig. 10.6(b)) reflects a mix of surface roughness and molecular diffusivities, and typical values over *Z. mays* are 0·30 s cm^{-1} for O_3 (Wesely *et al.*, 1978),

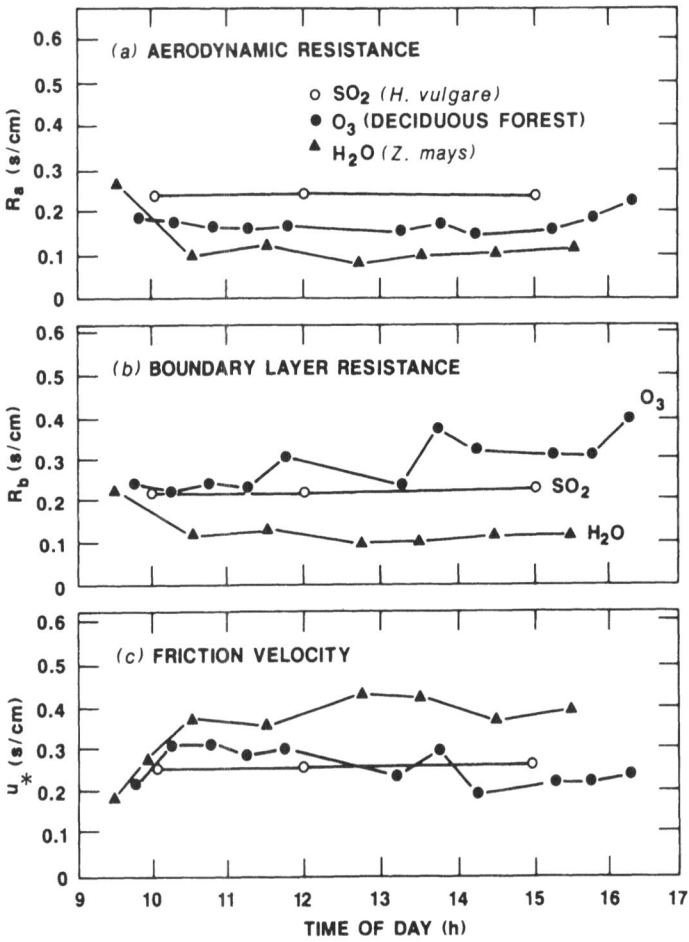

Fig. 10.6. Pattern of (a) aerodynamic resistance, (b) boundary layer resistance and (c) friction velocity as a function of the time of day for three gases above three different plant canopies. Data sources are Wesely *et al.* (1978) for *Z. mays*, Verma *et al.* (1986) for a deciduous forest, and Fowler and Unsworth (1979) for *H. vulgare*.

$0.20\,s\,cm^{-1}$ for SO_2 over *H. vulgare* (Fowler and Unsworth, 1979), and $0.10\,s\,cm^{-1}$ for H_2O over a deciduous forest (Verma *et al.*, 1986). Theoretical values for the combination of $R_a + R_b$ range from $2.0\,s\,cm^{-1}$ over low-stature crops under calm conditions to $0.20\,s\,cm^{-1}$ over tall crops under greater turbulence (Fowler, 1985).

Canopy surface resistance (R_c) is most often the primary resistance to the deposition of the common pollutant gases (Wesely, 1983) with the exception of HNO_3 (Fowler, 1980; Huebert and Roberts, 1985). Unfortunately, few field studies of dry deposition have been accompanied with measurements of stomatal resistance (R_s) so that the specific role of stomatal and nonstomatal factors in governing deposition under field conditions can not be clearly resolved. Leuning *et al.* (1979*b*) demonstrated that R_s for shaded and sunlit leaves of *Z. mays* were of the order of 10 to $16\,s\,cm^{-1}$ and 4 to $6\,s\,cm^{-1}$, respectively, whereas R_c values for the entire plant canopy ranged from 2 to $3\,s\,cm^{-1}$. The range in R_c values was associated with the degree of canopy closure and crop development and, in general, decreased as the leaf area index increased. Canopy resistances are typically gas-specific; Wesely *et al.* (1978) reported midday R_c values ranging from $0.30\,s\,cm^{-1}$ for O_3 to $1.25\,s\,cm^{-1}$ for H_2O.

Differences in R_c under field conditions are also associated with species-specific attributes that influence the affinity of the leaf surface for pollutant gases (Fig. 10.7). Typical diurnal variations of R_c for O_3 for *Z. mays, Pinus taeda* (loblolly pine), and well-watered *Glycine max* (soybean) demonstrate a sixfold difference between the R_c for the two herbaceous species, and the variation is likely to be associated with differences in species' intrinsic aspects of stomatal physiology and plant metabolism.

Surface wetness can have a substantial effect on the deposition of many pollutant gases to vegetation, and the case is particularly well documented for SO_2. Deposition to surfaces wetted with dew changes over time, being initially at a maximum as a function of the gas high solubility of the gas in H_2O.

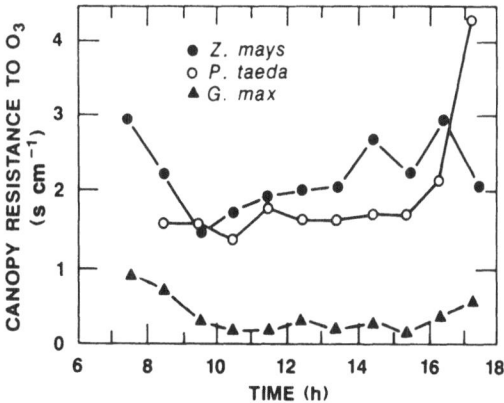

Fig. 10.7. Canopy resistance to O_3 for three different plant canopies under field conditions (Wesely, 1983).

TABLE 10.2
Selected physical and chemical properties of some common gases

Chemical	Molecular weight	Molecular diameter[a]	Solubility in H_2O[b]	Air diffusivity ratio[c]	V_d[d] Leaf	V_d[d] Teflon
Nontoxic gases						
N_2	28·01	0·260	1·04	0·80	—	—
O_2	32·00	0·267	2·18	0·75	—	—
H_2O	18·02	0·225	∞[e]	1·00	—	—
CO_2	44·01	0·378	76·5	0·64	—	—
Pollutant gases						
O_3	48·0	0·36	21·9	0·61	0·2–0·7	0
H_2O_2	34·01	0·294	∞	0·73	na[e]	na
NO	30·0	0·263	3·28	0·77	0·01–0·1	na
NO_2	46·01	0·366	decomp[e]	0·63	0·1–0·8	~0
HNO_3	63·01	0·427	∞	0·53	0·5–5·0	~0
NH_3	17·03	0·234	52789	1·03	0·2–0·6	na
PAN	121·05	~0·688	na	0·37	0·1–0·6	na
SO_2	64·06	0·393	3559	0·53	0·2–3·0	0·1
H_2S	34·08	0·266	195·1	0·73	0·2–0·4	0·06

[a] Expressed in units of nanometers and calculated from Bowen (1958).
[b] Expressed in units of μmol cm^{-3} and calculated from Weast (1986).
[c] The ratio is the square root of the quotient of the molecular weight of H_2O divided by the molecular weight of another gas.
[d] Expressed in units of cm s^{-1} and obtained from McMahon and Denison (1979), Sehmel (1980), van Aalst and Diederer (1985), and P. J. Hanson and G. E. Taylor, Jr (personal communication).
[e] ∞ = soluble in all proportions; decomp = soluble and decomposes in H_2O; na = not available.

Subsequently, deposition declines as the droplets are acidified, which decreases the water solubility of SO_2 (Fowler, 1978). Rain does not exhibit a comparable effect on SO_2 deposition because the droplets are commonly in equilibrium with the atmospheric pollutant concentrations prior to deposition. Since pollutant gases differ in H_2O solubility by several orders of magnitude (Table 10.2), differences in the sink capacity of dew for pollutant gases exist (Pierson *et al.*, 1986).

10.5.2 Individual leaf

The deposition of pollutant gases to an individual leaf is dependent on a mix of physical and chemical factors depicting the physicochemical properties of the trace gas and the effectiveness of the surface for deposition. Theoretical considerations using compound molecular weights or actual measures of binary diffusivity coefficients in air (Andrussow, 1969) have shown that the diffusion of a gas through the leaf's boundary layer and stomata can be estimated by numerical analogy to H_2O vapor exchange (assumes equivalent pathlengths of diffusion in the stagnant gas phase) using either measured binary diffusivity

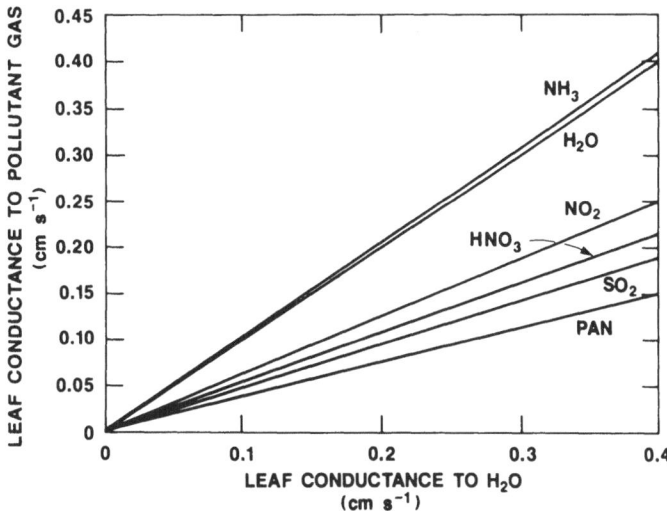

Fig. 10.8. Relationship between leaf diffusive conductance to H_2O and stomatal conductance to pollutant gases that differ in their physicochemical properties.

coefficients (Ball, 1987) or ratios of molecular weights (Gaastra, 1959), and it is argued that the former is a more accurate procedure (Farquhar and Sharkey, 1982). The relationship between stomatal conductance to H_2O vapor (reciprocal of stomatal resistance) and several pollutant gases (Fig. 10.8; Table 10.2) indicates that many pollutant gases diffuse at a slower rate than that of effluxing H_2O vapor molecules.

Identifying the specific physicochemical characteristics of pollutant gases governing deposition can provide an *a priori* means of predicting the reactivity of a gas with individual leaves. An empirically derived model, based on the parameters of molecular diameter and H_2O solubility (Taylor *et al.*, 1983), explained much of the variation in deposition of five sulfur-containing gases. Hill (1971) and Bennett and Hill (1973) measured uptake of several pollutant gases by an alfalfa canopy under controlled conditions and found that deposition rates among gases was in the following order: $HF > SO_2 > O_3 > NO_2 > PAN > NO$. Additional observations of trace gas uptake to plant surfaces under a range of controlled laboratory and/or field conditions would be useful in more fully characterizing the sites of deposition and developing a process-level model based on the principal physicochemical processes governing the flux rates.

The properties of the plant surface controlling pollutant deposition rates are either morphologically or physiologically based characteristics. The major morphological characteristics (i.e., leaf shape, surface features [trichomes, salt glands], total surface area) are commonly perceived as being univariant over periods of days to weeks in duration, corresponding to changes in plant ontogeny. Conversely, the more physiologically based characters relate to stomatal conductance. These factors change over time spans measured in seconds or minutes and reflect the dynamic nature of stomata and internal limitations to pollutant deposition.

The degree to which stomatal aperture influences H_2O and CO_2 exchange between the free atmosphere and the leaf interior is well documented (Raschke, 1975; Mansfield, 1986) and has been modeled extensively (e.g. Farquhar and Wong, 1984). Stomata have similarly been shown to exhibit considerable control over trace gas deposition to individual leaves, including NH_3 (Aneja *et al.*, 1986), NO_2 (Hill, 1971), O_3 (Taylor *et al.*, 1982), PAN (Garland and Penkett, 1976), and SO_2 (Winner and Mooney, 1980). Because stomata are the variable sites of resistance through which pollutants must diffuse before reaching sites of deposition in the leaf interior, an understanding of the environmental controls over stomatal aperture is essential before internal deposition can be predicted. Light (Chambers *et al.*, 1985), leaf water potential (Stewart and Dwyer, 1983), and vapor pressure deficit (Meidner 1987) play major roles in controlling stomatal aperture under normal field conditions and are reported to influence trace gas deposition. Less commonly, CO_2 levels (Raschke, 1975), extremes of temperature (Chambers *et al.*, 1985), and the presence of phytotoxic trace gases affect stomatal function (Mansfield and Freer-Smith, 1984).

Although stomatal conductance is probably the source of most physiologically based control over pollutant deposition to individual leaves, it is not the only site of control. The presence of additional mesophyll or cell-level resistances to pollutant deposition has been proposed (Taylor and Tingey, 1982; Tingey and Taylor, 1982; Garsed, 1985). Although changing resistances to gaseous pollutant deposition at the cell level are not completely understood, they include pollutant-induced changes in membrane permeability, altered enzyme activities, changing solute concentrations in the apoplasm (e.g., hydrogen ion, sulfate, nitrate), and efficacy of free-radical scavengers (Tingey and Taylor, 1982; Tingey and Olszyk, 1985).

10.6 MODELING DEPOSITION TO INDIVIDUAL LEAVES AND PLANT CANOPIES

10.6.1 Plant canopy

At the level of the plant canopy, the two methodological approaches to investigate trace gas deposition focus on the canopy operating either as a single layer ("big-leaf" model) or consisting of layers (multi-layer model) (Hicks *et al.*, 1986). Both methodologies assume a one-dimensional framework for transfer and a relatively flat, horizontally homogeneous surface. Multi-layer models address surface exchange processes involving the bi-directional fluxes of trace gases, whereas the "big leaf" model accommodates solely unidirectional trace gas flux.

The resistance network for the big-leaf model (Fig. 10.9) was initially developed by Sinclair *et al.* (1976) and identifies the primary resistances to deposition as the aerodynamic (R_a), diffusive boundary layer (R_b), and canopy

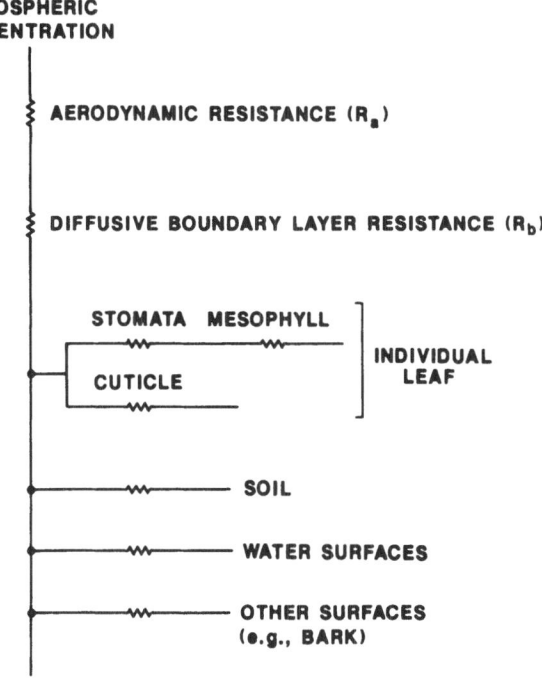

Fig. 10.9. Resistance analogy used to investigate the deposition of pollutant gases to terrestrial landscapes (Baldocchi *et al.*, 1987).

or surface resistance (R_c). The aerodynamic resistance, which is influenced by wind speed, surface roughness, and atmospheric stability (tendency for air parcel buoyancy to be increased or suppressed) is expressed as

$$R_a = (ku^*)^{-1}[\ln((z - d)/z_0) - \Psi_h],$$

where k is von Karman's constant, u^* is the friction velocity (cm s^{-1}), z is height (cm), z_0 is the surface roughness length (cm), d is the zero plane displacement, and Ψ_h is the integral form of the diabatic stability correction for heat and mass transfer (Wesely and Hicks, 1977; Hosker, 1986). The u^* is related to momentum transfer and computed from measurements of the wind speed, whereas d and z_0 are a function of the canopy's structure and roughness. Theoretically, the ranges of z_0 and d relative to that of canopy height are 4 to 15% and 40 to 90%, respectively (Shaw and Pereira, 1982).

The quasi-laminar boundary layer is introduced because the resistance to mass and energy transfer differs from that of momentum (Thom, 1975). Near the leaf surface, mass and energy transfer are controlled by molecular properties of the fluid, whereas the rate of momentum transfer is governed by bluff-body effects of the various surface elements. The excess resistance is a function of wind speed, surface properties, and molecular diffusivity of the trace gas.

The surface resistance for deposition is a composite term, consisting of stomatal and mesophyll resistances operating in series with each other and in parallel with the resistance occurring at the cuticle (R_{cut}), soil/litter (R_{soil}), surface moisture content (R_{wet}), and any miscellaneous surface component (e.g., bark) (R_{misc}). Stomatal resistance is driven by photosynthetically active radiation, temperature, vapor pressure deficit, CO_2 concentration, leaf water potential, leaf age and position, and the concentration of the pollutant (Jarvis, 1976). The procedure for making computations of the canopy's stomatal resistance is based on a canopy radiative transfer model that accounts for the distribution of sunlit and shaded leaves and the incident radiation on those leaves (Norman, 1980; Baldocchi *et al.*, 1987).

The mesophyll resistance term addresses those processes operating in the substomatal cavity and intercellular space of the leaf interior and principally dictated by the solubility of the gas in water and the internal cell surface area (O'dell *et al.*, 1977). This analysis takes into account Henry's Law coefficient, cell surface area, the trace gas diffusivity coefficient, depth of the mesophyll tissue, and stomatal frequency. The cuticular resistance term is a function of the chemical characteristics of the trace gas, and multiple physical and chemical features of the leaf surface including the surface area, presence of surface waxes and exudates, and the amount of pubescence (Hosker and Lindberg, 1982).

The alternative model of canopy deposition, the multi-layer model (Waggoner, 1975), incorporates many of the same pathways as the 'big-leaf' model but segregates the canopy into discrete layers (Baldocchi, 1988; Meyers, 1987). Differences between the two methodologies occur in the treatment of the aerodynamic and leaf boundary layer resistances. Conceptually, flux divergence at any layer in the canopy is balanced by the source/sink strength:

$$dF(z)/dz = -a(z)c(z)/(r_{al}(z) + r_{sl}(z))$$

where a is the leaf area density, r_{al} is the leaf boundary layer resistance, r_{sl} is the leaf surface resistance (function of stomata, cuticle, and mesophyll resistances), and c is the concentration of the trace gas at height z. The effectiveness of the soil surface as a deposition site is accounted for by using the empirically derived Freundlich equation, in which trace gas deposition is governed by Van der Waals and dipole forces, hydrophobic bonding, charge transfer and hydrogen bonding, ligand and ion exchange, and chemisorption bonding (Elrick *et al.*, 1975). Total canopy deposition is computed by integrating the individual-layer fluxes. Mathematically, this scheme yields only one equation with two unknowns (F and c). To solve for F and c, a first- or higher-order closure scheme must be adopted; first-order closure is accomplished by assuming

$$F(z) = -K(z)\, dc(z)/dz$$

and by prescribing a relationship for K, the eddy exchange coefficient for c. A higher-order turbulence closure model for trace gas deposition is based on the governing conservation equations for mass, momentum, and energy (Meyers, 1987). It improves on the multilayer resistance model because it is not constrained by the assumption that the turbulent transfer is driven by mean gradient and a time-averaged eddy exchange coefficient (Finnigan and Raupach, 1987).

At the level of the plant canopy, deposition of a specific pollutant gas cannot be accurately characterized with a univarant deposition velocity (V_d) since this parameter is composed of many environmental, physiological, and morphological variables. The net outcome is that the V_d for even a single trace gas varies substantially in space (vertically and horizontally within a canopy) and time (diurnally, seasonally, annually). The relationship between V_d and some of these controlling variables can be investigated with the "big leaf" model (see subsequent section for discussion) that incorporates both meteorological and physiological parameters (Baldocchi, *et al.*, 1987). The theoretical response curves for O_3 deposition to a canopy of *G. max* and *Quercus* spp. are shown in Fig. 10.10 as a function of radiation regime (PAR) and leaf water potential. In both species, increasing PAR from 50 to 375 W m^{-2} increased V_d of O_3, with the rate of increase being most pronounced (1) at the lower levels of PAR and (2) for the hardwood versus the herbaceous species. In the case of leaf water potential, the model illustrates pronounced reductions in V_d with

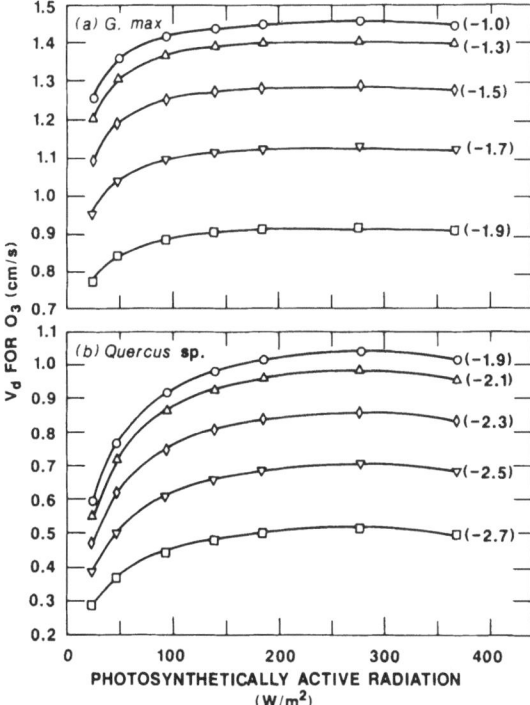

Fig. 10.10. Relationship between V_d for O_3 in a *G. max* canopy (a) and *Quercus sp.* canopy (b) as a function of leaf water potential and photosynthetic active radiation (PAR). The parenthetical data are leaf water potentials in units of MPa (D. D. Baldocchi, personal communication).

more negative water potential values, and the rate of decline exhibits the species-specific attributes associated with critical values that affect stomatal closure.

10.6.2 Individual leaf

At the level of an individual leaf, the analysis of trace gas deposition is commonly achieved using analogue resistance models as developed by Gaastra (1959) for CO_2 assimilation and subsequently applied to several gaseous air pollutants (Mansfield, 1973; Bennett and Hill, 1973). The approach is based on a mechanistic understanding of the physicochemical and physiological processes governing transfer to the leaf surface and leaf interior and assumes an analogy between the diffusive paths of H_2O vapor and that of the trace gas (Fig. 10.11). This approach quantifies the influence of boundary layer and stomatal resistance to pollutant deposition and any residual resistance term is attributed to physiological and/or biochemical processes affecting the chemical potential gradient of the gas or the diffusive resistance of the pathway. It is commonplace to refer to this residual term as the mesophyll resistance although it may not be physically located on or within the mesophyll cells of the leaf interior.

The application of resistance analogy must recognize some of the unique aspects of trace gas deposition to individual leaves that are not equal to those of effluxing H_2O molecules (Fig. 10.11). These include (1) variable deposition to the leaf surface as a function of genotype and environmental conditions; (2) existence of a residual resistance term; (3) potential re-emission of the trace gas by metabolically active tissues of the leaf interior; (4) gas-phase interactions in the substomatal cavity and intercellular space that may either consume the trace gas or physically react with it, thereby altering the net direction of diffusive flow (e.g., diffusiophoresis) or changing the effective diffusive coefficient; and (5) diffusive pathlengths that are not equivalent.

Analogue resistance models tend to overestimate the role of stomatal resistance to CO_2 assimilation (Farquhar and Sharkey, 1982). Several alternative methodologies have been offered that suggest that the stomatal limitation is not as great as previously assumed, implicating mesophyll factors in governing the rate of CO_2 assimilation. These approaches also offer some promising alternatives for modeling trace gas deposition to individual leaves.

Whereas the analogue resistance model focuses on individual leaves, the processes operating within the leaf interior in both the gas and liquid phases have only recently been addressed through finite element, modeling techniques that account for the diffusion and assimilation of gases such as CO_2 from the stomata into the palisades tissues in the leaf interior (Parkhurst, 1986). The modeling effort treats individual stomata as point sources for gas molecules and addresses the chemical gradient in the gas phase of the leaf interior. The further development and application of this type of modeling approach in conjunction with the two-layer stagnant film model is a possible alternative technique to investigate these leaf interior processes and provide a means of linking the disciplines of atmospheric chemistry and plant biochemistry.

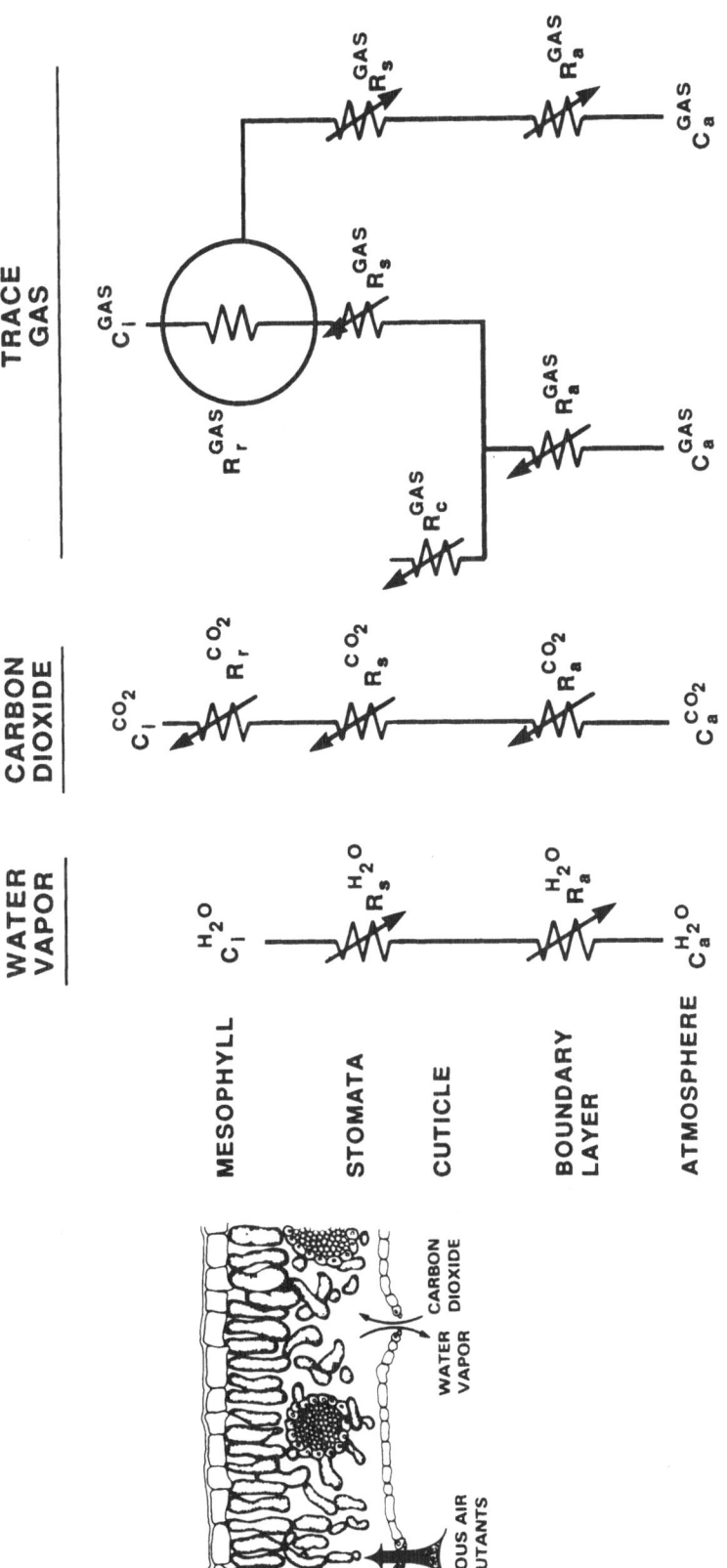

Fig. 10.11. The analogue resistance approach as originally developed for H_2O and CO_2 (Gaastra, 1959) and subsequently applied to gaseous pollutants (Bennett *et al.*, 1973; Unsworth *et al.*, 1976; Taylor and Tingey, 1982).

10.7 RELATIONSHIP BETWEEN DEPOSITION AND PLANT INJURY

The justification for characterizing pollutant flux into the leaf interior rests with the toxicological principle that biological responses are better defined by the biologically effective dose of the toxic chemical (i.e. pollutant dose reaching physiologically active sites of phytotoxicity) than by the administered dose (O'Flaherty, 1986). Relationships between administered dose and biological response are often nonlinear, reflecting the lack of proportionality between effective and administered dose. Curvilinear relationships often indicate capacity-limited or saturable systems, a feature common throughout toxicological literature and one whose reaction kinetics are well characterized empirically and conceptually (O'Flaherty, 1981). Nonlinear responses are common in air pollution studies. For example, Black and Unsworth (1979) observed that the responsiveness of net CO_2 assimilation in *Vicia faba* to low level SO_2 exposure is highly flux-dependent, exhibiting first-order reaction kinetics at fluxes less than $4 \, \text{nmol cm}^{-2} \text{h}^{-1}$ (Fig. 10.12). At higher fluxes, the response of CO_2 assimilation was typical of second-order reaction kinetics, achieving an asymptote indicative of saturation at fluxes greater than $8 \, \text{nmol cm}^{-2} \text{h}^{-1}$.

The issue of effective versus administered dose has received considerable attention in both applied and basic issues in air pollution effects on terrestrial vegetation. Notable examples of applied research include (1) defining the physiological and ecological features of ambient air quality in lieu of more arbitrary criteria based on meteorological or human health concerns (Adams

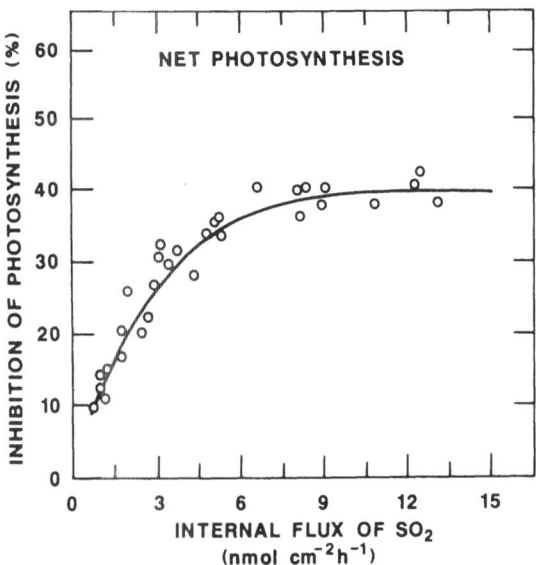

Fig. 10.12. Relationship between the internal deposition of SO_2 and the inhibition of net photosynthesis in *Vicia faba* (Black and Unsworth (1979), reproduced by permission from *Nature* **282**; 68–9, courtesy of Black. Copyright Macmillan Magazines Ltd).

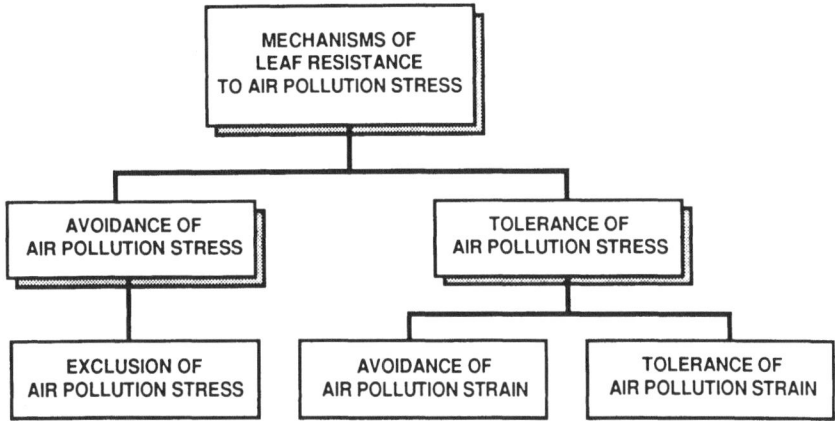

Fig. 10.13. Levitt's framework (1980) for investigating the physiological mechanisms underlying variation in plant response to air pollution stress (Taylor, 1978).

and Taylor, in press; Taylor and Norby, 1985); (2) extending site-specific and species-specific dose–response relationships to other environmental conditions (Fowler and Cape 1982); (3) developing methods to gauge pollutant phytotoxicity for projected emissions from advanced energy technologies (Taylor, 1983); and (4) selection of the most important pollutants within a mixture for further studies of interactive effects.

From a more basic perspective, the conceptual role for characterizing relationships between administered and effective dose are relevant in identifying physiological and biochemical processes underlying dissimilar plant responses to air pollution stress as a function of genotype, environment, and multiple pollutant interactions. Levitt's (1980) analysis is a valuable framework because it identifies two basic mechanisms underlying resistance: stress avoidance and stress tolerance (Fig. 10.13). In stress avoidance, the plant avoids stresses by excluding the pollutant from physiological sites of toxicity. An example of stress avoidance is stomatal closure, precluding deposition of pollutant molecules on cells of the leaf interior (Amiro *et al.*, 1984; Winner and Mooney, 1980). In stress tolerance, pollutant deposition is unchanged, and metabolic processes enable the plant to tolerate the stress (Taylor and Tingey, 1982; Tingey and Olszyk, 1985). Mechanisms of stress tolerance are related either to dissimilar biochemical thresholds for incipient physiological effects or dissimilar repair or compensatory processes. Reich (1987) concluded that mechanisms of stress avoidance were principally responsible for differences in the responsiveness of CO_2 assimilation to O_3 stress as a function of life forms (e.g., hardwoods, conifers, crops). In the case of environmental variables such as vapor pressure deficit, soil water availability, and air temperature, both stress avoidance and stress tolerance mechanisms are operative (McLaughlin and Taylor, 1981; Norby and Kozlowski, 1982; Taylor *et al.*, 1985; Tingey *et al.*, 1982).

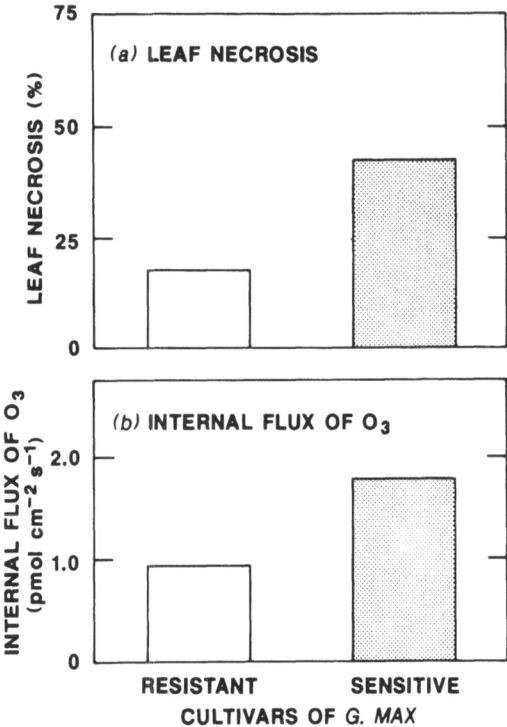

Fig. 10.14. Relationship between foliar injury by O_3 (a) and internal O_3 flux (b) in resistant and sensitive cultivars of *G. max*. The cultivars were exposed to 0·053 ppm or 104 $\mu g\,m^{-3}\,O_3$ for 4 h (Taylor *et al.*, 1982).

The rationale for initially characterizing pollutant deposition rates in physiological and biochemical investigations of stress physiology is demonstrated by the differential responses of *G. max* cultivars to O_3 stress (Taylor *et al.*, 1982). Following exposure to equivalent levels of O_3 in the atmosphere (0·053 ppm or 104 $\mu g\,m^{-3}$ for 4 h), the sensitive cultivar exhibited at least twice as much foliar injury as the resistant cultivar (Fig. 10.14(a)). Whereas this variation was clearly genetically determined, the analysis of O_3 deposition (Fig. 10.14(b)) indicated that in the sensitive cultivar internal O_3 flux into the leaf interior was nearly twice that of its resistant counterpart. This distinct difference between the atmospherically administered dose versus the physiologically effective dose suggests that analyses of the physiological bases underlying differential responses to air pollution stress should not focus on intrinsic metabolic and biochemical differences without initially measuring pollutant flux into the leaf interior.

Whereas the rationale for characterizing pollutant deposition to individual leaves is well documented, the issue of investigating the physical site of deposition has not been adequately resolved. Given the potential for pollutant gas deposition to different locations within the leaf interior (see previous discussion), it is proposed that incipient physiological effects are tissue-specific for each pollutant (Fig. 10.5). For example, if the deposition site of highly reactive and water-soluble gases (e.g., SO_2, HNO_3; Table 10.2) is the substomatal cavity, incipient effects on stomatal physiology are likely. For less chemically reactive and insoluble pollutant gases (e.g., O_3, NO_2, and NO), the principal sites of deposition might be deeper within the leaf interior (i.e., photosynthetically active mesophyll tissue). Consequently, interactive effects on CO_2 assimilation between two pollutant gases would be a consequence of pollutant-specific sites of action on stomatal conductance (SO_2) and CO_2 carboxylation within the mesophyll tissue (O_3). These examples suggest that researchers analyzing pollutant gas deposition and its relationship to injury must extend their efforts beyond characterizing deposition per se and focus also on the foliar sites of deposition on the leaf surface and among tissues of the leaf interior.

10.8 CONCLUSION

The exchange of trace gases between the atmosphere and biosphere is a subject of fundamental importance in the environmental sciences (Mooney *et al.*, 1987). As a subset of this subject, the deposition of pollutant gases to terrestrial vegetation in agricultural and forested landscapes is a central issue in a variety of applied (e.g., crop loss assessment) and basic (e.g., biogeochemistry) issues. Because deposition is a prerequisite to a biological response, analyses of deposition play a major role in characterizing the physiological and biochemical mechanisms underlying plant responses to air pollution stress. The basic tenet of these studies is the concept of dose as developed in the toxicological literature and focusing on the difference between administered and effective pollutant dose.

In the plant sciences, a unique feature of deposition research is the linkage required between the disciplines of atmospheric chemistry, whole-plant physiology, and plant biochemistry. Whereas pollutant deposition is commonly approached at the level of the individual leaf, subsequent analyses typically interface with the disciplines of atmospheric chemistry/meteorology (scaling up) and biochemistry (scaling down). In conjunction with its prominence in a variety of applied and basic issues, this interdisciplinary feature dictates that atmosphere–canopy exchange processes, in general, and deposition of pollutant gases, specifically, will continue to be a challenging and productive research area in the environmental sciences.

ACKNOWLEDGMENTS

G.E.T. Jr and P.J.H. acknowledge support from the Office of Health and Environmental Research, United States Department of Energy and the Electric Power Research Institute under contract DE-ACO5-84OR21400 with Martin

Marietta Energy Systems, Inc. ESD Publication No. 3077, Environmental Sciences Division, Oak Ridge National Laboratory. D.D.B. acknowledges support from the National Oceanic and Atmospheric Administration and the United States Department of Energy as a contribution to the National Acidic Precipitation Assessment Program, under the auspices of Task Group II (Atmospheric Chemistry).

REFERENCES

Adams, M. B. and G. E. Taylor, Jr. (1988). Effects of ozone on forest in the northeastern United States. In *Proceedings of Conference on Ozone Risk Communication*, ed. by E. J. Calabrese. Chelsea, Michigan, Lewis Publishers (in press).

Amiro, B. D., T. J. Gillespie, and G. W. Thurtell. (1984). Injury response of *Phaseolus vulgaris* to ozone flux density. *Atmos. Environ.*, **18**, 1207–15.

Andrussow, L. (1969). Diffusion. In *Landolt-Bornstein Zahlenwerte und Funktionen aus Physik, Chemie, Astronomie, Geophysik, und Technik*, 6th Edition, ed. by H. Borchers, H. Hausen, K.-H. Hellwege, K. Schafer, and E. Schmidt, 513–701. Berlin, Springer-Verlag.

Aneja, V. P., H. H. Rogers, and E. P. Stahel. (1986). Dry deposition of ammonia at environmental concentrations on selected plant species. *J. Air Pollut. Control Assoc.*, **36**, 1338–41.

Atkinson, C. J., W. E. Winner, and H. A. Mooney. (1986). A field portable gas-exchange system for measuring carbon dioxide and water vapor exchange rates of leaves during fumigation with SO_2. *Plant Cell Environ.*, **9**, 711–19.

Baldocchi, D. D. (1988). A multi-layer model for estimating sulfur dioxide deposition to a deciduous oak forest canopy. *Atmos. Environ.* **22**, 869–84.

Baldocchi, D. D., B. B. Hicks, and P. Camera. (1987). A canopy stomatal resistance model for gaseous deposition to vegetated surfaces. *Atmos. Environ.*, **21**, 91–101.

Ball, J. T. (1987). Calculations related to gas exchange. In *Stomatal function*, ed. by E. Zieger, G. D. Farquhar, and I. R. Cowan, 445–76. Stanford, Stanford University Press.

Bennett, J. H. and A. C. Hill. (1973). Absorption of gaseous pollutants by a standardized plant canopy. *J. Air Pollut. Control Assoc.*, **23**, 203–6.

Bennett, J. H. and A. C. Hill. (1975). Interactions of air pollutants with canopies of vegetation. In *Responses of plants to air pollution stress*, ed. by J. B. Mudd and T. T. Kozlowski, 273–306. New York, Academic Press.

Bennett, J. H., A. C. Hill, and D. M. Gates. (1973). A model for gaseous pollutant sorption by leaves. *J. Air Pollut. Control Assoc.*, **23**, 957–62.

Black, V. J. and M. H. Unsworth. (1979). Resistance analysis of sulphur dioxide fluxes to *Vicia faba*. *Nature*, **282**, 68–9.

Bolin, B. (1974). On the exchange of carbon dioxide between the atmosphere and the sea. *Tellus*, **12**, 326–8.

Bowen, H. J. M. (1958). *Tables of interatomic distances and configurations in molecules and ions*. Special Publication No. 11. London, The Chemical Society.

Broecker, W. S. and T.-H. Peng. (1974). Gas exchange rates between air and sea. *Tellus*, **26**, 21–35.

Businger, J. A. (1986). Evaluation of the accuracy with which dry deposition can be measured with current micrometeorological techniques. *J. Climate Appl. Meteorol.*, **25**, 1100–24.

Chambers, J. L., T. H. Hinckley, G. S. Cox, C. L. Metcalf, and R. G. Aslin. (1985).

Boundary line analysis and models of leaf conductance for four oak-hickory forest species. *For. Sci.*, **31**, 437–50.

Cowan, I. R. (1977). Stomatal behavior and environment. In *Advances in botanical research*, ed. by R. D. Preston, 117–228. New York, Academic Press.

Danckwerts, P. V. (1970). *Gas–liquid reactions*. New York, McGraw-Hill.

deCormis, L. (1968). Dégagement d'hydrogene sulfure pae des plants soumise à une atmosphère contestant de l'anhydride sulfureux. *C.r. Acad. Sci. Paris*, **268**, 683–5.

Elrick, D. E., P. H. Groenevelt, and T. J. M. Blom. (1975). Problems of chemical reaction and biological processes in soils. In *Heat and mass transfer in the biosphere. I. Transfer processes in the plant environment*, ed. by D. A. deVries and N. H. Afgan, 537–48. New York, Scripta Book Company.

Farquhar, G. D. and T. D. Sharkey. (1982). Stomatal conductance and photosynthesis. *Ann. Rev. Plant Physiol.*, **33**, 317–45.

Farquhar, G. D. and S. C. Wong. (1984). An empirical model of stomatal conductance. *Aust. J. Plant Physiol.*, **11**, 191–210.

Farquhar, G. M., P. M. Firth, R. Wetselaar, and B. Weir. (1980). On the gaseous exchange of ammonia between leaves and the environment. *Plant Physiol.*, **66**, 710–16.

Finnigan, J. J. and M. R. Raupach. (1987). Transfer processes in plant canopies in relation to stomatal characteristics. In *Stomatal function*, ed. by E. Zeiger, G. D. Farquhar, and I. R. Cowan, 385–430. Stanford, Stanford University Press.

Fowler, D. (1978). Dry deposition of SO_2 on agricultural crops. *Atmos. Environ.*, **12**, 269–73.

Fowler, D. (1980). Removal of sulfur and nitrogen compounds from the atmosphere in rain and by dry deposition. In *Proceedings of the international conference on the ecological impact of acid precipitation*, ed. by O. Drablos and A. Tollan, 22–32. Oslo, Norway, Swedish National Science Foundation.

Fowler, D. (1985). SO_2 deposition. In *Sulfur dioxide and vegetation*, ed. by W. E. Winner, H. A. Mooney, and R. A. Goldstein, 75–95. Stanford, Stanford University Press.

Fowler, D. and J. N. Cape. (1982). Air pollutants in agriculture and horticulture. In *Effects of gaseous air pollution in agriculture and horticulture*, ed. by M. H. Unsworth and D. P. Ormrod, 13–26. London, Butterworths Scientific.

Fowler, D. and M. H. Unsworth. (1979). Turbulent transfer of sulphur dioxide to a wheat crop. *Quart. J. R. Meteorol. Soc.*, **105**, 767–83.

Gaastra, P. (1959). Photosynthesis of crop plants as influenced by light, carbon dioxide, temperature, and stomatal diffusive resistance. *Meded. Landbouwhogesch Wageningen* **59**, 1–68.

Gaffney, J. S., G. E. Streit, W. D. Spall, and J. H. Hall. (1987). Beyond acid rain. *Environ. Sci. Technol.*, **21**, 519–24.

Galbally, I. E. (1971). Ozone profiles and ozone fluxes in the atmospheric surface layer. *Quart. J. R. Meteorol. Soc.*, **97**, 18–29.

Garland, J. A. and S. A. Penkett. (1976). Absorption of peroxy acetyl nitrate and ozone by natural surfaces. *Atmos. Environ.*, **10**, 1127–31.

Garsed, S. G. (1985). SO_2 uptake and transport. In *Sulfur dioxide and vegetation*, ed. by W. E. Winner, H. A. Mooney, and R. A. Goldstein, 75–95. Stanford, Stanford University Press.

Garsed, S. G. and D. J. Read. (1977). Sulphur dioxide metabolism in soybean, *Glycine max* var Bioloxi. I. The effect of light and dark on the uptake and translocation of $^{35}SO_2$. *New Phytol.*, **78**, 111–19.

Hallgren, J. E., S. Linder, A. Richter, E. Troeng, and L. Granat. (1982). Uptake of SO_2 in shoots of Scots pine: Field measurements of net flux of sulphur in relation to stomatal conductance. *Plant Cell Environ.*, **5**, 75–83.

Harris, G. C., J. K. Cheesbrough, and D. A. Walker. (1983). Measurement of CO_2 and H_2O vapor exchange in spinach leaf discs. *Plant Physiol.*, **71**, 102–7.

Hicks, B. B. (1986). Measuring dry deposition: a re-assessment of the state of the art. *Water, Air, Soil Pollut.*, **30**, 75–9.

Hicks, B. B., M. L. Wesely, S. E. Lindberg, and S. M. Bromberg. (1986). *Proceedings of the NAPAP Workshop on Dry Deposition*, 25–27 March 1986, Harpers Ferry, West Virginia. National Oceanic and Atmospheric Administration, Atmospheric Turbulence and Diffusion Division, P.O. Box 2456, Oak Ridge, TN 37831.

Hill, A. C. (1971). Vegetation: a sink for atmospheric pollutants. *J. Air Pollut. Control Assoc.*, **21**, 341–6.

Hosker, R. P. (1986). Practical application of air pollutant deposition models—current status, data requirements and research needs. *Proceedings international conference on air pollutants and their effects on terrestrial ecosystems*, 546–67. New York, John Wiley Publishers.

Hosker, R. P. and S. E. Lindberg. (1982). Review: Atmospheric deposition and plant assimilation of gases and particles. *Atmos. Environ.*, **5**, 889–910.

Huebert, B. J. and C. H. Roberts. (1985). The dry deposition of nitric acid to grass. *J. Geophys. Res.*, **90**(D1), 2085–91.

Hutchinson, G. L. (1972). Air containing nitrogen-15 ammonia: Foliar absorption by corn seedlings. *Science*, **175**, 759–61.

Jarvis, P. J. (1976). The interpretation of the variations in leaf water potential and stomatal conductance found in canopies in the field. *Phil. Trans. R. Soc. London, Ser. B.*, **273**, 593–610.

Jarvis, P. G. and K. G. McNaughton. (1986). Stomatal control of transpiration: Scaling up from the leaf to region. *Adv. Ecol. Res.*, **15**, 1–49.

Legge, A. H., D. R. Jacques, R. G. Amundson, and R. S. Walker. (1977). Field studies of pine, spruce, and aspen periodically subjected to sulphur gas emissions. *Water, Air, Soil, Pollut.*, **8**, 105–29.

Lendzian, K. J. (1982). Permeability of plant cuticles to gaseous air pollutants. In *Gaseous air pollutants and plant metabolism*, ed. by M. J. Koziol and F. R. Whatley, 77–81. London, Butterworths Scientific.

Lenshow, D. H., R. Pearson, and B. B. Stankov. (1982). Measurement of ozone vertical flux to ocean and forest. *J. Geophys. Res.*, **87**, 8833–7.

Leuning, R. (1983). Transport of gases into leaves. *Plant Cell Environ.*, **6**, 181–94.

Leuning, R., M. H. Unsworth, H. H. Neumann, and K. M. King. (1979a). Ozone fluxes to tobacco and soil under field conditions. *Atmos. Environ.*, **13**, 1155–63.

Leuning, R., H. H. Neumann, and G. W. Thurtell. (1979b). Ozone uptake by corn (*Zea mays* L.): A general approach. *Agric. Meteorol.*, **20**, 115–35.

Levitt, J. (1980). *Responses of plants to environmental stresses. Vol. II. Water, radiation, salt and other stresses*. New York, Academic Press.

Lindberg, S. E. and G. M. Lovett. (1985). Field measurements of dry deposition rates of particles to inert and foliar surfaces in a forest. *Environ. Sci. Technol.*, **19**, 228–44.

Liss, P. J. (1971). Exchange of SO_2 between the atmosphere and natural waters. *Nature*, **233**, 327–9.

Liss, P. S. and P. G. Slater. (1974). Flux of gases across the air–sea interface. *Nature*, **247**, 181–4.

Liss, P. S. and P. G. Slater. (1976). Mechanism and rate of gas transfer across the air–sea interface. In *Atmosphere–surface exchange of particulate and gaseous pollutants*, ed. by R. J. Engelmann and G. A. Sehmel, 354–66. Technical Information Center, US Department of Energy.

Mansfield, T. A. (1973). The role of stomata in determining the responses of plants to air pollutants. In *Current advances in plant sciences*, ed. by H. Smith, 13–20. New York, Pergamon Press.

Mansfield, T. A. (1986). Porosity at a price: The control of stomatal conductance in relation to photosynthesis. In *Photosynthetic mechanisms and the environment*, ed. by J. Barber and N. R. Baker, 419–52. New York, Elsevier Science Publishers B.V.

Mansfield, T. A. and P. H. Freer-Smith. (1984). The role of stomata in resistance mechanisms. In *Gaseous air pollutants and plant metabolism,* ed. by M. J. Kozial and F. R. Whatley, 131–46. London, Butterworths Scientific.

McLaughlin, S. B. and G. E. Taylor, Jr. (1981). Relative humidity: Important modifier of pollutant uptake by plants. *Science,* **212,** 167–8.

McMahon, T. A. and P. J. Denison. (1979). Empirical atmospheric deposition parameters—a survey. *Atmos. Environ.,* **13,** 571–85.

Mehlhorn, H. and A. R. Wellburn. (1987). Stress ethylene formation determines plant sensitivity to ozone. *Nature,* **327,** 417–18.

Meidner, H. (1987). The humidity response of stomata and its measurement. *J. Exp. Bot.,* **38,** 877–82.

Meyers, T. P. (1987). The sensitivity of modeled SO_2 fluxes and profiles to stomatal and boundary layer resistances. *Water, Air, Soil, Pollut.,* **35,** 261–78.

Mooney, H. A., P. M. Vitousek, and P. A. Matson. (1987). Exchange of materials between terrestrial ecosystems and the atmosphere. *Science,* **238,** 926–32.

Nobel, P. S. (1974). *Introduction to biophysical plant physiology.* San Francisco, W. H. Freeman.

Nobel, P. S. and D. B. Walker. (1985). Structure of leaf photosynthetic tissue. In *Photosynthetic mechanisms and the environment,* ed. by J. Barber and N. R. Baker, 501–36. New York, Elsevier Science Publishers B.V.

Norby, R. J. and T. T. Kozlowski. (1982). Relative sensitivity of three species of woody plants to SO_2 at high and low exposure temperature. *Oecologia,* **41,** 33–6.

Norman, J. M. (1980). Interfacing leaf and canopy light interception models. In *Predicting photosynthesis for ecosystem models,* ed. by J. D. Hesketh and J. W. Jones, 49–67. Boca Raton, CRC Press.

O'dell, R. A., M. Taheri, and R. L. Kabel. (1977). A model for uptake of pollutants by vegetation. *J. Air Pollut. Control Assoc.,* **27,** 1104–9.

O'Flaherty, E. J. (1981). *Toxicants and drugs; kinetics and dynamics.* New York, John Wiley.

O'Flaherty, E. J. (1986). Dose dependent toxicity. *Com. Toxic.,* **1,** 23–34.

Okano, K., T. Fukuzawa, T. Tazaki, and T. Totsuka. (1986). [15]N dilution method for estimating the absorption of atmospheric NO_2 by plants. *New Phytol.,* **102,** 73–84.

Olszyk, D. M. and D. T. Tingey. (1985). Interspecific variation in SO_2 flux: Leaf surface versus internal flux, and components of leaf conductance. *Plant Physiol.,* **79,** 949–56.

Parkhurst, D. F. (1986). Internal leaf structure: A three-dimensional perspective. In *On the economy of plant form and function,* ed. by T. J. Givnish, 215–49. New York, Cambridge University Press.

Parkinson, K. J. and H. L. Penman. (1970). A possible source of error in the estimation of stomatal resistance. *J. Exp. Bot.,* **21,** 405–9.

Petit, C., M. Ledoux, and M. Trinite. (1977). Transfer resistances to SO_2 capture by some plane surfaces, water and leaves. *Atmos. Environ.,* **11,** 1123–6.

Pierson, W. R., W. W. Brachaczek, R. A. Gorse, S. M. Japar, and J. M. Norbeck. (1986). On the acidity of dew. *J. Geophys. Res. D.,* **91,** 4083–96.

Raschke, K. (1975). Stomatal action. *Ann. Rev. Plant Physiol.,* **26,** 309–40.

Reich, P. B. (1987). Quantifying plant response to ozone: a unifying theory. *Tree Physiol.,* **3,** 63–91.

Russi, H. (1986). Unsetzungsrate von Ozon mit fluchtigen Terpenen im Oberflachenbereich von Koniferennadeln. *Z. Naturforsch.,* **41C,** 421–5.

Sehmel, G. A. (1980). Particle and gas dry deposition: A review. *Atmos. Environ.* **14,** 983–1011.

Sestak, A., J. Catsky, and P. J. Jarvis. (1971). *Plant photosynthetic production: Manual of methods,* The Hague, Dr. W. Junk, NU Publishers.

Shaw, R. H. and A. R. Pereira. (1982). Aerodynamic roughness of a plant canopy: A numerical experiment. *Agric. Meteorol.,* **26,** 51–65.

Sinclair, T. R., C. E. Murphy, and K. R. Knoerr. (1976). Development and evaluation of simplified models for simulating canopy photosynthesis and transpiration. *J. Appl. Ecol.*, **13**, 813–30.

Slinn, W. G. N. (1987). Transpiration's inhibition of air pollution fluxes to substomatal cavities. In *18th conference agricultural and forest meteorology*, 277–80. W. Lafayette, American Meteorological Society.

Smith, S. V. (1985). Physical, chemical and biological characteristics of CO_2 gas flux across the air–water interface. *Plant Cell Environ.*, **8**, 387–98.

Stewart, D. W. and L. M. Dwyer. (1983). Stomatal response to plant water deficits. *J. Theor. Biol.*, **104**, 655–66.

Taylor, G. E., Jr. (1978). Plant and leaf resistance to gaseous air pollution stress. *New Phytol.*, **80**, 523–34.

Taylor, G. E., Jr. (1983). The significance of the developing energy technologies of coal conversion to plant productivity. *Hortscience*, **18**, 684–9.

Taylor, G. E., Jr and D. T. Tingey. (1982). Sulfur dioxide flux into leaves of *Geranium carolinianum* L.: Evidence for a nonstomatal or residual resistance. *Plant Physiol.*, **72**, 237–44.

Taylor, G. E., Jr and R. J. Norby. (1985). The significance of elevated levels of ozone on natural ecosystems of North America. In *Evaluation of the scientific basis for ozone/oxidants standards*, ed. by S. D. Lee, 152–75. Pittsburgh, Air Pollution Control Association.

Taylor, G. E., Jr, D. T. Tingey, and H. C. Ratsch. (1982). Ozone flux in *Glycine max* (L.) Merr.: Sites of regulation and relationship to leaf injury. *Oecologia*, **53**, 179–86.

Taylor, G. E., Jr, S. B. McLaughlin, D. S. Shriner, and W. J. Selvidge. (1983). The flux of sulfur-containing gases to vegetation. *Atmos. Environ.*, **17**, 789–96.

Taylor, G. E., Jr, W. J. Selvidge, and I. J. Crumbly. (1985). Temperature effects on plant response to sulfur dioxide in *Zea mays*, *Liriodendron tulipifera*, and *Fraxinus pennsylvanica*. *Water, Air, Soil, Pollut.*, **24**, 405–18.

Thom, A. S. (1975). Momentum, mass, and heat exchange of plant communities. In *Vegetation and the atmosphere, Vol. 1. Principles*, ed. by J. L. Monteith, 57–109. London, Academic Press.

Tingey, D. T. and G. E. Taylor, Jr. (1982). Variation in plant response to ozone: A conceptual model of physiological events. In *Effects of gaseous air pollution in agriculture and horticulture*, ed. by M. H. Unsworth and D. P. Ormrod, 113–38. London, Butterworths Scientific.

Tingey, D. T. and D. M. Olszyk. (1985). Intraspecific variability in metabolic responses to SO_2. In *Sulfur dioxide and vegetation*, ed. by W. E. Winner, H. A. Mooney, and R. A. Goldstein, 178–208: Stanford, Stanford University Press.

Tingey, D. T., G. L. Thutt, M. L. Gumpertz, and W. E. Hogsett. (1982). Plant water status influences ozone sensitivity of bean plants. *Agric. Environ.*, **7**, 243–354.

Unsworth, M. H., P. V. Biscoe and V. Black. (1976). Analysis of gas exchange between plants and polluted atmospheres. In *Effects of air pollutants on plants*, ed. by T. A. Mansfield, 5–16. New York, Cambridge University Press.

Unsworth, M. H., A. S. Heagle and W. W. Heck. (1984). Gas exchange in open-top field chambers. I. Measurement and analysis of atmospheric resistances to gas exchange. *Atmos. Environ.*, **18**, 373–80.

van Aalst, R. M. and H. S. M. A. Diederer. (1985). Removal and transformation processes in the atmosphere with respect to SO_2 and NO_x. In *Interregional air pollution modelling*, ed. by S. Zwerver and J. van Ham, 83–145. New York, Plenum Press.

Verma, S. B., D. D. Baldocchi, D. E. Anderson, D. R. Matt, and R. J. Clement. (1986). Eddy fluxes of CO_2, water vapor, and sensible heat over a deciduous forest. *Bound. Layer Meteor.*, **36**, 71–91.

Waggoner, P. E. (1975). Micrometeorological models. Vol. 1. In *Vegetation and the atmosphere,* ed. by J. L. Monteith, 205–28. London, Academic Press.

Weast, R. C. (1986). *CRC Handbook of chemistry and physics.* Boca Raton, Florida, CRC Press, Inc.

Wesely, M. L. (1983). Turbulent transport of ozone to surfaces common in the eastern half of the United States. In *Trace atmospheric constituents; properties, transformations, and fates,* ed. by S. E. Schwartz, 345–69. New York, John Wiley.

Wesely, M. L. and B. B. Hicks. (1977). Some factors that affect the deposition rates of sulfur dioxide and similar gases to vegetation. *J. Air Pollut. Control Assoc.,* **27,** 1110–26.

Wesely, M. L., J. L. Eastman, D. R. Cook, and B. B. Hicks. (1978). Daytime variations of ozone eddy fluxes to maize. *Bound. Layer Meteor.,* **15,** 361–73.

Wesely, M. L., J. A. Eastman, D. H. Stedman, and E. D. Yalvac. (1983). An eddy-correlation measurement of NO_2 flux to vegetation and comparison to O_3 flux. *Atmos. Environ.,* **16,** 815–20.

Winner, W. E. and H. A. Mooney. (1980). Ecology of SO_2 resistance: I. Effects of fumigations on gas exchange of deciduous and evergreen shrubs. *Oecologia,* **44,** 290–5.

11

BIOCHEMICAL MECHANISMS OF POLLUTANT STRESS

ROBERT L. HEATH

University of California, Riverside, California, USA

11.1 INTRODUCTION

A discussion of biochemical mechanisms in this symposium is critical if one wishes to develop a concept of how O_3 exposure induces alterations in a plant's metabolism, which ultimately lower a plant's final yield or market value. Furthermore, an understanding of what is and is not known will allow future research to be formulated. In keeping with this concept, I hope to describe several possible scenarios for the development of yield reduction due to primary and secondary biochemical responses. I will concentrate upon O_3, as it was the major oxidant investigated within the NCLAN studies.

The major problem with a biochemical description of the events leading to impaired biomass production is that the integrated networks of biochemical pathways are highly complex. Changes in one pathway are often compensated for by other pathways. Tracing a sequence of perturbations from an initial site of attack of an oxidant, or its products, to the ultimate disruption of normal *homeostasis* is extremely difficult in one paper. I must, therefore, rely upon the many excellent reviews of biochemical pathways and only briefly describe certain biochemical events. Also, I wish to emphasize two major areas which, I believe, will be extensively involved in understanding some of the results obtained through the NCLAN studies.

1. The extracellular protection and interruption of "chain reactions" of oxidants which influence the possible reaction products of O_3 with water (the biochemical solvent) and with several important biochemical groups external to the cell.
2. The spread of the oxidant-induced stress disruption into the cell through

the membrane, in which the above initial events would result in altered, intracellular biochemical pathways (e.g., photosynthesis).

In these discussions it must be clear that several terms of injury are probably not interchangeable. In the past, biochemical investigations often did not consider the types of injury that were occurring. I believe more appropriate terms include the following:

(A) *Acute stress* (high level of oxidants for short time periods). Cellular and tissue death lead to a decline of *total* photosynthetic tissue.

(B) *Chronic stress* (lower levels of oxidants for short periods, but possibly with repeated exposures). Some small localized areas of cellular death, but principally, the altered biochemical states within tissue lead to the inability of the plant to respond properly to existing conditions and to further stress.

(C) *"Accelerated" senescence* (extremely low levels for prolonged time periods). This may be an acceleration of natural aging, which alters the "set points" of control mechanisms and leads to inaccurate internal responses. Loss of measured productivity occurs when the lower biosynthetic capacity is integrated over a long period of growth. Unfortunately, small losses in yield (over long time periods) cannot be detected easily by a single time point measurement of one or more biochemicals or processes.

Therefore, I will define *acute stress* to mean cellular death, and thus, no productivity nor repair is possible. However, the dead cells in the tissue may, by the release of reactive biochemicals into extracellular spaces, react to eliminate some of the oxidants reaching the tissue, thereby reducing the effective dose (Heath, 1987). In this paper I will focus on *chronic stress*, since "accelerated" senescence is presently difficult to quantify (Thomson *et al.*, 1987).

These concepts are diagrammed in Fig. 11.1, emphasizing the initial site (or sites) of interaction and reactions of O_3 with the cell components. This interaction leads to a partial disruption of the normal orderly progress of biochemistry. Clearly, there exist many types of stress adaptations (Levitt, 1972), which can feedback to stabilize the disruption. If the disruption is extreme, and hence, unstable, it leads to a "runaway" situation whereby adaptation can not compensate (acute stress), and cell or tissue death may result. However, if adaptation is successful, a new, stable biochemical state is established. This state represents a new level of metabolic activity, which may be lower in net productivity since it is not optimized for the existing environment (chronic stress).

11.2 AQUEOUS CHEMICAL REACTIONS

The entry of pollutants into the interior of the leaf tissue has been previously discussed (Heath, 1980; Tingey and Taylor, 1982) and has been summarized in

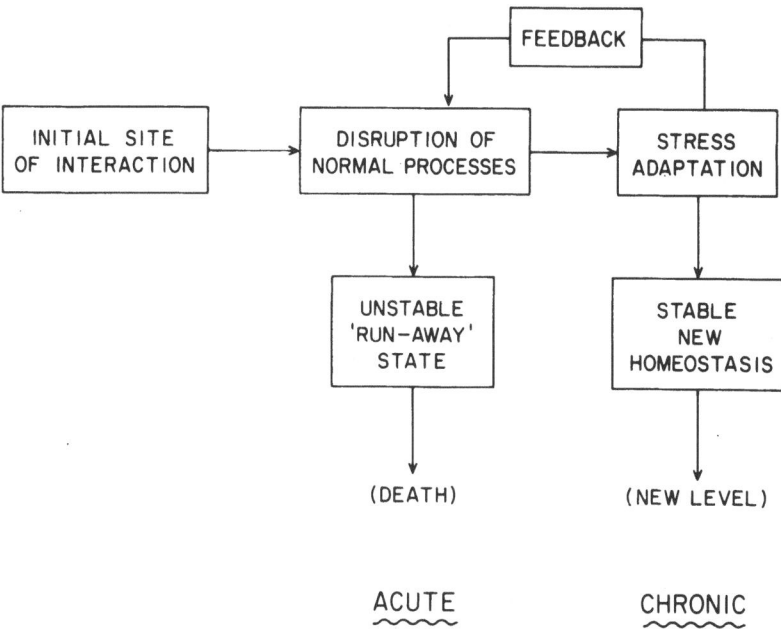

Fig. 11.1. Diagram of the general processes leading to ozone injury of green plants.

this symposium by Taylor *et al.* (1988). I will assume that O_3 has moved into the intracellular space and, aside from being in a hydrated atmosphere, has not yet been altered (however, refer to the next section). First, the O_3 must pass from the atmosphere into a very thin layer of water surrounding each individual cell, which can be estimated to be no thicker than the cell wall or 2 to $3\,\mu m$ for a typical wall (Rogers and Perkins, 1968). Ozone dissolves readily into water with a Bunsen coefficient of about $0.21\,ml\,O_3\,ml^{-1}$ water at 30°C (if O_3 behaves as a perfect gas; Thorp, 1954). This translates into an O_3 concentration of water of $0.009\times$ (O_3 partial pressure) moles l^{-1}, or for $0.1\,ppm$ $(51\,\mu g\,O_3\,m^{-3})$, $9\times10^{-10}\,moles\,l^{-1}$. This concentration is very low, but it is in an extremely thin sheath of water around the cell wall; therefore, the diffusion path is very short (for a $3\,\mu m$ wall thickness, O_3 is at a level of $5\times10^{-16}\,moles\,cm^{-2}$ cell surface). It is important to understand the reactivity of O_3 in this wall region in order to predict possible sites of initial O_3 attack and possible products of O_3 breakdown. In this section, I wish to describe likely reactions of O_3 with: (1) the aqueous solution within the wall, (2) unsaturated fatty acids, and (3) sulfhydryl groups. Other biochemical species are present (Mudd, 1973), but the chemicals listed above are those most likely to be associated with the site of initial attack.

The ionic environment, including H^+ (measured as pH), interacts with the varied reactions involving O_3. The cell wall, of course, surrounds the plant cell and the walls of the cells lining the substomatal cavity, where the majority of the O_3 reactions are thought to take place, are only primary walls (Esau, 1965). These cell walls somewhat restrict the movement of molecules by size, but do not impede the movements of water or ions (Carpita *et al.*, 1979). Furthermore, the primary wall has many specific characteristics, including various side groups of sugars and phosphate groups, which could affect O_3 chemistry and/or its reaction products (Bailey, 1978). This microenvironment needs to be further understood if we hope to characterize the primary site of O_3 injury.

The pH in this microenvironment can influence transport of ions (Lefebvre and Gillet, 1973; Tromballa, 1974) and sugars (Schwab and Komor, 1978). If the pH of the solution within the wall was grossly altered by O_3 chemistry, some observed initial events (membrane permeability changes, described later) might be explained merely by an altered environment within the wall region rather than any chemical modification of the membrane or its components. For example, in unicellular algae, the efflux and influx of a Rb^+ tracer (as marker of K^+ movements) are increased and the rate of deoxyglucose uptake is slowed by a rise in pH, similar to the effects of O_3 upon these transport properties (Heath, 1987). Thus, some of the early events in O_3 injury may be due to a pH change caused by O_3. Once the O_3 is removed, this pH imbalance could readily be reversed by the cell (but not instantaneously) without further injury to the cell. However, the pH difference would alter the "set point" of the cell's metabolism because the cell would sense an external pH, altered by the presence of O_3. A relatively large pH change could be caused by a small amount of O_3 reaction since the pH change would occur within the small wall space.

While O_3 reactions in organic solutions are well understood and useful for organic synthesis (Bailey, 1978), the picture is not so clear when the solvent, in particular water, can participate in O_3 chemistry. In fact, some confusion has arisen in the past because insufficient attention was paid to the diversity of O_3 reactions. Ozone can break down in aqueous solution leading to several types of products (Heath, 1987).

Weiss (1935) first described some O_3 reactions in water. Even with numerous subsequent investigations, it is doubtful that the final picture has emerged. Yet, some conclusions are summarized in Fig. 11.2. Ozone in acidic solutions is reasonably stable; its breakdown can hardly be measured (Shechter, 1973; Heath, 1979; Staehelin and Holgné, 1982). As the solution becomes more alkaline, the reaction with hydroxide ion becomes significant and leads to an autocatalytic or "cyclic" reaction (Staehelin and Holgné, 1982, 1985), in which the first products are the peroxyl-radical ($HO_2\cdot$) and superoxide (O_2^-). Superoxide anion can be formed from peroxyl-radicals by deprotonation through a pH equilibrium reaction ($pK = 4.8$). The peroxyl-radicals have been long noted as initiators of unsaturated fatty acid peroxidation (Pryor *et al.*, 1982).

These particular reaction kinetics also occur upon the entry of O_3 into an aqueous solution, since O_3 is both solubilized by water and reacts with water. For example, the solubility of O_3 often is listed as greater in acidic solutions (Selm, 1959; Staehelin and Holgné, 1985). In a sense, it is because an alkaline environment catalyzes the breakdown of O_3; the measured steady-state

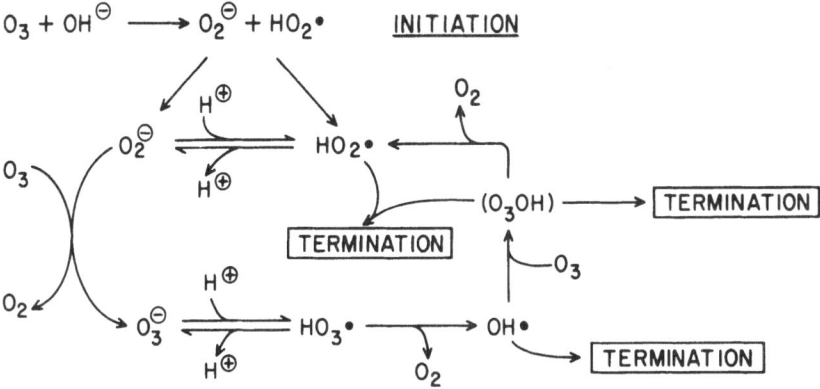

Fig. 11.2. Reaction scheme of ozone interactions with an aqueous solvent. Adapted from Staehelin and Holgné (1985) and Heath (1987). The top reaction is the initiation step of the total cycle, in which hydroxyl ion catalyses the production of both peroxyl-radical and superoxide. Several termination steps are shown without details, but involve the production of hydrogen peroxide from the hydroxyl radical (HO·).

concentration of O_3 is lowered at alkaline pH (Staehelin and Holgné, 1982; Gurol and Singer, 1982). Also, a greater pH change (release of hydroxyl ion or uptake of protons) is observed upon O_3 solubilization and subsequent reactions at lower pH, suggesting that a protonation of O_3 upon water solubilization releases a net hydroxyl ion to the solution (Selm, 1959; Gurol and Singer, 1982). This is large only at acidic pH, since the net pH shift with added O_3 declines at alkaline pH (Staehelin and Holgné, 1982).

Other reactions of the cycle (Fig. 11.2) are less understood with respect to products and kinetics (Staehelin and Holgné, 1982, 1985). However, it is believed that another O_3 molecule can react with superoxide anion to generate oxygen and an O_3 radical, which in turn, reacts with a proton. The resulting protonated O_3 radical (ozonide) decomposes rapidly, releasing oxygen to form a hydroxyl radical—a very reactive species, which has been detected by spin-trapping under certain circumstances (Pryor *et al.*, 1982; Grimes *et al.*, 1983). The cycle is postulated to continue with still another O_3 reaction, forming the protonated peroxyl radical. This radical can restart the cycle, thus inducing an autocatalysis. A variety of chemicals can either promote, presumably by proton donation, or inhibit/terminate the reactions by radical scavenging (Pryor *et al.*, 1982). Promotors found within the wall space include carboxylic acids (e.g., polyuronic acids), alcohols (e.g., sugars), and bicarbonate/CO_2; the biologically ubiquitous ion, phosphate, has been reported to be a scavenger (Staehelin and Holgné, 1982).

The description of O_3 attack upon the membranes often revolves around the ability of O_3 to break double bonds of unsaturated fatty acids (Criegee, 1975). This concept with respect to O_3 injury was discussed recently (Pryor *et al.*, 1982; Heath, 1984, 1987), so only a summary will be presented here.

Ozone, being a very polar molecule, is extremely hydrophilic, and under most circumstances, should not easily enter the hydrophobic regions of the membrane containing unsaturated fatty acids. (For example, oxygen enters charged micelles resulting in an increased internal concentration, but not to the extent that a gaseous hydrocarbon does; Matheson and King, 1978.) Thus, under chronic stress situations, O_3 reactions with fatty acids may be limited to only occasional reactions. For example, an artificial bilayer, which phosphatidylcholine forms in aqueous solution, protects the lipid within from oxidative attack by a hydroxy radical by excluding the reactive species from the organic phase, even though the radical reacts rapidly with each contacted phosphatidylcholine molecule (Barber and Thomas, 1978). Under chronic stress, investigations have shown that lipid alterations do not occur rapidly (Heath, 1984, 1987). Many observations have been conducted over the longer term (1 to 3 days), and, therefore, may represent metabolic shifts. Nonetheless, the entrance of O_3 into the membrane and reactions of O_3 with the double bonds of polyunsaturated fatty acids within an ordered structure are poorly understood, so further ideas are speculative at best.

Ozone can react with and alter critical sulfhydryls of the cell (Mudd et al., 1984) which could cause inactivation of some enzymes. It is known that sulfhydryls can be oxidized into sulfones chemically (Bailey, 1978) and within isolated enzymes (Mudd et al., 1984) but they cannot be reduced by biological thiol reagents. On the other hand, earlier work by Heath et al. (1974) suggested that the kinetics of O_3 reactions with sulfhydryls in an aqueous chemical system were not consistent with sulfone production, but rather appear to be the simple oxidation of the sulfhydryl to the disulfide bridge (see later section on ATPases).

This apparent contradiction can be understood if the types of reactions that can occur in solution are re-examined. Initial radicals produced by hydroxide ions (Staehelin and Holgné, 1982, 1985) will react rapidly with reduced compounds, such as sulfhydryls, to neutralize the radical, and thus turn, for example, the peroxyl-radical into a hydrogen peroxide molecule. Besides breaking the "cyclic" reaction, the reaction products, thiol radicals, can react with one another to produce a disulfide bridge. This reaction scheme would yield the observed second-order kinetics with respect to sulfhydryl concentration in aqueous solutions (Heath et al., 1974).

One would expect that the cell's response to these oxidative species would be to mobilize the antioxidants within the cell to reduce the oxidized species and to break all radical-induced "cyclic" reactions (Thompson et al., 1987). These responses, most probably, would lead to reversing many of the events of ozone-induced injury and prevent later stress-related events.

11.3 POLLUTANT TRANSPORT AND TRANSFORMATIONS WITHIN THE WALL

The question of pollutant dosage comes to the forefront in most discussions. To follow the lead of Taylor et al. (1988), one can calculate that for a stomatal conductance of water of $0.5\,\mathrm{cm\,s^{-1}}$, the conductance of O_3 is $0.31\,\mathrm{cm\,s^{-1}}$, which translates into a pollutant movement through the stomata of approximately 1.3×10^{-12} moles O_3 cm^{-2} s^{-1} for an external concentration of 0.1 ppm. However, the internal concentration of the pollutants must be zero to yield the maximum gradient. Naturally, if the internal concentration is zero, there could be no reactions within the cell. Similarly, if the internal concentration of the pollutant was the same as the outside, there would be no gradient to drive the movement of the pollutant into the leaf. Thus, the rate of O_3 delivery and the calculated internal O_3 concentrations—see previous section—are the maxima.

The surface area is based upon the leaf area, and the internal area or mesophyll area (A_{mes}) to leaf area (A) must be known to translate this rate into internal mesophyll cell surface area (Turrell, 1934; Nobel, 1974). For bean, an A_{mes}/A value of 12 is reasonable (Ticha and Catsky, 1977). So the dosage to total mesophyll cell surface area is 1.1×10^{-13} moles O_3 cm^{-2} s^{-1}. In terms of detoxification, the biochemical response must produce either reductive power or other metabolites, such as "antioxidants," at rates above this level.

The cell wall is often dismissed as a relatively inert container for the cell. Yet, the wall governs much of cell expansion (Cleland, 1975), and the cell membrane is intimately associated with the wall (Roland, 1973). A diagram of the wall system is shown in Fig. 11.3. The left side of the figure indicates the varied ion movements (H$^+$, K$^+$, and Ca^{2+}) with pumps or transport proteins indicated as circles, and passive leaks or pores as triangles. Both the H$^+$/K$^+$ and Ca^{2+} transporters require an energy source, most probably ATP. The

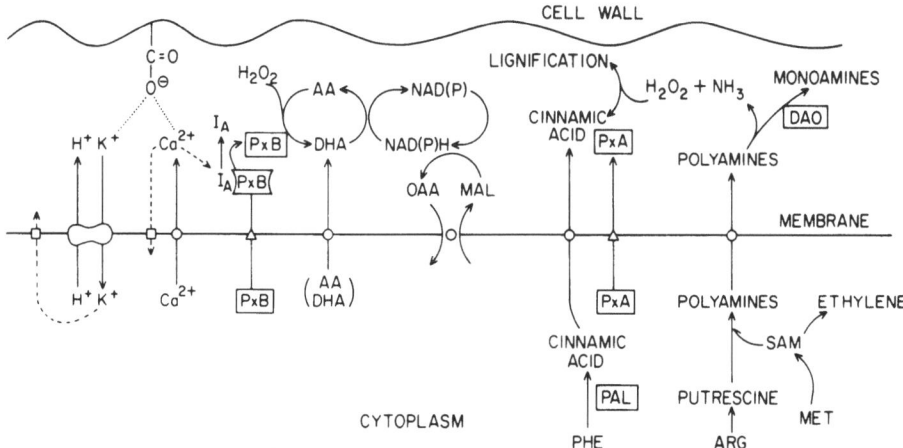

Fig. 11.3. A schematic of some of the reactions involved at the membrane surface and within the wall space. Ion movements with pumps or transport proteins are indicated as solid lines and circles, while passive leaks or pores are as dotted lines and triangles. P × A represents acidic or anionic peroxidases and P × B represents basic or cationic peroxidases. Ascorbic acid, dehydroascorbic acid, malate, and oxaloacetate, are identified as AA, DHA, MAL and OAA, respectively. Methionine, arginine, and phenylalanine are indicated as MET, ARG, and PHE. The enzymes, phenylalanine ammonia lyase, and diamine oxidase are shown as PAL and DAO, respectively. The ethylene intermediate, SAM, is the S-adenosyl-methionine. For additional discussion of the diagram, see text.

monovalent and divalent ions interact through passive ionic binding with the charged groups, polyuronic acid residues, of the wall, shown as a wavy line. The ionic interactions will be discussed in a later section in more detail.

The center of the diagram indicates the varied peroxidase reactions that have been measured. Two types of peroxidases (Px) have been designated (PxA—acidic or anionic—and PxB—basic or cationic), which are reported to move from the inside to the outside of the cell under certain conditions (Gaspar et al., 1985). Peroxidases catalyze two reactions: (1) the reaction of hydrogen peroxide with ascorbate to form dehydroascorbate (Thom and Maretzki, 1985) (which is regenerated by plasma membrane electron transport using a dehydrogenase (Gross and Janse, 1977) and a malate/oxaloacetate shuttle through the membrane); and (2) the reaction with cinnamic acid (from phenylalanine through phenylalanine ammonia lyase and hydrogen peroxide to form lignin within the wall). The acidic peroxidase (PxA) is thought to be involved with lignification and is actively transported out of the cell for the conversion of cinnamic acid to lignin (Ting, 1982). Some basic peroxidases (PxB) are maintained within the cell and some can be transported out after wounding (Lagrimini and Rothstein, 1987; Gaspar et al., 1985), presumably coupled to the synthesis of stress ethylene (Yang and Hoffman, 1984). Some isozymes are kept external to the cell but are inhibited by a low molecular weight compound (I_A). This inhibitor can be removed by excess Ca^{2+} or by lyso-phospholipids, produced by lipase activity (Gaspar et al., 1985). Thus, peroxidases can be activated by the release of free Ca^{2+}.

In addition, the formation of the polyamines and ethylene are shown on the right side of Fig. 11.3 with the carbon being derived from methionine and arginine. The transport of polyamines out of the cell is hypothetical, but must occur for the polyamines to function as intercell effectors (Smith, 1985). Diamine oxidase catalyzes the breakdown of polyamines and is localized within the wall space

As previously discussed, before any pollutant can enter the cell and interact with intracellular components, it must pass through the cell wall region. The wall's concentrated negative charges, which are thought to coordinate Ca^{2+} ions (Morvan et al., 1979), can influence the actual pH within the wall, but can equilibrate with all monovalent cations, including K^+ (Fig. 11.3). Any alkalinization, raising the pH within the wall (for example, through the introduction of O_3 (Selm, 1959)) causes a wall 'tightening' to occur, which would lead to a decline in cell growth. This is in opposition to the wall "loosening" induced by the H^+-pump, which is stimulated by the growth regulator, auxin (Cleland, 1975). The wall, during expansion, is believed to have an acidic pH (using pH-sensitive dyes, Kinraide et al. (1984) estimated the pH of the free wall space to be 5·1 for oat coleoptile). At this pH (5.1) the net effect, as O_3 breaks down, would be the production of both O_2^- and $HO_2^{\cdot}$ (Fig. 11.2).

In addition, O_3, by inducing an increased permeability, causes the loss of intracellular K^+ ions (Heath, 1987) which, in turn, must increase the concentration of K^+ within the wall region. The elevated K^+ levels would be expected to displace Ca^{2+} (Morvan *et al.*, 1979) and alter the wall's rigid structure (Cosgrove, 1987). Thus, there is another scenario by which an alteration of the wall structural capacity induced by O_3 involves no biochemical modifications but merely a localized shift of ions. This would also have profound effects on the ability for turgor pressure to induce cell growth.

While this is an interesting hypothesis (Heath, 1987), the magnitude of the ionic shifts is unknown. This question can be addressed if one makes some assumptions. Not surprisingly, the ability for walls to bind ions depends greatly upon pH. For speculative calculations, we will assume that the pH in the wall region is about 5. Titration of wall material isolated from *Lemna minor* L. (Morvan *et al.*, 1979) indicates that there are two sites—the important one can be titrated with a short equilibration time (1 h for capacity of 0.3 meq. g^{-1} fresh weight of wall material, with a pK of 3.2)—and is attributed to polyuronic acid. Using the polyuronic acid sites and assuming a density of wall material of 1.5 g ml^{-1}, there are about 2×10^{-4} eq. of sites cm^{-3} wall region. But the wall is nearly 2 pH units higher than the pK; therefore, the capacity is reduced by about 100. If the thickness of the wall region is about 3 μm, then the buffering capacity of the wall is about 2×10^{-10} eq. cm^{-2} surface area. This is a million-fold excess over the rate of delivery of O_3 to that region (see above), but is similar to what had been calculated by Heath (1987), using data from Nieboer and MacFarlane (1984), at a neutral pH.

We can calculate the rate of the initiation step of the O_3 cycle (Fig. 11.2). Using a rate coefficient of $140 M^{-1} s^{-1}$ (Staehelin and Holgné, 1985), the initiation rate would be $4.6 \times 10^{-12} M s^{-1}$, at $[O_3] = 0.1$ ppm. For a 3-μm wall, this would amount to a formation rate of O_2^- and $HO_2^{\cdot}$ of 1.4×10^{-18} moles $cm^{-2} s^{-1}$ (at pH 5.1). The initiation rate would be at only a millionth of the O_3 delivery rate, and O_2^- and $HO_2^{\cdot}$ would be formed at a rate of only 1.7 nM h^{-1}, possibly too low to be a problem for the cell (but see later section on wall enzymes).

Interestingly, Ca^{2+} plays little role in the alteration of the wall pH, but there is an interaction between the binding of Ca^{2+} and monovalent ions (Morvan *et al.*, 1979). The role of Ca^{2+} in cell metabolism is an increasingly researched area. While it has been long known that Ca^{2+} plays a role in metabolism, only recently have techniques been developed that allow its role to be fully investigated. For example, in the study of amino acid transport in bean root plasma membranes (Rickauer and Tanner, 1986), it has been shown that the depletion of extracellular Ca^{2+} makes the membrane potential less negative (measured by increased efflux) and greatly increases the net permeability of the membrane. This is similar to what is observed in O_3 injury (Heath, 1987) and in senescent tissue (Borochov and Faiman-Weinberg, 1984). Cramer *et al.* (1985) found that increasing the Na^+ concentration from 100 mM to 150 mM on the outside of root cells increased the $^{86}Rb^+$ efflux (preloaded and used as a measure of K^+) 3-fold at either high or low Ca^{2+} levels. A further increase of Na^+ to 200 mM caused only a small increase (20%) in efflux in the presence of high Ca^{2+} levels but induced still another 3-fold increase in efflux if the Ca^{2+} level was reduced by 25-fold. These investigators used a fluorescent probe of

Ca^{2+} to determine the relative Ca^{2+} level within the wall region of the individual cells. At high Ca^{2+}, the fluorescence level remained at 60% of the base level (with no Na^+ present) for Na^+ concentrations beyond 100 mM. At low Ca^{2+}, the fluorescence level showed a progressive decline as the Na^+ was increased (55% at 100 mM, 30% at 150 mM, 23% at 200 mM). Thus, there are sites for Ca^{2+} for which Na^+ can exchange easily and the ions do not increase K^+ efflux (up to 100 mM Na^+). Only at limiting Ca^{2+} does a higher concentration of Na^+ displace Ca^{2+} and cause an increase in K^+ efflux.

In terms of O_3 injury, any excess efflux of K^+ from O_3-exposed cells would increase the possibility of still more efflux, since the K^+ lost from the cell would then displace the Ca^{2+} in the wall space and thereby further increase the membrane's permeability to K^+ (by the above mechanism). In general, if the wall region is only about 1% of the total cell's volume (Rogers and Perkins, 1968), a loss of only a few percent of the total K^+ within the cell (about 350 mM) would raise the level of univalent ion within the wall to the levels quoted above. Thus, a loss of 1% of the cell's K^+, which is not critical to the inner workings of the cell, could induce a "run-away" or autocatalysis which may be injurious. The loss of a small amount of K^+ to the exterior of the cell induces a loss of much more K^+, due to the disruption of the Ca^{2+} balance outside the cell.

Ions are not the only components in the cell wall area; there are many enzyme systems, the functions of which are poorly understood (Gaspar et al., 1985; Smith, 1985; Majerus et al., 1986). Peroxidases are thought to be an important marker of O_3 injury (Curtis et al., 1976). Yet when total leaf tissue was used, the data were confusing. For example, Tingey et al. (1975) showed that peroxidase activity was initially depressed (within a day of exposure), but later showed a rise (two to three days later). Recently, Castillo and his coworkers (Castillo et al., 1984; Castillo, 1986; Castillo and Greppin, 1986) have clarified these results. There are several isozymes of peroxidase (Fig. 11.3), and when the total leaf extract is used, the average level of peroxidase changes irregularly because each existing isozyme is responding differently. Only later are new isozymes synthesized. Furthermore, if leaf tissue is greatly injured (acute stress), then some peroxidase enzymes may be irreversibly inactivated.

Under very low concentrations of O_3 (no visible injury), peroxidase activity seems to reflect a stress response. Ozone exposure causes the activity of basic (or cationic) isozymes located in the wall region to rise (for techniques, see Castillo et al., 1987; Heath and Castillo, 1988). Peroxidases can be activated by Ca^{2+} (Castillo et al., 1984). Presumably, the O_3-induced activation of peroxidases would be due to the availability of free Ca^{2+} within the wall. One such peroxidase experiment is shown in Table 11.1. The peroxidase level in the extracellular fluid of needles from Norway spruce, exposed to 0·16 ppm O_3 for 2 days (7 h per day), increased 3-fold, while the total peroxidase level increased only 30% (Castillo et al., 1987). Longer exposures yield greater increases. In other cases, different types of peroxidases can be assayed by varied substrates (Castillo and Greppin, 1986); ascorbate-linked peroxidases

TABLE 11.1

The activity of several enzymes and metabolites located within the cell wall space and cell of Norway spruce, as affected by ozone fumigation

	Intercellular fluid		Cell material	
Assay	Control	Ozone	Control	Ozone
Peroxidase[a]	43 ± 10	126 ± 30	32 ± 6	42 ± 8
Superoxide dismutase[b]	712 ± 4	841 ± 116	326 ± 66	386 ± 70
Ascorbic acid[c]	$2 \cdot 4 \pm 0 \cdot 5$	$1 \cdot 1 \pm 0 \cdot 1$	521 ± 55	452 ± 50
Ethylene[d]			$16 \cdot 2 \pm 2 \cdot 5$	$16 \cdot 8 \pm 1 \cdot 8$

[a] A_{470} mg-protein^{-1} min^{-1}.
[b] Units mg-protein^{-1}.
[c] µg g-Fresh Weight^{-1};
[d] nl g-Fresh Weight^{-1} h^{-1}.

Data are summarized from Castillo *et al.* (1987). The intercellular fluid, washed from the wall space of Norway spruce needles, was collected by vacuum infiltration with distilled water and subsequent centrifugal separation of the resulting fluid from the needles. The cell material had been from samples frozen in liquid nitrogen. The enzymes and metabolites were assayed by usual procedures (Castillo *et al.*, 1987). Ethylene production was measured after 24 h of incubating the needles in the presence of 1-aminocyclopropane-1-carboxylic acid (ACC, 1 mM). The control was filtered ambient air, and the ozone samples were those exposed for 2 days at 0·16 ppm for 7 h per day.

(cationic) increase within 3 h of exposure, while syringaldazine-linked peroxidases (anionic) show no increase for 24 h.

As we have seen, the interaction of air pollutants with the cell surface in the wall region of plant leaves can result in the generation of superoxide anions and peroxides (for example, O_2^-, H_2O_2 by free radical combinations in aqueous solutions during O_3 breakdown (Grimes *et al.*, 1983; Heath, 1987)). These other oxidants certainly are potential toxins; the hydroxyl radical (HO·) can attack membrane lipids and proteins (Slater, 1984; Wolf *et al.*, 1986; Thompson *et al.*, 1987).

The chemical reaction rates of several important biochemicals with superoxide (in basic solutions) or the peroxyl radical (in acidic solutions) have been tabulated by Bielski (1983) and are given in Table 11.2. While ascorbate reacts rapidly, the ascorbate radical, formed by the first one-electron reaction, reacts even more rapidly with both compounds. Thus, ascorbate (or its radical) will be very effective in removing both oxygen radicals from any solution. In contrast, superoxide can not easily react with either linolenic acid or α-tocopherol in an alkaline environment, yet the peroxyl-radical rapidly reacts

TABLE 11.2

Reaction rate coefficients for superoxide and peroxy-radical
with varied antioxidants

Antioxidant	Rate coefficient $(M^{-1}s^{-1})$	
	Acid solution HO_2· species	Basic solution O_2^- species
Ascorbate	$1·6 \times 10^4$	5×10^4
Ascorbate radical	5×10^9	$2·6 \times 10^8$
Linolenic acid	$1·2 \times 10^3$	0
α-Tocopherol	2×10^5	6

Data were taken from Bielski (1983), and are for the species
in water or 95% ethanol (for tocopherol and linolenic acid).

with both in acid. While α-tocopherol reacts more rapidly than linolenic acid
on an absolute scale, the concentration of α-tocopherol in membranes is 100 to
1000 times lower than that of linolenic acid (Boguth, 1969), and so the real
rates may be comparable within a membrane.

Arguments have been used that peroxidases react with hydrogen peroxide
produced by O_3 and thereby protect the cell. This is probably not so since
H_2O_2 seems to be present with the wall (Sagisaka, 1976) and is required for
lignification (Ting, 1982). It is possible that added O_3 raises the level of H_2O_2
above a toxic threshold, but little is known regarding this postulated
mechanism. Nonetheless, investigators have tried employing antioxidants,
including ascorbate, to prevent O_3 injury (Dass and Weaver, 1968). Interestingly, these researchers found that some of these treatments also returned the
O_3-induced, high peroxidase activity back to normal. One compound has
recently gained prominence as an 'antiozonide'—N-[2-(2-oxo-1-
imidazolidinyl)ethyl]-N'-phenylurea, or EDU. Bennett et al. (1984) have
summarized the findings. In essence, either a foliar spray or a soil addition of
EDU will protect against O_3 injury, without noticeable stomatal closure.
Furthermore, EDU increases total cellular superoxide dismutase and catalase
activity in addition to increasing the protein and RNA content of the tissues.
Lee and Chen (1982) showed that EDU has cytokinin-like activity but is many
times less effective at comparable concentration. Both EDU and cytokinin
prevent senescence, but EDU was more effective against O_3-induced early
senescence. Hence, EDU may be a partial 'antisenescent' plant effector rather
than a true 'antiozonide'.

Along this line, Matters and Scandalios (1987) showed that maize seedlings
exposed to as much as $0·75$ ppm O_3 for 8 h showed no increase in either
superoxide dismutase or catalase activity, even 4 days postexposure. Further,
the activity of superoxide dismutase (the isozymes were not separated) in
needles from Norway spruce shows only a small increase upon O_3 exposure (at
much lower O_3 levels than those previously described), while the level of
ascorbate from within the wall space, declined by more than 50% (Castillo et

al., 1987 see Table 11.1). The levels of both superoxide dismutase and ascorbate in the total cell extract showed little change. Thus, O_3 does not seem to be an obvious inducer of some protectants.

The ascorbic acid levels have been correlated previously with oxidant stress, but in an indirect manner. Mehlhorn *et al.* (1986) showed that ascorbate levels in spruce rose after a long-term exposure to low levels of O_3 (at a near background of 0.037 ppm for one summer), although the rise was small (3–5%). Greater percentage increases in levels of α-tocopherol and glutathione were noted. However, the researchers measured only total amounts of compounds at the end of the exposure period. No doubt, stimulation of biochemical pathways could have already occurred (presumably induced as a response to the oxidants). Ozone does increase antioxidant chemicals in the short term; for example, Lee *et al.* (1984) clearly showed that ascorbate levels rose nearly 2-fold when assayed 2 days postexposure.

The polyamines have been implicated in retardation of senescence and the stabilization of membrane function. There is an interrelation between Ca^{2+} and polyamines, in that they can partially substitute for each other in keeping the membrane semipermeable (Smith, 1985). Some have argued that the polyamines should be classed as a growth regulator since they have been shown to stimulate growth in certain systems and are found in high concentrations in tissue that have the greatest growth rates (Smith, 1985).

Polyamines can induce an activation of the membrane-bound ATPases, an inhibition of the peroxidases located in the membrane-wall region, and an increase in ethylene production (upon the wounding of the tissue) by an inhibition of protein synthesis. All of the above events are in opposition to those observed in O_3-induced stress. Not surprisingly therefore, polyamines are observed to accumulate under general stress conditions (Smith, 1985), which is thought to be in partial response to an altered H^+ level within the cells. Ormrod and Beckerson (1986) measured the amount of chlorotic/necrotic tissue of tomatoes caused by O_3 exposure during and after feeding polyamines to the shoots (the roots were cut off, and the transpiration stream carried the polyamines at a concentration of 0.01 mM into the leaves). Putrescine [$H_2N—(CH_2)_4—NH_2$] and spermine [$H_2N—(CH_2)_3—HN—(CH_2)_4—NH_2$] reduced the amount of damaged tissue by greater than 50%. Unfortunately, there are only preliminary reports on the levels of endogenous polyamines after O_3 fumigation (Borland and Rowland, 1988; Villanueva *et al.*, 1988). The levels are observed to increase, presumably concurrent with an inhibition of the ATPases, an increased peroxidase activity, and ethylene release. Possibly, some protection of the tissue is being caused by the increased levels of polyamines.

Recently, we have examined the activity of diamine oxidase, which is located external to the cell in pinto bean plants (*Phaseolus vulgaris* L.), and found that it is inhibited (nearly 50%) by an O_3 exposure which caused no visible injury (0.2 ppm for 4 h) (Peters, Castillo and Heath, unpublished data). It is thought that the diamine oxidase prevents the loss of polyamines to the apoplastic space, where they could affect metabolism in cells some distance away from the originating cells (Smith, 1985). Under O_3 fumigation, the increased membrane permeability (presumably also for amines) and the inhibition of the diamine oxidase would contribute to the movement of polyamines (stimulated by O_3) from an injured cell to the surrounding cells. This would, in turn, affect neighboring cells, perhaps, to aid their defenses.

Both the polyamines and ethylene arise from the amino acid, methionine (Fig. 11.3; Yang and Hoffman, 1984). Methionine must be "activated" by adenosylation by ATP to form S-adenosyl-methionine (SAM). If SAM is decarboxylated, the resulting molecule builds spermidine and spermine from putrescine, which is derived from arginine (Smith, 1985). If the SAM group is removed, 1-aminocyclopropane-1-carboxylic acid (ACC) is formed, leading to ethylene production. Inhibitors that block the formation or breakdown of ACC (e.g., aminoethoxyvinylglycine (AVG), an inhibitor of ACC synthase, which is the final enzyme in the pathway of ethylene biosynthesis) would be expected to increase the formation of precursors of polyamines. Thus, one might predict that polyamine formation and ethylene release might be correlated.

Gaseous ethylene has been known to induce many physiological responses in green plants; indeed, it is now classified as a plant growth hormone or effector (Ting, 1982). The production of ethylene is, in part, controlled by levels of indoleacetic acid (IAA), also a plant hormone. Although the pathway of biosynthesis is now known, ethylene's mode of action is obscure. There are reports of ethylene binding to cell components (Sisler and Goren, 1981), presumably to induce biochemical responses. The pathway to eliminate ethylene seems to be by the natural escape of the gas, rather than any specific series of biochemical reactions. Certainly, ethylene is produced within 1 h of wounding of plant tissue, and its production is preceded by increases in the components of the biosynthetic pathway from methionine (Imaseki, 1985). Wounding is likewise involved with IAA and auxin changes. More interestingly, wounding plant tissue in the presence of a small amount of ethylene accentuates the production of peroxidase and polyphenol oxidase activities, which are induced by the wounding.

Photosynthesis can be inhibited by ethylene, which is probably due to ethylene-induced stomatal closure (Pallas and Kays, 1982); however, this interpretation has not always been accepted (Reid and Wample, 1985). It may be that ethylene production may cause a series of events, including an inhibition of photosynthesis, which mirror the later events associated with air pollution injury. Wounding plant tissue, with the concurrent loss of turgor pressure and release of metabolites, induces ethylene release (Imaseki, 1985). Thus, it is easily argued that the initial events of O_3 injury cause a general wounding response.

Actually, many researchers (Pell, 1988, for review) have observed such a release of ethylene (called stress ethylene) upon O_3 fumigation of plants. The rise of ethylene occurred relatively rapidly after the fumigation period and was dependent somewhat upon O_3 dose. Yet with cultivars differing in their sensitivity to O_3, ethylene release did not correspond directly to visible injury

(Adepipe and Tingey, 1979). The connection between ethylene and O_3 injury was strengthened by a recent report by Mehlhorn and Wellburn (1987) who found that the amount of visible injury could be reduced by a spray of AVG. The authors suggested that reduction of ethylene synthesis could directly reduce the occurrence of visible injury, yet the AVG-induced inhibition of ethylene could be forcing precursors into polyamine synthesis, leading to the observed protection. The need for more work in this area is clearly indicated.

11.4 MEMBRANE ALTERATIONS

Interactions of O_3, or its decomposition products, may be immediately external to, within, or inside the membrane depending upon how far O_3, and its products, can move before reaching reacting biochemicals. Radiation chemists have studied diffusion paths of reactive species and have defined a 'mean free path length' (mfp) for reactive species, which is the average distance that a molecule will travel before reacting with the solvent. For example, $HO^{\cdot}$ and singlet O_2 have a mfp in water of only 3·5 and 11·5 nm, respectively (Saran, 1988), which can be shortened by other reactive components within the solvent. Since the membrane is about 10 nm in width (Baker, 1978), one would think that none of the products (with the exception of superoxide) of Fig. 11.2 could penetrate into the cytoplasm, even through water-filled pores. However, disruption of normal membrane function, by whatever means, will alter the cell's normal metabolism.

Visible injuries observed in plant leaf surfaces exposed to photochemical air pollutant episodes are wilting, water-logging (the filling of intracellular spaces with water), and/or localized epidermal collapse (Bobrov, 1952), and are classed as acute injury. These damages to plant tissues are related to the loss of osmotically active intracellular materials, indicating impaired membrane function.

The effects of O_3 upon membrane function have been summarized by Heath (1980), Tingey and Taylor (1982), Lendzian and Unsworth (1983), and Mudd *et al.* (1984). More recently, Heath (1987) and Heath and Castillo (1988) discussed alterations of membrane function and emphasized possible biochemicals that could be altered *in vivo* and *in vitro,* as opposed to discussions of the alterations of isolated biochemicals (Mudd *et al.,* 1984).

Little doubt now remains that passive permeability is altered by O_3 exposure. The major questions now seem to revolve about the biochemical mechanism and its reversibility (Heath and Frederick, 1979; Heath, 1984, 1987). Ozone alters permeability by increasing the "leakiness" of the membrane (Evans and Ting, 1974; Heath, 1980, 1987) and by inhibiting pumps or transporters (Evans and Ting, 1973, 1974; Sutton and Ting, 1977; Dominy and Heath, 1985). Ozone (exposure of leaves or individual cells) may also alter other membrane elements: membrane potential may be depolarized (Heath, 1987); diffusion may be altered by an increase in membrane 'fluidity' (Pauls and Thompson, 1981); energy requirements may be impaired by a decline in cytoplasmic ATP levels (Pell and Brennan, 1973); and the internal levels of Ca^{2+} may be increased by inhibited pumps (Macklon, 1984; Heath and Castillo, 1988).

Besides permeability changes, alterations to the plasmalemma's K^+-dependent ATPase, thought to be the K^+ pump or transporter, has been demonstrated by Dominy and Heath (1985; also Heath, 1987). The ATPase is inhibited *in vivo* if the plant is exposed to O_3 before the plasma membrane is isolated, and is accomplished without added sulfhydryl reagent. The inclusion of a sulfhydryl reagent reverses the inhibition, indicating that the inactivation is caused by the oxidation of a protein sulfhydryl to a disulfide bridge. This is the second membrane-bound protein to have been observed to be altered by O_3 exposure.

The first membrane-bound protein observed to be altered by O_3 was the β-glucan synthase, thought to be involved in the synthesis of cellulose within the cell wall. Ordin *et al.* (1969) found that oat coleoptile sections exposed *in vitro* (the sections were floating in water into which a high concentration of O_3 was bubbled) grew more slowly (especially in the presence of auxin) and possessed lower levels of UDP-glucose-dependent cellulose synthase (now thought to be β-glucan synthase). The synthase is membrane-bound and requires lipid for activity (Wasserman and McCarthy, 1986), although, the ratio of lipid to protein is critical for stimulation (a high ratio can depress activity). Also, the activity can be stimulated by Ca^{2+} (Wasserman and McCarthy, 1986), and the ratio of cellulose:callose formation seems to be altered by the amounts of Ca^{2+} (Kauss and Jeblick, 1986). In fact, the relative concentrations of Mg^{2+} and Ca^{2+} play a role in shifting the products and the binding constant (K_m) for its substrate (UDP-glucose) (Hayashi *et al.*, 1987). Few have appreciated the possible interrelationship of O_3-induced inhibition of this enzyme (Ordin *et al.*, 1969) and its regulation by divalent ions.

Since O_3 exposure of plant material induces the modification of membrane permeability, other ions, such as Ca^{2+}, are affected (Heath and Castillo, 1988). Calcium is known to pass through membranes via a discrete channel (Reuter, 1983), and the characteristics of these channels could be functionally impaired by O_3 and its products.

Free calcium ion is recognized as a major intracellular regulator of many different cellular processes, acting as a second messenger and regulating many interrelated cellular functions of plants (such as cell elongation, cell division, protoplasmic streaming, enzyme secretion), as well as regulating the stomatal response (Schwartz, 1985). Calcium can act alone or by binding to receptor proteins, which seem to accentuate the properties of Ca^{2+}. For example, the Ca^{2+} binding protein, calmodulin, becomes activated and is then capable of modifying other enzymic activities (Hepler and Wayne, 1985).

Plant cells must utilize large amounts of metabolic energy to maintain a very low level of free Ca^{2+} in the cytoplasm (less than 10^{-6} M); the electrochemical potential gradient across both the plasmalemma and the tonoplast compared with the cytoplasm is high (Macklon, 1984). The low Ca^{2+} concentration within the cytoplasm is controlled by separate ATP-dependent, membrane-bound transporters which lead to both extrusion of Ca^{2+} out of the cell into the wall region and accumulation of Ca^{2+} within cellular organelles such as the mitochondrion, endoplasmic reticulum, and vacuole (Marmé, 1985; Rasi-Caldogno *et al.*, 1987).

By changing Ca^{2+} fluxes through the membrane surrounding the cytoplasm, air pollutants would induce an alteration in the cytosolic concentration of free Ca^{2+} and, consequently, be able to change the Ca^{2+}-dependent biochemical and physiological reactions (Morgan *et al.*, 1986). Preliminary evidence (Heath and Castillo, 1988) suggests that the permeability of the membrane to Ca^{2+} is increased as it is for other ions, and the extrusion pump is inhibited by O_3. Thus, the net effect of these events would be the raised level of intercellular Ca^{2+}. This, in turn, could lead to widespread alterations observed in O_3-induced injury (Heath, 1980, 1984). The largest Ca^{2+} concentrations associated with the cell are found in the cell wall (Demarty *et al.*, 1984); thus the effect of pollutants on cell *homeostasis* via an effect on the Ca^{2+} permeability should not be neglected.

Calcium also affects other extracellular processes, which have been closely associated with O_3-induced injury. Extracellular peroxidases have been found to rise with O_3 exposure, and are dependent upon free Ca^{2+} (refer to previous section and Fig. 11.3). Furthermore, in other stress-related alterations of metabolism, researchers have found that the increase in cytoplasm Ca^{2+} leads to an activation of phospholipases, which leads to an increase in the hydrolysis of membrane lipids (Gaspar *et al.*, 1985). The levels of free fatty acids rise, and the lipids must be replaced more rapidly. The altered lipid content observed previously over a longer time period and the appearance of lipid oxidation products (such as ethane and malondialdehyde, summarized by Heath (1984) and Pell (1988)) may be due to such mechanisms.

What is the cost of detoxification to the cell? As we have previously seen, the maximum rate of delivery of O_3 to a mesophyll cell would be approximately 10^{-13} moles cm^{-2} s^{-1}. The production of O_2^-/HO_2˙ from the solvent seems to be small (10^{-19} moles cm^{-2} s^{-1}) and easily handled by existing compounds, e.g., superoxide dismutase (SOD) in the cell wall. (There are 700 units of SOD per mg protein in spruce (Castillo *et al.*, 1987) and about 6 μg protein per 20 μl extracellular fluid (Peters, Castillo, and Heath, unpublished).) Therefore, for a 3-μm thick wall filled with fluid, there is about 1 unit of enzyme cm^{-2}, or 1 μmole of superoxide can be eliminated cm^{-2} s^{-1}, in excess of any production from O_3 (see above, 4×10^{-19} cm^{-2} s^{-1}).

But even if each O_3 molecule oxidized one sulfhydryl to one-half disulfide bridge, then one NADPH would be required to re-reduce it to sulfhydryl. Heath *et al.* (1985) showed that the respiratory rate of pinto bean leaves was about 8×10^{-11} moles CO_2 cm^{-2} s^{-1} or $6\cdot7 \times 10^{-12}$ moles CO_2 cm^{-2} s^{-1}. If 10% of the respiratory rate is actually being shunted through the pentose shunt from which CO_2 is produced, then approximately $1\cdot3 \times 10^{-12}$ moles of NADPH (2 NADPH produced/CO_2 released) is produced for reductive power. This is greater than the amount required for 'repair' of oxidation due to O_3 exposure by more than 10-fold. Thus, for the pentose shunt capacity to handle the increased demand for reducing power, the apparent respiratory rate must increase only 1% due to more carbon being routed through the pentose shunt. The respiratory rate has been observed to be stimulated occasionally by high levels of O_3 (Black, 1984), but not all the O_3 within the cells attacks the sulfhydryls, so it would seem that there is adequate NADPH production to detoxify the attacking molecules, at a concentration of 0·1 ppm O_3.

11.5 TURGOR PRESSURE AND ITS CONTROL OF METABOLISM

It has been clear for years that turgor pressure within the cell is critical for plant growth, but only recently has the sequence of events and its control begun to be elucidated (for review, see Cosgrove, 1987). Turgor is the hydraulic pressure induced by the accumulation of osmotically active solute molecules within the cell. The relationship between turgor and cell expansion is not a simple one, since the wall plays an important role in the expression of the turgor pressure. Further, turgor also seems to affect metabolic processes as well as metabolic processes affecting turgor (see below).

The yield threshold (Y, that level of pressure required for the initiation of cell expansion) of the wall is high but not unalterably fixed, since it is dependent upon the cellular structure, developmental age, genetic expression, and hormonal status (including the level of active auxin within the cell; Cleland, 1975; Cosgrove, 1987). Once the turgor pressure (P_t) exceeds the yield threshold, the cell's volume increases linearly (by a coefficient, m) in accordance with the excess turgor ($P_t - Y$, Van Volkenburgh and Cleland, 1984), at least over the physiological range.

More importantly, it has been observed that as the turgor pressure is changed rapidly in unicellular alga (*Chara corallina*) (increase or decrease of 3 to 4 bars within 2 to 3 s), the membrane potential depolarizes (within 2 to 3 s) by as much as 80 to 100 mV (Zimmermann and Beckers, 1978). Most of the effects easily observed are extreme, e.g., a gradual decline in the membrane breakdown potential (at about −800 mV) with increasing turgor (lowering by 50 mV by the application of 0·4 bars, Coster *et al.*, 1977), yet the influence of turgor upon the membrane is present. The above effects were observed within unicellular alga using a sophisticated pressure probe technique; however, less complex experiments with carrot storage tissue have shown that K^+ efflux can be lowered by high external osmotic pressure (to produce less turgid cells), but externally added Ca^{2+} can lower both the efflux with and without added external osmoticum (Enoch and Glinka, 1983).

Externally added osmoticum (to lower net cellular turgor) can affect the membrane potential of cells. Using broadbean (*Vicia faba* L. cv. Aguadulce) mesocarp cells, Li and Delrot (1987) demonstrated that the membrane potential became more negative (from −70 mV to −120 mV) with an increasing osmotic pressure up to a certain point (external osmotic strength of about −5 bars). Kinraide and Wyse (1986) likewise showed that an external osmoticum made the membrane potential more negative in sugar beet tap root, which was attributable to an inhibition of the H^+ pump. However, as the osmotic pressure was raised higher (and thus the net turgor pressure was lowered), the membrane potential became less negative (reaching −40 mV at −15 bars). (It is clear in one of their figures that the cells are in osmotic equilibrium at about −5 bars; external osmotic pressure causes a decrease in cell size (measured by apparent free space of retained inulin) and the apparent wall space increases.) Thus, the water potential of the cell has an effect upon the membrane characteristics, consistent with the proton pump of the membrane (required for "loosening" the wall) acting as a transducer of the cell water potential state.

Using the pressure probe, Westgate and Steudle (1985) have estimated the time of propagation of a turgor pressure pulse across a maize leaf. They found that, not surprisingly, the time required for the turgor pressure of a cell to adjust to a pressure pulse in the xylem depended upon the distance of the cell from the midrib, but the time was still amazingly rapid (a half-time of 14 s for a cell located at a relative distance of 0·230 × the midrib diameter, increasing to 60 s for a distance of 0·776 ×). They also found that the path that a pressure pulse traveled (through the apoplastic or symplastic space) depended upon the manner in which the leaf water potential was altered. For pressure driven flow, the preferred path was apoplastic, while for osmotic driven flow, it was symplastic. Thus, one could predict that the interrelationships between high external K^+, due to leakage induced by O_3, and the path and speed of water movement, which affects the water potential of each cell, would be dependent upon the type of disturbance. With pressure driven flows (O_3-induced turgor collapse) where apoplastic movement predominates, the water movement within the walls would be influenced by the extracellular K^+. Initially with little intracellular K^+ loss under well-watered conditions, the cell walls would be expected to be the path of water transfer, but under dry conditions (low leaf water potential), passage of water through the cell might predominate. These two pathways are different and so could alter the manner in which O_3 injury was initially expressed.

Space limits a complete discussion of the effects of the cell's water status upon metabolism. Of interest, however, is the increase of glutathione reductase by water stress (Gamble and Burke, 1984). This mechanism may allow a greater availability of antioxidants such as glutathione (Mehlhorn *et al.*, 1986), under O_3-induced cell water loss.

11.5.1 Dependence of photosynthesis upon turgor pressure

The emphasis in NCLAN was upon productivity as measured by yield. The yield is dependent upon photosynthesis (although the exact dependence is controversial (Hesketh and Jones, 1980)). Thus, it seems natural to concentrate upon the interactions of the initial events of air pollutant injury with photosynthesis rather than other metabolic events.

Kaiser and his coworkers (1981*a*, *b*) have studied the effect of external osmoticum upon photosynthesis in several types of preparations. In leaf slices, the inhibition of photosynthesis (measured by CO_2 fixation) required high levels of osmoticum (for 50% inhibition, -25 to -40 bars of sorbitol; Kaiser and Heber, 1981). Further, when the osmoticum was eliminated, the photosynthetic rate was restored rapidly but not completely (Kaiser et al., 1981b). Berkowitz and Whalen (1985), using vacuum infiltration to eliminate air spaces within the leaf slices, found that photosynthesis was inhibited 14% by increasing the osmoticum from 0·33 to 0·5 M (equivalent to a pressure shift of -8 to -12 bars). More importantly, the inhibition of photosynthesis in leaf slices from plants grown on low K^+-nutrient solutions (which resulted in a lowering of the total K^+ from 240 to 74 μeq. g^{-1} F.W.) was greater (41%) under the same shift of osmoticum.

Berkowitz and Whalen (1985) did not observe the same sensitivity to osmotic potential as that observed by Kaiser and Heber (1981) when isolated, intact chloroplasts from K^+-deficient plants were used; although photosynthesis in intact chloroplasts was still inhibited by external osmoticum. Kaiser and Heber (1981) reported, however, that the osmotic stress caused a rapid shrinkage of the chloroplast volume and a generalized loss of metabolites. The volume soon recovered, and the loss of metabolites was slowed. Both starch degradation and the use of NADPH/ATP dependent pathways (dihydroxyacetone-3-phosphate to 3-phosphoglyceric acid and ribose-5-phosphate + CO_2 to 3-phosphoglyceric acid) were inhibited. Many enzymes, assayed in vitro, were also inhibited (but recovered their activities when the osmoticum was removed, Kaiser et al., 1981a).

Robinson (1985) suggested that the inhibition of photosynthesis in intact plastids by external osmoticum was not physiologically important since it depended upon the isolation procedure. He states, "Sudden exposure to osmotic stress apparently induces a transient change in permeability of the chloroplast envelope . . ." and this results in many of the observed effects due to loss of metabolites. For slow changes, Robinson believes that the chloroplast can adjust and maintain its photosynthetic rate. This may have been the cause of the inhibition of nitrite reductase observed in isolated chloroplasts by Behrens et al. (1985), where the reductase was inhibited by 40% upon a shift from -11 bars to -30 bars. In this case, electron transport was studied, and no change in its activity was noted with the osmotic stress. The authors concluded that the structural interaction of ferrodoxin with the reductase was disturbed.

In terms of O_3 injury, it seems unlikely that O_3 or its products could penetrate into an organelle to disrupt photosynthesis. The data of Coulson and Heath (1972) still relate to this problem. Ozone can inhibit the photosynthetic CO_2 fixation of isolated intact chloroplasts, but does not enter the plastid to inhibit the grana reactions. It is hard to imagine O_3 flowing through the cytoplasm unaltered to attack the envelope of the chloroplast. If, however, the chloroplast was in close proximity to the plasma membrane, chances of envelope alterations would be increased and it may be possible to inhibit CO_2 fixation, as described by Coulson and Heath (1972). But nevertheless, a water stress, induced by osmotic imbalance due to O_3 exposure, can alter the photosynthetic rate and this alteration can be influenced by cellular K^+. Most probably, the loss of K^+, metabolites, and turgor pressure induced by O_3 at the cell wall/membrane causes the photosynthetic alteration, which has been previously observed (Hällgren, 1984). These disruptions have important ramifications with respect to productivity and, undoubtedly, play a critical role in how the cell will be able to compensate for the losses and repair any injury.

Turgor pressure can also affect the plant processes. Water stress has been known to slow and even stop translocation from the source leaf to varied sinks (Hansom and Hitz, 1982). In fact, the altered microrhizosphere population observed after O_3 exposure to leaves may be more attributable to the loss of translocated sugar—the food source—to the organisms in the microrhizosphere than anything else (Manning *et al.*, 1971).

Miller (1988) describes the effects of pollutant exposure on translocation. However, recent studies on the effect of water potential status upon translocation and phloem flow are quite interesting. Sovonick-Dunford (1986) described an experiment in which the turgor pressure within the phloem of a white ash sieve tube was measured by a pressure probe, before and after the tube was cut. Initially, the relation between the flow and turgor pressure followed the standard relationship of a pressure threshold. Later, the flow slowed, and its solutes were diluted. The release of the turgor pressure when the sieve tube was cut, forced water from surrounding cells into the tube, which partially maintained the flow but altered the concentration of the solutes within the tube. The interrelationship between turgor and flow may explain the results of Porter *et al.* (1987) in which they describe the inhibition of transport of fixed carbon from the pedicel tissue of maize seeds, which was caused by increased osmotic pressure in the apoplastic space of the developing embryo (800 mM mannitol or other sugars caused a 70% inhibition in rate). The turgor pressure relationships have been altered by the application of extracellular osmotic material; thus the water relationships between the cells were changed such that the material flow was maintained but solute in the vessels was diluted.

Turgor pressure can easily alter both photosynthesis within the cell and the movement of its products to the developing portions of the remainder of the plant. Thus, altered turgor pressure should lower the net, measured productivity, but in a highly complex manner.

11.6 SUMMARY

It is obvious that there is much more to learn before a clear mechanism can be formulated to relate early biochemical events to final air pollutant injury associated with yield decline. Biochemical pathways are very interrelated, and we do not have sufficient knowledge of all the control and regulatory mechanisms in operation. As our understanding of the biochemistry of healthy plants progresses, we should be able to better understand O_3-induced pathology. In closing, I wish to offer a few goals for future research.

First, the major problem in interpreting biochemical changes induced by air pollutants is that gases do not enter the plant uniformly and therefore do not injure the plant uniformly. In addition, some areas of the leaf surface respond to the gases, even at a constant dose, in a nonuniform manner. At the present time, most investigators look at the leaf as a whole and report a change as a single datum point. But of course this is not so; there are areas heavily injured, which will ultimately become neocrotic; areas slightly injured, which will have distorted biochemistry; and areas uninjured, which are similar to the level in the control. When a measurement of the leaf is treated as a single datum, all of these responses are combined together. For a given biochemical reaction, the first area may yield an inhibition of the process; the second, a stimulation; and the third, no change. The average datum may be $+$, $-$, or 0 depending upon the relative amounts of the respective areas. We must be able to determine the areas of the leaf that are normal and should be treated as an 'internal control'. In this regard, we may have to adapt nondestructive sampling such as delayed light emission (Ellenson and Amundson, 1982) or chlorophyll fluorescence (Omasa *et al.*, 1987) in order to image the leaf and ascertain where injury has occurred. Unfortunately, current instrumentation is expensive and does not lend itself to routine monitoring.

The literature is filled with bits of biochemical data from many diverse groups of plants, and so a uniform model of injury is difficult to construct. A single plant type should be used for biochemical work, which is suited for biochemical analysis and does not contain noxious chemicals. For example, spinach has been used for photosynthetic work because organelles can be obtained with a minimum of injury. The species selected should have a large amount of known genetic diversity among its varieties to allow for genetic analysis and a known developmental sequence, which can be manipulated for analysis of growth within the environment. And finally, this most-perfect plant should be alterable by recombinant DNA techniques to test varied hypotheses at the molecular level.

We must understand the initial oxidant attack of cells to determine where to focus our attention. Measurement of the alterations of other pathways by oxidants may be merely due to a water stress induced by the loss of osmotically active material and, as such, is interesting as a general stress response. Certainly, as we understand more of the cell wall and membrane microenvironment and their interactions, it becomes easier to understand the initial biochemical events of oxidant injury.

REFERENCES

Adepipe, N. O. and D. T. Tingey. (1979). Ozone phototoxicity in relation to stress ethylene evolution and stomatal resistance in cowpea (*Vigna unquiculata*) cultivars. *Z. Pflanzenphysiol.*, **93**, 259–64.

Bailey, P. S. (1978). *Ozonation in organic chemistry: olefinic compounds*, Vol. 1, 493 pp. New York, Academic Press.

Baker, D. A. (1978). *Transport phenomena in plants*, 80 pp. London, Chapman & Hall.

Barber, D. J. W. and J. R. Thomas. (1978). Reactions of radicals with lecithin bilayers. *Rad. Res.*, **74**, 51–65.

Behrens, P. W., F. Xu, M. Werner, T. Hoffman, T. V. Marsho and A. B. Mackay. (1985). The effect of low osmotic potential on nitrite reduction in intact spinach chloroplasts. *Plant Physiol.*, **79**, 441–4.

Bennett, J. H., E. H. Lee, and H. E. Heggestad. (1984). Biochemical aspects of plant tolerance to ozone and oxyradicals: superoxide dismutase. In *Gaseous pollutants and plant metabolism*, ed. by M. J. Koziol and F. R. Whatley, 413–24. London, Butterworths Scientific.

Berkowitz, G. A. and C. Whalen. (1985). Leaf K^+ interaction with water stress inhibition of nonstomatal-controlled photosynthesis. *Plant Physiol.*, **79**, 189–93.

Bielski, B. H. J. (1983). Evaluation of the reactivities of HO_2/O_2^- with compounds of biological interest. In *Oxy-radicals and their scavenger systems: molecular aspects.* ed. by G. Cohen and R. G. Greenwald, Vol. 1, 1–7. New York, Elsevier Press.

Black, V. J. (1984). The effects of air pollutants on apparent respiration. In *Gaeous pollutants and plant metabolism*, ed. by M. J. Koziol and F. R. Whatley, 231–48. London. Butterworths Scientific.

Bobrov, R. A. (1952). The effect of smog on the anatomy of oat leaves. *Phytopathology*, **42**, 558–63.

Boguth, W. (1969). Aspects of the action of vitamin E. *Vit. Hormones*, **27**, 1–15.

Borland, A. M. and A. J. Rowland. (1988). Pollution and polyamines. In *Air pollution and plant metabolism*, ed. by S. Schulte-Hosted, L. Blank, N. Darrall and A. R. Wellburn, (in press). Berlin, De Gruyter.

Borochov, A. and R. Faiman-Weinberg. (1984). Biochemical and biophysical changes of plant protoplasmic membranes during senescence. *What's New in Plant Physiol.*, **15**, 1–4.

Carpita, N., D. Sabularse, D. Montezinos, and D. P. Delmer. (1979). Determination of the pore size of plant walls of living plant cells. *Science*, **205**, 1144–7.

Castillo, F. J. (1986). Extracellular peroxidases as markers of stress? In *Molecular and physiological aspects of plant peroxidases*, ed. by H. Greppin, Cl. Penel, and Th. Gasper, 419–26. Genieve, Switzerland, Centre de Botanique, Université de Genieve.

Castillo, F. J. and H. Greppin. (1986). Balance between anionic and cationic extracellular peroxidase activities in *Sedum album* leaves after ozone exposure: analysis by high-performance liquid chromatography. *Physiol. Plant.*, **68**, 201–8.

Castillo, F. J., Cl. Penel, and H. Greppin. (1984). Peroxidase release induced by ozone in *Sedum album* leaves: involvement of Ca^{2+}. *Plant Physiol.*, **74**, 846–51.

Castillo, F. J., P. R. Miller, and H. Greppin. (1987). Extracellular biochemical markers of photochemical oxidant air pollutant damage to Norway spruce. *Experientia*, **43**, 111–15.

Cleland, R. E. (1975). Auxin-induced hydrogen ion excretion: correlation with growth, and control by external pH and water stress. *Planta*, **127**, 233–42.

Cosgrove, D. J. (1987). Wall relaxation and the driving forces for cell expansive growth. *Plant Physiol.*, **84**, 561–4.

Coster, H. G. L., E. Steudle, and U. Zimmerman. (1977). Turgor pressure sensing in plant cell membranes. *Plant Physiol.*, **58**, 636–43.

Coulson, C. L. and R. L. Heath. (1972). Inhibition of the photosynthetic capacity of isolated chloroplasts by ozone. *Plant Physiol.*, **53**, 32–8.

Cramer, G. R., A. Lauchli, and V. S. Polito. (1985). Displacement of Ca^{2+} by Na^+ from the plasmalemma of root cells: a primary response to salt stress. *Plant Physiol.*, **79**, 207–11.

Criegee, R. (1975). Mechanism of ozonolysis. *Angewandte Chemie* (Intl. Ed.), **14**, 745–60.

Curtis, C. R., R. K. Howell, and D. F. Kremer. (1976). Soybean peroxidases from ozone injury. *Environ. Pollut.*, **11**, 189–94.

Dass, H. C. and G. M. Weaver. (1968). Modification of ozone damage to *Phaseolus vulgaris* by antioxidants, thiols, and sulfhydryl reagents. *Can. J. Plant Sci.*, **48**, 569–74.

Demarty, M., C. Morvan, and M. Thellier. (1984). Calcium and cell wall. *Plant Cell Environ.*, **7**, 399–411.

Dominy, P. J. and R. L. Heath. (1985). Inhibition of the K^+-stimulated ATPase of the plasmalemma of Pinto bean leaves by ozone. *Plant Physiol.*, **77**, 43–5.

Ellenson, J. L. and R. G. Amundson. (1982). Delayed light imaging for the early detection of plant stress. *Science*, **215**, 1104–6.

Enoch, S. and Z. Glinka. (1983). Turgor-dependent membrane permeability in relation to calcium level. *Physiol. Plant.*, **59**, 203–7.

Esau, K. (1965). *Plant anatomy*, 2nd edn, 37 pp. New York, John Wiley.

Evans, L. S. and I. P. Ting. (1973). Ozone induced membrane permeability changes. *Am. J. Bot.* **60**, 155–62.

Evans, L. S. and I. P. Ting. (1974). Effect of ozone on ^{86}Rb-labeled potassium transport in leaves of *Phaseolus vulgaris* L. *Atmos. Environ.*, **8**, 855–61.

Gamble, P. E. and J. J. Burke. (1984). Effects of water stress on the chloroplast antioxidant system I. Alterations in glutathione reductase activity. *Plant Physiol.*, **76**, 615–21.

Gasper, Th., Cl. Penel, F. J. Castillo, and H. Greppin. (1985). A two-step control of basic and acidic peroxidases and its significance for growth and development. *Physiol. Plant.*, **64**, 418–23.

Grimes, H. D., K. K. Perkins and W. F. Boss. (1983). Ozone degrades into hydroxyl radical under physiological conditions: a spin trapping study. *Plant Physiol.*, **72**, 1016–20.

Gross, G. C. and C. Janse. (1977). Formation of NADH and hydrogen peroxide by cell wall-associated enzymes from *Forsythia* xylem. *Zeit. fur Pflanzenphysiol.*, **84**, 447–52.

Gurol, M. D. and P. C. Singer. (1982). Kinetics of ozone decomposition: a dynamic approach. *Environ. Sci. Technol.*, **16**, 377–83.

Hällgren, J.-E. (1984). Photosynthetic gas exchange in leaves affected by air pollutants. In *Gaseous pollutants and plant metabolism*, ed. by M. J. Koziol and F. R. Whatley, 147–60. London, Butterworths Scientific.

Hansom, A. D. and W. D. Hitz. (1982). Metabolic responses of mesophytes to plant water deficits. *Ann. Rev. Plant Physiol.*, **33**, 163–203.

Hayashi, T., S. M. Read, J. Bussell, M. Thelen, F.-C. Lin, R. M. Brown, Jr. and D. P. Delmer. (1987). UDP-glucose: (1-3)-β-glucan synthase from mung bean and cotton. *Plant Physiol.*, **83**, 1054–62.

Heath, R. L. (1979). Breakdown of ozone and formation of hydrogen peroxide in aqueous solutions of amine buffers exposed to ozone. *Toxicol. Letters*, **4**, 449–53.

Heath, R. L. (1980). Initial events in injury to plants by air pollutants. *Ann. Rev. Plant Physiol.*, **31**, 395–431.

Heath, R. L. (1984). Air pollutant effects on biochemicals derived from metabolism: organic, fatty, and amino acids. In *Gaseous pollutants and plant metabolism*, ed by M. J. Koziol and F. R. Whatley, 275–90. London, Butterworths Scientific.

Heath, R. L. (1987). The biochemistry of ozone attack on the plasma membrane of plant cells. *Rec. Adv. Phytochem.*, **21**, 29–54.

Heath, R. L. and P. E. Frederick. (1979). Ozone alteration of membrane permeability in *Chlorella*: I. Permeability of potassium ion as measured by [86]Rubidium tracer. *Plant Physiol.*, **64**, 455–9.

Heath, R. L. and F. J. Castillo. (1988). Membrane disturbances in response to air pollutants. In *Air pollution and plant metabolism*, ed by S. Schulte-Hosted, L. Blank, N. Darrall and A. R. Wellburn, 55–75. Berlin, De Gruyter.

Heath, R. L., P. E. Chimiklis, and P. Frederick. (1974). Role of potassium and lipids in ozone injury to plant membranes. In *Air pollution effects on plant growth*, ed. by W. M. Dugger, Jr., 58–75. Washington, DC, American Chemical Society.

Heath, R. L., R. T. Furbank, and D. A. Walker. (1985). Effects of polyethylene-glycol induced osmotic stress on transpiration and photosynthesis in Pinto bean leaf disc. *Plant Physiol.*, **78**, 627–9.

Hepler, P. C. and R. O. Wayne. (1985). Calcium and plant development. *Ann. Rev. Plant Physiol.*, **36**, 397–439.

Hesketh, J. D. and J. W. Jones. (1980). *Predicting photosynthesis for ecosystem models*, 273 pp. Boca Raton, FL, CRC Press.

Imaseki, H. (1985). Hormonal control of wound-induced responses. In *Plant hormones. Encyclopedia of Plant Physiology, New Series*, Vol. 11, 485–512. Berlin, Springer-Verlag.

Kaiser, W. M. and U. Heber. (1981). Photosynthesis under osmotic stress: effect of high solute concentrations on the permeability properties of the chloroplast envelope and on activity of stroma enzymes. *Planta*, **153**, 423–9.

Kaiser, W. M., G. Kaiser, P. K. Prachuab, S. G. Wildman, and U. Heber. (1981*a*). Photosynthesis under osmotic stress: inhibition of photosynthesis of intact chloroplasts, protoplasts, and leaf slices at high osmotic potentials. *Planta*, **153**, 416–22.

Kaiser, W. M., G. Kaiser, S. Schoner, and S. Neimanis. (1981*b*). Photosynthesis under osmotic stress: differential recovery of photosynthetic activities of stroma enzymes, intact chloroplasts, protoplasts, and leaf slices after exposure to high solute concentrations. *Planta*, **153**, 430–5.

Kauss, H. and W. Jeblick. (1986). Influence of free fatty acids, lysophosphatidylcholine, platelet-activating factor, acylcarnitine, and echinocandin B on 1,3-β-D-glucan synthase and callose synthesis. *Plant Physiol.*, **80**, 7–13.

Kinraide, T. B. and R. E. Wyse. (1986). Electrical evidence for turgor inhibition of proton extrusion in sugar beet taproot. *Plant Physiol.*, **82**, 1148–50.

Kinraide, T. B., I. A. Newman and B. Etherton. (1984). A quantitative simulation model for H^+-amino acid cotransport to interpret the effects of amino acids on membrane potential and extracellular pH. *Plant Physiol.*, **76**, 806–13.

Koziol, M. J. and F. R. Whatley (Eds). (1984). *Gaseous pollutants and plant metabolism.*, 466 pp. London, Butterworths Scientific.

Lagrimini, L. M. and S. Rothstein. (1987). Tissue specificity of tobacco peroxidase isozymes and their induction by wounding and tobacco mosaic virus infection. *Plant Physiol.*, **84**, 438–42.

Lee, E. H. and C. M. Chen. (1982). Studies on the mechanisms of ozone tolerance: cytokinin-like activity of *N*-[2-(2-oxo-1-imidazolidinyl)ethyl]-*N'*-phenylurea, a compound protecting against ozone injury. *Physiol. Plant.*, **56**, 486–91.

Lee, E. H., J. A. Jersey, C. Gifford and J. Bennett. (1984). Differential ozone tolerance in soybean and snapbeans: analysis of ascorbic acid in O_3-susceptible and O_3-resistant cultivars by high-performance liquid chromatography. *Environ. Exptl. Bot.*, **24**, 331–41.

Lefebvre, J. and C. Gillet. (1973). Effect of pH on the membrane potential and electrical resistance of *Nitella-flexilis* in the presence of calcium. *J. Exptl. Bot.*, **24**, 1024–30.

Lendzian, K. J. and M. H. Unsworth. (1983). Ecophysiological effects of atmospheric pollutants. In *Encyclopedia of plant physiol.*, New Series, Vol. IV, 466–502. Berlin, Springer-Verlag.

Levitt, J. (1972). *Responses of plants to environmental stress.*, 697 pp. New York, Academic Press.

Li, Z.-S. and S. Delrot. (1987). Osmotic dependence of the transmembrane potential difference of Broadbean mesocarp cells. *Plant Physiol.*, **84**, 895–9.

Macklon, A. E. S. (1984). Calcium fluxes at plasmalemma and tonoplast. *Plant Cell Environ.*, **7**, 407–13.

Majerus, P. W., P. W., T. M. Connolly, H. Deckmyn, T. S. Ross, T. E. Bross, H. Ishii, V. S. Bansal, and D. B. Wilson. (1986). The metabolism of phosphoinositide-derived messenger molecules. *Science*, **234**, 1519–26.

Manning, W. J., W. A. Feder, P. M. Papia, and I. Perkins. (1971). Influence of foliar ozone injury on root development and root surface fungi of Pinto bean. *Environ. Pollut.*, **1**, 305–12.

Marmé, D. (1985). The role of calcium in the cellular regulation of plant metabolism. *Physiol. Veg.*, **23**, 945–53.

Matheson, I. B. C. and A. D. King, Jr. (1978). Solubility of gases in micellar solutions. *J. Colloid Interface Sci.*, **66**, 464–9.

Matters, G. L. and J. G. Scandalios. (1987). Synthesis of isozymes of superoxide dismutase in Maize leaves in response to O_3, SO_2, and elevated O_2. *J. Exptl. Bot.*, **38**, 842–52.

Mehlhorn, H. and A. R. Wellburn. (1987). Stress ethylene formation determines plant sensitivity to ozone. *Nature*, **327**, 417–18.

Mehlhorn, H., G. Seufert, A. Schmidt, and K. J. Kunert. (1986). Effect of SO_2 and O_3 on production of antioxidants in conifers. *Plant Physiol.*, **82**, 336–38.

Miller, J. E. (1988). Effects on photosynthesis, carbon allocation, and plant growth associated with air pollutant stress. In *Assessment of crop loss from air pollutants*. Proceedings of the International Conference, Raleigh, North Carolina, USA, ed. by W. W. Heck, O. C. Taylor, and D. C. Tingey. London, Elsevier Applied Science.

Morgan, B. P., J. P. Luzio, and A. K. Campbell. (1986). Intracellular Ca^{2+} and cell injury: a paradoxical role of Ca^{2+} in complement membrane attack. *Cell Calcium*, **7**, 399–411.

Morvan, C., M. Demarty, and M. Thellier. (1979). Titration of isolated cell walls of *Lemna minor* L., *Plant Physiol.*, **63**, 1117–22.

Mudd, J. B. (1973). Biochemical effects of some air pollutants on plants. *Adv. Chem.*, **122**, 31–47.

Mudd, J. B., S. K. Banerjee, M. M. Dooley, and K. L. Knight. (1984). Pollutants and plant cells: effects on membranes. In *Gaseous pollutants and plant metabolism*, ed. by M. J. Koziol and F. R. Whatley, 105–16. London, Butterworths Scientific.

Nieboer, E. and J. D. MacFarlane. (1984). Modification of plant cell buffering capacities by gaseous air pollutants. In *Gaseous pollutants and plant metabolism*. ed. by M. J. Koziol and F. R. Whatley, 313–29. London, Butterworths Scientific.

Nobel, P. S. (1974). *Introduction of biophysical plant physiology*, 2nd edn, 363 pp. San Francisco, W. H. Freeman.

Omasa, K., K.-I. Shimazaki, I. Aiga, W. Larcher, and M. Onoe. (1987). Image analysis of chlorophyll fluorescence transients for diagnosing the photosynthetic system of attached leaves. *Plant Physiol.*, **84**, 748–52.

Ordin, L., M. A. Hall, and J. I. Kindinger. (1969). Oxidant-induced inhibition of enzymes involved in cell wall polysaccharide synthesis. *Arch. Environ. Health.* **18**, 623–6.

Ormrod, D. P. and D. W. Beckerson. (1986). Polyamines as antiozonants for tomato. *HortSci.*, **21**, 1070–1.

Pallas, J. E. and S. J. Kays. (1982). Inhibition of photosynthesis by ethylene—a stomatal effect. *Plant Physiol.*, **70**, 598–601.

Pauls, K. P. and J. E. Thompson. (1981). Effects of *in vitro* treatment with ozone on the physical and chemical properties of membranes. *Physiol. Plant.*, **53**, 255–62.

Pell, E. J. (1988). Secondary metabolism and air pollutants. In *Air pollution and plant metabolism*, ed. by S. Schulte-Hosted, L. Blank, N. Darrall, and A. R. Wellburn, 222–37. Berlin, De Gruyter.

Pell, E. J. and E. Brennan. (1973). Changes in respiration, photosynthesis, adenosine 5'-triphosphate, and total adenylate content of ozonated Pinto bean foliage as they relate to symptom expression. *Plant Physiol.*, **51**, 378–81.

Porter, G. A., D. P. Knievel, and J. C. Shannon. (1987). Assimilate unloading from Maize (*Zea mays* L.) pedicel tissue: I. Evidence for regulation of unloading by cell turgor. *Plant Physiol.*, **83**, 131–6.

Pryor, W. A., J. W. Lightsey, and D. G. Prier. (1982). The production of free radicals *in vivo* from the action of xenobiotics: the initiation of autoxidation of polyunsaturated fatty acids by NO_2 and O_3. In *Lipid peroxides in biology and medicine*, ed. by K. Yagi, 1–22. New York, Academic Press.

Rasi-Caldogno, F., M. C. Pugiarello, and M. I. De Michelis. (1987). The Ca^{2+}-transport ATPase of plant plasma membrane catalyzes an H^+/Ca^{2+} exchange. *Plant Physiol.*, **83**, 994–1000.

Reid, D. M. and R. L. Wample. (1985). Water relations and plant hormones. In *Plant hormones. Encyclopedia of plant physiology*, New Series, Vol. 11, 513–78. Berlin, Springer-Verlag.

Reuter, H. (1983). Calcium channel modulation by neurotransmitters, enzymes, and drugs. *Nature*, **301**, 569–74.

Rickauer, M. and W. Tanner. (1986). Effects of Ca^{2+} on amino acid transport and accumulation in roots of *Phaseolus vulgaris*. *Plant Physiol.*, **82**, 41–6.

Robinson, S. P. (1985). Osmotic adjustment by intact isolated chloroplasts in response to osmotic stress and its effect on photosynthesis and chloroplast volume. *Plant Physiol.*, **79**, 996–1002.

Rogers, H. J. and H. R. Perkins. (1968). The architecture of the plant cell wall. In *Cell walls and membranes.*, 90–113. London, E. & F. N. Spon.

Roland, J. (1973). The relationship between the plasmalemma and cell wall. *Int. Rev. Cytol.*, **36**, 45–91.

Sagisaka, S. (1976). The occurrence of peroxide in a perennial plant, *Populus gelrica*. *Plant Physiol.*, **57**, 308–9.

Saran, M., C. Michel and W. Bors. (1988). Reactivities of free radicals. In *Air pollution and plant metabolism*, ed. by S. Schulte-Hostede, L. Blank, N. Darrall, and A. R. Wellburn, Berlin, De Gruyter.

Schulte-Hostede, S., L. Blank, N. Darrall, and A. R. Wellburn. (1988). *Air pollution and plant metabolism*, (in press). Berlin, De Gruyter.

Schwab, W. G. W. and E. Komor. (1978). A possible mechanistic role of the membrane potential in proton-sugar cotransport of *Chlorella*. *FEBS Lett.*, **87**, 157–60.

Schwartz, A. (1985). Role of Ca^{2+} and EGTA on stomatal movements in *Commelina communis* L. *Plant Physiol.*, **79**, 1003–5.

Selm, R. P. (1959). Ozone oxidation of aqueous cyanide waste solution in stirred batch reactors and packed towers. *Adv. Chem.*, **21**, 66–77.

Schechter, H. (1973). Spectrophotometric method for determination of ozone in aqueous solutions. *Water Res.*, **7**, 729–39.

Sisler, E. C. and R. Goren. (1981). Ethylene-binding—the basis for hormone action in plants. *What's New in Plant Physiol.*, **12**, 37–40.

Slater, T. F. (1984). Free-radical mechanisms in tissue injury. *Biochem. J.*, **222**, 1–15.

Smith, T. A. (1985). Polyamines. *Ann. Rev. Plant Physiol.*, **36**, 117–43.

Sovonick-Dunford, S. (1986). Water relations parameters of White Ash sieve tubes. In *Phloem transport*, ed. by J. Cronshaw, W. J. Lucas, and R. T. Giaquinta, 187–91. New York, Alan R. Liss.

Staehelin, J. and J. Holgné (1982). Decomposition of ozone in water: rate of initiation by hydroxide ion and hydrogen peroxide. *Environ. Sci. Technol.*, **16**, 676–81.

Staehelin, J. and J. Holgné. (1985). Decomposition of ozone in water in the presence of organic solutes acting as promoters and inhibitors of radical chain reactions. *Environ. Sci. Technol.*, **19**, 1206–13.

Sutton, R. and I. P. Ting. (1977). Evidence for the repair of ozone induced membrane injury. *Am. J. Bot.*, **64**, 404–11.

Taylor, G. E., Jr, P. J. Hanson, and D. D. Baldocchi. (1988). Pollution deposition to individual leaves and plant canopies: sites of regulation and relationship to injury. In *Assessment of crop loss from air pollutants*, Proceedings of the International Conference, Raleigh, North Carolina, USA, ed. by W. W. Heck, O. C. Taylor, and D. C. Tingey, 227–57. London, Elsevier Applied Science.

Thom, M. and A. Maretzki. (1985). Evidence for a plasmalemma redox system in sugarcane. *Plant Physiol.*, **77**, 873–6.

Thompson, J. E., R. L. Legge, and R. F. Barber. (1987). The role of free radicals in senescence and wounding. *New Phytol.*, **105**, 317–44.

Thomson, W. W., E. A. Nothnagel, and R. C. Huffaker. (1987). *Plant senescence: its biochemistry and physiology*, 255 pp. Rockville, American Society of Plant Physiology.

Thorp, C. E. (1954). *Bibliography of ozone technology.*, Vols I and II. Chicago, IL, Armour Research Foundation, J. S. Swift & Co.

Ticha, I. and J. Catsky. (1977). Ontogenetic changes in the internal limitations to bean-leaf photosynthesis. *Photosynthetica*, **11**, 361–6.

Ting, I. P. (1982). *Plant physiology*, 499–502. Reading, Mass., Addison-Wesley.

Tingey, D. T. and G. E. Taylor, Jr. (1982). Variation in plant response to ozone: a conceptual model of physiological events. In *Effects of gaseous air pollution in agriculture and horticulture*, ed. by M. H. Unsworth and D. P. Ormrod, 113–38. London, Butterworths Scientific.

Tingey, D. T., R. C. Fites, and C. Wickliff. (1975). Activity changes in selected enzymes from Soybean leaves following ozone exposure. *Physiol. Plant.*, **33**, 316–20.

Tromballa, H. W. (1974). Der Einfluss des pH-werts auf Aufnahme und Abgabe von Natrium durch *Chlorella*. *Planta*, **117**, 339–48.

Turrell, F. M. (1934). The area of the internal exposed surface of dicotyledon leaves. *Am. J. Bot.*, **23**, 255–64.

Van Volkenburgh, E. and R. E. Cleland. (1984). Control of leaf growth by changes in cell wall properties. *What's New in Plant Physiol.*, **15**, 25–8.

Villanueva, V. R., M. Mardon, F. Mancelon, and A. Santerre. (1988). *Biochemical markers in polluted Picea trees. 1. Is putrescine a useful pollution biological marker?*, ed. by S. Schulte-Hostede, L. Blank, N. Darrall, and A. R. Wellburn (in press). Berlin, De Gruyter.

Wasserman, B. P. and K. J. McCarthy. (1986). Regulation of plasma membrane β-glucan synthase from red beet root by phospholipids. *Plant Physiol.*, **82**, 396–400.

Weiss, J. (1935). Investigations on the radical $HO_2\cdot$ in solution. *Trans. Faraday Soc.*, **31**, 668–81.

Westgate, M. E. and E. Steudle. (1985). Water transport in the midrib tissue of Maize tissue of Maize leaves: direct measurement of the propagation of changes in cell turgor across a plant tissue. *Plant Physiol.*, **8**, 183–91.

Wolf, S. P., A. Garner, and R. T. Dean. (1986). Free radicals, lipids, and protein degradation. *Trends in Biochem. Sci.*, **11**, 27–31.

Yang, S. F. and N. E. Hoffman. (1984). Ethylene biosynthesis and its regulation in higher plants. *Ann. Rev. Plant Physiol.* **35**, 155–89.

Zimmermann, U. and F. Beckers. (1978). Generation of action potentials in *Chara corallina* by turgor pressure changes. *Planta*, **138**, 173–9.

12

EFFECTS ON PHOTOSYNTHESIS, CARBON ALLOCATION, AND PLANT GROWTH ASSOCIATED WITH AIR POLLUTANT STRESS

JOSEPH E. MILLER

US Department of Agriculture, North Carolina State University, Raleigh, North Carolina, USA

12.1 INTRODUCTION

Over the past 30 to 50 years, extensive literature has accumulated concerning air pollutant effects on plants. The relative degree of phytotoxicity of the common pollutant gases is known, and the potential for effects on many botanical classes of plants is understood somewhat. Knowledge is at a point where attempts are being made to assess and predict the effects of current or projected air pollutant levels on the growth or productivity of vegetation. The most successful attempts have been made with crops because the most extensive data are available in this area. The National Crop Loss Assessment Network (NCLAN) has made noteworthy contributions in the areas of generating empirical, field-based, dose–response data and of economic assessments (Heck *et al.*, 1982; 1984; Adams *et al.*, 1985). But, even in the case of assessment of crop losses due to air pollutants, it has become apparent that insufficient data or understanding exist to make predictions with a high degree of confidence. The NCLAN effort has clearly identified the need for additional information in the form of dose–response data concerning the interaction of air pollutants with other environmental variables and the need to establish the functional relationships between air pollution stress and plant growth and yield.

This paper emphasizes the latter area, the physiological basis for air pollution-induced reductions in crop growth and yield. The topics of photosynthesis and carbon allocation were chosen because of the obviously close relationship they bear to growth and yield and, therefore, to assessment of crop losses. The review will be limited to consideration of air pollutant effects on crop species and will emphasize the gaseous pollutants ozone (O_3) and

287

sulfur dioxide (SO_2) because more is known about their effects on physiological processes and growth than for other air pollutants. The factors regulating gas uptake by plant tissues (e.g., stomatal and mesophyll conductance) will not be discussed because they are addressed in Chapter 10 (Taylor *et al.*). The biochemistry of photosynthetic carbon assimilation will not be emphasized except as it relates to the mechanisms controlling allocation.

12.2 REVIEW OF AIR POLLUTANT EFFECTS ON GROWTH, PHOTOSYNTHESIS, AND CARBON ALLOCATION

12.2.1 Growth and partitioning of biomass

The purpose of this paper is not to provide a comprehensive review of air pollutant effects on plant growth and yield, but rather to address the ways in which air pollutant stress modifies the processes of carbon allocation affecting growth and yield. A number of other reviews or books are available concerning effects on growth (e.g., Guderian, 1977; Heck *et al.*, 1982; Unsworth and Ormrod, 1982; Heck *et al.*, 1984; Koziol and Whatley, 1984; Roberts, 1984; Skarby and Sellden, 1984; Treshow, 1984; Guderian, 1985; Winner *et al.*, 1985; Heck *et al.*, 1986; Miller, 1987). The following discussion will deal with a few selected studies of the effects of air pollutant stress on partitioning of biomass. This information provides a more detailed look at the effects of pollutant stress on growth and can give some insight into the processes of allocation that the stress may affect.

The majority of the work concerning O_3 and SO_2 effects on partitioning of biomass among plant tissues or organs has been with pot-grown plants in controlled environments (Tables 12.1 and 12.2). The results vary depending on the species, the conditions of the study, and the characteristics of the pollutant exposures. A generalized summary of the findings from these studies is shown in Fig. 12.1. One significant generality from these results is that O_3 and SO_2 often suppress root growth more than shoot growth. Biomass partitioning to reproductive structures also seems to be reduced in some instances, although in several cases either an increase or no effect has been found. Increases in allocation to shoots have been attributed to the metabolic costs of repair of the stress damage (McLaughlin and Shriner, 1980) and the general precedence of shoots over roots as primary sinks during periods of stress or limited periods of photosynthate production (Silvius *et al.*, 1977; Chatterton and Silvius, 1979). Also, the partitioning of carbohydrate between sucrose and starch in the leaf may influence relative root and shoot development (Huber *et al.*, 1985). Thus, any air pollution-induced shift in sucrose-starch metabolism might affect carbohydrate allocation patterns. Regardless, the outcome seems to be preferential translocation of assimilate to shoot tissues or retention in the shoot under pollutant stress.

Growth analysis techniques have also been employed in some studies to investigate more specifically the effects on growth characteristics (Tables 12.1 and 12.2). For example, Endress and Grunwald (1985) found that chronic O_3

stress at concentrations similar to those in O_3-impacted areas reduced several indices of growth in soybeans (*Glycine max*). Among those parameters reduced were total plant dry weight; leaf area (LA); leaf area duration (LAD); relative growth rate (RGR) of leaves, stems, and roots; and net assimilation rate (NAR). Overall, these results indicate that O_3 altered the production and distribution of photosynthates within the plant. Reductions in RGR by O_3 at some periods in plant development have also been found with cotton (*Gossypium hirsutum*) (Oshima *et al.*, 1979), snap bean (*Phaseolus vulgaris*) (Blum and Heck, 1980), and sunflower (*Helianthus annuus*) (Shimizu *et al.*, 1980). Net assimilation rates are also often reduced by O_3 (Oshima *et al.*, 1979; Reich *et al.*, 1986; Shimizu *et al.*, 1980). However, Bennett and Runeckles (1977) found that a higher NAR was associated with reduced leaf size in annual ryegrass (*Lolium multiflorum*) treated with O_3, which they attributed to a compensation reaction to the stress.

Fewer studies have attempted to determine air pollutant effects on biomass partitioning or carbon allocation in plants under field conditions due to problems such as lack of environmental control, inherent field variability, and the difficulty in recovering roots from soil. Periodic sampling of plants throughout the growing season has been done in several open-top chamber studies. Kohut *et al.* (1986*b*) found that total above-ground biomass of 'Hodgson' soybeans responded to O_3 stress within 5 weeks after initiation of treatments, while the percentage of biomass found in leaves, stems, and pods did not respond until about 7 weeks. Most of the change was due to accelerated senescence of leaves from the O_3 stress. In a very comprehensive study over two growing seasons with 'Davis' soybeans, Flagler (1986) found that O_3 stress caused a reduction in main stem nodes, branches, leaf number, leaf size, total leaf area, and total dry weight. Root dry weights were reduced by the O_3 stress in 1 year of the study, but root/shoot ratios showed no obvious trends. While the O_3 treatments generally reduced pod weight, a greater proportion of the total dry weight was present in the pod fraction as O_3 increased. In another field study with 'Young' soybeans, root/shoot ratios were reduced by O_3 during the vegetative phase of growth (J. E. Miller, unpublished results). A field study with 'McNair-235' cotton also considered the effects of O_3 on total plant growth and biomass partitioning throughout an entire growing season (J. E. Miller, unpublished results). During vegetative growth, there was a tendency for leaf area ratio (LAR) and leaf weight ratio (LWR) to be elevated by O_3, although this reversed later in the growing season due to accelerated leaf drop as a result of O_3 stress. There was some tendency for O_3 to increase stem height and node formation, but not stem biomass. This resulted in a more spindly stem, which is probably more susceptible to lodging. Root/shoot and root/stem ratios showed a tendency to be reduced by increasing O_3 concentrations at some times during the season. Leaf area and LAD were reduced by O_3, which undoubtedly accounted for much of the seed cotton yield loss. However, the ratio of final seed cotton yield to LAD was reduced by increasing O_3 concentrations, which indicates that reductions in the efficiency of the leaves as well as reductions in the LA were responsible for the yield

TABLE 12.1

Summary of selected studies concerning O_3 effects on growth, biomass partitioning, and growth analysis of crop plants

Species	Conditions/treatment	Growth (biomass)				Partitioning		Growth analysis				Comments	References
		Shoot	Root	Rep.	Other	Root/shoot	Rep./shoot	RGR	NAR	LAR	Other		
Capsicum annuum 'M-75'	GH; 0·12 and 0·2 ppm, 3 h d⁻¹, 3 d wk⁻¹; 5% of growth period of 106 d	NE	NE	D	—	D	D	—	—	—	—	Plant height and no. leaves I at 0-20 ppm	Bennett *et al.* (1979)
Daucus carota 'Imperator 58'	GH; 0·19 and 0·25 ppm, 6 h d⁻¹, 1·5 times wk⁻¹; 6% of growth period of 108 d	NE	D	—	—	D	—	—	—	—	—	Height and no. leaves I	Bennett and Oshima (1976)
Festuca arundinacea 'Alta'	GH; 0·1-0·3 ppm, 6 h d⁻¹, 1 d wk⁻¹, 12 wk	D	D	—	Tiller (D)	—	—	—	—	—	—	—	Flagler and Youngner (1982)
Glycine max 'Corsoy'	GH; 0·070 and 0·097 ppm, 6 h d⁻¹, 4-5 d wk⁻¹, 10 wk, 341 h total	D	D	Seed (D)	—	—	—	D	D	I	LWR(I) SLA(D) LAD(D)	0·046 ppm O_3 often I growth	Endress and Grunwald (1985)
Glycine max 'Davis'	F, 0-T; 0·045-0·107 7 h d⁻¹, 16 wk	D	D	D	—	I	I	—	—	—	—	—	Flagler (1986) Heagle *et al.* (1987)
Glycine max 'Hodgson'	F, 0-T; 0·035-0·122 ppm, 7 h d⁻¹, 10 wk	D	—	D	—	—	D	—	—	—	—	Accelerated senescence of leaves	Kohut *et al.* (1986*b*)
Glycine max 'Hodgson'	GC; 0·05, 0·09 and 0·13 ppm, 6-8 h d⁻¹, 7 d wk⁻¹, 8·5 wk	D	D	Pod (D)	—	NE	NE	—	D	—	LA(D)	O_3 delayed onset of flowering	Amundson *et al.* (1986) Reich *et al.* (1986)
Glycine max 'Hood and Dare'	GH; 0·05 and 0·10 ppm, 8 h d⁻¹, 5 d wk⁻¹, 3 wk	D	D	—	—	D	—	D	—	—	—	—	Tingey *et al.* (1973*b*)

Species	Conditions									Growth analysis	Comments	Reference
Gossypium hirsutum 'Acala SJ-2'	GH; 0.25 ppm, 6 h d⁻¹, 2 d wk⁻¹	D	D	—	D	D	D	D	I	—	Leaf area and no. branch I following initial D	Oshima *et al.* (1979)
Gossypium hirsutum 'McNair 235'	F, O-T; 0.041–0.071 ppm, 12 h d⁻¹, 18 wk	D	D	—	D	—	I	—	(I/D)	LAD(D) SLA(I) LWR(I/D)	LAR and LWR I early in season and D later	Miller *et al.* (unpublished data)
Helianthus annuus 'Russian Mammoth'	GC; 0.1 and 0.2 ppm, 24 h d⁻¹, 12 d	D	D	—	D	D	D	D	I	LWR(I), SWR(D)	—	Shimizu *et al.* (1980)
Lolium multiflorum	GH; 0.03 and 0.09 ppm, 8 h d⁻¹, 6 wk	D	D	—	D	D	NE	I	D	SLA(D)	Authors concluded I in NAR offset D in LAR	Bennett and Runeckles (1977)
Petroselinum crispum 'Banquet'	GH; 0.2 ppm, 4 h d⁻¹, 2 d wk⁻¹, 8 wk	D	D	—	D	D	—	D	—	—	Root RGR D at early treatment	Oshima *et al.* (1978)
Phaseolus vulgaris 'BBL-290'	GC; 0.3 and 0.6 ppm, 1·5 h, 2 d	D	D	—	NE	D	D	D	—	AGR(D)	RGR of roots D more than tops	Blum and Heck (1980)
Solanum tuberosum 'Centennial Russet'	FC; variable filtration of ambient air (Riverside, CA), 3–42 ppm-h dose	D	D	Tuber (NE)	D	D	D	—	—	—	Ambient air (nonchamber) effects similar	Foster *et al.* (1983)
Trifolium incarnatum	GH; 0.03 and 0.09 ppm, 8 h d⁻¹, 6 wk	D	D	—	D	D	NE	NE	D	SLA(D)	—	Bennett and Runeckles (1977)
Triticum aestivum 'Vona'	F, O-T; 0.042–0.096 ppm, 7 h d⁻¹, 8–5 wk	D	D	—	D	D	D	—	—	—	—	Kohut *et al.* (1987)

Abbreviations: D (decrease), I (increase), NE (no effect), GH (greenhouse chamber), GC (growth chamber), O-T (open-top chamber), F (field), FC (field chamber), RGR (relative growth rate), NAR (net assimilation rate), LAR (leaf area ratio), SLA (specific leaf area), LWR (leaf weight ratio), SWR (stem weight ratio), AGR (absolute growth rate), LA (leaf area), LAD (leaf area duration), Rep. (reproductive).

TABLE 12.2

Summary of selected studies concerning SO_2 effects on growth, biomass partitioning, and growth analysis of crop plants

Species	Conditions/treatment	Growth (biomass) Shoot	Root	Rep.	Other	Partitioning Root/shoot	Rep./shoot	Growth analysis RGR	NAR	LAR	Other	Comments	References
Avena sativa 'Carolee'	GC; 0·4–1·6 ppm, 3 h d⁻¹, 4 d	D	D	Panicle no. (D)	—	D	—	—	—	—	—	Interactions with temp and RH studied	Heck and Dunning (1978)
Festuca arundinacea 'Alta'	GH; 0·1 ppm, 6 h d⁻¹, 1 d wk⁻¹, 12 wk	NE	D	—	—	D	—	—	—	—	—	No O_3 interactions	Flagler and Youngner (1982)
Glycine max 'Wells'	F-OA; 0·12–0·79 ppm, 4·7 h d⁻¹, 24 d	—	—	D	Chaff (D)	—	Harvest ratio (D)	—	—	—	—	—	Sprugel *et al.* (1980)
Helianthus annuus 'Russian Mammoth'	GC; 0·1 ppm, 24 h d⁻¹, 5 wk	NE	NE	—	Leaf (D)	—	—	D	—	I	LWR(I) SLA(NE)	Depression of flower bud development and stem elongation	Shimizu *et al.* (1980)
Nicotiana tabacum 'Samsun'	GC; 0·02 ppm, 24 h d⁻¹, 4 wk	D	D	—	—	D	—	—	—	—	LA(D)	—	Mejstrik (1980)
Phleum pratense 'Aberystuyth S48'	GC; 0·06 ppm, 24 h d⁻¹, 6 wk	NE	D	—	—	D	—	—	—	I	SLA(I), LWR(I)	—	Jones and Mansfield (1982)
Solanum tuberosum 'Centennial Russet'	FC; 0·1 ppm, 24 h d⁻¹, ~25 ppm-h dose	NE	NE	—	Tuber (D)	NE	—	—	—	—	—	$SO_2 \times O_3$ interaction only on tuber N content; tuber N(I) by SO_2	Foster *et al.* (1983)

Abbreviations: D (decrease), I (increase), NE (no effect), GH (greenhouse chamber), GC (growth chamber), F-OA (field-open air), FC (field chamber), F (field), O-T (open-top chamber), AGR (absolute growth rate), LAR (leaf area ratio), LWR (leaf weight ratio), NAR (net assimilation rate), RGR (relative growth rate), SLA (specific leaf area), SWR (stem weight ratio) Rep. (reproductive), LA (leaf area).

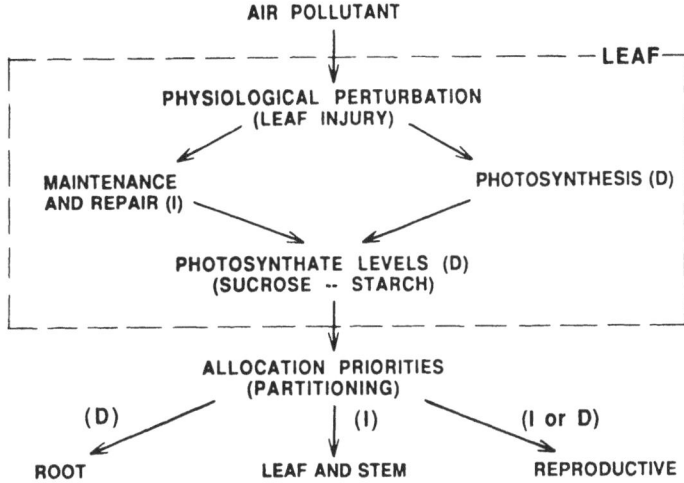

Fig. 12.1. Generalized summary of effects of air pollutants on carbon allocation in crop plants. Results may vary with the crop, environment, and nature of the air pollutant stress.

effects. Unsworth *et al.* (1984) came to similar conclusions in a field study with soybeans.

Clearly, air pollutants affect the partitioning of assimilated carbon as well as the overall growth (increase in total biomass) of crop plants. The mechanisms by which effects on growth are realized have received some attention. However, much less is known concerning air pollutant effects on the mechanisms controlling the allocation of carbon than for effects on growth alone, although these are interrelated processes. More research is required to define the influence of air pollutants on source–sink relationships and to explain the shifts in sink priorities that may occur. The following sections will provide an overview of some of the research that has been done on the major components of carbon assimilation and allocation, i.e., photosynthesis, respiration, and translocation, and will attempt to identify some of the deficiencies in knowledge.

12.2.2 Photosynthesis

The majority of the dry weight of plants is derived from photosynthetic CO_2 fixation and, thus, processes related to CO_2 fixation should be sensitive indicators of air pollutant stress. An understanding of air pollution-induced alterations of photosynthetic carbon fixation is essential in quantifying the carbon budget and relating the stress effects to growth and yield responses. One serious difficulty in relating existing studies of photosynthesis to plant productivity is that plant growth integrates carbon assimilation and metabolism over its entire life span, while measurements of photosynthesis or detailed studies of photosynthetic mechanisms usually have been short-term due to experimental limitations. Still, even with this limitation, studies of photosynthesis are one important part of understanding air pollutant stress.

12.2.2.1 Photosynthesis—O₃

There have been several studies of photosynthesis with field-grown crops using open-top chambers to control pollutant concentrations. The results for O_3 shown in Fig. 12.2 illustrate that photosynthesis rates may be reduced by concentrations near typical ambient levels. There is a surprising similarity in the results considering the diversity in species, duration of O_3 exposure, and experimental approach to the measurements. With the exception of the one set of measurements with wheat, all data were collected on single leaves. All measurements were taken intermittently during the season (with different frequencies in different experiments), so the data do not represent the total photosynthetic history of the plants throughout the growing season. But, it is clear that ambient levels of O_3 have the potential to reduce photosynthesis.

It is difficult to compare relative species sensitivity in these studies, since exposure durations (hours per day and days per experiment) and exposure conditions differed. Reich and Amundson (1985) presented a summary of their research on seven crop and tree species that were exposed to chronic O_3 stress under laboratory and/or field conditions. After converting to the total treatment dose (parts per million-hours, ppm-h) during exposure periods, they made a tentative comparison of their data. Red clover (*Trifolium pratense*), wheat (*Triticum aestivum*), soybean, and hybrid poplar (*Populus deltoides* × *trichocarpa*) appeared to be the most sensitive, and sugar maple (*Acer saccharum*) also appeared sensitive when the data were presented in this fashion. White pine (*Pinus strobus*) and red oak (*Quercus rubra*) were much

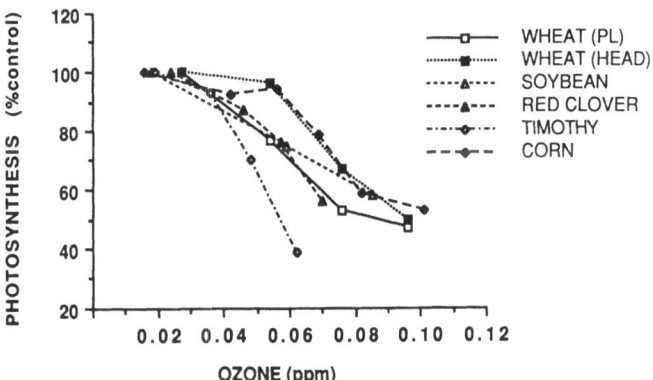

Fig. 12.2. Effects of O_3 on relative rates of photosynthesis of field-grown crop plants. Wheat (plant)—'Vona' (*Triticum aestivum*); O_3 treatment (tr) 7 h day^{-1}, 5 weeks; photosynthesis (Ps) measured on whole plant, anthesis-maturity (Amundson *et al.*, 1987). Wheat (head)—'Vona' (*Triticum aestivum*); O_3 tr 7 h day^{-1}, 5 weeks; Ps measured on head, anthesis-maturity (Amundson *et al.*, 1987). Soybean—'Young' (*Glycine max*); O_3 tr 12 h day^{-1}, 18 weeks; Ps measured on sixth leaf down from apex, measured weekly for 12 weeks, vegetative-reproductive maturity; well-watered plots (S. F. Vozzo and J. E. Miller, unpublished data). Red Clover—'Arlington' (*Trifolium pratense*); O_3 tr 7 h day^{-1}, 6 days week^{-1}, 3 weeks; Ps measurement is average of various age leaves (^{14}C fixation) (Kohut *et al.*, 1985). Timothy—'Champlain' (*Phleum pratense*); O_3 tr 12 h day^{-1}, 7 day week^{-1}; Ps measured on third leaf below inflorescence after 2–4 months of tr, ^{14}C fixation (Kohut *et al.*, 1986*a*). Corn—'FR23 × LH74' (*Zea mays*); O_3 tr 12 h day^{-1}, 7 day week^{-1}; Ps measured on ear leaf in late August (P. M. Irving, personal communication).

less sensitive. It is interesting that the four most sensitive species also had the highest inherent rates of photosynthesis. Probably, the higher gas exchange rates in these species resulted in greater O_3 flux into the leaf, which may be partially responsible for the greater apparent sensitivity of these species. Reich (1987) later presented the view that species exhibit a common response to O_3 at the cellular and biochemical levels and that differences in the inherent leaf diffusive conductance (controlling pollutant uptake) may be used to predict relative species sensitivity. It should be noted that the low and intermediate O_3 concentrations used in the studies of Reich and Amundson (1985) compared well with growing season concentrations found in much of North America and elsewhere. In the eastern and midwestern United States, cumulative seasonal O_3 doses of 30 to 50 ppm-h are common during daylight hours over an average growing season. Cumulative doses as low as 10 ppm-h caused substantial photosynthetic depression in five of the seven species in these studies.

In most studies of photosynthesis, measurements are made for a limited time after O_3 exposure and, thus, do not represent the complete photosynthetic history of the leaves throughout their development. However, Reich *et al.* (1986) did measure photosynthesis of soybean during leaf development. They found that chronic O_3 exposure reduced the capacity for photosynthesis throughout the developmental period of the leaves with the effects becoming greater as the leaves aged and had a longer history of exposure to O_3. This was partly due to reductions in leaf chlorophyll and acceleration of senescence. Other studies have also shown a reduction in the maximum photosynthetic efficiency due to O_3. For example, light saturation of fully developed leaves of snap bean that had been exposed to 0·072 ppm (141 μg m^{-3}) O_3 for 66 h over a period of approximately 18 days were studied by Coyne and Bingham (1978). The maximum photosynthetic rate at light saturation was reduced 18% by the O_3 treatment. Thus, O_3 apparently reduces maximum photosynthetic efficiency under conditions of chronic exposure.

Several conclusions concerning O_3 effects on photosynthesis that are pertinent to the understanding of O_3 effects on growth and yield can be drawn: (1) short-term exposures that do not cause visible injury may reduce photosynthesis, but the plants may recover within hours; (2) higher O_3 concentrations may cause cell damage, including the destruction of chlorophyll that permanently reduces the capacity for photosynthesis; and (3) chronic exposures at lower O_3 levels can reduce the ability of the plant to photosynthesize, probably by reducing chlorophyll, accelerating senescence, and impairing other physiological activities. The latter case is what is usually encountered in areas impacted by O_3.

12.2.2.2 *Photosynthesis—SO₂*

The inhibitory effects of SO_2 on photosynthesis have been documented in several reviews (e.g., Ziegler, 1975; Hällgren, 1978; Heath, 1980; Black, 1982). Species may vary in the degree of photosynthetic response to SO_2, as illustrated in Fig. 12.3. However, one should not make direct comparisons among species based on this summary of results, because the experimental and measurement protocols varied so widely among experiments. In a few

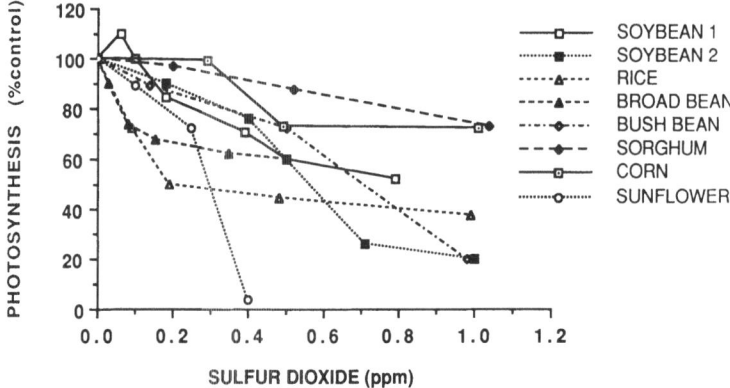

Fig. 12.3. Effects of SO$_2$ on relative photosynthetic rates of crop plants. Soybean 1—'Wells' (*Glycine max*); approximately six 4-h treatments (tr); field exp. (Muller *et al.*, 1979). Soybean 2—'Wayne' (*Glycine max*); one 2-h tr (Carlson, 1983). Rice—'Nihonmasari' (*Oryza sativa*); one 5-h tr (Katase *et al.*, 1983). Corn—'Golden Cross Bantam' (*Zea mays*); one 5-h tr (Katase *et al.*, 1983). Sorghum—(*Sorghum vulgare*); one 5-h tr (Katase *et al.*, 1983). Sunflower—'Russian Mammoth' (*Helianthus annuus*); one 5-h tr; (Katase *et al.*, 1983). Bush Bean—'Blue Lake 274' (*Phaseolus vulgaris*); one 6-h tr (Taylor and Selvidge, 1984). Broad Bean—'Dylan' (*Vicia faba*); one to three 8-h tr (Black and Unsworth, 1979).

cases, inhibition of photosynthesis was observed at low SO$_2$ concentrations (<0·1 ppm, 266 μg m^{-3}). For example, broad bean (*Vicia faba*) was especially sensitive with a 7% reduction at 0·035 ppm for 7 h (Black and Unsworth, 1979). In this study, the photosynthetic rates recovered to control values within an hour after termination of exposure, and visible injury did not occur (even at concentrations up to 0·175 ppm). In other studies with a variety of species, much higher concentrations have been required to cause significant inhibition of photosynthesis (e.g., corn (*Zea mays*) in Fig. 12.3). This is consistent with the findings that growth and yield of corn is quite resistant to SO$_2$ (Miller *et al.*, 1981). Other investigators have also noted a lack of a response of photosynthesis to moderate SO$_2$ concentrations. For example, Kohut *et al.* (1985) rarely found a response in red clover exposed to SO$_2$ concentrations ranging up to 0·9 ppm in open-top chambers. Occasionally, a stimulation at low SO$_2$ concentrations has been observed such as with soybean shown in Fig. 12.3. A variety of explanations for stimulation have been offered, such as increased sulfur nutrition, effects on stomatal gas exchange, and stimulation of energy-requiring repair processes. The reasons for the apparent variation in response among species is uncertain, although factors such as differential gas uptake and metabolic differences are both likely involved. Environmental factors during plant growth and during the exposure period most certainly are involved to some extent when different experiments are compared. For example, interactive effects of relative humidity (RH) and SO$_2$ on physiological processes have been observed. Effects of SO$_2$ are generally greater when RH is high, probably because of greater SO$_2$ uptake at the high RH (Barton *et al.*, 1980).

Despite the considerable information available, it is usually not possible to predict the effect of SO_2 on photosynthesis. The majority of investigations indicate that SO_2 exposure depresses net photosynthesis, although some workers report temporary enhancement at low concentrations (e.g., Muller *et al.*, 1979). Sometimes these enhanced rates can be correlated with increased stomatal conductance, and depressed photorespiration may also occur. The input of sulfur, as a nutrient, may also be a factor in longer-term stimulations. Depressions of photosynthetic rates may occur within minutes to hours after the initiation of exposure, are often reversible, and may not be accompanied by major visible injury, at least at low concentrations (Black and Unsworth, 1979). Recurrent exposure to low concentrations may cause the destruction of chlorophyll and hasten leaf senescence (Irving and Miller, 1981). At higher concentrations, responses often are not reversible and often are associated with the appearance of visible injury.

12.2.2.3   Photosynthesis—$O_3 \times SO_2$

Reports of synergistic interactions of O_3 and SO_2 on growth (e.g., Tingey and Reinert, 1975; Heggestad and Bennett, 1981) have prompted investigators to look to photosynthesis for a possible explanation. Ormrod *et al.* (1981) and Black *et al.* (1982) studied the response of broad bean to O_3 with or without SO_2 added at a concentration of 0·04 ppm for 4 h. The results suggested a synergistic response to concentrations of O_3 between approximately 0·06 to 0·15 ppm and an additive or antagonistic response to O_3 concentrations above that level. Furukawa and Totsuka (1979) observed a synergistic decrease in photosynthesis of sunflower at concentrations of 0·20 ppm each of SO_2 and O_3. Chevone and Yang (1985) observed that either 0·20 ppm O_3 or 0·70 ppm SO_2 administered separately for 2 h did not significantly affect photosynthetic rates of soybean, but the combination of the two gases at the same concentrations caused a 70% reduction. With soybean, Le Sueur-Brymer and Ormrod (1984) tested the effects of SO_2 (0·30 ppm) and O_3 (0·067 ppm) separately and combined. Each pollutant separately tended to reduce photosynthesis during 5 days of fumigation, and the combination caused a further decline, but the interactive effect of the two gases was not statistically significant. Currently, no research has conclusively tied synergistic reductions in photosynthesis due to SO_2 and O_3 to synergistic growth reductions.

12.2.2.4   Comparison of photosynthesis with growth and yield

There have been only several studies in which air pollutant effects on photosynthesis were successfully studied in conjunction with estimates of growth and/or yield. These data are presented as reductions in measured photosynthesis rates versus reductions in biomass or yield (Fig. 12.4). Reductions in whole-plant photosynthesis of wheat due to O_3 did correspond extremely well with grain yield reductions, whereas the measurements of wheat head photosynthesis tended to underestimate yield reductions (Amundson *et al.*, 1987). With field-grown soybeans, it was found that photosynthesis of single leaves in the upper part of the plant canopy was reduced proportionally more by O_3 than bean yield in well-watered (WW) plots, while in

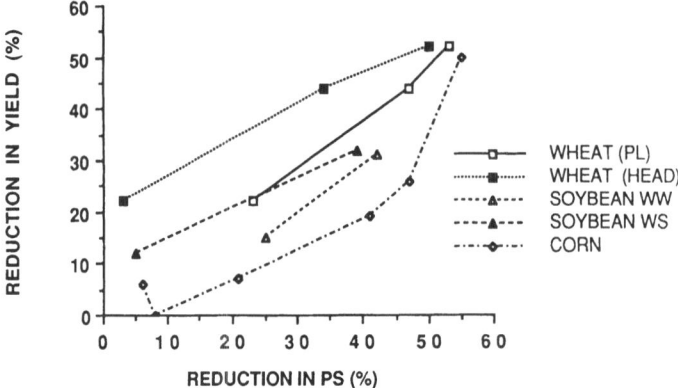

Fig. 12.4. Relationship of photosynthesis and yield reductions in crops exposed to O_3. (See Fig. 12.2 for description of treatments.) 'Vona' wheat (Amundson *et al.*, (1987)). 'Young' soybean; well-watered (WW) and water-stressed (WS) (Vozzo and Miller, unpublished data). 'FR23 × LH74' field corn (Irving, personal communication).

water-stressed (WS) plots yield was reduced more than photosynthesis at lower O_3 concentrations and slightly less at higher concentrations (S. F. Vozzo and J. E. Miller, unpublished results).

These comparisons, as presented, are not meant to suggest that photosynthetic rates alone be used to estimate yield or biomass reductions due to pollutant stress. First of all, in all of the studies, the measurements represent only a small part of the total photosynthetic history of the plants. In some cases, only a single leaf or part of a plant was measured. The photosynthetic rates of individual leaves or other photosynthetic tissues on a plant differ due to factors such as canopy shading, stage of development, and length of exposure to O_3. The earlier discussion of O_3 effects on LA and LAD are obviously quite relevant to the relationship between photosynthesis and yield. Air pollutants may reduce the leaf surface available for photosynthesis throughout the growth period of the plant, as well as reducing the photosynthetic efficiency of the tissues. Measurements of photosynthesis also do not account for other physiological aspects of carbon allocation and growth such as respiration, translocation, hormonal relationships, and metabolism in general. Also, when economic yield is the biomass measurement (such as soybean or wheat seed), the possible effects on partitioning between vegetative and reproductive tissues must be considered.

12.2.3 Respiration

Respiratory processes in plants are also quite sensitive to air pollution, and knowledge in this area is required to understand the response of the carbon budget of the plants to air pollution stress. Measurement of net CO_2 exchange during the light period expresses the balance between CO_2 uptake due to photosynthesis and respiratory CO_2 release. This measurement alone does not help to identify the complete physiological basis of the effect nor does it account for respiration during dark periods. Independent measurements of dark respiration (true respiration) and photorespiration are required to quantify the total carbon exchange.

Not surprisingly, research has shown that air pollutants will affect respiration if injury occurs. For example, Todd (1958) determined that rates of respiration in pinto bean were stimulated by O_3 when visible injury occurred. Similarly, the increase in respiration in leaf discs of bean after 24 h exposure to O_3 led Pell and Brennan (1973) to propose that enhanced respiration was a consequence of cellular injury. Stimulated respiration rates and uncoupling of oxidative phosphorylation were also reported by MacDowall and Ludwig (1962) after the appearance of visible damage due to high concentrations of O_3. MacDowall (1965) later reported that low concentrations of O_3 resulted in an inhibition of respiration and a reduction in mitochondrial phosphorylation in the absence of visible injury.

There is also evidence that exposure to low concentrations of O_3 can result in a stimulation of respiration in the absence of visible injury. For example, Todd and Propst (1963) found that O_3 caused up to a three-fold stimulation in the respiration of coleus (*Coleus spp.*) and tomato (*Lysopersicum esculentum*) in the absence of visible injury, while Dugger and Ting (1970) noted that one of the first measurable changes after the start of an O_3 exposure period was an increase in respiration rates.

The respiratory function of tissues other than leaves may be affected by O_3 as well. Hofstra *et al.* (1981) reported that the respiratory activity of roots of bean was very sensitive to O_3 exposure of the leaves. Root growth was inhibited, and a reduction occurred in the CO_2 evolution from the root system within 24 h of the beginning of exposure to 0·15 ppm O_3. Inhibition of root activity occurred before the appearance of visible foliar injury. It is likely that the effects in the roots were caused by reduced photosynthesis resulting in reduced carbohydrate supplies transported to the roots.

As with O_3, SO_2 has been found to have a variety of effects on respiration rates. Some researchers have found little effect on respiration from SO_2 (Thomas and Hill, 1937; Sij and Swanson, 1974; Furukawa *et al.*, 1980; Takemoto and Noble, 1982). However, the majority of work indicates some effect of SO_2 on respiratory processes, and both inhibitions (Taniyama *et al.*, 1972) and stimulations (Koziol and Jordan, 1978; Black and Unsworth, 1979) of dark respiration have been reported. Undoubtedly, the results depend on the experimental conditions and the species. As with O_3, the respiratory effects may or may not be accompanied by visible injury. Respiratory rates may return to control values following exposure as long as concentrations are relatively low. This may reflect the capacity of plants to detoxify sulfite or repair damage incurred by exposure to toxic concentrations.

There are fewer studies of air pollutant effects on photorespiration than dark respiration despite the importance of this process in C-3 plants, possibly due to the greater difficulties in making quantitative measurements. Assessments of pollutant-induced changes in this process are of particular importance since rates of photorespiration probably are significantly higher than those of dark respiration (Zelitch, 1971). Several studies have investigated pool sizes of metabolites of the glycolate pathway in response to O_3. Ito *et al.* (1984) found pool sizes of glycine and serine (and ^{13}C incorporation into them) in bean were increased by O_3 exposure, which they interpreted as evidence that carbon flow through the pathway was stimulated. Johnson (1984) found a similar buildup of glycine and serine with O_3 exposure of white clover (*Trifolium repens*). However, another possible interpretation of these results would be that the metabolism or translocation of the two amino acids was blocked.

Several workers have reported apparent inhibitions of photorespiration by SO_2. For example, Furukawa *et al.* (1980) found that photorespiratory rates of sunflower plants were inhibited by exposure to 1·5 ppm SO_2 for 30 min. Koziol and Cowling (1978) reported that when ryegrass (*Lolium perenne*) was exposed to SO_2, an increase in $^{14}CO_2$ photoassimilation was observed with a concomitant increase in ^{14}C-labelled glycine and serine. They attributed this to an inhibition of the photorespiratory cycle. Libera *et al.* (1974) demonstrated that one of the photorespiratory intermediates, glycolic acid, accumulated in leaves of spinach (*Spinacia oleracea*) exposed to a high concentration of SO_2, which they attributed to a blocking of glycolate oxidation. It has been shown that sulfite may inhibit glycolate oxidase *in vitro*, although treatment of tobacco (*Nicotiana tabacum*) plants with 1·3 ppm SO_2 for 18 h caused apparent *de novo* synthesis of glycolate oxidase (Spedding and Thomas, 1973; Soldatini and Ziegler, 1979).

The type or degree of respiratory response varies depending on the species, the pollutant, the characteristics of the pollutant exposure, the environmental conditions, and probably the measurement technique. Pollutant effects on respiration deserve more study because of the relationships among dark respiration, photorespiration and photosynthesis in determining the energy status and reserves of the plant, which ultimately control growth. As with photosynthesis, air pollutant effects on respiration must be quantified throughout the diurnal cycles. While this could be done on a whole-plant basis, a more complete understanding will be obtained if it is done for individual plant tissues as they function as sources or sinks.

12.2.4 Translocation

Air pollution-induced alterations in the partitioning of biomass among plant parts has led a number of investigators to consider the processes related to translocation of carbohydrates as a possible explanation. Photoassimilated ^{14}C-, ^{13}C-, and ^{11}C-labelled CO_2 have often been used as tools to address these questions. McLaughlin and McConathy (1983) found an increased retention of photoassimilated ^{14}C in leaves of O_3- and SO_2-exposed bean, whereas developing pods contained less label than controls. Blum *et al.* (1983) reported that after 6 days of exposure of white clover to O_3 concentrations up to 0·10 ppm, ^{14}C allocation to roots increased, but at 0·15 ppm O_3 the allocation to roots was sharply decreased with a concomitant increase in allocation to developing leaves. Okano *et al.* (1984) used $^{13}CO_2$-labelled kidney bean and found a reduction in ^{13}C transport to roots and an increase in transport to young leaves caused by 0·2 ppm O_3. Interestingly, translocation from trifoliate leaves (primarily supplying the young growing leaves) was stimulated, whereas that from the primary leaves (supplying roots) was reduced. Okano *et al.* (1985) later found that 2·0 ppm NO_2 stimulated photosynthesis and increased translocation to sink tissues, whereas the combination of NO_2 plus 0·2 ppm O_3 caused a greater reduction in translocation to roots and lower stem tissues than O_3 alone. The physiological basis of the effects of O_3 on translocation from source leaves is not apparent from this work. However, Pell and Weissberger (1976) did find that paraveinal cells (believed to be involved in the transport of photosynthate from the palisade parenchyma to the phloem) were the most ozone-sensitive cells in soybean leaves, which may help to explain cases in which translocation was reduced. The reasons for increased translocation to certain tissues may be a consequence of increased sink size or activity, although this is an incomplete explanation from the standpoint of physiological mechanisms.

A few studies have shown that translocation from source leaves is especially sensitive to SO_2. In a study with ryegrass, Koziol and Cowling (1978) found that the leaves retained more photoassimilated ^{14}C with increasing concentrations of SO_2 up to 0·15 ppm. Noyes (1980) reported that the translocation of ^{14}C from $^{14}CO_2$-labelled bean leaves was reduced 39, 44, or 66% by a 2-h exposure to SO_2 at 0·1, 1·0, or 3·0 ppm, respectively. The reduction in translocation at 0·1 ppm was not accompanied by a reduction in photosynthesis, indicating that translocation may be more sensitive than photosynthesis to SO_2. Teh and Swanson (1982) reported a 45% decrease in ^{14}C-assimilate translocation in bean exposed to 2·9 ppm SO_2 for 2 h, but, while photosynthesis recovered within 2 h after termination of the exposure, translocation did not. McLaughlin and McConathy (1983) found that the disruption of translocation patterns was persistent for a week after discontinuing pollution exposures. Controlled field studies by Milchunas *et al.* (1982) showed that translocation of ^{14}C-photosynthate from source leaves to developing leaves of bluestem (*Agropyron smithii*) was stimulated by 12% at a mean seasonal exposure level of 0·08 ppm SO_2, although translocation to roots was not markedly affected. Jones and Mansfield (1982) found both reduced root growth and reduced ^{14}C translocation to roots of timothy (*Phleum pratense*) exposed to 0·06 ppm SO_2 for six weeks. Gould (1986) also found that mixtures of SO_2 and NO_2 reduced the amount of assimilate translocated to roots whereas more was translocated to young developing tissues. Griffith and Campbell (1987) also determined that SO_2 affected the export of photoassimilated ^{14}C from snap bean leaves. Leaves of plants exposed to 0·43 ppm SO_2 exported more ^{14}C than controls, while those treated with 0·87 ppm SO_2 exported less.

In several of the above studies, the increased retention of carbon in the leaves or reduced translocation to other tissues was attributed to reduced phloem loading. For example, Noyes (1980) found an accumulation of ^{14}C-labelled photoassimilate in the veins surrounding the sieve tubes with SO_2 treatment. Several recent studies have more directly implicated phloem (sieve-tube) loading and reloading as sites of the SO_2 effect on translocation.

Using $^{11}CO_2$, Minchin and Gould (1986) found evidence that phloem loading was reduced by SO_2 in C-3 species such as wheat and beans. In a subsequent study, Gould (1986) speculated, based on the application of a mass balance flow model, that SO_2 decreased the reloading of labelled assimilate as it was transported along the phloem. However, it seems that more than inhibition of phloem loading or reloading may be involved, since in some cases translocation from source leaves is stimulated by SO_2 (Milchunas *et al.*, 1982). Certainly, sink strength (demand for assimilate), will also play a role.

12.2.5 Carbohydrate pools

Air pollutant-induced reductions in photosynthesis would seem to suggest that soluble carbohydrate pools and storage carbohydrates would be consistently decreased, but a survey of the available research indicates a range of responses (Koziol, 1984). Carbohydrate concentrations within the tissues of plants are the result of a number of interrelated processes, and effects on other facets of carbohydrate metabolism and translocation can tend to modify the expected outcome. For example, low concentrations of SO_2 may elevate photosynthetic rates and decrease photorespiration, as well as possibly inhibit translocation. Elevation of sugar and starch levels at relatively low concentrations or with short exposure durations of SO_2 has been observed. Koziol and Jordan (1978) measured an increase in free sugars in leaf, stem, and root tissues of red kidney bean exposed for 24 h to approximately 0·8 ppm SO_2 or less. Starch levels were also elevated except in roots, where slight reductions occurred. At higher SO_2 concentrations (4 ppm and above), the sugar and starch levels were generally slightly decreased or unaffected. In a study with ryegrass, Koziol and Cowling (1980) also found an increase in free carbohydrates in leaf tissue after 29 days of exposure to 0·02 ppm SO_2, but little effect at 0·15 ppm. After an additional 22 days of exposure, 0·02 ppm SO_2 caused a slight decline. Fructosans were reduced, especially after the second harvest. Photosynthesis and dark respiration of the plants from this experiment were generally unaffected by SO_2 (Cowling and Koziol, 1978). Koziol (1980) reported similar findings with free sugars in soybean leaves exposed to less than 0·1 ppm SO_2 for 34 days. In general, the tendency for elevation of free carbohydrates in leaf tissues due to SO_2 is consistent with the observations of SO_2-induced leaf retention of photoassimilate and inhibition of phloem loading. However, in cases where elevation of carbohydrates was found in stem and root tissues, other explanations such as possible increases in photosynthesis must be considered. Koziol and Jordan (1978) noted that carbohydrate depletion in kidney bean at higher SO_2 concentrations was correlated with reduced photosynthesis and increased respiration.

With O_3 stress, the pattern of effects on carbohydrate pools may be somewhat different than with SO_2, although too few data are available to make a conclusive statement. A decrease in reducing sugars of soybean leaf tissue immediately following O_3 exposure was noted by Tingey *et al.* (1973*a*), although the levels returned to normal within 2 days following termination of the exposure. Blum *et al.* (1982) found a reduction in total nonstructural carbohydrate (TNC) in the shoot of ladino clover (*Trifolium repens*) in response to O_3 but not in the roots. Pell *et al.* (1980) observed an increase in reducing sugars in potato (*Solanum tuberosum*) tubers from ozonated plants. In a study of O_3 and SO_2 effects on forage quality of tall fescue (*Festuca arundinaceae*), Flagler and Younger (1985) found that O_3, but not SO_2, reduced shoot TNC concentrations. Rebbeck (1987), in a study of a ladino clover/tall fescue pasture mixture, found that starch levels of clover were suppressed by increasing O_3 concentrations, while starch and fructosan in fescue were not affected. The total soluble sugars were generally unaffected in both species. In a field experiment with cotton, J. E. Miller (unpublished results) found that the effects of O_3 on the concentration and partitioning of reducing sugars, sucrose, and starch in the plant tissues were dependent on the stage of plant growth. Overall, there was a tendency for O_3 to reduce the soluble carbohydrates and starch in leaves, stems, and roots. Starch levels were the most consistently reduced throughout the growing season, especially in stems and roots. The results for cotton plants at early reproductive growth are shown in Fig. 12.5.

The relationship of pool levels of metabolites such as carbohydrates to plant development is often difficult to interpret because it represents a balance between various anabolic and catabolic reactions, as well as translocation within the plant. Part of the difficulty in relating carbohydrate pools to plant growth and yield is due to the lack of comprehensive data on any particular plant/pollutant combination over the life cycle of the plant. Generalities are difficult because the patterns of carbon assimilation and allocation differ among species and among different growth stages within a species. Indirect effects of the pollutants on carbohydrate partitioning and the influence of environment are also likely of importance. For example, O_3 and SO_2 stress may affect cellular water relations and influence osmoregulatory processes, which can cause a redistribution of carbohydrates among the various metabolic pools. Since carbohydrate pools are intimately involved in plant growth, it is clear that further research is needed in this area to understand their relationship to air pollutant stress.

12.3 RESEARCH NEEDS

A considerable amount of research has been conducted concerning air pollution stress effects on plant growth and a limited number of physiological processes such as photosynthetic gas exchange. Much less has been done with respect to air pollution effects on the mechanisms controlling the allocation of photosynthate, even though it is of primary importance as a determinant of the economic yield of crops (Gifford *et al.*, 1984). A few studies have shown that photosynthate distribution is affected by air pollutants (e.g., Blum *et al.*, 1983; McLaughlin and McConathy, 1983), but little is known concerning the mechanisms of the effects. For this reason, the discussion of research needs will focus on some of the processes involved in carbon allocation to illustrate

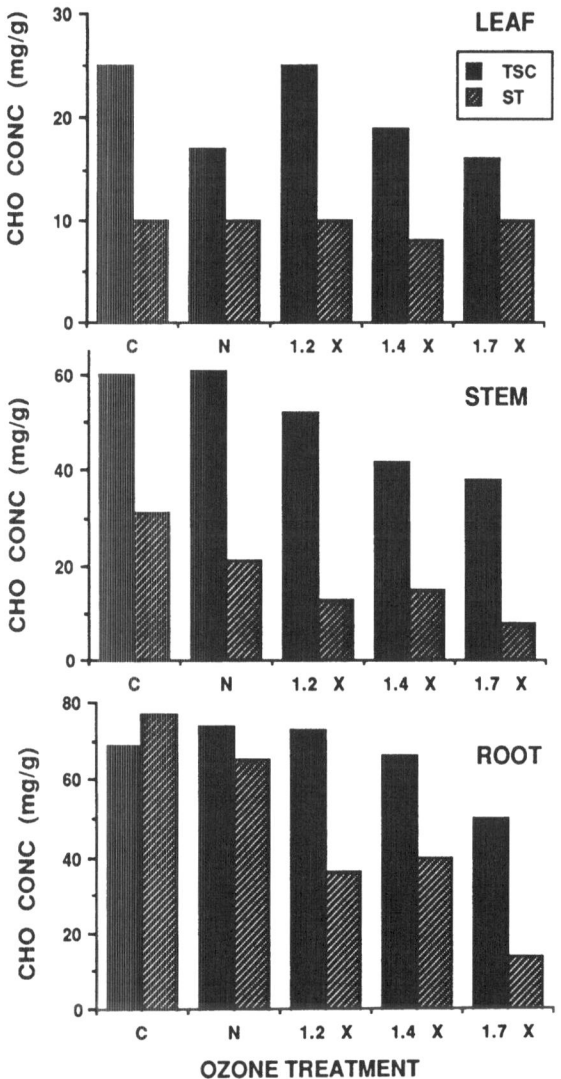

Fig. 12.5. Effect of O_3 on total soluble carbohydrates (TSC) and starch (ST) of field-grown 'McNair-235' cotton (*Gossypium hirsutum*) in open-top chambers during early reproductive development. Ozone treatment levels: C, charcoal-filtered air (0·02 ppm, 12 h d^{-1} seasonal mean); N, non-filtered air (0·041 ppm); 1·2, 1·2 × ambient air (0·050 ppm); 1·4, 1·4 × ambient air (0·60 ppm); 1·7, 1·7 × ambient air (0·071 ppm) (J. E. Miller, unpublished results).

some potentially productive areas for research that are needed to further the understanding of air pollution affects on plant productivity.

First of all, to appreciate the relative magnitude of the flow of carbon in a crop plant, the estimates of Gifford *et al.* (1984) are presented (Fig. 12.6). Of the total CO_2 that is photoassimilated during the life cycle of a cereal crop, it was estimated that 15 to 20% is lost due to photorespiration and approximately 34 to 40% due to true respiration (catabolism). This results in 40 to 51% of the total in dry matter, of which 10 to 23% (of the total carbon fixed) is thought to

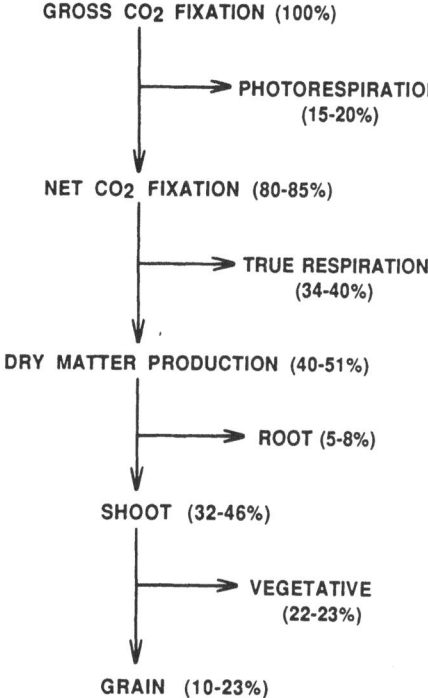

GROSS CO₂ FIXATION (100%)

→ PHOTORESPIRATION
(15-20%)

NET CO₂ FIXATION (80-85%)

→ TRUE RESPIRATION
(34-40%)

DRY MATTER PRODUCTION (40-51%)

→ ROOT (5-8%)

SHOOT (32-46%)

→ VEGETATIVE
(22-23%)

GRAIN (10-23%)

Fig. 12.6. Carbon allocation in a temperate cereal crop (data from Gifford *et al.*, 1984). Values in parentheses are percentages of the total carbon photoassimilated.

go to grain production. It is easy to envision the compounding effects of stresses such as air pollution, even when effects on total carbon assimilation are ignored. For example, a 10% increase in the total amount of carbon lost to respiration might result in a 15 to 30% reduction in grain yield if the relative partitioning of carbon among the plant tissues remains the same.

A diagram of a generalized pathway for movement of photoassimilated carbon from a source to a sink is presented in Fig. 12.7. The initial products of photosynthesis, triose phosphates, serve not only as the initial pool of metabolites for synthesis of translocatable or storage carbon (carbohydrates), but they are also involved in biochemical regulation of allocation among carbohydrate pools in the leaf. Sucrose is synthesized in the cytoplasm from triose phosphates released from the chloroplasts. This pathway is highly regulated, and two enzymes appear to play key roles in the control of sucrose synthesis: sucrose phosphate synthase (SPS) and fructose-1,6-bisphosphatase (Fru-1,6-BPase) (Harbron *et al.*, 1981). Higher SPS activity has been associated with increased partitioning of carbon into sucrose, increased translocation of assimilates from leaves, and decreased buildup of starch in leaves (Huber *et al.*, 1985). The Fru-1,6-BPase enzyme also plays a key role in the control of sucrose synthesis partially by means of the regulatory metabolite fructose-2,6-bisphosphate (Fru-2,6-BP), which is a potent inhibitor of Fru-1,6-BPase activity. The concentration of Fru-2,6-BP increases when photosynthesis

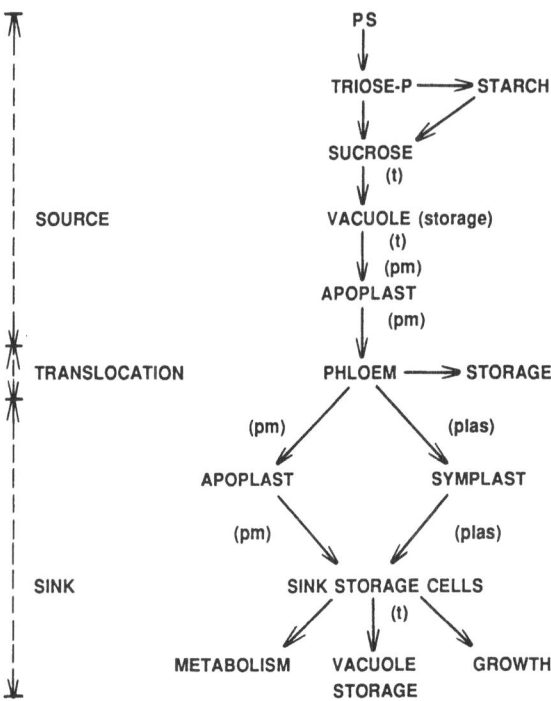

Fig. 12.7. Generalized pathway for movement of photoassimilated carbon from a source leaf to sinks. Abbreviations: t (tonoplast), pm (plasma membrane), and plas (plasmodesmata).

is less than maximal (Stitt *et al.*, 1984) or when sucrose accumulation occurs due to a restriction in export (Huber and Bickett, 1984). An involved set of metabolic controls involving the triose phosphates, Fru-2,6-BP, Fru-1,6-BPase, and SPS can result in the formation of starch rather than sugar. The importance of this metabolic regulation becomes apparent when it is recognized that root growth in some species occurs primarily in the daytime with sucrose exported from source leaves, while the primary source of sucrose for growth of stems and expanding leaves is mobilization of starch reserves in source leaves at night (Huber, 1983). Thus, the partitioning of carbon between sucrose and starch in source leaves may represent one mechanism affecting the relative growth of root and shoot tissues. It is interesting to note that hydrolysis of starch in leaves of several plant species was found to be reduced by O_3 (Hanson and Stewart, 1970). These mechanisms have received little attention with respect to the effects of air pollutants. Evidence has been obtained, however, suggesting that Fru-1,6-BPase may be inhibited by SO_2, probably as a result of H_2O_2 production due to sulfite oxidation in the chloroplast (Alscher *et al.*, 1987).

Prior to translocation, some of the sucrose in the source leaf may be stored in the vacuole as a buffer against short-term changes in the balance between photosynthesis and export (Geiger *et al.*, 1983) (Fig. 12.7). Sucrose being exported from the mesophyll cells is probably released into the apoplast prior to phloem loading. Thus, the tonoplast and plasma membrane both represent possibilities for control points of allocation, although their role in this capacity is not clear, especially with respect to air pollutant effects on the control. Membrane integrity and function are known to be affected by air pollutants such as O_3 (R. L. Heath, Chapter 11, this volume), although the exact relationship to carbon allocation is not certain.

The loading of sucrose into the phloem is an active process requiring energy, and high concentrations of SO_2 have been found to reduce this process (Minchin and Gould, 1986). Gould (1986) also found that lower concentrations of SO_2 and NO_2 (0·1 ppm) reduced phloem loading in wheat. Other evidence also suggests that phloem loading may be affected by air pollutants. For example, it is known that air pollutants may increase retention of photoassimilate in leaves (e.g., Noyes, 1980; Blum *et al.*, 1983; McLaughlin and McConathy, 1983; Okano *et al.*, 1984). But, the air pollution-induced increases in transport of photosynthate to young developing leaves that have been found in some cases (e.g., Jones and Mansfield, 1982; Okano *et al.*, 1984) illustrate that inhibition of phloem loading alone does not account for all effects on allocation.

The path of translocation itself (i.e., the phloem) is another possible site for air pollutant effects on allocation. But, Gifford and Evans (1981) concluded that the movement of sucrose through phloem is not limiting to the allocation of carbon, which might suggest that the transport path would not be the most likely site for air pollution effects. Others do not assume such a passive role for the phloem in the control of allocation, and Gould (1986) does suggest that SO_2 may reduce the transport velocity of ^{11}C-labelled assimilate in wheat. The translocation path may also represent a sink for storage of exported carbon (Fig. 12.7). This stored carbohydrate may buffer changes in sucrose concentrations in the phloem when export rates from sources are low (Minchin *et al.*, 1984), as might occur under conditions of air pollutant stress. Gould (1986) suggested that SO_2 inhibits active reloading of phloem along the transport path in wheat without affecting passive unloading.

The mechanisms controlling phloem unloading in sink regions are not well understood despite their obvious importance as a factor influencing allocation patterns. One problem is that the process seems to differ depending on the type of sink. Both unloading into the apoplast and through the symplast (via plasmodesmata) are known to occur (Wyse, 1985) (Fig. 12.7). A sucrose carrier and the involvement of metabolic energy have been implicated, as well as osmotic control (Griffith and Wyse, 1983; Patrick, 1983). In cases where apoplastic unloading occurs, transport of the sucrose or its products into the sink tissues involves transport across the plasmalemma. In some tissues, this may be by way of a sucrose–proton cotransporter (Lichtner and Spanswick, 1981), while in other cases, if concentrations are high enough, transport down a concentration gradient may occur (Saftner *et al.*, 1983). In some tissues, the sucrose or resynthesized sucrose may be stored in the vacuole of the sink tissues, requiring transport across the tonoplast. Again, little direct information is available concerning air pollutant effects on phloem unloading at sink tissues, although Gould (1986) does suggest that unloading to wheat heads was increased by high SO_2 concentrations.

Not all assimilate is immediately metabolized or translocated to its final sink. Even sucrose may exist in highly mobile and less mobile pools, i.e., the cytoplasmic and vacuolar fractions. The type, location, and quantity of storage products may have quantitative and qualitative effects on partitioning among sink tissues. Storage pools may be highly transitory or longer term. These temporary storage pools often seem to promote biochemical stability of the sink during changing conditions (Franceschi, 1985) and thus are of considerable interest in studies of air pollution stress effects on allocation. Long-term storage and remobilization may be important in crop plants during the stages of reproductive development or seed fill when the demand for assimilate is high, but the production is low due to senescence of leaves.

From this limited discussion, it can be seen that there are a number of areas that are relatively unexplored from the standpoint of understanding air pollutant effects on carbon allocation, that relate to plant growth and yield. Certainly, there are many other potential sites for air pollutant effects on carbon allocation, and the above discussion is meant only to illustrate a few of the areas for which information is sparse or lacking. For example, the entire area of intermediary metabolism has been treated very lightly in this discussion, but it too is an integral part of the control of allocation.

When attempting to relate air pollution effects on physiological processes involved in carbon allocation to growth and yield, there are a wide range of possible approaches that a researcher can take. Obviously, the exact approach chosen will be largely dictated by the specific goals of the research or the hypotheses to be tested. One choice that must be made is whether to emphasize the details of the physiological function of individual tissues or cells or to emphasize the performance of the entire plant. If the main purpose of the work is to establish a correlative relationship between some physiological function (such as photosynthesis) and the growth of a plant, the most successful approach probably will be to study the integrated function of the entire plant (in this example, whole-plant photosynthesis). On the other hand, if the goal is to gain a physiological understanding of the mechanisms of the pollutant effects, then it is necessary to study the details of tissue or cell function. The data from the first case will have a more immediate application in a practical sense (for example, model development to relate physiological measurements to yield for predictive purposes). The second case will have primary application in the longer term when more fundamental knowledge is needed to attain goals such as: (1) diagnosis based on physiological processes; (2) the development of resistant varieties or other preventive measures; and (3) the development of physiological process models to refine predictive capabilities.

REFERENCES

Adams, R. M., S. A. Hamilton, and B. A. McCarl. (1985). An assessment of the economic effects of ozone on U.S. agriculture. *J. Air Pollut. Control Assoc.,* **35,** 938–43.

Alscher, R., M. Franz, and C. W. Jeske. (1987). Sulfur dioxide and chloroplast metabolism. In *Phytochemical effects of environmental compounds,* ed. by J. A. Saunders, L. Kosak-Channing and E. E. Conn, 1–28. New York, Plenum Press.

Amundson, R. G., R. M. Rabe, A. W. Schoettle, and P. B. Reich. (1986). Response of soybean to low concentrations of ozone: II. Effects on growth, biomass allocation and flowering. *J. Environ. Qual.,* **15,** 161–7.

Amundson, R. G., R. J. Kohut, A. W. Schoettle, and P. B. Reich. (1987). Correlative reductions in whole-plant photosynthesis and yield of winter wheat caused by ozone. *Phytopathology,* **77,** 75–9.

Barton, J. R., S. B. McLaughlin, and R. K. McConathy. (1980). The effects of SO_2 on components of leaf resistance to gas exchange. *Environ. Pollut.,* **21,** 255–65.

Bennett, J. P. and R. J. Oshima. (1976). Carrot injury and yield response to ozone. *J. Am. Soc. Hortic. Sci.,* **101,** 638–9.

Bennett, J. P. and V. C. Runeckles. (1977). Effects of low levels of ozone on growth of crimson clover and annual ryegrass. *Crop. Sci.,* **17,** 443–5.

Bennett, J. P., R. J. Oshima, and L. F. Lippert. (1979). Effects of ozone on injury and dry matter partitioning in pepper plants. *Environ. Exp. Bot.,* **19,** 33–9.

Black, V. J. (1982). Effects of sulphur dioxide on physiological processes in plants. In *Effects of gaseous air pollution in agriculture and horticulture,* ed. by M. H. Unsworth and D. P. Ormrod, 67–91, London, Butterworths Scientific.

Black, V. J. and M. H. Unsworth. (1979). Effects of low concentrations of sulfur dioxide on net photosynthesis and dark respiration of *Vicia faba. J. Exp. Bot.,* **31,** 473–83.

Black, V. J., D. P. Ormrod, and M. H. Unsworth. (1982). Effects of low concentration of ozone, singly, and in combination with sulphur dioxide on net photosynthesis rates of *Vicia faba* L. *J. Exp. Bot.,* **33,** 1302–11.

Blum, U. and W. W. Heck. (1980). Effects of acute ozone exposures on snap bean at various stages of its life cycle. *Environ. Exp. Bot.,* **20,** 73–85.

Blum, U., G. R. Smith, and R. C. Fites. (1982). Effects of multiple O_3 exposures on carbohydrate and mineral contents of ladino clover. *Environ. Exp. Bot.,* **22,** 143–54.

Blum, U., E. Mrozek, Jr, and E. Johnson. (1983). Investigation of ozone (O_3) effects on ^{14}C distribution in ladino clover. *Environ. Exp. Bot.,* **23,** 369–78.

Carlson, R. W., 1983. The effect of SO_2 on photosynthesis and leaf resistance at varying concentrations of CO_2. *Environ. Pollut., Ser. A,* **39,** 309–22.

Chatterton, N. J. and J. W. Silvius. (1979). Photosynthate partitioning into starch in soybean leaves. I. Effects of photoperiod versus photosynthetic period duration. *Plant Physiol.,* **63,** 749–53.

Chevone, B. I. and Y. S. Yang. (1985). CO_2 exchange rates and stomatal diffusive resistance in soybean exposed to O_3 and SO_2. *Can. J. Plant Sci.,* **65,** 267–74.

Cowling, D. H. and M. J. Koziol. (1978). Growth of ryegrass (*Lolium perenne* L.) exposed to SO_2. *J. Exp. Bot.,* **29,** 1029–36.

Coyne, P. I. and G. E. Bingham. (1978). Photosynthesis and stomatal light responses in snap beans exposed to hydrogen sulfide and ozone. *J. Air Pollut. Control Assoc.,* **28,** 1119–23.

Dugger, W. M. and I. P. Ting. (1970). Air pollutant oxidants—their effects on metabolic processes in plants. *Ann. Rev. Plant Physiol.,* **21,** 215–34.

Endress, A. G. and C. Grunwald. (1985). Impact of chronic ozone on soybean growth and biomass partitioning. *Agric. Ecosys. Environ.,* **13,** 9–23.

Flagler, R. B. (1986). *Effects of ozone and water deficit on growth, yield, and nitrogen metabolism of soybeans.* PhD thesis, North Carolina State University, Raleigh, NC.

Flagler, R. B. and V. B. Youngner. (1982). Ozone and sulfur dioxide effects on tall fescue: I. Growth and yield responses. *J. Environ. Qual.,* **11,** 386–9.

Flagler, R. B. and V. B. Youngner. (1985). Ozone and sulfur dioxide effects on tall fescue: II. Alteration of quality constituents. *J. Environ. Qual.,* **14,** 463–6.

Foster, K. W., H. Timm, C. K. Labanauskas, and R. J. Oshima. (1983). Effects of ozone and sulfur dioxide on tuber yield and quality of potatoes. *J. Environ. Qual.*, **12**, 75–9.

Franceschi, V. R. (1985). Temporary storage and its role in partitioning among sinks. In *Phloem transport*, ed. by J. Cronshaw, W. J. Lucas, and R. T. Giaquinta, 399–409. New York, Alan R. Liss, Inc.

Furukawa, A. and T. Totsuka. (1979). Effects of NO_2, SO_2 and ozone alone and in combination on net photosynthesis in sunflower. *Environ. Control Biol.*, **17**, 161–6.

Furukawa, A., T. Natori, and T. Totsuka. (1980). The effects of SO_2 on net photosynthesis in sunflower leaf. In *Studies on the effects of air pollutants on plants and mechanisms of phytotoxicity*, Vol. 11, 1–8. Research Report from the National Institute for Environmental Studies, Yatabe, Japan.

Geiger, D. R., B. J. Ploeger, T. C. Fox, and B. R. Fondy. (1983). Sources of sucrose translocated from illuminated sugar beet source leaves. *Plant Physiol.*, **2**, 964–70.

Gifford, R. M. and L. T. Evans. (1981). Photosynthesis, carbon partitioning and yield. *Ann. Rev. Plant Physiol.*, **32**, 485–509.

Gifford, R. M., J. H. Thorne, W. D. Hitz, and R. T. Giaquinta. (1984). Crop productivity and photoassimilate partitioning. *Science*, **25**, 801–8.

Gould, R. P. (1986). *The effects of gaseous sulphur dioxide and nitrogen dioxide on carbon allocation in plants*. PhD thesis, Lancaster University, UK.

Griffith, S. and R. E. Wyse. (1983). Some characteristics of the phloem unloading system in stem segments of *Vicia faba* L. *Plant Physiol.* (Suppl.), **72**, 138.

Griffith, S. M. and W. F. Campbell. (1987). Effects of sulfur dioxide on nitrogen fixation, carbon partitioning, and yield components in snapbean. *J. Environ. Qual.*, **16**, 77–80.

Guderian, R. (1977). *Air pollution, phytotoxicity of acidic gases and its significance in air pollution control, Ecological Studies*, Vol. 22, Berlin, Springer-Verlag.

Guderian, R. (1985). *Air pollution by photochemical oxidants, Ecological Studies*, Vol. 52. New York, Springer-Verlag.

Habron, S., C. Foyer, and D. Walker. (1981). The purification and properties of sucrose-phosphate synthetase from spinach leaves: The involvement of this enzyme and fructose bisphosphate in the regulation of sucrose biosynthesis. *Arch. Biochem. Biophys.*, **212**, 237–46.

Hällgren, J. E. (1978). Physiological and biochemical effects of sulfur dioxide on plants. In *Sulfur in the environment: Part II Ecological impacts*, ed. by J. O. Nriagu, 163–209. New York, John Wiley.

Hanson, G. P. and W. S. Stewart. (1970). Photochemical oxidants: effect on starch hydrolysis in leaves. *Science*, **168**, 1223–4.

Heagle, A. S., R. B. Flagler, R. P. Patterson, V. M. Lesser, S. R. Shafer, and W. W. Heck. (1987). Injury and yield response of soybean to chronic doses of ozone and soil moisture deficit. *Crop. Sci.*, **27**, 1016–24.

Heath, R. L. (1980). Initial events in injury to plants by air pollutants. *Ann. Rev. Plant Physiol.*, **31**, 395–431.

Heath, R. L. (1988). Biochemical mechanisms of pollutant stress. In *Assessment of crop loss from air pollutants*, Proceedings of an International Conference, Raleigh, NC, ed. by W. W. Heck, O. C. Taylor, and D. T. Tingey, 259–85. London, Elsevier Applied Science.

Heck, W. W. and J. A. Dunning. (1978). Response of oats to sulfur dioxide; interactions of growth temperature with exposure temperature and humidity. *J. Air Pollut. Control Assoc.*, **28**, 241–6.

Heck, W. W., O. C. Taylor, R. Adams, G. Bingham, J. Miller, E. Preston, and L. Weinstein. (1982). Assessment of crop loss from ozone. *J. Air Pollut. Control Assoc.*, **32**, 353–62.

Heck, W. W., W. W. Cure, J. O. Rawlings, L. J. Zaragoza, A. S. Heagle, H. E. Heggestad, R. J. Kohut, L. W. Kress, and P. J. Temple. (1984). Assessing impacts of ozone on agricultural crops: II. Crop yield functions and alternative exposure statistics. *J. Air Pollut. Control Assoc.*, **34**, 810–17.

Heck, W. W., A. S. Heagle, and D. S. Shriner. (1986). Effects on vegetation: native, crops, forests. In *Air pollution*, Vol. 6, ed. by A. S. Stern, 247–50. New York, Academic Press.

Heggestad, H. E. and J. H. Bennett. (1981). Photochemical oxidants potentiate yield losses in snap beans attributable to sulfur dioxide. *Science*, **213**, 1008–10.

Hofstra, G., A. Ali, R. T. Wukasch, and R. A. Fletcher. (1981). The rapid inhibition of root respiration after exposure of bean (*Phaseolus vulgaris* L.) plants to ozone. *Atmos. Environ.*, **15**, 483–7.

Huber, S. C. (1983). Relation between photosynthetic starch formation and dry weight partitioning between the shoot and root. *Can. J. Bot.*, **61**, 2709–16.

Huber, S. C. and D. M. Bickett. (1984). Evidence for control of carbon partitioning by fructose-2,6-bisphosphate in spinach leaves. *Plant Physiol.*, **74**, 445–7.

Huber, S. C., P. S. Kerr, and W. Kalt-Torres. (1985). Regulation of sucrose formation and movement. In *Regulation of carbohydrate partitioning in photosynthetic tissue*, Proceedings of the 9th annual symposium in botany, Riverside, CA, USA, ed. by R. Heath and J. Preiss, 199–214. Baltimore, Waverly Press.

Irving, P. M. and J. E. Miller. (1981). Productivity of field-grown soybeans exposed to acid rain and sulfur dioxide alone and in combination. *J. Environ. Qual.*, **10**, 473–8.

Ito, O., F. Mitsumori, and T. Totsuka. (1984). Effects of NO_2 and O_3 alone or in combination on kidney bean plants. III. Photosynthetic CO_2 assimilation observed by ^{13}C nuclear magnetic resonance. In *Studies on effects of air pollutant mixtures of plants*, No. 66, Part 2, 27–36. Yatabe, Japan, The National Institute for Environmental Studies.

Johnson, E. L. (1984). *Effect of ozone on photosynthetic pathways in white clover* (*Trifolium repens cv Tillman*). PhD thesis, North Carolina State University, Raleigh, NC.

Jones, T. and T. A. Mansfield. (1982). Studies on dry matter partitioning and distribution of ^{14}C labelled assimilates in plants of *Phleum pratense* exposed to SO_2 pollution. *Environ. Pollut., Ser. A*, **28**, 199–208.

Katase, M., T. Ushijima, and T. Tazaki. (1983). The relationship between absorption of sulphur dioxide (SO_2) and inhibition of photosynthesis in several plants. *Bot. Mag.*, **96**, 1–13.

Kohut, R. J., R. G. Amundson, and J. A. Laurence. (1985). Effects of O_3 and SO_2 on the physiology and yield of a crop of red clover and timothy. In *National Crop Loss Assessment Network (NCLAN) 1984 Annual Report*, 69–94. Corvallis, OR, US Environmental Protection Agency, EPA/600/3-86/041.

Kohut, R. J., R. G. Amundson, and J. A. Laurence. (1986*a*). Effects of O_3 and SO_2 on the physiology and yield of a crop of red clover and timothy. In *National Crop Loss Assessment Network (NCLAN) 1985 Annual Report*, 133–62. Corvallis, OR, US Environmental Protection Agency, EPA/600/3-87/031.

Kohut, R. J., R. G. Amundson, and J. A. Laurence. (1986*b*). Evaluation of growth and yield of soybean exposed to ozone in the field. *Environ. Pollut.*, **4**, 219–34.

Kohut, R. J., R. G. Amundson, J. A. Laurence, L. Colavito, P. van Leuken, and P. King. (1987). Effects of ozone and sulfur dioxide on yield of winter wheat. *Phytopathology*, **77**, 71–4.

Koziol, M. J. (1980). *Effects of prolonged exposure to SO_2 on the growth and carbohydrate metabolism of soyabean and ryegrass*. PhD thesis, University of Oxford, Oxford, UK.

Koziol, M. J. (1984). Interactions of gaseous pollutants with carbohydrate metabolism. In *Gaseous air pollutants and plant metabolism*, ed. by M. J. Koziol and F. R. Whatley, 251. London, Butterworths Scientific.

Koziol, M. J. and D. W. Cowling. (1978). Growth of ryegrass (*Lolium perenne* L.) exposed to SO_2. II. Changes in the distribution of photoassimilated ^{14}C. *J. Exp. Bot.*, **29**, 1431–9.

Koziol, M. J. and D. W. Cowling. (1980). Growth of ryegrass (*Lolium perenne* L.) exposed to SO_2. *J. Exp. Bot.*, **31**, 1687–99.

Koziol, M. J. and C. F. Jordan. (1978). Changes in carbohydrate levels in red kidney bean (*Phaseolus vulgaris* L.) exposed to sulfur dioxide. *J. Exp. Bot.*, **29**, 1037–43.

Koziol, M. J. and F. R. Whatley (eds). (1984). *Gaseous air pollutants and plant metabolism*. London, Butterworths Scientific.

Le Sueur-Brymer, N. M. and D. P. Ormrod. (1984). Carbon dioxide exchange rates of fruiting soybean plants exposed to ozone and sulfur dioxide singly or in combination. *Can. J. Plant Sci.*, **64**, 69–75.

Libera, W., I. Ziegler, and H. Ziegler. (1974). The action of sulfite on the HCO$_3$-fixation and the fixation pattern of isolated chloroplasts and leaf tissue slices. *Z. Pflanzenphysiol.*, **74**, 420–33.

Lichtner, F. T. and R. M. Spanswick. (1981). Electrogenic sucrose transport in developing soybean cotyledons. *Plant Physiol.*, **67**, 869–74.

MacDowall, F. D. H. (1965). Stages of ozone damage to respiration of tobacco leaves. *Can. J. Bot.*, **43**, 419–27.

MacDowall, H. and R. A. Ludwig. (1962). Some effects of ozone on tobacco leaf metabolism. *Phytopathology* (Abst.), **52**, 740.

McLaughlin, S. B. and D. S. Shriner. (1980). Allocation of resources to defense and repair. *Plant Dis.*, **5**, 407–31.

McLaughlin, S. B. and R. K. McConathy. (1983). Effects of SO$_2$ and O$_2$ on allocation of ^{14}C-labeled photosynthate in *Phaseolus vulgaris*. *Plant Physiol.*, **73**, 630–5.

Mejstrik, V. (1980). The influence of low SO$_2$ concentrations on growth reduction of *Nicotiana tabacum* L. cv. Samsun and *Cumulus sativus* L. cv. Unikat. *Environ. Pollut.*, **21**, 73–6.

Milchunas, D. G., W. R. Laurenroth, and J. L. Dodd. (1982). The effect of SO$_2$ on C-14 translocation in *Agropyron smithii* Rydb. *Environ. Exp. Bot.*, **22**, 81–92.

Miller, J. E. (1987). Effects of ozone and sulfur dioxide stress on growth and carbon allocation in plants. In *Phytochemical effects of environmental compounds*, ed. by J. A. Saunders, L. Kosak-Channing and E. E. Conn, 55–100. New York, Plenum Press.

Miller, J. E., H. J. Smith, and W. Prepejchal. (1981). Evidence for extreme resistance of field corn to intermittent sulfur dioxide stress. In ANL-81-85, Part III, 27–9. Argonne, IL, Argonne National Laboratory.

Minchin, P. E. H. and R. Gould. (1986). Effect of SO$_2$ on phloem loading. *Plant Sci.*, **43**, 179–83.

Minchin, P. E. H., K. G. Ryan, and M. R. Thorpe. (1984). Further evidence of apoplastic unloading into the stem of bean: Identification of the phloem buffering pool. *J. Exp. Bot.*, **35**, 1744–53.

Muller, R. N., J. E. Miller, and D. G. Sprugel. (1979). Photosynthetic response of field-grown soybeans to fumigations with sulphur dioxide. *J. Appl. Ecol.*, **16**, 567–76.

Noyes, R. D. (1980). The comparative effects of sulfur dioxide on photosynthesis and translocation in bean. *Physiol. Plant Pathol.*, **16**, 73–9.

Okano, K., O. Ito, G. Takeba, A. Shimizu, and T. Totsuka. (1984). Alteration of ^{13}C-assimilate partitioning in plants of *Phaseolus vulgaris* exposed to ozone. *New Phytol.*, **97**, 155–63.

Okano, K., O. Ito, G. Takeba, A. Shimizu, and T. Tosuka. (1985). Effects of O$_3$ and NO$_2$ alone or in combination on the distribution of ^{13}C-assimilate in kidney bean plants. *Jpn J. Crop Sci.*, **54**, 152–9.

Ormrod, D. P., V. J. Black, and M. H. Unsworth. (1981). Depression of net photosynthesis in *Vicia faba* L. exposed to sulphur dioxide and ozone. *Nature*, **291**, 585–6.

Oshima, R. J., J. P. Bennett, and P. K. Braegelmann. (1978). Effect of ozone on growth and assimilate partitioning in parsley. *J. Am. Soc. Hortic. Sci.*, **103**, 348–50.

Oshima, R. J., P. K. Braegelmann, R. B. Flager, and R. R. Teso. (1979). The effects of ozone on the growth, yield and partitioning of dry matter in cotton. *J. Environ. Qual.*, **8**, 575–9.

Patrick, J. W. (1983). Photosynthate unloading from seed coats of *Phaseolus vulgaris* L. General characteristics and facilitated transfer. *Z. Pflanzenphysiol.*, **11**, 9–18.

Pell, E. J. and E. Brennan. (1973). Changes in respiration, photosynthesis, adenosine, 5'-triphosphate, and total adenylate content of ozonated pinto bean foliage as they relate to symptom expression. *Plant Physiol.*, **51**, 378–81.

Pell, E. J. and W. C. Weissberger. (1976). Histopathological characterization of ozone injury to soybean foliage. *Phytopathology*, **66**, 856–61.

Pell, E. J., W. C. Weissberger, and J. J. Speroni. (1980). Impact of ozone on quantity and quality of greenhouse-grown potato plants. *Environ. Sci. Technol.*, **14**, 568.

Rebbeck, J. (1987). *The effects of ozone and soil moisture on the growth and energy reserves of ladino clover and tall fescue.* PhD thesis, North Carolina State University, Raleigh, NC.

Reich, P. B. (1987). Quantifying plant response to ozone: A unifying theory. *Tree Physiol.*, **3**, 63–91.

Reich, P. B. and R. G. Amundson. (1985). Ambient levels of ozone reduce net photosynthesis in tree and crop species. *Science*, **30**, 566–70.

Reich, P. B., A. W. Schoettle, R. M. Raba, and R. G. Amundson. (1986). Response of soybean to low concentrations of ozone: I. Reductions in leaf and whole plant net photosynthesis and leaf chlorophyll content. *J. Environ. Qual.*, **15**, 31–6.

Roberts, T. M. (1984). Long-term effects of sulfur dioxide on crops: An analysis of dose–response relations. *Philos. Trans. R. Soc., London B*, **305**, 299–316.

Saftner, R. A., J. Daie, and R. W. Wyse. (1983). Sucrose uptake and compartmentation in sugar beet root tissue discs. *Plant Physiol.*, **72**, 1–6.

Shimizu, H., S. Motohashi, H. Iwaki, A. Furukawa, and T. Totsuka. (1980). Effects of chronic exposures to ozone on the growth of sunflower plants. *Environ. Control Biol.*, **19**, 137–47.

Sij, J. W. and C. A. Swanson. (1974). Short-term kinetic studies on the inhibition of photosynthesis by sulfur dioxide. *J. Environ. Qual.*, **3**, 103–7.

Silvius, J. E., R. R. Johnson, and D. B. Peters. (1977). Effects of water stress on carbon assimilation and distribution in soybean plants at different stages of development. *Crop Sci.*, **17**, 713–16.

Skarby, L. and G. Sellden. (1984). The effects of ozone on crops and forests. *Ambio*, **13**, 68–72.

Soldatini, G. F. and I. Ziegler. (1979). Induction of glycolate oxidase by SO$_2$ in *Nicotiana tabacum*. *Phytochemistry*, **18**, 21–2.

Spedding, D. J. and W. J. Thomas. (1973). Effect of sulphur dioxide on the metabolism of glycolic acid by barley (*Hordeum vulgare*) leaves. *Aust. J. Biol. Sci.*, **26**, 281–6.

Sprugel, D. G., J. E. Miller, R. N. Muller, H. J. Smith and P. B. Xerikos. (1980). Sulfur dioxide effects on yield and seed quality in field-grown soybeans. *Phytopathology*, **70**, 1129–33.

Stitt, M., B. Herzog, and H. W. Heldt. (1984). Control of photosynthetic sucrose synthesis by fructose-2,6-bisphosphate. I. Coordination of CO$_2$ fixation and sucrose synthesis. *Plant Physiol.*, **75**, 548–53.

Takemoto, B. K. and R. D. Noble. (1982). The effects of short-term SO$_2$ fumigation on photosynthesis and respiration in soybean *Glycine Max*. *Environ. Pollut.*, **28**, 67–74.

Taniyama, T., H. Arikado, Y. Iwata, and K. Sawanka. (1972). Studies on the mechanism of injurious effects of toxic gases on crop plants. On photosynthetic and dark respiration of rice plant fumigated with SO$_2$ for long period. *Proc. Crop Sci. Soc. Jpn*, **41**, 120–5.

Taylor, Jr, G. E. and W. J. Selvidge. (1984). Phytotoxicity in bush bean of five sulfur-containing gases released from advanced fossil energy technologies. *J. Environ. Qual.*, **13**, 224–30.

Taylor, G. E., Jr, P. J. Hanson, and D. D. Baldocchi. (1988). Pollutant deposition to individual leaves and plant canopies: Sites of regulation and relationship to injury. In *Assessment of crop loss from air pollutants*, Proceedings of an International

Conference, Raleigh, NC, ed. by W. W. Heck, O. C. Taylor, and D. T. Tingey, 227–57. London, Elsevier Applied Science.

Teh, K. H. and C. A. Swanson. (1982). Sulfur dioxide inhibition of translocation in bean plants. *Plant Physiol., 69*, 88–92.

Thomas, M. D. and G. K. Hill. (1937). Relation of sulphur dioxide in atmosphere to photosynthesis and respiration in alfalfa. *Plant Physiol., 12*, 309–83.

Tingey, D. T. and R. A. Reinert. (1975). The effect of ozone and sulphur dioxide singly and in combination on plant growth. *Environ. Pollut., 9*, 117–25.

Tingey, D. T., R. C. Fites, and C. Wickliff. (1973a). Ozone alteration of nitrate reduction in soybean. *Physiol. Plant., 29*, 33–8.

Tingey, D. T., R. A. Reinert, C. Wickliff, and W. W. Heck. (1973b). Chronic ozone or sulfur dioxide exposures, or both, affect the early vegetative growth of soybean. *Can. J. Plant Sci., 53*, 875–9.

Todd, G. W. (1958). Effect of ozone and ozonated 1-hexene on respiration and photosynthesis of leaves. *Plant Physiol., 33*, 416–20.

Todd, G. W. and B. Propst. (1963). Changes in transpiration and photosynthetic rates of various leaves during treatments with ozonated hexene or ozone gas. *Physiol. Plant., 16*, 57–65.

Treshow, M. (ed.). (1984). *Air pollution and plant life.* New York, John Wiley.

Unsworth, M. H. and D. P. Ormrod (eds). (1982). *Effects of gaseous air pollution in agriculture and horticulture.* London, Butterworths Scientific.

Unsworth, M. H., V. M. Lesser, and A. S. Heagle. (1984). Radiation interception and the growth of soybeans exposed to ozone in open-top field chambers. *J. Appl. Ecol., 21*, 1059–79.

Winner, W. E., H. A. Mooney, and R. A. Goldstein (eds). (1985). *Sulfur dioxide and vegetation: Ecology, physiology and policy issues.* Stanford, CA, Stanford University Press.

Wyse, R. E. (1985). Sinks as determinants of assimilate partitioning: possible sites for regulation. In *Phloem transport,* ed. by J. Cronshaw, W. J. Lucas, and R. T. Giaquinta, 197–209. New York, Alan R. Liss, Inc.

Zelitch, I. (1971). *Photosynthesis, photorespiration and plant productivity,* 130–213. New York, Academic Press.

Ziegler, I. (1975). The effect of SO_2 pollution on plant metabolism. *Residue Rev., 56*, 79–105.

Session V

ABIOTIC AND BIOTIC
INTERACTIVE STRESS FACTORS

(Leonard H. Weinstein, *Chairman*; Richard A. Reinert,
Co-Chairman)

13

PROBLEMS OF CROP LOSS ASSESSMENT WHEN THERE IS EXPOSURE TO TWO OR MORE GASEOUS POLLUTANTS

T. A. MANSFIELD
University of Lancaster, Lancaster, UK

and

D. C. McCUNE
Boyce Thompson Institute, Ithaca, New York, USA

13.1 INTRODUCTION

If the effects of two or more pollutants could always be estimated simply from a knowledge of how each acts alone, there would be no need for a lengthy paper on this topic. It is because pollutants actually occur as mixtures and because the effects of combinations are sometimes greater or sometimes less than expected, in experimental exposures of crops in realistic situations, that we must give careful consideration to this subject. Indeed, one could regard the body of information presently available from experiments on single pollutants merely as a first step in understanding the real effects of air pollution on crops.

Several comprehensive reviews of the joint action of different air pollutants on the growth of crops have appeared in recent years, and it is not our intention simply to restate or to update those reviews. Instead, we will try to use them as a foundation for an in-depth discussion of the possible importance of particular combinations of gaseous pollutants (neglecting therefore, aerosols, precipitation, and suspended solids) because these have been the subject of much investigation, and also because space does not allow a complete treatment of interactions among all contaminants of the atmosphere.

Even with this limited set of mixtures of pollutants, it would be possible to occupy the space we are allowed by listing the results of experiments that used them. Such an approach to the topic is unprofitable because the picture that emerges is very complex, and it is not possible to reach many useful conclusions about the magnitude of effects on particular crops. In fact, few of the experimental studies described in the literature are relevant to crops in the field because the concentrations of pollutants employed were relatively high and the durations of exposures very short. An alternative approach to the subject is to consider some known effects of particular combinations of pollutants that may be encountered in realistic situations and to give special attention to possible modes of action. It may eventually be possible to predict the consequences of exposures to complex mixtures if we can identify the primary causes of injury.

13.2 CONTEXT

13.2.1 Origin of the problem

The joint action of two or more air pollutants arises under two general cases. The first occurs when two or more pollutants are emitted from the same source, or at least from two sources that are virtually the same with reference to the occurrence of pollutants in the impacted area. The simplest assumption is that the pollutants occur concurrently and in a constant proportion (independent of concentration) at any site. The second case occurs when pollutants from two separate sources affect the same site and the interaction of consecutive exposures to each, as well as the simultaneous action of mixtures, must be considered—as when a peak concentration from a point-source is embedded in a broader peak from an area-source, or when the pollutants from several sources interact and undergo atmospheric transformations.

We have chosen to give special consideration in this review to mixtures of sulfur dioxide (SO_2) and nitrogen dioxide (NO_2), SO_2 and ozone (O_3), and combinations involving hydrogen fluoride (HF). This emphasis can be justified by the volume of literature available but does not imply that other combinations of pollutants are unimportant. (Concentrations of SO_2, NO_2, or O_3 are expressed in ppb with 0·01 ppm assumed equal to, respectively, 26·2, 18·8, or 19·6 $\mu g\,m^{-3}$, all at 25°C and 760 mmHg. Concentration of HF is expressed as μg of fluoride (F) per m^3 with 0·01 ppm of HF equivalent to 7·7 $\mu g\,m^{-3}$.)

Combinations of the primary pollutants SO_2 and NO_2 have been chosen for study by several researchers because the two gases do occur together in many situations. They are often produced at the same time by the same sources: for example, combustion processes which use sulfur-containing fossil fuels. Some nitrogen oxides (NO_x) may come from nitrogenous contaminants of the fuel, but the bulk of it is the result of heat-induced combination of atmospheric O_2 and N_2 to form NO, which is subsequently oxidized to NO_2. Coal-fired electricity generation can lead to substantial emissions of $SO_2 + NO_2$ (Ormrod, 1982). The interactive effects of SO_2 and NO_2 may be important in two situations: (1) in or near urban areas, or close to industrial emissions, where short episodes of high concentrations of the two gases may be expected; and (2) in rural locations situated near urban areas, where prolonged exposures to low concentrations of both gases may occur. We shall direct most attention to effects of long-term exposures because these are probably most important in relation to crops in the field.

There have been many experiments on the joint action of SO_2 and O_3, but often they have involved concentrations of SO_2 higher than those expected in most polluted environments, whereas those of O_3 have been more realistic. A comprehensive review of the experiments reported prior to 1981 concerning effects of these two pollutants on growth (Guderian and Rabe, 1981) showed that among 24 experiments, 20 involved SO_2 concentrations of 0·100 ppm or higher (of which 14 were above 0·400 ppm) and none had employed concentrations below 0·05 ppm. Ozone concentrations were, on the other hand, more realistic, being between 0·025 and 0·150 ppm in 17 experiments. The tacit assumption in such experiments is that importance should be attached to peak concentrations of SO_2 occurring against a background of elevated O_3. (Concern with SO_2 from an isolated, single source has been predominant in North America, whereas in Northwest Europe it has been with low, continuous levels of SO_2.) This assumption may apply near major sources of SO_2, but it is worth noting that such sources (those involving combustion) often emit NO, which can remove O_3 from the atmosphere as it is converted to NO_2. Chronic effects of the two gases have received relatively little attention, although there is considerable information on the co-occurrence of SO_2 and O_3 in rural areas (Lefohn and Tingey, 1984). The diurnal cycle of the two pollutants may be different, with the peak of SO_2 occurring earlier in the day than that of O_3, but their co-occurrence in industrialized parts of the world is commonplace.

Hydrogen fluoride is thought to be the most phytotoxic of the primary air pollutants (Weinstein, 1977), but the effects of HF are geographically less widespread than those of pollutants such as SO_2, NO_x, and O_3. The processes generating HF do, nevertheless, usually produce other pollutants, and consequently any interactions between HF and other pollutants need careful examination. Future increases in the use of coal for power generation will call attention to this problem.

13.2.2 Application of the results

The role of joint action of gaseous pollutants in crop loss assessment needs to be considered with an awareness of the regulatory environment in which the information will be applied. In the formulation of secondary air quality standards appropriate to the occurrence of two or more pollutants, three options are available to the regulatory process (Jordan and Zaragosa, 1984). The first is to regard the effect of the mixture as being equal to the sum of the effects of each component considered independently. The second is to regard the effects of the pollutant under review as being modified by other pollutants that may be present (as though these were environmental factors). The third is to formulate a separate standard for each combination. In judging which approach is most appropriate, a model should be at hand that supplies (1) criteria sufficient to decide when effects of mixtures must be considered, and (2) quantitative relationships between exposures and effects for determining marginal costs and benefits expected with the control of one or more pollutants. Consequently, the accuracy, effectiveness, and cost of estimations of effects, predictions of risks, and regulatory actions for combinations of

pollutants rest upon an understanding of the joint action of their components
with the plant.

13.2.3 Possibilities for joint action

An estimation of the effects of a mixture of pollutants on an assemblage of
plants solely as a combination of the effects that each component produced
alone rests upon establishing the validity of three assumptions concerning the
independence of response, tolerance, and action.

Independence of response assumes that the likelihood of an effect of one
pollutant on a plant is independent of the occurrence of an effect of a second
pollutant on another plant. Violations of this assumption could occur when an
effect on a plant alters the effective dose received by another plant, its
tolerance to the pollutant, or directly or indirectly, its resistance to abiotic or
biotic stress. The conditions that would lead to a violation of this assumption
are determined by the composition and the structural and functional relationships in the assemblage. Although of questionable validity for long-term effects
in forests, the assumption should be valid for crops and more certainly for
plants subjected to experimental exposures.

Independence of tolerance assumes that, in the assemblage, phenotypic
tolerance to one pollutant is distributed independently of phenotypic tolerance
to another. (Tolerance is defined here as the dose below which no response is
elicited in an individual.) For example, in a set of plants subjected to two
pollutants, each at the median effective dose, a response in 75% of the plants
would be expected under independence. If the tolerances followed a bivariate
normal distribution with a correlation coefficient of 0.5 or -0.5, the expected
response would be about 67% or 85%, respectively. Comparisons among
cultivars of *Phaseolus vulgaris* L. (Beckerson *et al.*, 1979) indicate that
tolerances to O_3, SO_2, and their combination are associated, but evidence with
Pinus strobus L. suggests that the genetically determined response to $SO_2 + O_3$
may be complex (Costonis, 1973; Houston and Stairs, 1973). Although direct
evidence is lacking, climatic and edaphic factors could also affect tolerances to
make this assumption of questionable validity in the field (McCune *et al.*,
1984) but valid in experiments where single cultivars are exposed under
controlled, uniform conditions.

Independence of action, wherein it is assumed that action of one pollutant is
not affected by the presence or action of another pollutant, has been the focus
of research. Nevertheless, the interpretation of results solely in this respect
rests upon the other assumptions being valid.

13.3 RESPONSES OF CROPS

13.3.1 Reductions in yield

Although there have been many experiments exploring the joint action of
SO_2 and O_3 or NO_2, the results are very difficult to interpret because so many
different kinds of response have been reported. We shall not attempt a
comprehensive review of this topic because one was recently compiled (Kohut,

1985). There have been few additional research contributions to change the extent of our knowledge, and earlier reviews can also be recommended (Reinert *et al.*, 1975; Ormrod, 1982; Runeckles, 1984).

Some of the earliest experiments with short-term exposures (Engle and Gabelman, 1966; Menser and Heggestad, 1966) provided evidence of synergistic effects of SO_2 and O_3, i.e., concentrations that did not produce visible injury on their own did so when the two gases were applied together. The types of foliar injury caused by SO_2 and O_3 are sufficiently different for them to be distinguished, and it was noted in these and some other early experiments that mixtures below the injury threshold concentrations for both gases can produce symptoms characteristic of O_3 injury. Further investigations have shown that not only quantitative but also qualitative changes in symptom expression result from the joint action of the two pollutants (Lewis and Brennan, 1978; Miller and Davis, 1981).

Tingey *et al.* (1971) described the foliar injury in six different species caused by 4-h exposures to mixtures of SO_2 and NO_2. The concentrations of the individual gases in the mixtures ranged from 0·050 to 0·250 ppm. There was no visible injury to leaves caused by NO_2 alone below 2·000 ppm or by SO_2 below 0·500 ppm in these short fumigations, but severe injury did develop during exposure to the much lower concentrations in the mixtures of the two. For example, in tobacco, 0·100 ppm SO_2 + 0·100 ppm NO_2 caused injury to 26% of the total leaf area, from visual estimates made 2 days after exposure. This paper includes a careful description of the nature of foliar injury to the six species (pinto bean, oats, radish, soybean, tobacco, tomato), and it was noted that the flecking and necrotic bleaching normally associated with ozone pollution could be caused by mixtures of SO_2 and NO_2. The experiments also revealed unexpected dose–response characteristics, with reductions in effects at particular concentrations of the two pollutants. Since this pioneering study, there have been many further reports of greater-than-additive effects of SO_2 and NO_2.

Although NO_x pollution can be phytotoxic in its own right, there is very little evidence to suggest that the concentrations found in the open air cause much direct damage to plants. It is only in special circumstances, when NO_x concentrations become exceptionally high, that the phytotoxicity of NO and NO_2 causes concern (Law and Mansfield, 1982). Nevertheless, the importance of ambient levels of NO_2 in combination with SO_2 is indicated by a series of long-term fumigations of grasses undertaken at Lancaster in northern England (e.g., Ashenden and Mansfield, 1978; Ashenden, 1979; Ashenden and Williams, 1980; Whitmore and Mansfield, 1983). The plants were grown at near-ambient temperatures and with natural illumination, and were exposed to SO_2 and NO_2 concentrations of 0·060 to 0·070 ppm (weekly averages). The percentage reductions in growth, when plants were fumigated over winter and harvested in early spring, usually showed evidence of a synergistic effect of the two pollutants, i.e., a statistically significant interaction (Table 13.1). Treatment with NO_2 alone often stimulated growth, but the same concentration of NO_2 combined with SO_2 caused greater damage than SO_2 on its own.

When the growth of *Poa pratensis* L. was studied for longer periods of time

TABLE 13.1

Percentage reductions (relative to control) of all growth parameters measured for *Lolium multiflorum* and *Phleum pratense* after being exposed for 140 days to atmospheres containing (a) 0·068 ppm NO_2, (b) 0·068 ppm SO_2, (c) 0·068 ppm NO_2 + 0·068 ppm SO_2 (from Ashenden and Williams, 1980).

	NO_2	SO_2	$SO_2 + NO_2$	Effect
Lolium multiflorum				
Number of tillers	17	23	32	
Number of leaves	18	27	40	
Leaf area	1[a]	22	43	S
Dry weight of green leaves	10	28	65	S
Dry weight of dead leaves and "stubble"	5	3[a]	28	
Dry weight of roots	35[a]	7[a]	58	S
Phleum pratense				
Number of tillers	6	33	55	
Number of leaves	10[a]	29	68	S
Leaf area	30[a]	11	82	S
Dry weight of green leaves	14[a]	25	84	S
Dry weight of dead leaves and "stubble"	12	47	64	
Dry weight of roots	1[a]	58	92	

[a] Increase.
S = Synergistic effects of pollutants (i.e., a statistically significant interaction).

(up to 11 months) with SO_2, NO_2, or $SO_2 + NO_2$, sequential harvests showed that more-than-additive effects of SO_2 and NO_2 occurred during late winter, but during rapid growth in the summer there were no interactive effects of the two pollutants (Whitmore and Freer-Smith, 1982; Whitmore and Mansfield, 1983). If the effects of $SO_2 + NO_2$ are plotted against time, the dramatic change in the response in the period from April to May can be seen clearly (Fig. 13.1). These observations suggest that the damaging effects of $SO_2 + NO_2$ might be associated with the slower metabolism of plants in the winter, or that there might be some interaction with cold stress. The plants in these experiments were exposed to natural frosts in winter, and there is evidence from work with SO_2 (but not with $SO_2 + NO_2$) that grasses and cereals become more sensitive to frost during fumigation (Davison and Bailey, 1982; Baker *et al.*, 1982).

Field-grown soybeans were exposed to SO_2 and NO_2 alone and in combination (Irving *et al.*, 1982) with the zonal-air-pollution (ZAP) system (Miller *et al.*, 1980) used to release the two gases into the experimental plots. The concentrations used were relatively high (0·140 to 0·420 ppm for SO_2, and 0·070 to 0·370 ppm for NO_2), but the exposures were of short duration and were a reasonable simulation of peak exposures that can occur in the vicinity of major sources of the two pollutants. The experiment was conducted over two growing seasons, and on each occasion there was a significant $SO_2 + NO_2$

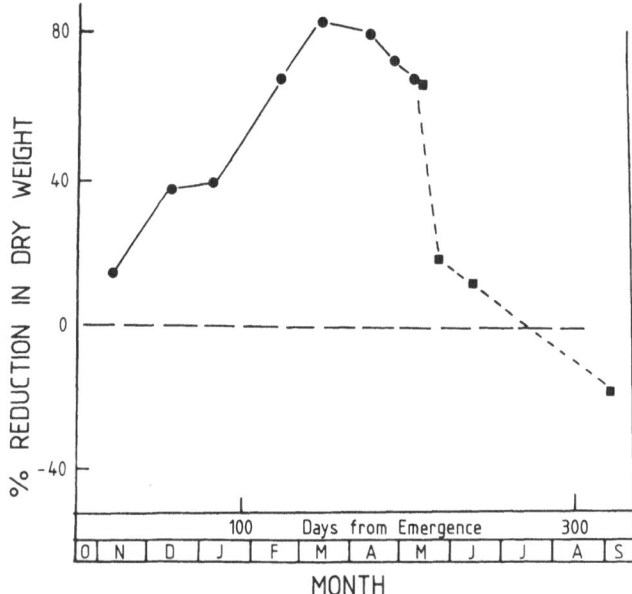

Fig. 13.1. Percentage reductions (compared with controls in clean air) in the growth of the grass *Poa pratensis* in 0·062 ppm SO_2 + 0·062 ppm NO_2. ● represents harvest of the whole plant (i.e., root and shoot) and ■ represents harvest of the shoots only (From Whitmore and Freer-Smith (1982), reproduced by permission of Macmillan Magazines Ltd.)

interaction. Decreases in yield occurred as a result of treatment with both pollutants even when there was no effect of either SO_2 or NO_2 alone. Ambient concentrations of O_3 were present during these experiments and it is possible that O_3 may have contributed to the observed effects. Nevertheless, synergistic effects of SO_2 and NO_2 have also been found in experiments on soybean in open top chambers from which ambient O_3 was excluded (Amundson and MacLean, 1982).

13.3.2 Dose–response relationships

The concentrations or durations of exposure may determine the character of the response to a combination of pollutants. In soybean with O_3 and SO_2, Heagle and Johnston (1979) found synergistic, additive, or antagonistic effects depending on both concentration and duration. They attached importance to the amount of injury caused by each pollutant on its own. Synergism was usually found when the doses were such that the pollutants on their own had little effect, but there was antagonism when their individual effects were severe. More recently, no effect of SO_2 by itself or interactively with linear or higher degree components of the O_3 dose–response function was found with winter wheat (Kohut *et al.*, 1987).

When the responses of plants to a number of factors are interdependent, experiments are often conducted with a full set of treatments in a factorial design. A few such experiments have been performed using SO_2 and O_3 (e.g., Gardner and Ormrod, 1976) but for most researchers such an approach cannot be undertaken because of the limitations imposed by experimental facilities. To overcome this problem, the use of response-surface techniques has been recommended (Ormrod *et al.*, 1984) as being more efficient in estimating certain parameters than a full factorial design. The value of this approach is clearly illustrated by the data (dry weight of leaves of garden peas) presented in their paper. Figure 13.2 depicts two of the response surfaces obtained. In Fig. 13.2(a) the changes in leaf area of radish in response to SO_2 and O_3 are shown. Here the "isoeffect lines" are equally spaced along the O_3 axis and run parallel to the SO_2 axis. There is clearly a linear response to O_3, and SO_2 had no detectable effect on this response. Figure 13.2(b) shows the response surface that is obtained when the effects of SO_2 and O_3 are additive at low concentrations, but at higher concentrations of O_3, the presence of SO_2 seems to reduce the threshold at which a particular level of injury occurs. Another value of the technique is that an axis normal to the contours indicates the weights (relative effectiveness) to be assigned each component in representing or approximating the dose–response relationship with a single pollutant-variate. Although such a simplification may not be valid globally, it could represent a satisfactory local solution over a domain of regulatory concern.

The dose–response relationship of $SO_2 + NO_2$ has been determined for *Poa pratensis* in studies in which both the gas concentrations and the durations of exposure had been varied (Whitmore, 1985). The results in Fig. 13.3 show a clear relationship, with a stimulation in dry weight at low doses of the pollutants, and an abrupt change to a negative impact as dose was increased. Dose in this case was a sum of the concentrations of the components. A similar result is also obtained with mixtures of SO_2 and O_3 for a quadratic relationship between dose and mass of stalks (but not roots) of soybean (Tingey *et al.*, 1973) wherein the dose variate can be represented as a weighted sum of the concentration of each pollutant and a non-monotonic relationship is present (Fig. 13.4). Linear relationships were also found between a weighted sum of

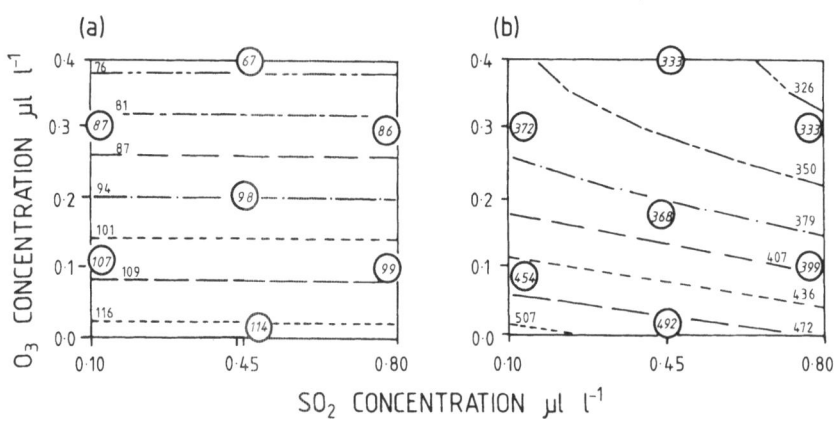

Fig. 13.2. Examples of contour plots illustrating the responses of leaf area in radish (a) and leaf dry weight in peas (b) to combinations of O_3 and SO_2. ((From Ormrod *et al.* (1984), reproduced by permission of the American Society for Plant Physiologists.)

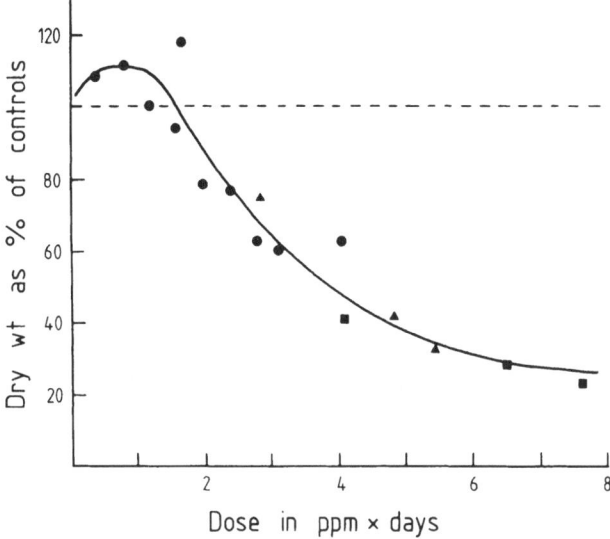

Fig. 13.3. Dose–response curve for the effect of mixtures of SO_2 and NO_2 on the dry mass of *Poa pratensis*. Controls were exposed to 0·007 ppm of each gas and the effects of higher concentrations ($\bullet = 0·040$ ppm, $\blacktriangle = 0·070$ ppm, and $\blacksquare = 0·100$ ppm) are shown relative to the controls. Periods of fumigation varied from 4 to 50 days. (From Whitmore (1985), reproduced by permission of the New Phytologist Trust.)

SO_2 and O_3 concentrations and total mass of radish (Tingey *et al.*, 1971), stomatal conductance of garden peas (Olszyk and Tingey, 1986), and potassium efflux from leaf discs of petunia (Elkiey and Ormrod, 1979). These data were obtained using seedling plants under controlled conditions in laboratory fumigation systems. Such clear dose–response relationships cannot be assumed for plants of different ages growing in more variable conditions. Nevertheless, the characterization of the relationships, and the very clear indication in some cases that positive and negative responses to a pollutant mixture can be dependent on total dose, represents an advance in our understanding and a

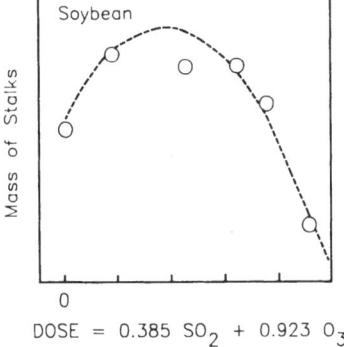

Fig. 13.4. Dose–response curve for the effect of mixtures of SO_2 and O_3 on the dry mass of soybean stalks. Plants were exposed for $8\,h\,d^{-1}$, $5\,d\,wk^{-1}$. Dose is the weighted sum of the concentrations of SO_2 at 5 or 20 pphm and O_3 at 5 or 10 pphm (Tingey *et al.*, 1973).

sound foundation for future biochemical and physiological evaluations of the mechanism of action of the pollutants.

13.3.3 Allocation

A general conclusion from studies of the action of mixtures of SO_2 and NO_2 is that there are disproportionate effects on roots and shoots. The allocation of dry matter to the roots appears to be more greatly affected than that to the newly developing, above-ground portions of the plants. Effects on root growth represent a particularly good example of the unexpected synergism that can occur between two pollutants. It has been well established that SO_2 on its own can appreciably change the allocation of photosynthetic products to roots. Like SO_2, NO and NO_2 can inhibit photosynthesis (Capron and Mansfield, 1976) but generally they do not seem to induce a reallocation of dry matter away from roots. In *Poa pratensis*, Whitmore and Mansfield (1983) found that SO_2 had a much greater influence on shoot:root ratio than did NO_2. Nevertheless, it seems unlikely that NO_x has any direct inhibitory effect on phloem translocation.

When SO_2 and NO_2 are applied simultaneously, the inhibition of root growth is usually greater than that caused by SO_2 alone, and it is often very severe. Two examples of effects on grasses are found in Table 13.1. Pande and Mansfield (1985) found that there was a significant effect of $SO_2 + NO_2$ on root growth in barley (Table 13.2) and that the inhibition appeared to be linearly related to the concentration of the two gases (Fig. 13.5). Another example is found in *Dactylis glomerata* L. (Fig. 13.6) where during a 140-day period of exposure, NO_2 alone had no appreciable effect on the allocation of dry matter among green leaves, dead leaves and stubble, and roots; however, NO_2 substantially altered the perturbation produced by SO_2 (Ashenden, 1979).

TABLE 13.2

Effects of NO_2 and SO_2 alone and in combination on the growth of spring barley (c.v. Patty). Percentage changes relative to controls in clean air are shown. Fumigation began 2 days after germination and continued for 20 days (from Pande and Mansfield, 1985)

	0·100 ppm NO_2	0·100 ppm SO_2	0·100 ppm 0·100 ppm $NO_2 + SO_2$
Number of tillers	No change	9·6 ↓	12·3 ↓
Number of fully			
expanded leaves	4.6 ↑	6·9 ↓	21·5 ↓
Root dry weight	18·1 ↑	20·2 ↓ [a]	50·0 ↓ [a]
Stubble dry weight	1·7 ↓	4·9 ↓	35·1 ↓ [a]
Leaf dry weight	1·3 ↓	18·3 ↓ [a]	45·4 ↓ [a]
Leaf area	2·4 ↓	6·3 ↓	44·8 ↓ [a]

↑ = Increase above controls.
↓ = Decrease below controls.
[a] Significantly different from controls.

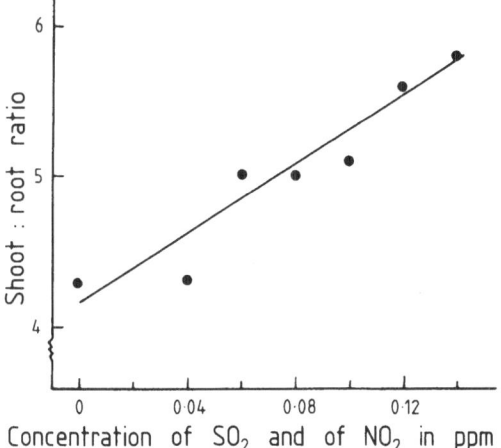

Fig. 13.5. Changes in shoot:root ratio in *H. vulgare* (spring barley) as a result of fumigation with SO_2 and NO_2. Seedlings were fumigated for 2 weeks, beginning 3–4 days after germination. The concentrations of both gases were equal on a unit volume basis (i.e., in ppb) (from Pande and Mansfield, 1985).

Tingey *et al.* (1973) performed one of the few experiments using long-term exposures to low concentrations of both SO_2 and O_3. They exposed soybean seedlings to 0·050 ppm SO_2 + 0·050 ppm O_3, and to a small range of concentrations of each pollutant, for 8 h daily, 5 days each week, beginning 3 days from germination. The fresh and dry weights of roots were significantly affected by the $SO_2 + O_3$ treatment, and the roots responded more than the above-ground portions. The result was a significant increase in the shoot:root ratio. The apparent synergism between O_3 and 0·050 ppm SO_2 occurred in spite of the resistance of soybean to SO_2 alone. Under the conditions of these experiments, 0·200 ppm SO_2 had no effect on the growth of roots or shoots. It is generally agreed that soybean is relatively insensitive to SO_2 as compared with O_3.

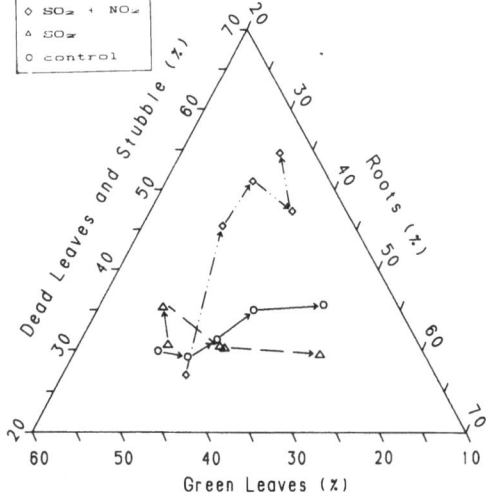

Fig. 13.6. Changes in the distribution of dry mass of *Dactylis glomerata* over successive 28-day intervals for plants exposed to: no detectable SO_2 or NO_2 (or to NO_2 at 6·8 pphm) (○); to SO_2 at 6·8 pphm (△); or to SO_2 and NO_2 both at 6·8 pphm (◇) (after Ashenden, 1979).

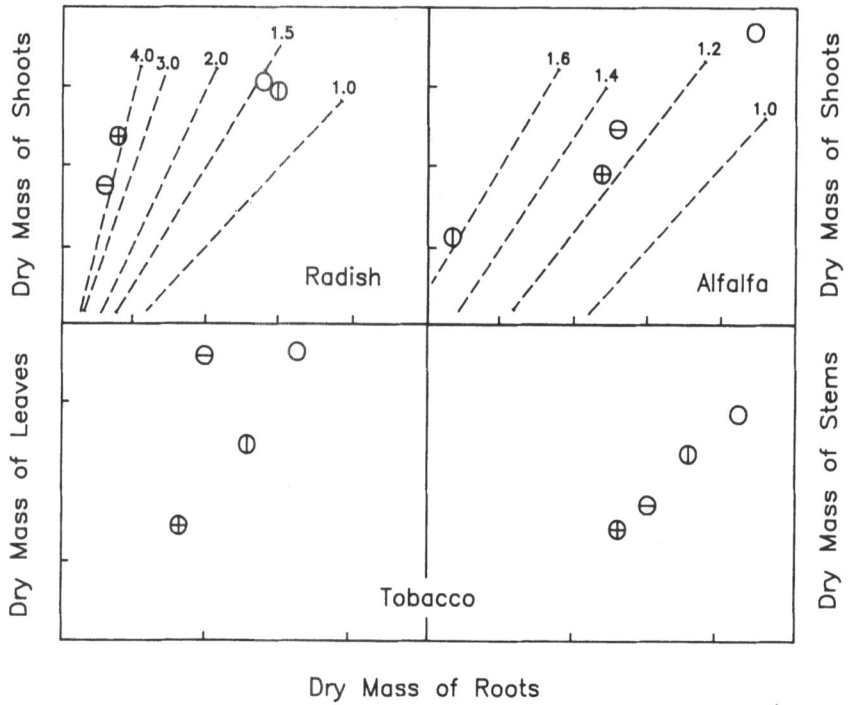

Dry Mass of Roots

Fig. 13.7. Effects of SO_2 and O_3, singly or in combination, on the dry mass of roots relative to the dry mass of shoots of radish and alfalfa or to the dry mass of stems or leaves of tobacco. Exposure of radish was a single 4-h period with O_3 and SO_2 each at 0·45 ppm; for alfalfa and tobacco, exposures were 40 h wk^{-1} with O_3 and SO_2 each at 0·05 ppm. Treatments are denoted by: $\bigcirc$ = control; $\oplus$ = SO_2; $\ominus$ = O_3; $\oplus$ = O_3 and SO_2. Dashed lines indicate values for shoot:root ratios (after Tingey and Reinert, 1975).

These observations suggested that the effect of O_3 under field conditions might be enhanced by the presence of SO_2, but no significant interactions were observed in soybeans exposed to lower levels of both pollutants in a field fumigation system (Reich and Amundson, 1984).

The difficulty of drawing general conclusions about the action of this combination of pollutants is well illustrated by the variable responses of roots and shoots. Tingey and Reinert (1975) studied the effects of chronic exposures to SO_2 and O_3 on three different species and found that the growth responses were species-dependent (Fig. 13.7). In alfalfa, the response of the plant could be represented by a weighted sum of the root and shoot wherein the effect of SO_2 reduced the mass of both fractions, produced a disproportionate effect on root growth and appeared to be negated by the presence of O_3 which by itself had a similar but lesser effect (Fig. 13.8). In radish, the effect of O_3 was a reduction in a weighted sum of root and shoot; SO_2 increased the relative mass of roots in the absence of O_3 but had the opposite effect in the presence of O_3. In tobacco, the main effect of O_3 was a reduction in mass of roots, whereas SO_2 tended to reduce foliar mass with a substantially greater effect in the presence of O_3. These findings illustrate that major differences can occur

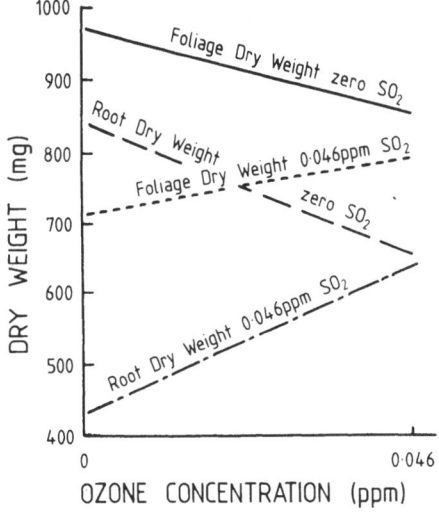

Fig. 13.8. The effects of chronic exposure to O_3 and/or SO_2 on the growth of alfalfa. Means for foliage dry weights and root dry weights are based on 60 and 20 observations, respectively. Alfalfa was exposed $40\,h\,wk^{-1}$ (from Tingey and Reinert, 1975).

between one species of crop and another in their responses to $SO_2 + O_3$, not only in production but also in allocation.

The different responses of roots and shoots to a mixture of SO_2 and O_3 suggest a partial explanation of the variable reports of the types of response (i.e., more-than-additive, additive, and less-than-additive). First, it is possible that some of the apparent discrepancies have arisen because different investigators have used different parameters to determine plant growth or performance. If roots are not included when dry mass increments are determined, then clearly the effects will be underestimated if the main response resides in the roots. A change in the pattern of translocation may sometimes lead to an increase in the growth of some components and a decrease in others. Second, interactions between pollution effects and environmental factors are likely to arise when pollution causes a change in the growth form of the plants. Ability to withstand soil moisture stress is an obvious factor to consider when it is known that pollution causes a reduction in the biomass of the root system.

These considerations do not, however, provide a convincing explanation of the range of responses that has been reported. A study of the literature leads inevitably to the conclusion that the reactions of different species to $SO_2 + O_3$ are highly individualistic. Even within a species there is not always agreement about the type of response expected. The duration and timing of exposures, the concentrations of the two pollutants and the age and condition of the plants may be major determining factors. Assessments of crop losses as a result of a coincidence of SO_2 and O_3 are, therefore, difficult to contemplate from the body of data that is presently available.

13.3.4 Translocation

The responses of root growth to SO_2/O_3 mixtures suggests that a possible common mode of action for their effects is an apparent interference with the process of translocation, and a frequently observed consequence of fumigation with the two gases separately is a reduced allocation of assimilates to the roots. It has been well established that SO_2 on its own can appreciably change the allocation of photosynthates to roots. Noyes (1980) found that a dose of SO_2 insufficient to inhibit photosynthesis in *Phaseolus vulgaris* reduced the translocation of ^{14}C-labeled assimilates out of a fumigated leaf by 39%. The quantitative changes in photosynthesis and translocation during SO_2 fumigation showed quite different dose–response characteristics. Teh and Swanson (1982) came to the same conclusion and, like Noyes, suggested that SO_2 inhibited directly the process of phloem loading. Jones and Mansfield (1982) showed that the distribution of ^{14}C-labeled assimilates was affected by SO_2 pollution in the grass *Phleum pratense* L. The growth of roots in this species can be inhibited by SO_2 even at doses that are apparently "tolerated", i.e., that have no effect on growth in terms of dry mass (Jones and Mansfield, 1982; Mansfield and Jones, 1985). Thus, there are physiological consequences of SO_2 pollution that are not revealed by studies of net assimilation or dry matter accumulation.

Oshima *et al.* (1978, 1979) found comparable changes in assimilate partitioning in parsley and cotton exposed to O_3 pollution. Root dry mass was decreased by 43% in 0·200 ppm O_3 but there was little effect on the leaves. Okano *et al.* (1984) conducted tracer experiments with $^{13}CO_2$ to examine the effects of O_3 on the distribution of newly formed assimilates from source leaves in *Phaseolus vulgaris*. They found that in response to 0·200 ppm O_3, the labeled assimilates moving to non-photosynthetic portions of the plants decreased by 53%, whereas those going to young leaves decreased by only 28%. (On the other hand, NO_2 increased assimilation and thereby translocation from the leaf.)

13.4 MECHANISMS

13.4.1 Exclusion

The action of one agent in modifying the uptake into foliage of another can be suggested as a possible basis for interactions because SO_2, NO_x, and O_3 have all been found to affect stomatal behavior and hence the diffusive conductance of leaves (Mansfield and Freer-Smith, 1984).

This seems likely to be the case with a pollutant as highly toxic as HF (McCune, 1986) where the effects of fluoride can be cumulative and the accumulation of fluoride by foliage can often be of as much practical significance as effects on growth or yield. All the possible kinds of effects that other pollutants might have on the accumulation of fluoride have been reported, viz. increases, decreases, or no change. The variation among species in the response of fluoride uptake to the presence of other pollutants (see Mandl *et al.*, 1975; McCune, 1986) is likely to reflect the highly variable manner in which stomata respond. Low concentrations of SO_2 normally cause stomatal opening whereas higher concentrations cause closure. With respect to fluoride toxicity, this could lead to unexpected dose–response relationships, with a larger accumulation of fluoride in the presence of low concentrations of SO_2, and a smaller accumulation when SO_2 concentrations are higher. There is evidence that stomatal conductance is increased by HF at $1 \cdot 5\ \mu g\ m^{-3}$ (Amundson *et al.*, 1982), but whether HF decreases stomatal conductance at concentrations above $5\ \mu g\ m^{-3}$ is equivocal, and HF has no effect on SO_2-induced changes at or above this concentration (Bonte *et al.*, 1983). A study of the joint action of HF and NO_2 on *Zea mays* L. (Amundson *et al.*, 1982) found that changes in leaf conductance could partially explain the negative interaction between the two gases on plant growth and on fluoride accumulation. NO_2 was found to inhibit stomatal closure in the dark, but the presence of HF overcame this inhibition. This effect of HF was suggested as the cause of a reduction in damage (reduced dry weight), which was found when NO_2 was applied singly. Moreover, the interactive effect of NO_2 did not depend upon its own concentration but upon that of HF: there was no effect of NO_2 on fluoride accumulation at $1 \cdot 5\ \mu g\ m^{-3}$ but there was at $0 \cdot 5\ \mu g\ m^{-3}$. Because stomatal conductance is of such importance in determining the entry of gases to the internal tissues of leaves, pollutants that have a substantial effect on stomata are likely to exert a major influence on the action of a cumulative toxic agent such as fluoride. Moreover, when each component of a mixture can increase or decrease (depending on concentration) stomatal conductance, it is possible for one pollutant to increase the effect of a second while the second decreases the effect of the first. Consequently, the terms antagonism and synergism must be used rather carefully.

Research by Amundson and Weinstein (1981) suggested that changes in stomatal conductance may be important in determining the effects of short exposures to high concentrations of SO_2 and NO_2. Some of the earliest experiments with short-term exposures by Engle and Gabelman (1966) and Menser and Heggestad (1966) provided evidence of synergistic effects of SO_2 and O_3, i.e., concentrations that did not produce visible injury on their own did so when the two gases were applied together. The types of foliar injury caused by SO_2 and O_3 are sufficiently different for them to be distinguished, and it was noted in these and some other early experiments that mixtures below the injury threshold concentrations for both gases can produce symptoms characteristic of O_3 injury. The data were consistent with the stimulation of stomatal opening by SO_2, allowing greater access of O_3 to sensitive sites. A careful study of changes in stomatal conductance in response to O_3 and for SO_2 in petunia hybrids did not, however, indicate that stomatal responses had a major controlling influence (Elkiey and Ormrod, 1979, 1980). In barley, there was more leaf injury after exposure to O_3 alone than after exposure to $O_3 + SO_2$, but the antagonistic effect of the two pollutants could not be explained in terms of a lower stomatal conductance reducing pollutant uptake (Ashmore and Önal, 1984). Pratt *et al.*, (1983) found that the accumulation of sulfur in the leaves of soybean was reduced in the presence of O_3 (Fig. 13.9), which suggested that in this case the well documented inhibitory effect of O_3 on stomatal opening was predominant. Earlier,

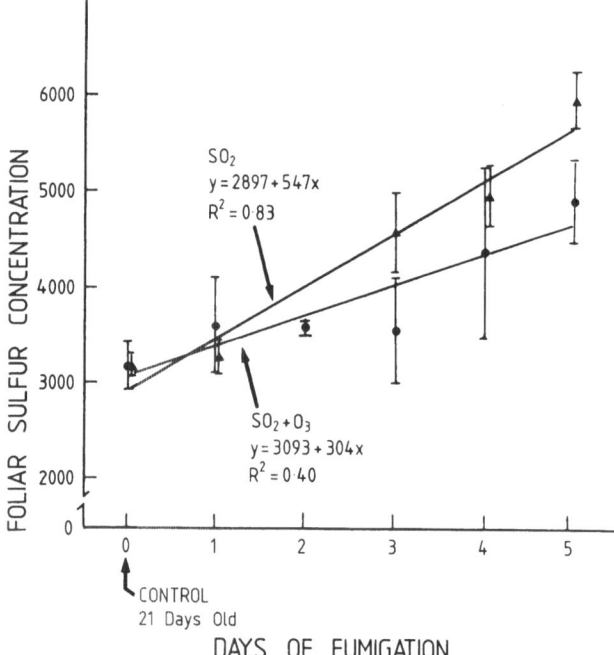

Fig. 13.9. Foliar sulfur concentration of soybean cv. Hodgson first trifoliate leaves fumigated for up to 5 consecutive days with $1050\,\mu g\,m^{-3}\,SO_2$ for $2\,h\,d^{-1}$, or $1050\,\mu g\,m^{-3}\,SO_2 + 196\,\mu g\,m^{-3}$ ($2\,h\,d^{-1}$). The slopes of the two lines are significantly different at $p = 0.05$. Each point represents the mean of three observations in three replications. Similar results were obtained with unifoliate leaves (from Pratt *et al.*, 1983).

Beckerson and Hofstra (1979) had found that while $0.150\,ppm\,SO_2$ tended to stimulate stomatal opening, the same concentration of SO_2 in combination with O_3 caused more stomatal closure than SO_2 on its own. Despite the reduced accumulation of foliar sulfur when O_3 was present with SO_2, the loss of chlorophyll was appreciably greater (Fig. 13.10). These findings raise important questions about the combined action of the two pollutants because of the implication that there can be more damage even when the uptake of both pollutants is reduced (if less sulfur accumulates, as in Fig. 13.9, then it must be assumed that there is a reduced entry of O_3 into the leaves because the diffusive pathways for SO_2 and O_3 are similar).

Studies of the joint action between pollutants thus draw particular attention to the considerable effects some gases may have on the entry of others into leaves. This leads us to suggest that future studies of the combined action of pollutants should involve measurement of the rates of deposition of the individual components to the plant leaves. Ideally a detailed analysis of the absorption (i.e., uptake into internal tissues) and of sorption onto the surface should be undertaken. This will provide a preliminary indication of the level at which interactions are occurring. Only when such effects are defined will it be possible to begin to quantify the importance of interactions at metabolic levels.

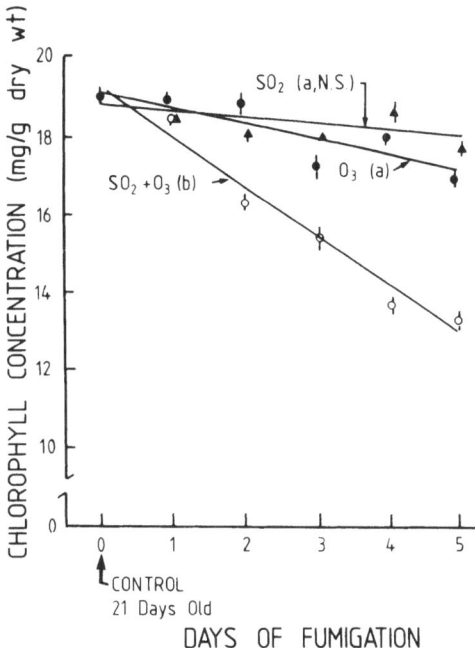

Fig. 13.10. Relationship between chlorophyll concentration of first trifoliate leaves of soybean cv. Hodgson and successive days of fumigation with O_3 ($196 \mu g\,m^{-3}\,2\,h\,d^{-1}$). SO_2 ($1050 \mu g\,m^{-3}$ $2\,h\,d^{-1}$), or the two pollutants simultaneously (same concentrations, $2\,h\,d^{-1}$). Each point is the mean of 16 observations in t replications (NS-regression line not significantly different from 0 slope: different lower case letters indicate significantly different slope ($p = 0.05$, F-test); slopes for the lines; $SO_2 = -0.16$, $O_3 = -0.40$, $SO_2 + O_3 = -1.25$). (From Pratt *et al.*, 1983.)

13.4.2 Detoxification

The activity of the enzyme nitrite reductase is considered to be critical for the metabolic utilization of the solution products of NO_x after its entry into the leaf via the stomata. Yoneyama and Sasakawa (1979) used $^{15}NO_2$ to show that it was converted into nitrite and nitrate, and that after reduction of these ions, the ^{15}N was incorporated into amino acids. Nitrite reductase activity appears to be confined to plastids and it must be regarded as essential to the detoxification of NO_x, because nitrite ions are highly toxic and do not normally accumulate in cells.

The most detailed study at the biochemical level has been by A. R. Wellburn and his colleagues (Wellburn, 1982, 1984). Much of the work has been performed on pasture grasses, fumigated using the same equipment as that employed in the growth studies described previously. Wellburn (1982) used plastid preparations from leaves of *Lolium perenne* L. that had been exposed to relatively high concentrations of SO_2, NO_2 or $SO_2 + NO_2$ for 5 to 15 days. Fumigation with NO_2 caused a statistically significant increase in the activity of nitrite reductase within 7 to 9 days, but no such increase was observed if NO_2 was accompanied by SO_2 (Fig. 13.11).

Longer term fumigations (20 weeks) of *L. perenne* with 0.068 ppm of the

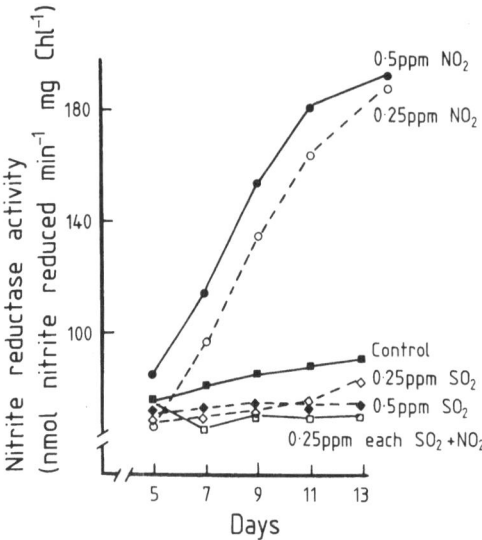

Fig. 13.11. Levels of nitrite reductase activity in extracts from *Lolium perenne* L. laminae (6 weeks old from selfing, before experiment started) which have been exposed to clean air or various combinations of SO_2-polluted and/or NO_2-polluted air for 5 to 15 days. (From Wellburn (1982), reproduced by permission of Butterworths, courtesy of Wellburn.)

two gases separately or in combination led to similar conclusions (Fig. 13.12). These studies were performed on two cultivars (S23 and S24) and on clonal material known to possess some degree of SO_2 resistance. One was the Helmshore clone well known to be resistant to chronic injury (Bell and Mudd, 1976), and another, called S23 Bell resistant, had been selected from seedlings of the S23 cultivar by Dr J. N. B. Bell. The results for the two cultivars show a very large inhibition of nitrite reductase formation when SO_2 was present with NO_2. The same trends were evident in the two resistant clones but the magnitude of the inhibition appeared to be less. Similar results were obtained from 20-week exposures to three other grasses (*Dactylis glomerata, Phleum pratense, Poa pratensis*) to the same pollution treatments (Fig. 13.12).

Wellburn concluded from these extensive studies that the presence of SO_2 had prevented the induction of the additional nitrite reductase activity that usually occurs in the presence of NO_2. In consequence, the plants are likely to have been unable to detoxify the solution products of NO_2, and the damage caused normally by SO_2 would be accompanied by injury caused by nitrite and perhaps other compounds formed from this reactive ion.

One of the mechanisms suggested as underlying the synergistic effect of O_3 and NO_2 on carbon assimilation in bean leaves (Okano *et al.*, 1984) is a reduction in nitrite reductase activity induced by exposure to ozone (Leffler and Cherry, 1974) wherein the magnitude of the reduction and its recovery depended upon the species of plant and concentration of O_3.

Nitrite reductase activity is closely linked to the photosynthetic electron transport chain from which it receives the essential supply of reductant. It is possible that leaf cells that contain no chloroplasts, or only poorly developed chloroplasts, may be specially vulnerable when NO_2 is absorbed via the

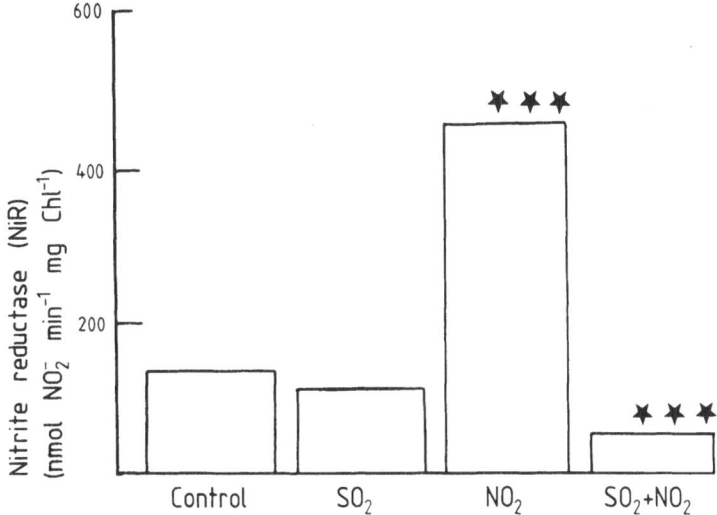

Fig. 13.12. Nitrate reductase activity in *Lolium* S23 exposed to clean air or to air containing SO_2 and/or NO_2 (0·068 ppm) over a period of 20 weeks (asterisks indicate the mean significant difference from the control). (From Wellburn (1982), reproduced by permission of Butterworths, courtesy of Wellburn.)

stomata. Some recent studies have suggested that epidermal cells may be in this category (Wright *et al.*, 1987).

Whether the light-dependent emission of H_2S and the increased levels of free thiols (predominantly reduced glutathione) following exposures to SO_2 could be regarded as similar detoxification mechanisms is an interesting problem. The latter has been viewed as a means of avoidance of sulfate toxicity in the leaf (Rennenberg, 1984) but other stresses of which O_3 is one (Guri, 1983; Mehlhorn *et al.*, 1986), also induce higher levels of reduced glutathione. Consequently, detoxification mechanisms may include more than just the solution products of the pollutant itself.

13.4.3 Action in common

Two basic considerations in a mechanistic explanation of joint action are to what extent the two agents have the same sites of action and how each may influence the action of the other. The different species of active oxygen—singlet oxygen ($^1O_2$), the superoxide (O_2^-) and perhydroxyl (HOO$\cdot$) radicals, hydrogen peroxide (H_2O_2), and the hydroxyl (OH$\cdot$) radical—and their reactions constitute a possible nexus for the actions of O_3, SO_2, and the natural metabolic reactions of the photosynthetic plant cell.

Reactions of O_3 in and with components of the cell produce these species. In the case of SO_2, we have some information on the action of solution products such as bisulfite and sulfite, of which it is known that the aerobic oxidation can generate free radicals, such as superoxide (evidence reviewed by Peiser and Yang, 1985). Superoxide leads to the formation of the very reactive hydroxyl radical, and reactions of these oxidative species with macromolecules can lead to serious damage to cellular functions, e.g., changes in membrane permeability and fluidity and loss of enzyme function (Thompson *et al.*, 1987). The involvement of free radicals in the injury to cells caused separately by SO_2 and O_3 suggests an obvious basis for a synergistic interaction: the two pollutants have common sites of action because they give rise to the same toxic products.

Inasmuch as photosynthetic processes also produce active oxygen, the occurrence of oxidative injury in cells is normally regulated or moderated by specific enzymes, particularly superoxide dismutase, catalase, and those of the ascorbate–glutathione–NADPH cycle, and compounds, such as α-tocopherol and carotenes. A small increment in production of free radicals by pollutants may be scavenged effectively by the normal enzyme systems available for this function, but if radical production exceeds the cell's capacity for scavenging them, then the damage may abruptly appear. Nevertheless, an increasing body of data indicates that the cell's response to O_3 and SO_2 can comprise an increase in the levels of antioxidants and activity of scavenging systems, although to a different extent according to the pollutant (Mehlhorn *et al.*, 1986). Such regenerative responses could constitute a means by which one pollutant affects the action of another.

Osswald and Elstner (1986, 1987) suggested that the reaction between O_3 in the atmosphere and endogenously produced ethylene might be important in forest decline. Mehlhorn and Wellburn (1987) found that O_3 fumigations only damaged pea plants when they were producing endogenous ethylene, and they suggested that reaction of O_3 with ethylene could produce water-soluble, oxidative, free radicals. This could be one (perhaps the major) basis for O_3 injury to plant cells, which is known to involve damage to membranes (Mudd *et al.*, 1984). Unsaturated fatty acids in plasma membranes are likely to be especially vulnerable to any free radicals produced when ethylene emitted from cells reacts with O_3 near the cell surface. It is also known that ethylene production is increased by exposure to SO_2 and is metabolically linked with the metabolism of sulfur-containing compounds (see review by Bucher, 1984). Consequently, ethylene could provide another mechanism for coupling the actions of O_3 and SO_2 at different sites within the foliar tissues.

These speculative proposals are examples of several that can be made to account for a combined action of SO_2 and O_3 on plants. It should also be noted that certain components of these speculations, e.g., active oxygen and ethylene, are not specific to the action of pollutants but are connected to the plant's response to natural stresses. We suggest that future investigation of this topic should involve a more robust testing of realistic hypotheses set up to explain some of the effects already well established from experimental fumigations. In this way, we are likely to progress toward a better understanding of the mechanisms involved, and to provide a foundation from which crop loss assessments can be made.

13.5 PREDISPOSITION TO STRESS

Some recent studies indicate that $SO_2 + NO_2$ could predispose the plant to climatic stress via their reduction of the structural and functional integrity of epidermal cells of the leaf (Wright *et al.*, 1987). It was found both in broadleaved trees and in grasses that the ability of leaves to conserve water during periods of water stress was reduced after exposure to small doses of $SO_2 + NO_2$ at concentrations as low as 0·010 to 0·020 ppm. A similar effect occurred with SO_2 alone, but the magnitude was much greater when SO_2 was accompanied by NO_2. This appeared to be a good example of SO_2/NO_2 synergism in affecting an essential physiological process.

An example of a readily observed change in the capacity for water conservation is shown in Fig. 13.13. Leaves of silver birch (*Betula pendula*) were put under severe stress by being excised from the plant, and simply being allowed to dry out. They were weighed at intervals to indicate the rate at which water was being lost. Leaves from plants grown in clean air showed good control of water loss under these extreme conditions; the stomata closed quickly and much of the original fresh weight was retained, even after several hours. Leaves that had been in $SO_2 + NO_2$ for 30 days behaved quite differently and lost water at a much more rapid rate. Even 0·020 ppm $SO_2 +$ 0·020 ppm NO_2 had a significant effect on water conservation. Similar changes in the drying curves were also found for leaves of the grass *Phleum pratense*, and in this case even 0·010 ppm $SO_2 + 0·010$ ppm NO_2 had an appreciable effect.

Studies of frozen hydrated leaves of birch under the scanning electron microscope have revealed patches of damaged epidermal cells, accompanied by wide-open stomata. When epidermal cells are damaged, there can be

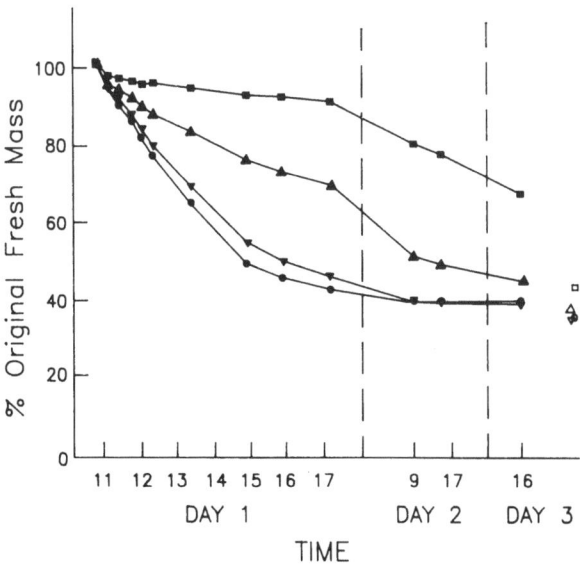

Fig. 13.13. Fresh weight for excised leaves measured over time from clonal plants of *Betula pendula* grown for 30 days in clean air (■), 0·020 ppm $SO_2 + 0·020$ ppm NO_2 (▲), 0·040 ppm $SO_2 + 0·040$ ppm NO_2 (▼), 0·060 ppm $SO_2 + 0·060$ ppm NO_2 (●). Each point is a mean of nine replicates, one leaf from one tree. The open symbols show the final dry mass, after oven drying at 80°C (from Wright *et al.* 1987).

TABLE 13.3

Effects on the mean relative growth rate (RGR) and net assimilation rate (NAR) for *Phleum pratense* exposed continuously for 40 days to SO_2 and NO_2 in combination, followed by a period of 23 days in clean air during which water was either supplied daily or was withheld (from Wright *et al.*, 1987)

Treatment $SO_2 + NO_2$ (ppm)	$RGR\ g\,g^{-1}\,d^{-1}$ Watered	Unwatered	$NAR\,g\,cm^{-2}\,d^{-1}$ Watered	Unwatered
0 control	0·024	0·013	0·216	0·101
0·030	0·021	0·008	0·211	0·096
0·060	0·028	0·006[a]	0·183	0·076[a]
0·090	0·027	0·005[a]	0·162[a]	0·037[a]

[a] Indicates significant differences from control at the $P < 0.05$ level.

mechanical forces in the epidermis that prevent stomatal closure so that transpiration continues even when a leaf is under very severe water stress.

Observations of this kind on detached leaves give no indication of how such changes may affect the water relations of whole plants. Studies of the gas exchange of common birch (*Betula pubescens*) showed that rates of transpiration were enhanced after fumigation with $SO_2 + NO_2$, but that rates of photosynthesis were unaffected. When water was withheld from the plants, those previously exposed to $SO_2 + NO_2$ displayed severe effects of drought much more quickly than the clean air controls. Comparable differences in water consumption and drought resistance were observed in *Phleum pratense*. The data in Table 13.3 show an after-effect of fumigation with $SO_2 + NO_2$, which was revealed most clearly in the plants that were subjected to water stress.

Whether the potentiation by O_3 of the SO_2-induced loss of magnesium and calcium from the crowns of spruce in throughfall (Arndt *et al.*, 1985) has a similar causal basis, the fact remains that predisposition to or potentiation of natural climatic stresses may play a significant role in the joint action of air pollutants under ambient conditions.

13.6 CONCLUSIONS AND RECOMMENDATIONS

Although there is now an extensive amount of literature available on the combined action of SO_2 and NO_2, it is still not possible to come to firm conclusions about effects on particular crops. Nevertheless, the subject has progressed to a stage that permits us to discuss some of the issues that need to be faced in future research. The discussion is assisted by the advances that have been made in explaining the action of the two pollutants, which can be summarized as follows:

(1) More-than-additive effects of SO_2 and NO_2 have been observed in both short-term and long-term experiments. They do not invariably occur but

it seems likely that they are sufficiently important to require special consideration when the economic effects of SO_2 and of NO_2 are considered.

(2) During long-term fumigations under the conditions and stresses of ambient weather, considerable variations in response occur. Effects of $SO_2 + NO_2$ may be especially important in cool weather when light intensities are low.

(3) The translocation of assimilates to roots, which is affected by SO_2 as a single pollutant, is subject to greater inhibition in $SO_2 + NO_2$ exposures even though NO_2 on its own does not appear to influence translocation directly.

(4) The ability of leaves to detoxify the nitrite ions entering cells when NO_x is taken up from the atmosphere appears to be severely inhibited by SO_2. Whether this results from the direct action of SO_2 or competition for reductant, it may be a primary cause of synergistic action between the two pollutants.

(5) Damage to epidermal cells has been observed when leaves are exposed to mixtures of SO_2 and NO_2. This can prevent complete closure of stomata and affects the ability of plants to conserve water under conditions of drought, and may be especially significant if growth of roots has already been reduced by the mixture.

Publications on the joint action of $SO_2 + O_3$ on plants are much more numerous than on any other pollutant combination, and the experimental work that has led to this extensive literature obviously has been very costly to perform. Unfortunately, there are few useful conclusions that can be drawn about the likely impact of this combination of pollutants on crops. It is clear that the funds for research must be directed differently if we are to make real progress towards assessment of crop losses in situations where SO_2 and O_3 are coincident.

Noticeably absent from much of the literature is any attempt to understand the mechanisms behind the observed effects. Although the occurrence of more-than-additive effects of SO_2 and O_3 are probably of most importance in relation to crop loss assessment, less-than-additive effects also need to be taken into account because they imply that losses could increase with the reduced atmospheric level of either or both pollutants. Both types of action have been reported frequently enough from experimental studies to suggest that they may occur from time to time in the field. Unless we have some mechanistic understanding there is little hope of predicting when and where they may occur.

In the light of these observations it is suggested that future research be directed as follows:

(1) To establish the effects of $SO_2 + NO_2$ on plants that are exposed to low temperatures and low light intensities during part of their growing period. Such effects are more likely to be important in temperate regions where there are many diverse sources of the two pollutants.

(2) To discover whether mixtures of pollutants enhance the plant's sensitivity to drought (and also to frost, since there are factors shared between the two regarding their effects on plants). Very few experiments have been specifically designed to examine pollution/drought interactions even though these may be of vital importance in determining the scale of the impact of pollution in the field. The simultaneous occurrence of disturbed stomatal behavior and reduced root growth requires special attention.

(3) To elucidate the effect of SO_2 on the ability of plants to reduce nitrite to ammonia. More careful study of this area of metabolism is needed because there are differences among plant species in the location of the reduction of nitrate and nitrite. In some the process occurs not in the leaves but in the roots, whereas in others the entire process is believed to take place in leaves (Smirnoff *et al.*, 1984).

(4) To determine more accurately and extensively the role of stomatal mechanisms in the joint action of mixtures. One controlling factor common to all gaseous pollutants is uptake into foliage. An additional factor to be considered, of which nothing is known, is reciprocal actions of pollutants on conductance through the mesophyll tissues of the leaf.

To evaluate the mechanisms behind injury, we need far more detailed physiological and biochemical studies, so that we can begin to define the nature of the disturbances caused by the pollutants. While it is possible that the damage caused by two pollutants in combination will be quite different from that produced by the gases individually, it is more likely that the basic disturbances are the same, owing to common sites and processes being affected by common intermediates, and that the synergism results from an amplification of their impact.

REFERENCES

Amundson, R. G. and D. C. MacLean. (1982). Influence of oxides of nitrogen on crop growth and yield: an overview. *Stud. Environ. Sci.*, **21**, 501–10.

Amundson, R. G. and L. H. Weinstein. (1981). Joint action of sulfur dioxide and nitrogen dioxide on foliar injury and stomatal behavior in soybean. *J. Environ. Qual.*, **10**, 204–6.

Amundson, R. G., L. H. Weinstein, P. van Leuken, and L. J. Colavito (1982). Joint action of HF and NO_2 on growth, fluorine accumulation, and leaf resistance in Marcross sweet corn. *Environ. Exp. Bot.*, **22**, 49–55.

Arndt, U., G. Seufert, J. Bender and H. J. Jäger (1985). Untersuchungen zum Stoffhaushalt von Waldbäumen aus belasteten Modellökosystemen in Open-Top-Kammern. *VDI-Ber.*, **560**, 783–803.

Ashenden, T. W. (1979). The effects of long-term exposures to SO_2 and NO_2 pollution on the growth of *Dactylis glomerata* L. and *Poa pratensis* L. *Environ. Pollut.*, **18**, 249–58.

Ashenden, T. W. and T. A. Mansfield (1978). Extreme pollution sensitivity of grasses when SO_2 and NO_2 are present in the atmosphere together. *Nature (London)*, **273**, 142–3.

Ashenden, T. W. and I. A. D. Williams (1980). Growth reductions in *Lolium multiflorum* Lam. and *Phleum pratense* L. as a result of SO₂ and NO₂ pollution. *Environ. Pollut., Ser. A*, **21**, 131–9.

Ashmore, M. R. and M. Önal. (1984). Modification by sulphur dioxide of the responses of *Hordeum vulgare* to ozone. *Environ. Pollut., Ser. A*, **36**, 31–43.

Baker, C. K., M. H. Unsworth, and P. Greenwood. (1982). Leaf injury on wheat plants exposed in the field in winter to SO₂. *Nature (London)*, **299**, 149–51.

Beckerson, D. W. and G. Hofstra. (1979). Response of leaf diffusive resistance of radish, cucumber and soybean to O₃ and SO₂ singly or in combination. *Atmos. Environ.*, **13**, 1263–8.

Beckerson, D. W., G. Hofstra, and R. Wukasch. (1979). The relative sensitivity of 33 bean cultivars to ozone and sulfur dioxide singly or in combination in controlled exposures and to oxidants in the field. *Plant Dis. Rep.*, **63**, 478–82.

Bell, J. N. B. and C. H. Mudd. (1976). Sulphur dioxide resistance in plants: case study of *Lolium perenne*. *Semin. Ser.-Soc. Exp. Biol.*, **1**, 87–103.

Bonte, J., C. Bonte, and L. de Cormis. (1983). Effect of simultaneous action of sulfur dioxide and hydrogen fluoride on stomatal movement in *Zea mays* and *Pelargonium hortorum*. *Fluoride*, **16**, 220–8.

Bucher, J. B. (1984). Emissions of volatiles from plants under air pollution stress. In *Gaseous air pollutants and plant metabolism*, ed. by M. J. Koziol and F. R. Whatley, 399–412. London, Butterworths Scientific.

Capron, T. M. and T. A. Mansfield. (1976). Inhibition of net photosynthesis in tomato in air polluted with NO and NO₂. *J. Exp. Bot.*, **27**, 1181–6.

Costonis, A. C. (1973). Injury to eastern white pine by sulfur dioxide and ozone alone and in mixtures. *Eur. J. For. Pathol.*, **3**, 50–5.

Davison, A. W. and I. F. Bailey. (1982). SO₂ pollution reduces the freezing resistance of ryegrass. *Nature (London)*, **297**, 400–2.

Elkiey, T. and D. P. Ormrod. (1979). Leaf diffusion resistance responses of three petunia cultivars to ozone and/or sulfur dioxide. *J. Air Pollut. Control Assoc.*, **29**, 622–5.

Elkiey, T. and D. P. Ormrod. (1980). Sorption of ozone and sulfur dioxide by *Petunia* leaves. *J. Environ. Qual.*, **9**, 93–5.

Engle, R. L. and W. H. Gabelman. (1966). Inheritance and mechanism for resistance to ozone damage in onion, *Allium cepa* L. *Proc. Am. Soc. Hortic. Sci.*, **89**, 423–30.

Gardner, J. O. and D. P. Ormrod. (1976). Response of the Rieger begonia to ozone and sulphur dioxide. *Sci. Horticul. (Amsterdam)*, **5**, 171–81.

Guderian, R. and R. Rabe. (1981). *Effects of photochemical oxidants on plants*. Brussels, Commission of the European Communities.

Guri, A. (1983). Variation in glutathione and ascorbic acid content among selected cultivars of *Phaseolus vulgaris* prior to and after exposure to ozone. *Can. J. Plant Sci.*, **63**, 733–7.

Heagle, A. S. and J. W. Johnston. (1979). Variable responses of soybeans to mixtures of ozone and sulfur dioxide. *J. Air Pollut. Control Assoc.*, **29**, 729.

Houston, D. B. and G. R. Stairs. (1973). Genetic control of sulfur dioxide and ozone tolerance in eastern white pine. *For. Sci.*, **19**, 267–71.

Irving, P. M., J. E. Miller, and P. B. Xerikos. (1982). The effect of NO₂ and SO₂ alone and in combination on the productivity of field-grown soybeans. *Stud. Environ. Sci.*, **21**, 521–31.

Jones, T. and T. A. Mansfield (1982). Studies on dry matter partitioning and distribution of ¹⁴C-labelled assimilates in plants of *Phleum pratense* exposed to SO₂ pollution. *Environ. Pollut., Ser. A*, **28**, 199–207.

Jordan, B. and L. J. Zaragoza. (1984). Role of pollutant mixture studies in establishing national air quality standards. US Environmental Protection Agency, Res. Dev. [Rep.] EPA-600/3-84-037, 1–15.

Kohut, R. (1985). The effects of SO₂ and O₃ on plants. In *Sulfur dioxide and vegetation*, ed. by W. W. Winner, H. A. Mooney, and R. A. Goldstein, 296–312. Stanford University Press, Stanford.

Kohut, R. J., R. G. Amundson, J. A. Laurence, L. Colavito, P. van Leuken and P. King. (1987). Effects of ozone sulfur dioxide on yield of winter wheat. *Phytopathology*, **77**, 71–4.

Law, R. M. and T. A. Mansfield. (1982). Oxides of nitrogen and the greenhouse atmosphere. In *Proceedings of the 32nd Easter School of Agricultural Science University of Nottingham: Effects of Gaseous Air Pollutants in Agriculture and Horticulture*, ed. by M. H. Unsworth and D. P. Ormrod, 93–112. London, Butterworths Scientific.

Leffler, H. R. and J. H. Cherry. (1974). Destruction of enzymatic activities of corn and soybean leaves exposed to ozone. *Can. J. Bot.*, **52**, 1233–8.

Lefohn, A. S. and D. T. Tingey. (1984). The co-occurrence of potentially phytotoxic concentrations of various gaseous air pollutants. *Atmos. Environ.*, **11**, 2521–6.

Lewis, E. and E. Brennan. (1978). Ozone and sulfur dioxide mixtures cause a PAN-type injury to petunia. *Phytopathology*, **68**, 1011–14.

Mandl, R. H., L. H. Weinstein, and M. Keveny. (1975). Effects of hydrogen fluoride and sulphur dioxide alone and in combination on several species of plants. *Environ. Pollut.*, **9**, 133–43.

Mansfield, T. A. and P. H. Freer-Smith. (1984). The role of stomata in resistance mechanisms. In *Gaseous pollutants and plant metabolism*, ed. by M. J. Koziol and F. R. Whatley, 131–46. London, Butterworths Scientific.

Mansfield, T. A. and T. Jones. (1985). Growth/environment interactions in SO_2 responses of grasses. In *Sulfur dioxide and vegetation*, ed. by W. A. Winner, H. A. Mooney and R. A. Goldstein, 332–46. Stanford University Press, Stanford.

McCune, D. C. (1986). Hydrogen fluoride and sulfur dioxide. In *Adv. Environ. Sci Technol., Vol. 18: Air pollutants and their effects on the terrestrial ecosystem*, ed. by A. Legge and S. Krupa, 305–24. New York, Wiley-Interscience.

McCune, D. C., D. P. Ormrod and R. A. Reinert. (1984). Effects of pollutant mixtures on vegetation. US Environmental Protection Agency, Res. Dev. [Rep.] EPA-600/3-84-037, 46–82.

Mehlhorn, H. and A. R. Wellburn. (1987). Stress ethylene formation determines plant sensitivity to ozone. *Nature (London)*, **327**, 417–18.

Mehlhorn, H., G. Seufert, A. Schmidt, and K. J. Kunert. (1986). Effect of SO_2 and O_3 on production of antioxidants in conifers. *Plant Physiol.*, **82**, 336–8.

Menser, H. A. and H. E. Heggestad. (1966). Ozone and sulfur dioxide synergism: injury to tobacco plants. *Science*, **153**, 424–5.

Miller, C. A. and D. D. Davis. (1981). Response of pinto bean plants exposed to O_3, SO_2, or mixtures at varying temperatures. *HortScience*, **16**, 548–50.

Miller, J. E., D. G. Sprugel, R. N. Muller, H. J. Smith, and P. B. Xerikos. (1980). Open-air fumigation system for investigating sulfur dioxide effects on crops. *Phytopathology*, **70**, 1124–8.

Mudd, J. B., S. K. Banerjee, M. M. Dooley, and K. L. Knight. (1984). Pollutants and plant cells: effects on membranes. In *Gaseous air pollutants and plant metabolism*, ed. by M. J. Koziol and F. R. Whatley, 105–16. London, Butterworths Scientific.

Noyes, R. D. (1980). The comparative effects of sulfur dioxide on photosynthesis and translocation in bean. *Physiol. Plant Pathol.*, **16**, 73–9.

Okano, K., O. Ito, G. Takeba, A. Shimizu and T. Totsuka. (1984). Effects of NO_2 and O_3 alone and in combination on kidney bean plants. V. ^{13}C-assimilate partitioning as affected by NO_2 and/or O_3. *Res. Rep. Natl Inst. Environ. Stud. Jpn*, **66**, 49–57.

Olszyk, D. M. and D. T. Tingey. (1986). Joint action of O_3 and SO_2 in modifying plant gas exchange. *Plant Physiol.*, **82**, 401–5.

Ormrod, D. P. (1982). Air pollutant interactions in mixtures. *Proceedings of the 32nd Easter School of Agricultural Science, University of Nottingham: Effects of Gaseous*

Air Pollutants in Agriculture and Horticulture, ed. by M. H. Unsworth and D. P. Ormrod, 307–31. London, Butterworths Scientific.

Ormrod, D. P., D. T. Tingey, M. L. Gumpertz, and D. M. Olsyzk. (1984). Utilization of a response-surface technique in the study of plant responses to ozone and sulfur dioxide mixtures. *Plant Physiol.,* **75,** 43–8.

Oshima, R. J., J. P. Bennett, and P. K. Braegelmann. (1978). Effect of ozone on growth and assimilate partitioning in parsley. *J. Am. Soc. Hortic. Sci.,* **103,** 348–50.

Oshima, R. J., P. K. Braegelmann, R. B. Flagler and R. R. Teso. (1979). The effects of ozone on the growth, yield, and partitioning of dry matter in cotton. *J. Environ. Qual.,* **8,** 474–9.

Osswald, W. F. and E. F. Elstner. (1986). Fichtenerkrankungen in der Hochlagen der Bayerischen Mittelgebirge. *Ber. Dtsch. Bot. Ges.,* **99,** 313–39.

Osswald, W. F. and E. F. Elstner. (1987). Investigations on spruce decline in the Bavarian forest. *Free Radical Res. Commun.,* **3,** 185–92.

Pande, P. C. and T. A. Mansfield. (1985). Responses of spring barley to SO_2 and NO_2 pollution. *Environ. Pollut., Ser. A,* **38,** 87–97.

Peiser, G. and S. F. Yang. (1985). Biochemical and physiological effects of SO_2 on nonphotosynthetic processes in plants. In *Sulfur dioxide and vegetation,* ed. by W. A. Winner, H. A. Mooney and R. A. Goldstein, 148–61. Stanford University Press, Stanford.

Pratt, G. C., K. W. Kromroy, and S. V. Krupa. (1983). Effects of ozone and sulphur dioxide on injury and foliar concentrations of sulphur and chlorophyll in soybean *Glycine max. Environ. Pollut., Ser. A,* **32,** 91–9.

Reich, P. B. and R. G. Amundson. (1984). Low level O_3 and/or SO_2 exposure causes a linear decline in soybean yield. *Environ. Pollut., Ser. A,* **34,** 345–55.

Reinert, R. A., A. S. Heagle, and W. W. Heck. (1975). Plant responses to pollutant combinations. In *Responses of Plants to Air Pollution,* ed. by J. B. Mudd and T. Kozlowski, 159–77. New York, Academic Press.

Rennenberg, H. (1984). The fate of excess sulfur in higher plants. *Ann. Rev. Plant. Physiol.,* **35,** 121–53.

Runeckles, V. C. (1984). Impact of air pollutant combinations on plants. In *Air pollution and plant life,* ed. by M. Treshow, 239–58. Chichester, John Wiley.

Smirnoff, N., P. Todd, and G. R. Stewart. (1984). The occurrence of nitrate reduction in the leaves of woody plants. *Ann. Bot. (London)* **54,** 363–74.

Teh, K. H. and C. A. Swanson. (1982). Sulfur dioxide inhibition of translocation in bean plants. *Plant Physiol.,* **69,** 88–92.

Thompson, J. E., R. L. Legge, and R. F. Barber. (1987). The role of free radicals in senescence and wounding. *New Phytol.,* **105,** 317–44.

Tingey, D. T. and R. A. Reinert. (1975). The effect of ozone and sulphur dioxide singly and in combination on plant growth. *Environ. Pollut.,* **9,** 117–25.

Tingey, D. T., R. A. Reinert, J. A. Dunning, and W. W. Heck. (1971). Vegetation injury from the interaction of nitrogen dioxide and sulfur dioxide. *Phytopathology,* **61,** 1506–11.

Tingey, D. T., R. A. Reinert, C. Wickliff, and W. W. Heck. (1973). Chronic ozone or sulfur dioxide exposures, or both, affect the early vegetative growth of soybean. *Can. J. Plant Sci.,* **53,** 875–9.

Weinstein, L. H. (1977). Fluoride and plant life. *J. Occup. Med.,* **19,** 49–78.

Wellburn, A. R. (1982). Effects of SO_2 and NO_2 on metabolic function. *Proceedings of the 32nd Easter School of Agricultural Science, University of Nottingham: Effects of Gaseous Air Pollutants in Agriculture and Horticulture,* ed. by M. H. Unsworth and D. P. Ormrod, 169–87. London, Butterworths Scientific.

Wellburn, A. R. (1984). The influence of atmospheric pollutants and their cellular products upon photophosphorylation and related events. In *Gaseous air pollutants and plant metabolism,* ed. by M. J. Koziol and F. R. Whatley, 203–221. London Butterworths Scientific.

Whitmore, M. E. (1985). Relationship between dose of SO_2 and NO_2 mixtures and growth of *Poa pratensis*. *New Phytol.*, **99**, 545–53.

Whitmore, M. E. and P. H. Freer-Smith. (1982). Growth effects of SO_2 and/or NO_2 on woody plants and grasses during spring and summer. *Nature (London)*, **300**, 55–7.

Whitmore, M. E. and T. A. Mansfield. (1983). Effects of long-term exposures to SO_2 and NO_2 on *Poa pratensis* and other grasses. *Environ. Pollut., Ser. A*, **31**, 217–35.

Wright, E. A., P. W. Lucas, D. A. Cottam, and T. A. Mansfield. (1987). Physiological responses of plants to SO_2, NO_x and O_3: implications for drought resistance. CEC COST Workshop, Lökeberg, Sweden. Brussels, Commission of the European Communities.

Yoneyama, T. and H. Sasakawa. (1979). Transformation of atmospheric NO_2 absorbed in spinach leaves. *Plant Cell Physiol.*, **20**, 263–6.

14

DROUGHT STRESS APPLIED DURING THE REPRODUCTIVE PHASE REDUCED OZONE-INDUCED EFFECTS IN BUSH BEAN

THOMAS J. MOSER,[a] DAVID T. TINGEY,[b] KENT D. RODECAP,[a]
DEBRA J. ROSSI[a] and C. SCOTT CLARK[a]

[a] *Northrop Services Inc., Corvallis, Oregon, USA*
[b] *US Environmental Protection Agency, Corvallis, Oregon, USA*

14.1 INTRODUCTION

Plant response to O_3 is influenced by environmental conditions and the stage of plant development. Drought stress is a major factor affecting plant yield and modifies plant response to O_3. Soil moisture stress, especially during the critical periods of flowering and early pod set, reduces yield in legumes (Dubetz and Mahalle, 1969; Maurer *et al.*, 1969; Sionit and Kramer, 1977). Photosynthate allocation to reproductive organs is not substantial until pod initiation (Fischer and Turner, 1978), thus plant stress during reproductive development can suppress yield. Similarly, plant exposure to O_3 during reproductive development reduces yield. Kohut and Laurence (1983) reported that O_3 exposure during the pod-filling period reduced kidney bean yield. Reductions in yield were also reported in soybean exposed to enhanced O_3 concentrations during flowering and pod filling (Kohut *et al.*, 1986).

Field studies have demonstrated that O_3-induced foliar injury was greater on crops grown in moist soils than on those grown in drier soils, and that the intensity of injury was proportional to the amount of irrigation water applied (Walker and Vickery, 1961; Dean and Davis, 1967; Markowski and Grzesiak, 1974; Schwartz *et al.*, 1983). Controlled O_3 exposures in greenhouse chambers confirmed these field observations. Plants that were drought stressed prior to O_3 exposure showed little or no foliar injury compared to well-watered plants (Khatamian *et al.*, 1973; Olszyk and Tibbitts, 1981).

Although the moderating effect of drought stress on acute O_3 injury under both field and greenhouse conditions is well documented, there is little information on potential interactions of drought stress and chronic O_3 exposure on crop growth and yield. In studies conducted in controlled-environment chambers, Amundson *et al.* (1986) found that soybean growth was less impaired by O_3 when plants experienced drought stress. In field studies with three soybean cultivars, Heggestad *et al.* (1985) reported that one cultivar (Corosy) was partially protected by drought stress in the presence of above-ambient O_3 concentrations. However, all three soybean cultivars revealed synergistic yield responses to drought stress when exposed to near ambient O_3 levels. In one year, the growth and yield of field-grown, drought-stressed cotton was less impaired by O_3 than that of well-watered plants, but drought stress did not moderate the O_3 impact in the second year (Temple *et al.*, 1985). Similar year-to-year variations of water-stress influence on plant response to O_3 have been observed in field-grown soybeans (Heagle *et al.*, 1987).

Ozone impacts plant species only after diffusing through stomata (Rich *et al.*, 1970; Tingey *et al.*, 1982; Tingey and Hogsett, 1985). Consequently, environmental factors that influence stomatal behavior are important in controlling plant response. Plant water status, as controlled by soil water availability and evaporative demand, influences stomatal behavior and therefore can be expected to influence plant response to O_3. The mechanism for the drought-stress-induced protection from O_3 has not been clearly explained but appears to involve stomatal control that reduces O_3 uptake. The stomata of drought-stressed plants opened to a lesser degree, closed earlier during the day, and closed more rapidly in the presence of O_3 (MacDowall, 1965; Dean and Davis, 1967; Olszyk and Tibbitts, 1981; Tingey and Hogsett, 1985).

Accepting that stress during the reproductive phase of plant development reduces potential yield and that drought stress, in particular, will influence plant water status and stomatal behavior, information addressing the interaction of drought stress and O_3 on crop growth and yield is required. The objective of this research was to evaluate the effects of a moderate drought stress, applied during flowering and pod-filling phases, on growth and yield response of bush bean to simulated ambient O_3 exposures.

14.2 MATERIALS AND METHODS

14.2.1 Plant culture

Beans (*Phaseolus vulgaris* L. cv Bush Blue Lake 290) were grown in open-top field exposure chambers during the growing seasons of 1984 through 1986. Two crops were grown during each year (mid-May to late July; August through September). Bush beans were sown into polyvinyl chloride sleeves (500 ml volume; three seeds per sleeve) containing Promix BX[1] and were topped with approximately 1 cm of vermiculite to reduce surface evaporation. Seedlings were maintained in a greenhouse until 7 to 8 days after sowing, then they were thinned to one plant per sleeve and transferred into plant cultural systems (Fig. 14.1) in open-top field exposure chambers. The plant cultural systems incorporate the water-table, root-screen method for controlling plant water status, in which roots are constrained with Nitex[1] nylon cloth (20-μm mesh openings) above a uniform conductive medium (Smither's Oasis,[1] a commercial florist's foam) and a water table of known height.

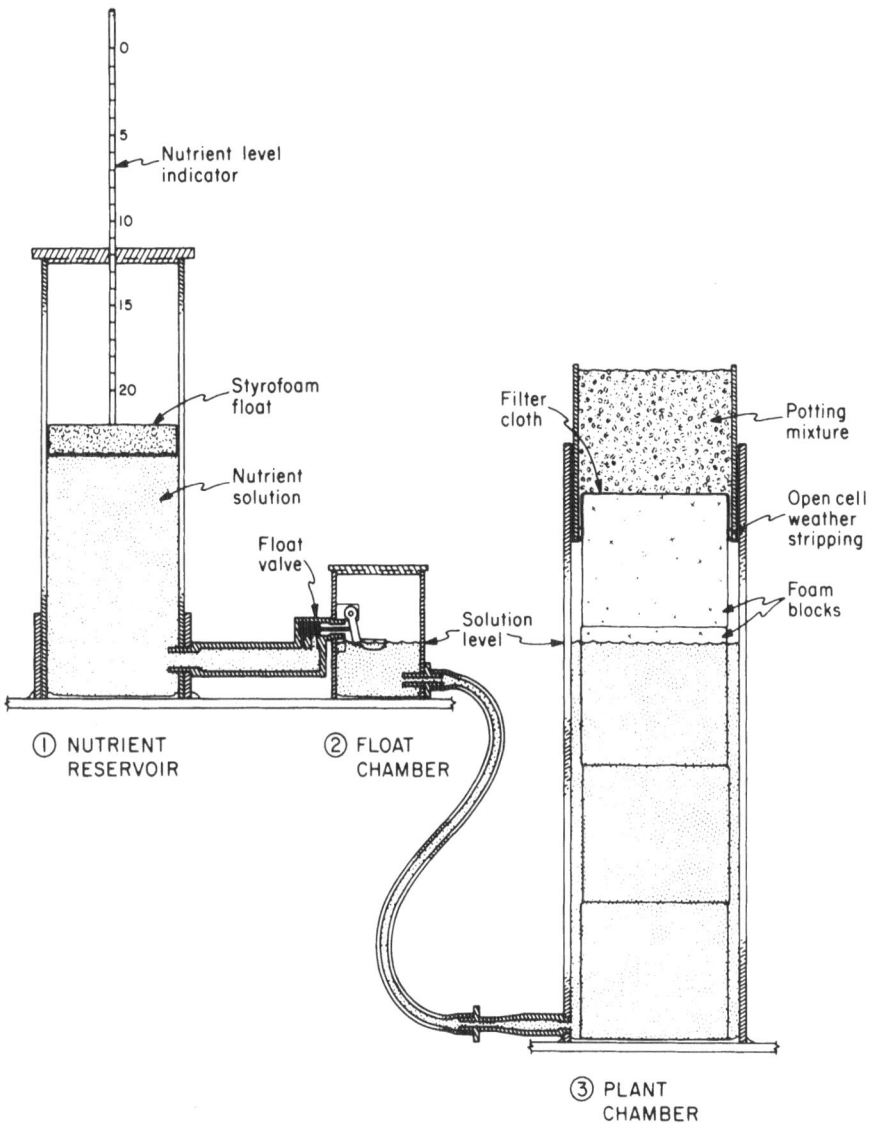

Fig. 14.1. Cross-sectional illustration of the plant cultural system used to control drought stress. In these field studies, the nutrient reservoirs were located on adjustable platforms outside the open-top chambers. Additional details of the plant cultural system are contained in Snow and Tingey (1985) and Tingey *et al.* (1987).

Drought stress is controlled by (1) the hydraulic conductivity of the florist's foam, (2) the distance between the roots and the water table, and (3) the transpiration rate of the plant (Snow and Tingey, 1985; Tingey *et al.*, 1987). Bush beans grown in cultural systems and maintained in open-top field chambers displayed typical physiological responses to changing environmental conditions and grew and matured normally (Tingey *et al.*, 1987). The cultural system not only controlled plant water status, but also provided the roots with a constant supply of North Carolina State University Phytotron nutrient solution (Downs and Hellmers, 1975). The volume of nutrient solution used by the plants was monitored daily. The plants were grown to maturity in open-top exposure chambers that were modified with a rain exclusion cap 30 cm above the chamber top (Hogsett *et al.*, 1985).

14.2.2 Experimental measurements

Bush beans were harvested when approximately 50% of the pods on the control plants were Commercial Grade 4 or larger. At harvest, the plants were separated into leaves, stems, roots, and pods. The pods from each plant were separated by Commercial Grades 1 to 5 (Duncan *et al.*, 1960) and then counted. Plant tissues were dried to constant weight in a forced-draft oven (72 h at 70°C), and then the oven-dry weights of leaves, stems, roots, and pods/commercial grade were measured. Total plant leaf area was measured with a LI-COR 3100 area meter. Foliar injury (on a per plant basis) was estimated weekly. To estimate leaf abscission, the number of leaves on each plant was counted weekly. In addition, the number of days to anthesis was monitored for each plant. For one year's growing season (two crops), the number of flowers, pods, and racemes was monitored each week. For all six crops, diurnal leaf water potentials were measured approximately every second day on the primary, first, and fourth trifoliate leaves with L-51A-SF leaf hygrometers using a Wescor HP-115 microprocessor-controlled microvolt-meter. The hygrometers were insulated according to the methods of Brown and Tanner (1981) and Savage *et al.* (1983). Leaf hygrometers were calibrated immediately before and after each year's study.

14.2.3 Drought stress treatments

Three drought stress treatments were used: (1) no drought stress (NDS), (2) early reproductive drought stress (ERDS), and (3) late reproductive drought stress (LRDS). Each drought stress treatment included three to four plants and was replicated in two chambers for each O_3 treatment during each crop's growing season. In the NDS treatment, the water table was maintained approximately 3 cm below the root–screen interface throughout the season. The ERDS and LRDS treatments, each approximately 14 days in duration, were applied at anthesis and at approximately pod fill, respectively, by lowering the water table from 3 to 16 cm below the root–screen interface. At the same time that the water table level of ERDS-treatment plants was returned to the control level, the water table of LRDS plants was lowered and maintained until harvest.

14.2.4 Ozone exposure treatments

Bush beans were exposed to three O_3 treatments (control and two episodic regimes) in open-top field exposure chambers. There were two chamber replicates for each O_3 treatment during each crop's growing season. To simulate the episodic occurrence of O_3, the procedures of Lefohn *et al.* (1986) were used to construct exposure regimes that incorporated the temporal concentration dynamics and frequency of O_3 occurrence that is typical of rural areas in the midwestern United States. Examples of low and high exposure regimes (30 days in duration) in the 1985 and 1986 replicates are illustrated (Fig. 14.2). The 30-d regimes were repeated until plant harvest. The high-episodic exposure regime used in 1984 (Tingey *et al.*, 1986) had a greater mean and seasonal O_3 dose than the high-episodic regime used in 1985 and 1986.

Ozone generation, exposure control, air sampling, quality assurance, and data acquisition procedures have been previously described (Hogsett *et al.*, 1985). Chamber O_3 concentrations were monitored at canopy height with UV photometers. Prior to use, the photometers were calibrated with a transfer standard (McElroy, 1979) over the range of requested O_3 concentrations for the study. The transfer standard was calibrated with a dedicated UV standard operated and calibrated in accordance with the procedures of Paur and McElroy (1979). Both standards were subjected to periodic performance audits, and weekly zero and span checks were performed on the monitors to ensure accuracy. At least once during each crop growing season, the O_3 loss within the sample line of each chamber was determined. Because O_3 loss was minimal (typically within 4 to 5%), no correction was applied to the data.

The O_3 exposure indices for each treatment and replicate are listed in Table 14.1. Because there is no single, universally accepted index to describe O_3 exposures, several seasonal means as well as the cumulative exposure and an O_3 exposure index (each hourly O_3 concentration was raised to the exponential power 2·5 and summed) are listed for comparison. This O_3 index was selected because it cumulates the concentration and places greater emphasis on peak concentrations, which are more important biologically in characterizing plant responses to O_3 (US Environmental Protection Agency [EPA], 1986). The exposure index is identical to the Impact Index proposed by Larsen *et al.* (1983), except that the concentration was raised to the 2·5 power.

14.2.5 Data analysis

The data from the six crops were combined and analyzed using a regression approach, utilizing a randomized block design with a split-plot structure (SAS, 1985). Crops were treated as blocks, while chambers and drought stress treatments were treated as main plots and subplots, respectively. To achieve the proper error terms, chamber/replication means were analyzed for O_3 effects, while chamber/replication × drought stress means were analyzed for drought stress and O_3 × drought stress effects. The O_3 model included factors for crops, O_3 (linear), and a quadratic O_3 term, when significant. The drought stress and interaction model included terms for crop growing season, chamber/replication (includes linear and quadratic O_3 terms), drought stress, and O_3 × drought stress. Bonferroni multiple comparison techniques were used to assess significance levels for more specific treatment comparisons. Prior to analysis, data were transformed to correct for unequal variances. A natural log transformation was used on plant growth and yield variables, a square-root transformation was used on pod and leaf number variables, and an

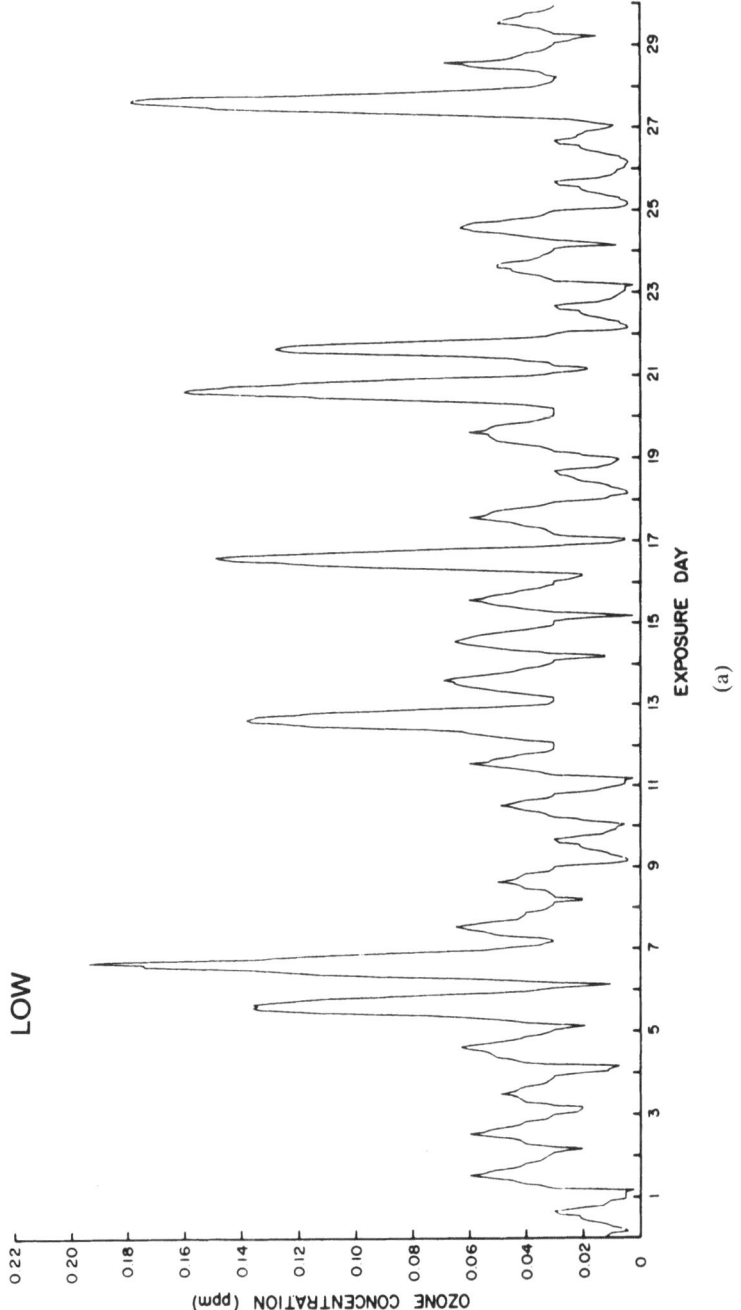

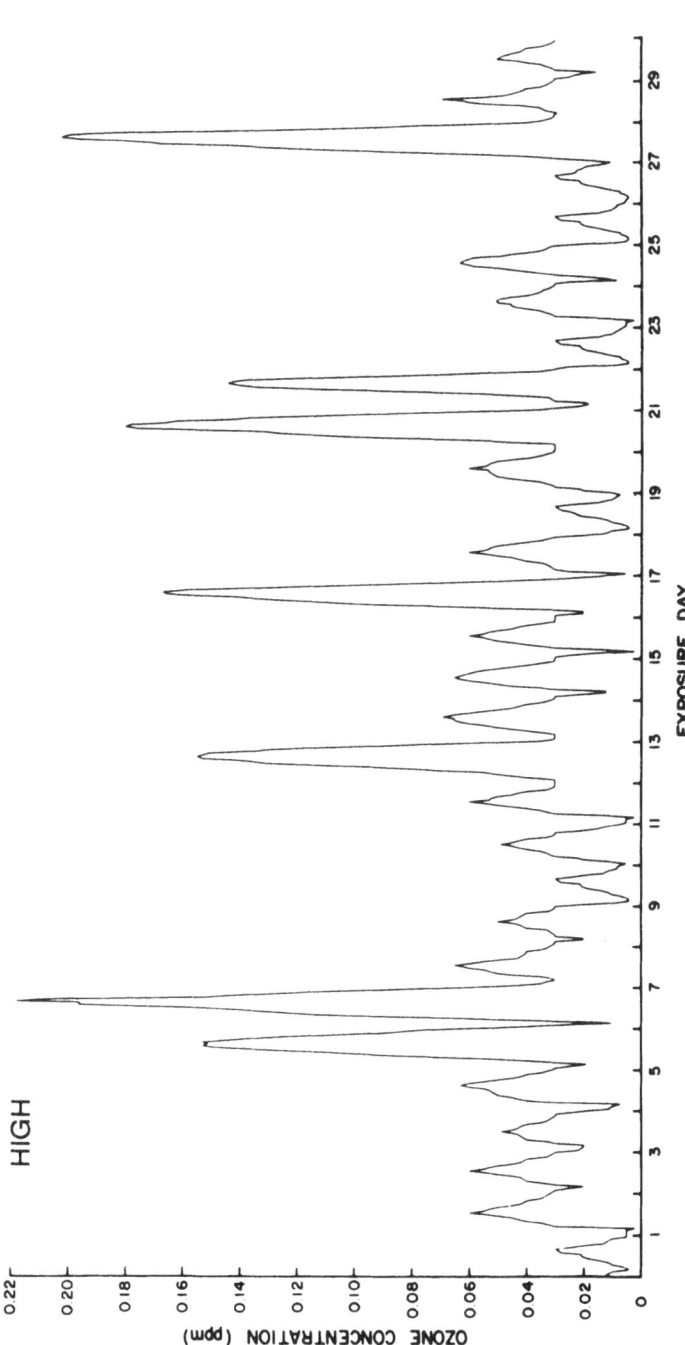

Fig. 14.2. (a) The low- and (b) the high-episodic-O_3 exposure regimes used in the drought stress/O_3 interaction studies on bush bean during the 1985 and 1986 growing seasons. The high-episodic exposure profile used in 1984 had a greater mean and seasonal dose than the high-episodic exposure profile used in 1985 and 1986 (refer to Table 14.1 for comparison of O_3 exposure statistics between the growing season replications). The 30-day exposure regime was repeated to attain the necessary exposure duration.

TABLE 14.1

Ozone exposure summary statistics (ozone values averaged over chamber replication for each year-season combination

Ozone treatment	Year-season	7-h seasonal mean (ppm)	12-h seasonal mean (ppm)	24-h seasonal mean (ppm)	Cumulative (ppm)	Ozone exposure index[a] (ppm)	Days of plant exposure
Control	1984-early	0·007	0·008	0·005	9·2	0·010	66
	1984-late	0·006	0·006	0·004	6·4	0·004	64
	1985-early	0·007	0·008	0·006	8·0	0·009	62
	1985-late	0·009	0·009	0·007	10·8	0·023	62
	1986-early	0·015	0·018	0·016	23·3	0·082	59
	1986-late	0·012	0·013	0·011	15·7	0·063	59
Low episodic	1984-early	0·067	0·065	0·050	79·7	1·918	66
	1984-late	0·055	0·055	0·041	62·7	1·107	64
	1985-early	0·063	0·059	0·043	63·6	1·371	62
	1985-late	0·064	0·061	0·044	63·2	1·479	62
	1986-early	0·062	0·060	0·044	61·9	1·599	59
	1986-late	0·052	0·051	0·038	52·8	1·099	59
High episodic	1984-early	0·091	0·088	0·068	108·0	3·266	66
	1984-late	0·085	0·083	0·063	96·6	2·657	64
	1985-early	0·063	0·061	0·044	65·8	1·687	62
	1985-late	0·065	0·062	0·046	67·4	1·777	62
	1986-early	0·068	0·066	0·051	72·6	1·969	59
	1986-late	0·054	0·054	0·040	56·6	1·332	59

[a] Ozone exposure index was calculated by raising each hourly O_3 concentration to the exponential power 2·5 and then summing these values.

arc sine square-root transformation was used on proportion and percent foliar injury variables.

14.3 RESULTS

14.3.1 Plant growth and yield

In the NDS treatment, O_3 significantly ($p < 0.001$) decreased pod, leaf, stem, and root biomass (Fig. 14.3). In NDS plants, O_3 reduced pod yield to a greater degree than the other plant tissues. For example, at a mean O_3 exposure index of 1·8 (approximates a cumulative seasonal O_3 exposure of 73 ppm-h), pod yield was reduced an average of 54·6%, while leaf, stem, and root biomass were reduced an average of 37·8, 30·1 and 40·3%, respectively. Yield reductions were directly related to a decrease in the number of pods produced ($p < 0.001$) as well as the production of fewer commercial Grade 5 pods ($p < 0.0001$) (Table 14.2). Ozone-induced yield reductions were not the result of delayed anthesis, nor a reduction in the number of nodes per plant or a reduction in the rate of flower formation (data not presented). However, prior to harvest, O_3-treated plants had fewer pods (data not presented), which

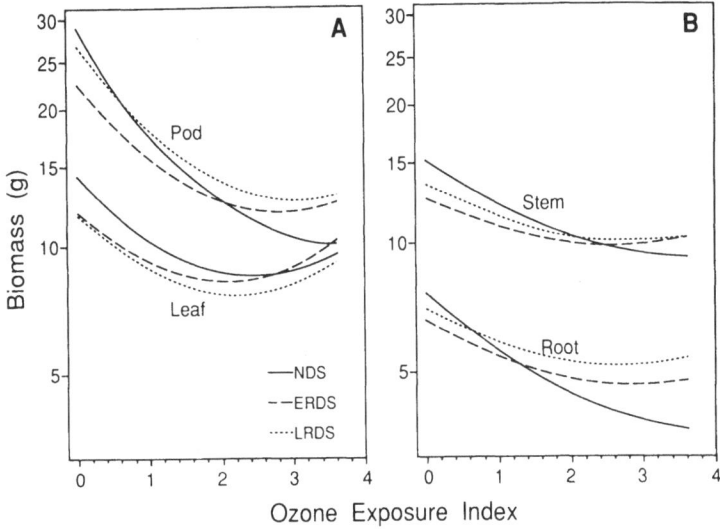

Fig. 14.3. Predicted response of bush bean biomass under three drought stress treatments as a function of the O_3 exposure index. R-squared values for the regression models for (A) pod and leaf biomass (B) stem and root biomass as a function of the O_3 exposure index were 0·88, 0·86, 0·86, and 0·76, respectively.

could have been the result of reduced success in formation of pods from existing flowers or of increased pod abortion.

Both drought stress treatments reduced plant growth and pod yield (Fig. 14.3). Diurnal measurements of leaf water potential revealed that plants subjected to drought stress during their reproductive development were under moderate stress. Midday leaf water potentials (mean (standard error)) over the six crops were −0·59 (0·07), −0·86 (0·10), and −1·25 (0·14) MPa for the NDS, ERDS, and LRDS treatments, respectively; while predawn leaf water potentials (mean (standard error)) were −0·27, (0·03), −0·53 (0·09), and −0·48

TABLE 14.2

The effects of ozone and drought stress on the predicted number of commercial grade pods

Ozone exposure index	Total Pods (Grade 1–5)[a]				Grade 5 Pods			
	NDS	ERDS	LRDS	SE[b]	NDS	ERDS	LRDS	SE
Control	40·3	30·0	33·8	(2·5)	21·8	17·5	14·4	(1·5)
0·9	32·8	26·5	28·7	(1·7)	15·5	14·5	12·4	(1·0)
1·8	27·4	24·2	25·4	(1·8)	10·4	11·7	10·6	(1·1)
2·6	23·4	23·1	23·0	(2·7)	6·6	9·5	9·1	(1·7)

[a] NDS = no drought stress; ERDS = early reproductive drought stress; LRDS = late reproductive drought stress.
[b] The standard error (SE) for the means associated with the three drought stress treatments within each ozone exposure index are listed in parentheses.

(0·05) MPa for the NDS, ERDS, and LRDS treatments, respectively. Drought stress during the early reproductive phase reduced plant growth and yield more than did stress during the late reproductive phase (Fig. 14.3). In the absence of O_3, the ERDS treatment reduced pod, leaf, stem, and root biomass an average of 22% ($p = 0·01$), 16% ($p = 0·12$), 17% ($p = 0·03$), and 13% ($p = 0·09$), respectively; while the LRDS treatment reduced pod, leaf, stem, and root biomass an average of 9% ($p = 0·39$), 18% ($p = 0·09$), 12% ($p = 0·15$) and 8% ($p = 0·31$), respectively.

The moderate levels of drought stress during the reproductive phase reduced the impact of O_3 on bush bean growth and yield. As illustrated (Fig. 14.3), the regression lines relating the various plant tissue biomasses, as a function of O_3, were less negative for ERDS and LRDS treatment plants than for the NDS treatment plants. Overall, the ERDS treatment was more effective (the slope was flatter) than the LRDS treatment in reducing the effects of O_3. For example, at the O_3 exposure index of 1·8 ppm, the ERDS treatment reduced the effect of O_3 on pod, leaf, stem, and root biomass an average of 12% ($p = 0·06$), 8% ($p = 0·35$), 10% ($p = 0·15$), and 13% ($p = 0·04$), respectively; whereas, the LRDS treatment reduced the effect of O_3 on pod, leaf, stem, and root biomass an average of 9% ($p = 0·18$), 4% ($p = 0·57$), 9% ($p = 0·26$), and 18% ($p = 0·02$), respectively.

14.3.2 Leaf area, foliar injury and leaf abscission
At harvest, total plant leaf area was significantly reduced ($p = 0·0001$) by O_3. For example, at an O_3 exposure index of 1·8, NDS plants exhibited a 32% reduction in leaf area when compared to nonfumigated plants (Fig. 14.4). Both

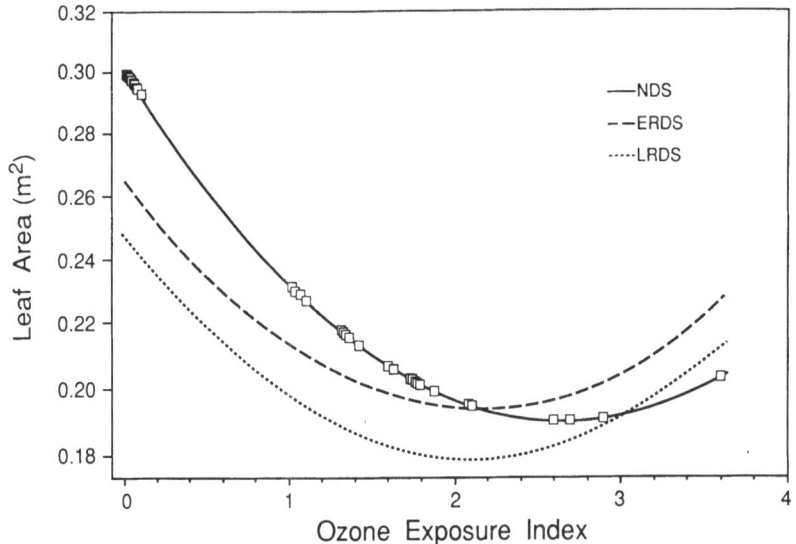

Fig. 14.4. Predicted response of bush bean total leaf area at harvest under three drought stress treatments as a function of the O_3 exposure index. The *R*-squared value for the regression model of total plant leaf area as a function of the O_3 exposure index was 0·84. Points are shown along the NDS curve in this figure to illustrate the location of the O_3 exposure index values used to generate the lines in Figs 14.3, 14.4, and 14.7.

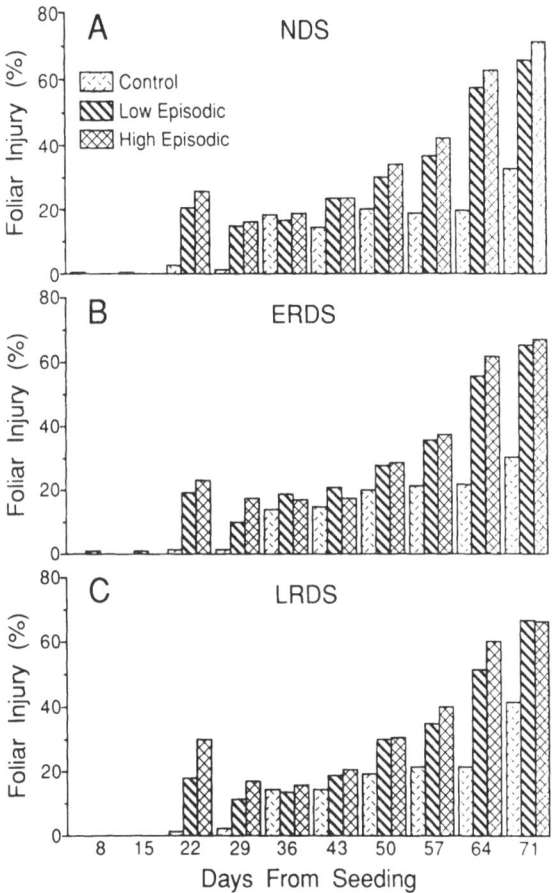

Fig. 14.5. Canopy foliar injury of bush beans under the three drought stress treatments (A) NDS, (B) ERDS, (C) LRDS in response to control, low- and high-episodic-O_3 exposure profiles during an early growing season replication.

the ERDS and LRDS treatments moderated, to a similar degree, the O_3-induced reductions in total plant leaf area.

Ozone exposure resulted in characteristic foliar injury symptoms, while the nonfumigated plants displayed only normal chlorosis and necrosis reflective of leaf aging. Ozone-induced foliar injury typically diverged from nonfumigated plants shortly following anthesis (45–50 d after seeding), and increased in intensity during reproductive development (Fig. 14.5). Ozone injury was not significantly reduced by either reproductive-phase drought stress treatment (Fig. 14.5). The O_3-induced reduction in total plant leaf area was, in part, related to increased leaf abscission at elevated O_3 levels. As illustrated in Fig. 14.6, the rate of leaf abscission in NDS treatment plants typically began diverging from nonfumigated controls at approximately 45 to 50 days from seeding, which corresponds closely with the occurrence of anthesis. When compared to nonfumigated plants, leaf abscission at final harvest was approximately two times greater in the plants exposed to the high-episodic-O_3 regime.

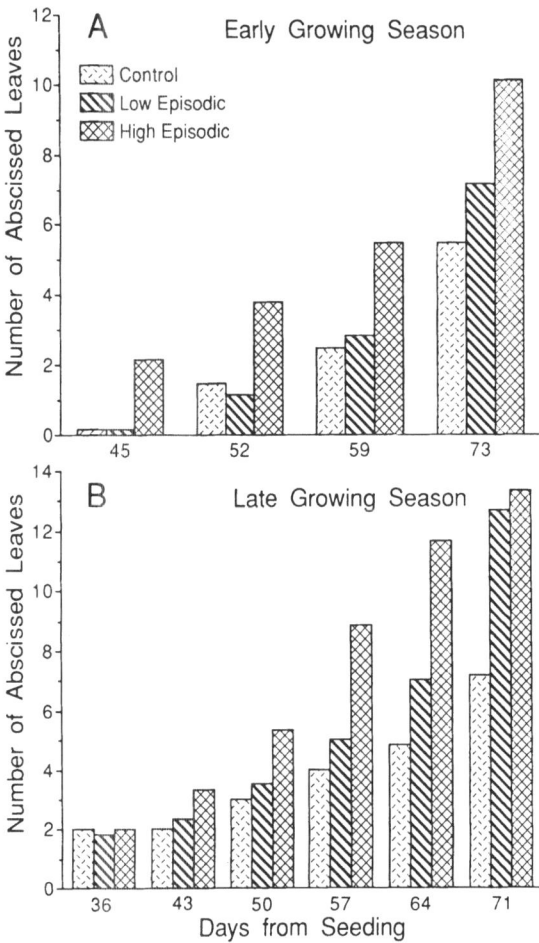

Fig. 14.6. Leaf abscission of nondrought-stressed bush beans in response to control, low- and high-episodic-O_3 exposure profiles during (A) an early and (B) a late growing season replication.

14.3.3 Dry-matter partitioning

In addition to the effects on growth and yield, O_3 also altered dry-matter partitioning (Fig. 14.7). In NDS treatment plants, the proportion of dry matter in the pods decreased ($p = 0.0001$) with increasing O_3, while the proportion of dry matter in the leaf ($p < 0.03$) and stem ($p = 0.001$) tissue increased. At an O_3 exposure index of 1·8, the proportion of dry matter in the NDS treatment plant pods was reduced an average of 7·9%, while dry-matter partitioning to the leaf and stem tissues was increased an average of 2·5 and 5·5%, respectively. Minimal changes in dry-matter partitioning to the roots were observed with increasing O_3. As illustrated in the regression lines, drought stress reduced the impact of O_3 on tissue dry-matter partitioning. As with plant growth and biomass effects, the ERDS treatment appeared to be more

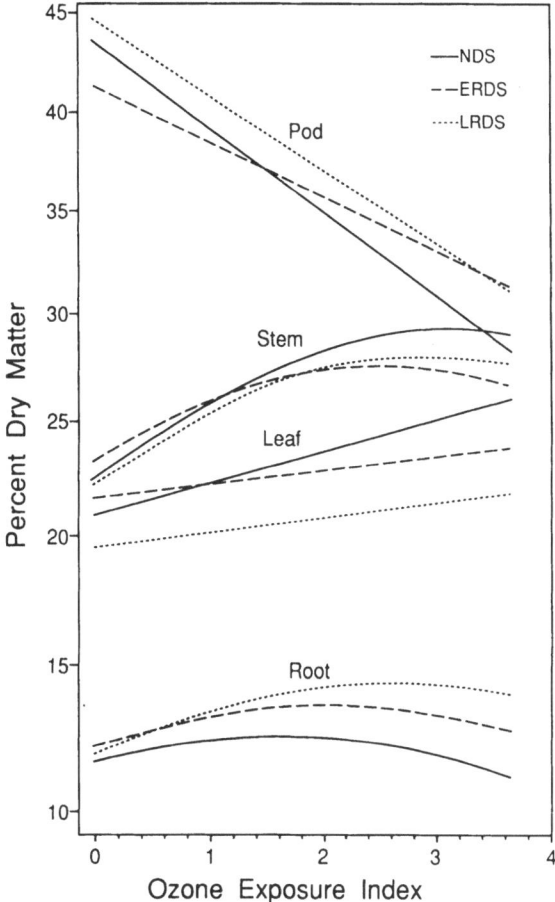

Fig. 14.7. Predicted response of dry-matter partitioning in bush bean under three drought stress treatments as a function of the O_3 exposure index. *R*-squared values for percent dry matter partitioning in pod, leaf, stem, and root tissue as a function of the O_3 exposure index were 0·82, 0·81, 0·82 and 0·76, respectively.

effective than the LRDS treatment in reducing O_3 effects, especially in the pod ($p = 0·18$) and stem ($p = 0·06$) tissues.

Ozone also affected dry-matter distribution among the five commercial pod grades (Table 14.3). In the NDS plants, increasing O_3 significantly increased the amount of dry matter in the smaller grades ($p < 0·002$) and Grade 4 pods ($p < 0·02$), while decreasing the amount of dry matter in Grade 5 pods ($p < 0·02$). Both reproductive-phase drought stress treatments reduced the effects of O_3 on pod dry-matter distribution, especially in the larger pod grades. For example, dry-matter distribution in Grade 4 pods of the LRDS plants was less affected by increasing O_3 than in NDS or ERDS plants ($p < 0·01$ and $p = 0·12$, respectively). However, dry-matter distribution in Grade 5 pods of the LRDS plants was less affected by increasing O_3 than in NDS or ERDS plants ($p < 0·01$ and $p = 0·14$, respectively).

358

TABLE 14.3

The effects of ozone and drought stress on predicted percent dry-matter distribution among commercial grade pods

Ozone exposure index	Grade 1–3 Pods[a]				Grade 4 Pods				Grade 5 Pods			
	NDS	ERDS	LRDS	SE[b]	NDS	ERDS	LRDS	SE	NDS	ERDS	LRDS	SE
Control	3·9	2·5	5·2	(1·7)	11·1	10·1	19·9	(4·5)	85·0	87·4	74·9	(5·4)
0·9	6·9	4·6	6·8	(1·1)	17·4	12·8	20·3	(3·0)	75·7	82·6	72·9	(3·5)
1·8	9·7	6·4	8·1	(1·2)	24·1	15·8	20·1	(3·3)	66·2	77·8	71·8	(3·9)
2·6	11·8	7·4	8·4	(1·8)	31·6	19·0	20·0	(5·0)	56·6	73·6	71·6	(5·9)

[a] NDS = no drought stress; ERDS = early reproductive drought stress; LRDS = late reproductive drought stress.
[b] The standard error (SE) for the means associated with the three drought-stress treatments within each ozone exposure index are listed in parentheses.

14.4 DISCUSSION

Legumes are known to be very sensitive to O_3 (US EPA, 1986) and the common bean is one of the most O_3-sensitive ones (Hill *et al.*, 1970). Bush Blue Lake (BBL 290), grown primarily for processing purposes in the northwest United States, is among the most O_3-sensitive of the bean cultivars (Butler and Tibbitts, 1979; Heggestad *et al.*, 1980). Ozone-induced yield reductions have been reported in legumes after chronic exposures were conducted in the field (Kohut and Laurence, 1983; Reich and Amundson, 1984; Heggestad *et al.*, 1985; Kohut *et al.*, 1986; Heagle *et al.*, 1987). Yield reductions in response to increased O_3 have been generally associated with pollutant impacts on various yield components: (1) reduction in the total number of pods per plant, (2) decrease in the number of filled pods per plant, (3) fewer seeds per pod, and (4) reduction in pod seed weight (Endress and Grunwald, 1985). In open-top field chamber studies conducted over a 3-year period, BBL 290 was the most sensitive cultivar to ambient levels of photochemical oxidants, with an average reduction in yield of 14% (Heggestad *et al.*, 1980).

Bush bean yield is determined by the number of pods per plant and the weight per pod. These yield components are primarily dependent on (1) the number of nodes per plant, which potentially determine the number of inflorescences produced, (2) the rate of flower and pod formation and abortion, (3) net seasonal photosynthesis by the plant canopy, and (4) allocation of photosynthate to reproductive organs.

In the present study, O_3 reduced pod biomass more than leaf, stem, and root biomass. The greater sensitivity of pod yield to O_3 may be related to a greater sensitivity to stress during reproductive development. Hass (1970) and Toivonen *et al.* (1982) reported that O_3 injury in white bean coincided with the onset of flowering. Preliminary data with BBL 290 demonstrated that plants with flowers removed developed significantly less O_3-induced injury than plants that were allowed to develop normally and produce pods.

Ozone-induced yield reductions were, in part, the direct result of a decrease in the total number of pods present at harvest. It was observed that fewer pods were present 2 to 3 weeks prior to plant harvest, yet it is uncertain whether the reduction in total harvested pods was a result of a decrease in pod set or an increase in pod abortion. Ozone did not delay the date of anthesis, nor did it reduce the number of nodes per plant or the rate of flower formation. Although low O_3 concentrations (0·05 to 0·10 ppm (131 to 262 $\mu g\,m^{-3}$)) have been reported to impair the fertilization process in tobacco (Feder, 1968) and corn (Mumford *et al.*, 1972), O_3 is not believed to have impaired fertilization in bush bean. *Phaseolus vulgaris* is a highly self-pollinated species, where pollination occurs prior to flower opening during the white bud stage (Webster *et al.*, 1977). Thus, the unopened flower is likely to inhibit O_3 diffusion into the flower during the period of fertilization.

In addition to fewer pods, O_3 also affected pod maturation as reflected by the reduction in the number of Grade 5 pods produced. These results are consistent with those of Kohut and Laurence (1983) who reported that O_3-induced yield reductions in field-grown red kidney beans were due to a reduction in the number of pods produced and a decrease in the number of mature pods.

The observed yield reductions are believed to be the result of a reduction in photosynthetic rate and/or canopy photosynthetic surface area and the alteration of normal partitioning of dry matter among plant organs. Other investigators have shown that O_3 reduces crop photosynthesis (Hill and Littlefield, 1969; Reich and Amundson, 1985; Reich et al., 1986). Leaf photosynthesis of greenhouse-grown bush beans (BBL 290) was reduced 15% after an 11-h exposure to $0·10$ ppm O_3 (McLaughlin and McConathy, 1983). In the present study, O_3 concentrations equal to or greater than $0·10$ ppm occurred approximately 8 and 14% of the time over the growing season in the low and high episodic treatments, respectively. In growth chamber studies, whole-plant photosynthesis was reduced 10% in soybeans after an 8-week exposure ($6·8 \, h \, d^{-1}$) to $0·05$ ppm O_3 (Reich et al., 1986).

A reduction in canopy photosynthetic surface area will also result in less photosynthate assimilated for plant growth and storage. In the present study, an O_3 exposure index value of $1·8$ (an approximate cumulative seasonal O_3 exposure of 73 ppm) decreased total leaf area at harvest by 32%. This O_3-induced reduction in leaf area may be the result of impaired leaf expansion, producing smaller leaves, or increased foliar injury, resulting in premature leaf abscission. Shortly following anthesis (approximately 50 days from seeding), plants in the high episodic O_3 treatment revealed an approximate 20% increase in plant foliar injury and a loss of two to three leaves due to premature abscission. The combination of these effects reduced viable canopy photo-synthetic surface area, and resulted in a reduction of potential photosynthate available for yield. Hofstra et al. (1978) demonstrated that the antioxidant ethylene-diurea (EDU) when applied after peak flowering decreased O_3-induced foliar injury and delayed leaf abscission, resulting in a 36% increase in navy bean yield.

Ozone may also affect yield by altering the partitioning of dry matter between plant parts (Oshima et al., 1979; Okano et al., 1984; Endress and Grunwald, 1985). During vegetative growth, much of the photosynthate is allocated to roots. Okano et al. (1984) reported that O_3 increased the proportion of photosynthate partitioned to the growing leaves at the expense of the root and stem in 14-day-old kidney beans. In contrast, the present study with mature plants revealed substantial increases in dry-matter partitioning to the leaves and stems with increasing O_3, but minimal alteration of dry-matter partitioning to the roots.

The onset of flowering in plants is characterized by shifts in photosynthate allocation patterns that generally favor the development of reproductive organs. Expanding leaves are the primary source of photosynthates for reproductive organs. Once flowering occurs, a greater proportion of photo-synthate is allocated to the reproductive organs. Ozone altered the normal allocation pattern of photosynthate in bush beans, because there was an increased retention of photosynthate by the leaves at the expense of allocation to the pods. In addition to the increased retention by leaves, photosynthate was also redirected to the stems. At an O_3 exposure index of 1·8, dry-matter allocation to the pods was reduced an average of 7·9%, while the fraction of dry matter in the leaves and stems increased an average of 2·5% and 5·5%, respectively. McLaughlin and McConathy (1983) reported that retention of ^{14}C-labeled photosynthate in bush bean (BBL 290) leaves increased while plants were exposed to O_3. This increased retention of assimilates by leaves was accompanied by decreased allocation to pods. The cause of altered allocation patterns of photosynthate in O_3-stressed plants is unclear. However, McLaughlin and McConathy (1983) suggested three possible mechanisms: (1) decreased translocation due to physical or biochemical blockage of the phloem transport system, (2) increased retention in foliage for repair of injured leaf tissue, and (3) reduction in photosynthetic rate, resulting in a greater demand by leaves for the limited pool of photosynthate produced. Another explanation for the accumulation of dry matter in the leaves and stems of O_3-exposed beans is that the reproductive sink for photosynthate was less demanding due to the production of fewer pods on these plants. Ciha and Brun (1978) demonstrated that leaf and stem nonstructural carbohydrate concentrations increased in soybeans with their pods removed.

Moderate drought stress during the early reproductive stage reduced the effects of O_3 on bush bean growth, yield, and dry-matter partitioning more than drought stress during the late reproductive stage. The reduced effect of O_3 provided by drought stress during bush bean reproductive development is assumed to be the result of decreased stomatal conductance, which reduces O_3 uptake (Rich and Turner, 1972; Tingey and Hogsett, 1985). Preliminary transpiration data from the present study substantiate this assumption, as transpiration rates in drought-stressed plants were reduced relative to the nondrought-stressed plants during the last 3 or 4 weeks of the growing season. In addition, the differences between the ERDS and LRDS treatments in moderating the effects of O_3 relative to the NDS treatment illustrate the importance of the stage of plant development when drought stress occurs. Drought stress/O_3 interaction studies conducted in open-top exposure chambers with cotton (Temple *et al.*, 1985) and soybean (Heagle *et al.*, 1987) revealed large year-to-year variation in yield responses to O_3 and drought stress. These variations are likely the consequence of a lack of control of drought stress due to the irrigation methods used in these studies, thus resulting in stress occurring at slightly different plant developmental stages. The plant culture system used in the present bush bean study allows for greater reproducibility and control of drought stress at specific plant developmental stages.

ACKNOWLEDGEMENTS

The authors gratefully acknowledge Dr William E. Hogsett, Steven R. Holman, and Grady E. Neely for their design and operation of the Field

Exposure Research Facility. We also recognize the efforts of Emily H. Bates, Laura W. Joos and Casey D. Tillery for assisting in the collection of experimental data and Dr E. Henry Lee for statistical consultation. The work upon which this publication is based was performed pursuant to contract No. 68-03-3246 with the US Environmental Protection Agency.

NOTE

1. Mention of trade or commercial products does not constitute endorsement or recommendation for use.

REFERENCES

Amundson, R. G., R. M. Raba, A. W. Schoettle and P. B. Reich. (1986). Response of soybean to low concentrations of ozone: II. Effects on growth, biomass allocation, and flowering. *J. Environ. Qual.*, **15**, 161–7.

Brown, P. W. and C. B. Tanner. (1981). Alfalfa water potential measurement: A comparison of the pressure chamber and leaf dew-point hygrometers. *Crop Sci.*, **21**, 240–4.

Butler, L. K. and T. W. Tibbitts. (1979). Variation in ozone sensitivity and symptom expression among cultivars of *Phaseolus vulgaris* L. *J. Am. Soc. Hortic. Sci.*, **104**, 208–10.

Ciha, A. J. and W. A. Brun. (1978). Effect of pod removal on nonstructural carbohydrate concentration in soybean tissue. *Crop Sci.*, **18**, 773–6.

Dean, C. R. and D. R. Davis. (1967). Ozone and soil moisture in relation to the occurrence of weather fleck on Florida cigar-wrapper tobacco in 1966. *Plant Dis. Rep.*, **51**, 72–5.

Downs, R. J. and H. Hellmers. (1975). *Environmental and the experimental control of plant growth*, 145. New York, Academic Press.

Dubetz, S. and P. S. Mahalle. (1969). Effect of soil water stress on bush beans *Phaseolus vulgaris* L. at three growth stages. *J. Am. Soc. Hortic. Sci.*, **94**, 479–81.

Duncan, A. A., R. W. Every, and I. C. MacSwan. (1960). Commercial production of bush snap beans in Oregon. Extension Bulletin 787, 1–20. Oregon State College, Corvallis, OR, Federal Cooperative Extension Service.

Endress, A. G. and C. Grunwald. (1985). Impact of chronic ozone on soybean growth and biomass partitioning. *Agric. Ecosys. Environ.*, **13**, 9–23.

Feder, W. A. (1968). Reduction in tobacco pollen germination and tube elongation, induced by low levels of ozone. *Science*, **160**, 1122.

Fischer, R. A. and N. C. Turner. (1978). Plant productivity in the arid and semiarid zones. *Ann. Rev. Plant Physiol.*, **29**, 277–317.

Hass, J. H. (1970). Relation of crop maturity and physiology to air pollution incited bronzing of *Phaseolus vulgaris*. *Phytopathology*, **60**, 407–10.

Heagle, A. S., R. B. Flagler, R. P. Patterson, V. M. Lesser, S. R. Shafer and W. W. Heck. (1987). Injury and yield response of soybean to chronic doses of ozone and soil moisture deficit. *Crop Sci.*, **27**, 1016–24.

Heggestad, H. E., A. S. Heagle, J. H. Bennett, and E. J. Koch. (1980). The effects of photochemical oxidants on the yield of snap beans. *Atmos. Environ.*, **14**, 317–26.

Heggestad, H. E., T. J. Gish, E. H. Lee, J. H. Bennett, and L. W. Douglass. (1985). Interaction of soil moisture stress and ambient ozone on growth and yields of soybeans. *Phytopathology*, **75**, 472–7.

Hill, A. C. and N. Littlefield. (1969). Ozone. Effect on apparent photosynthesis, rate of transpiration, and stomatal closure in plants. *Environ. Sci. Technol.*, **3**, 52–6.

Hill, A. C., H. E. Heggestad, and S. N. Linzon. (1970). Ozone. In *Recognition of air pollution injury to vegetation: A pictorial atlas*, ed. by J. C. Jacobson and A. C. Hill, B1–B32. Pittsburgh, Air Pollution Control Association.

Hofstra, G., D. A. Littlejohns, and R. T. Wukasch. (1978). The efficacy of the antioxidant ethylene-diurea (EDU) compared to carboxin and benomyl in reducing yield lossed from ozone in navy bean. *Plant Dis. Rep.*, **62**, 350–2.

Hogsett, W. E., D. T. Tingey, and S. R. Holman. (1985). A programmable exposure control system for determination of the effects of pollutant exposure regimes on plant growth. *Atmos. Environ.*, **19**, 1135–45.

Khatamian, H., N. O. Adepipe, and D. P. Ormrod. (1973). Soil-plant-water aspects of ozone phytotoxicity in tomato plants. *Plant Soil*, **38**, 531–41.

Kohut, R. J. and J. A. Laurence. (1983). Yield response of red kidney bean *Phaseolus vulgaris* to incremental ozone concentrations in the field. *Environ. Pollut.*, **32**, 233–40.

Kohut, R. J., R. G. Amundson, and J. A. Laurence. (1986). Evaluation of growth and yield of soybean exposed to ozone in the field. *Environ. Pollut.*, **41**, 219–34.

Larsen, R. I., A. S. Heagle, and W. W. Heck. (1983). An air quality data analysis system for interrelating effects, standards and needed source reductions: Part 7: An O_3–SO_2 leaf injury mathematical model. *J. Air Pollut. Control Assoc.*, **33**, 198–207.

Lefohn, A. S., W. E. Hogsett, and D. T. Tingey. (1986). A model for developing ozone exposures that mimic ambient conditions in agricultural areas. *Atmos. Environ.*, **20**, 361–6.

MacDowall, F. D. H. (1965). Predisposition of tobacco to ozone damage. *Can. J. Plant Sci.*, **45**, 1–12.

Markowski, A. and S. Grzesiak. (1974). Influence of sulphur dioxide and ozone on vegetation of bean and barley plants under different soil moisture conditions. *Bull. Acad. Pol. Sci. Ser. Sci. Biol.*, **22**, 875–87.

Maurer, A. R., D. P. Ormrod, and N. J. Scott. (1969). Effect of five soil water regimes on growth and composition of snap beans. *Can. J. Plant Sci.*, **49**, 271–8.

McElroy, F. F. (1979). *Transfer standards for calibration of air monitoring analyzers for ozone*. Research Triangle Park, NC, US Environmental Protection Agency, Environmental Monitoring and Support Laboratory. EPA-600/4-79-056.

McLaughlin, S. B. and R. K. McConathy. (1983). Effects of SO_2 and O_3 on allocation of ^{14}C-labeled photosynthate in *Phaseolus vulgaris*. *Plant Physiol.*, **73**, 630–5.

Mumford, R. A., H. Lipke, D. A. Laufer, and W. A. Feder. (1972). Ozone-induced changes in corn pollen. *Environ. Sci. Technol.*, **6**, 427–30.

Okano, K., O. Ito, G. Takeba, A. Shimizu, and T. Totsuka. (1984). Alteration of ^{13}C-assimilate partitioning in plants of *Phaseolus vulgaris* exposed to ozone. *New Phytol.*, **97**, 155–63.

Olszyk, D. M. and T. W. Tibbitts. (1981). Stomatal response and leaf injury of *Pisum sativum* L. with SO_2 and O_3 exposures. II. Influence of moisture stress and time of exposure. *Plant Physiol.*, **67**, 545–9.

Oshima, R. J., P. K. Braegelmann, R. B. Flagler, and R. R. Teso. (1979). The effects of ozone on the growth, yield, and partitioning of dry matter in cotton. *J. Environ. Qual.*, **8**, 474–9.

Paur, R. J. and F. F. McElroy. (1979). *Technical assistance document for the calibration of ambient ozone monitors*. Research Triangle Park, NC, US Environmental Protection Agency, Environmental Monitoring and Support Laboratory. EPA-600/4-79-057.

Reich, P. B. and R. G. Amundson. (1984). Low level O_3 and/or SO_2 exposure causes a linear decline in soybean yield. *Environ. Pollut.*, **34**, 345–55.

Reich, P. B. and R. G. Amundson. (1985). Ambient levels of ozone reduce net photosynthesis in tree and crop species. *Science*, **230**, 566–70.

Reich, P. B., A. W. Schoettle, R. M. Raba, and R. G. Amundson. (1986). Response of soybean to low concentrations of ozone: I. Reductions in leaf and whole plant net photosynthesis and leaf chlorophyll content. *J. Environ. Qual.*, **15**, 31–6.

Rich, S. and N. C. Turner. (1972). Importance of moisture on stomatal behavior of plants subjected to ozone. *J. Air Pollut. Control Assoc.*, **22**, 718–21.

Rich, S., P. E. Waggoner, and H. Tomlinson. (1970). Ozone uptake by bean leaves. *Science*, **169**, 79–80.

SAS. (1985). The REG Procedure. In *SAS User's Guide: Statistics. Version 5 Edition*, Chapter 31. Cary, NC, SAS Institute Inc.

Savage, M. J., H. H. Wiebe, and A. Cass. (1983). *In situ* field measurement of leaf water potential using thermocouple psychrometers. *Plant Physiol.*, **73**, 609–13.

Schwartz, H. F., M. Ballarin, and R. L. Riggle. (1983). Shade, irrigation and variety effects upon bronzing response of dry beans in Colorado. *Ann. Rep. Bean Improv. Coop.*, **26**, 20–1.

Sionit, N. and P. J. Kramer. (1977). Effect of water stress during different stages of growth of soybean. *Agr. J.*, **69**, 274–8.

Snow, M. D. and D. T. Tingey. (1985). Evaluation of a system for the imposition of plant water stress. *Plant Physiol.*, **77**, 602–7.

Temple, P. J., O. C. Taylor, and L. F. Benoit. (1985). Responses of cotton yield to ozone as mediated by soil moisture and evapotranspiration. *J. Environ. Qual.*, **14**, 55–60.

Tingey, D. T. and W. E. Hogsett. (1985). Water stress reduces ozone injury via a stomatal mechanism. *Plant Physiol.*, **77**, 944–7.

Tingey, D. T., G. L. Thutt, M. L. Gumpertz, and W. E. Hogsett. (1982). Plant water status influences ozone sensitivity of bean plants. *Agric. Environ.*, **7**, 243–54.

Tingey, D. T., K. D. Rodecap, E. H. Lee, T. J. Moser, and W. E. Hogsett. (1986). Ozone alters the concentration of nutrients in bean tissue. *Angew. Botanik*, **60**, 481–93.

Tingey, D. T., T. J. Moser, D. F. Zirkle, and M. D. Snow. (1987). A plant cultural system for monitoring evapotranspiration and physiological responses under field conditions. In *Proceedings of the International Conference on Measurement of Soil and Plant Water Status*. Logan, UT, Utah State University.

Toivonen, P. M. A., G. Hofstra, and R. T. Wukasch. (1982). Assessment of yield losses in white bean due to ozone using the antioxidant EDU. *Can. J. Plant Path.*, **4**, 381–6.

US Environmental Protection Agency. (1986). *Air quality criteria for ozone and other photochemical oxidants.* Research Triangle Park, NC, US Environmental Protection Agency, Environmental Criteria and Assessment Office. EPA-600/8-78-004.

Walker, E. K. and L. S. Vickery. (1961). Influence of sprinkler irrigation on the incidence of weather fleck on flue-cured tobacco in Ontario. *Can. J. Plant Sci.*, **41**, 281–7.

Webster, B. D., C. L. Tucker, and S. P. Lynch. (1977). A morphological study of the development of reproductive structures of *Phaseolus vulgaris* L. *J. Am. Soc. Hortic. Sci.*, **102**, 640–3.

15

EFFECTS OF AIR POLLUTANTS ON INTERACTIONS BETWEEN PLANTS, INSECTS, AND PATHOGENS

WILLIAM J. MANNING and KEVIN D. KEANE

University of Massachusetts, Amherst, Massachusetts, USA

15.1 INTRODUCTION

"Effects of air pollutants on vegetation are like icebergs: the more substantial effects are often below what is easily observed on the surface." (Laurence *et al.*, 1983)

This quotation admirably summarizes and describes the current state of knowledge of air pollution effects on plants. We know a great deal about the direct or primary effects of air pollutants on symptom expression, growth, and yield. This is the part of the iceberg that is obvious and readily seen. Much of this information, however, comes from experiments involving the interaction of a plant with a single air pollutant, under carefully controlled experimental conditions.

In comparison, we know very little about the indirect or secondary effects of air pollutants. Indirect effects can be many and may be subtle (Fig. 15.1) and yet have pronounced effects on how plants interact with herbivorous insects and pathogens. These effects relate to the much larger part of the iceberg that is not easily detected. As a result, the indirect effects of air pollutants on plants have received little attention.

During the past few years, it has become evident that the indirect effects of air pollutants on plants have been underestimated and that they may actually be more important than primary effects, especially when yield losses are determined.

It is clear that we must begin to look at the interactive effects of all the factors that affect plants if we are to begin to understand the full magnitude of the effects of air pollutants on plants (Heck, 1982). At present, we know little

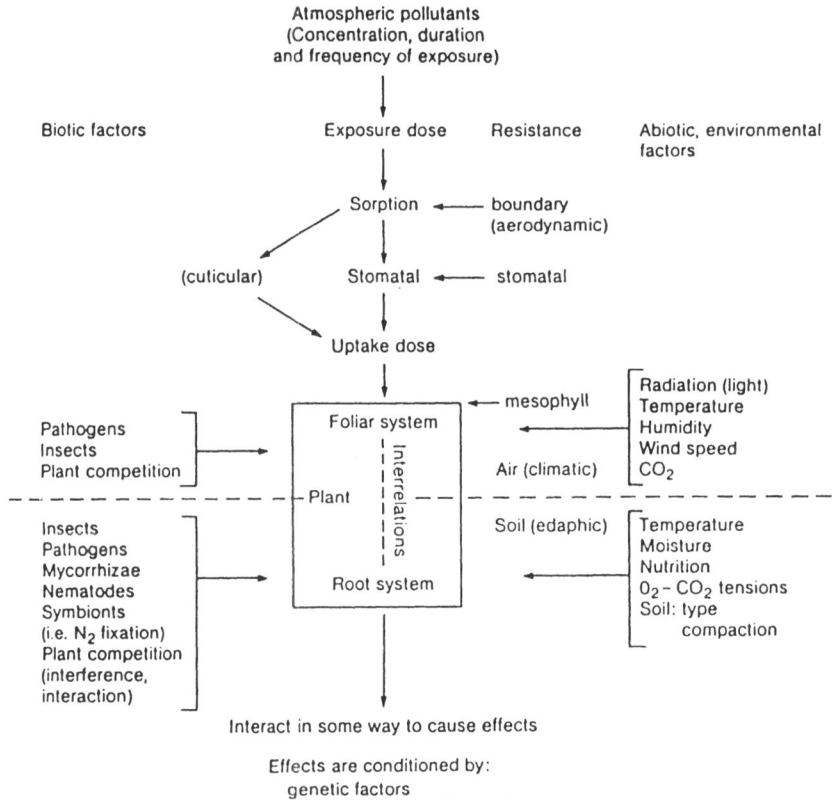

Fig. 15.1. Biotic and abiotic factors that can affect the response of plants to an air-pollutant exposure dose (from Heck, 1982).

about how air pollution-mediated changes in plants affect plant–insect or plant–pathogen interactions and plant yields (Laurence and Weinstein, 1981).

Insects and pathogens associated with plants are known to be affected by air pollutants, particularly the prevalent phytotoxic gases ozone (O_3), sulfur dioxide (SO_2), and nitrogen dioxide (NO_2). Sometimes the effect on the insect or pathogen is a direct and obvious one, but usually it is an indirect one that is expressed through a change in the physical or biochemical nature of the plant.

Our purpose here is to review the effects of the major air pollutants O_3, SO_2, and NO_2 on interactions between plants, insects, and pathogens, in relation to plant yield effects. Different and/or more inclusive reviews have been previously published (Alstad *et al.*, 1982; Flückiger and Braun, 1986; Heagle, 1973, 1982; Hughes, 1988; Hughes and Laurence, 1984; Huttunen, 1984; Laurence, 1981; Laurence *et al.*, 1983; Manning, 1975; Smith, 1981).

15.2 INTERACTIONS BETWEEN PLANTS AND INSECTS

Plants and herbivorous insects maintain a state of balanced coadaptation that is mutually essential. Insects depend on plants for nutrition, for growth and reproduction, and for behavioral cues. Pollutant effects on host nutritional quality, resulting from changes in photosynthesis, respiration, and translocation, can greatly affect insects. Changes in plant morphology, leaf toughness, and secondary metabolites also affect plant susceptibility to insects. Virtually any change in a plant can have some effects on associated herbivorous insects (Hughes, 1988; Hughes and Laurence, 1984).

15.2.1 Observations

There have been very many observations in nature regarding changes in insect populations along gradients away from air pollution sources. Causal relationships have been proposed between air pollutants and changes in insect populations (Alstad *et al.*, 1982; Laurence *et al.*, 1983; Przybylski, 1979). European foresters and scientists have been suggesting since 1865 that air pollutants influence outbreaks of insects in forests. Gerlach even coined the word "Rauch-russler" (smoke weevils) for the weevils observed on smoke-damaged spruce in Saxony (Flückiger and Braun, 1986).

Most of the reports in the literature on air pollution–plant–insect interactions are anecdotal accounts of field observations. Entomologists have been slow to investigate these interactions under controlled experimental conditions. Most research papers are of recent origin and report only from short-term experiments. Results from these experiments indicate that insects generally develop better when they feed on plants exposed to air pollutants, and often prefer such plants. Unless concentrations are very high, air pollutants seem to have few direct effects on insects.

15.2.2 Experimental results

There has been considerable recent interest in the effects of air pollutants on aphids (McNeill *et al.*, 1987). Ambient London air was found to consistently increase biomass of the black bean aphid (*Aphis fabae*). Sulfur dioxide and NO_2 stimulated the mean relative growth rate (MRGR) of *A. fabae* by 6·5% and 8·4% respectively, after 3 days. There were no direct effects of SO_2 and NO_2 on *A. fabae*, and it was concluded that the effects were plant mediated (Dohmen *et al.*, 1984). The MRGR of aphids (*Macrosiphon rosae*) feeding on rosebushes (*Rosa* sp.) in ambient air in Munich was about 20% higher than for those in carbon-filtered air (Dohmen, 1985). Aphid numbers were found to have increased by a factor of 4·4 times when green apple aphid (*Aphis pomi*) numbers were determined on hawthorns (*Crataegus* sp.) growing in carbon-filtered and ambient air chambers along motorways in Switzerland (Braun and Flückiger, 1985; Flückiger and Braun, 1986). Low SO_2 doses increased MRGRs for pea aphids (*Acyrthosiphon pisum*) on pea plants (*Pisum sativum*) in growth cabinets. Higher concentrations had deleterious effects on *A. pisum*. It was suggested that aphid MRGRs could be used to detect fluxes and short-lived physiological changes in plants (Warrington, 1987). Long-term exposure of pea aphids to low SO_2 in growth cabinets resulted in a 19% increase in the rate of nymph production (Warrington *et al.*, 1987). Recent work at Lancaster University in the UK indicates that O_3 as well as SO_2

enhances MRGRs of the cereal aphid (*Rhopalosiphum padi*) on spring barley (*Hordeum vulgare*). SO_2 enhanced MRGRs of the green spruce aphid (*Elatobium abietinum*) on Sitka spruce (*Picea sitchensis*) (personal communication from S. Warrington). In Germany, 0·08 ppm O_3 (156 μg m^{-3}) for 3 days decreased growth rate of *A. fabae* about 15% on broad beans (*Vicia faba*) (personal communication from P. Dohmen). Flückiger also observed a decrease of *A. fabae* on broad beans in Switzerland during an episode of high O_3 concentrations (personal communication from W. Flückiger). In an open-air fumigation system in southern England, the highest SO_2 concentration (0·057 ppm [150 μg m^{-3}], arithmetic mean) significantly increased numbers of grain aphids (*Sitobion avenae*) (McLeod, 1988; McLeod and Roberts, 1987).

Tomato pinworms (*Keiferia lycopersicella*) developed faster and survived better on tomato plants (*Lycopersicon esculentum*) injured by O_3 than on noninjured plants grown in carbon-filtered air. Survival during pupation, sex ratios of adults, and female longevity and fecundity were not affected. Ozone was thought to react chemically with plant proteins and amino acids to result in release of large amounts of soluble nitrogenous compounds. This improved host plant suitability for *K. lycopersicella* (Trumble *et al.*, 1987).

Total grasshopper density, and that specifically of *Melanoplus sanguinipes*, were determined in a northern mixed grass prairie, as influenced by an open-field SO_2 fumigation system. Grasshopper density decreased with increases in SO_2 concentration, and the effect was uniform and not species specific (McNary *et al.*, 1981).

Third instar larvae of the gypsy moth (*Lymantria dispar*) preferred to feed on O_3-treated oak (*Quercus* sp.) foliage, and expressed a feeding preference in response to a gradient of O_3 concentration. Leaves exposed to 0·15 ppm O_3 were preferred over those exposed to 0·12 ppm. Those exposed to 0·09 ppm were preferred less than nonexposed control leaves (Jeffords and Endress, 1984).

The effects of SO_2, and to a lesser extent O_3, on the growth, development, fecundity, longevity, feeding behavior, and feeding preference of the Mexican bean beetle (MBB, *Epilachna varivestris*) on bean (*Phaseolus vulgaris*) (preferred host) and soybean (*Glycire max*) have been extensively investigated. More is known about air pollutant effects on MBB than on any other insect.

When MBB larvae were fed bean leaves that had been exposed to 0·15 ppm SO_2 for 7 days, they were not adversely affected. Adult females, however, showed a significant feeding preference for SO_2-treated leaves (Hughes *et al.*, 1981). With soybean leaves, however, MBB larvae developed faster and grew larger, and adult females were more fecund when the leaves had previously been exposed to SO_2. Adult females also showed marked feeding preference for SO_2-treated leaves (Hughes *et al.*, 1982). The same increases in larval growth and adult female fecundity were found when MBB fed on caged soybean plants exposed to SO_2 in the field (Hughes *et al.*, 1983). MBB pupal weights were found to increase with increasing SO_2 concentrations up to 0·30 ppm, and then decreased at higher concentrations (Hughes *et al.*, 1985). A relationship between reduced glutathione (GSH) concentrations in SO_2-stressed soybean leaves and MBB larval growth rates was determined. The GSH concentrations varied with SO_2 in the same way that MBB responded. Plant GSH concentration was suggested as a useful predictor of MBB larval growth for plants exposed to SO_2 (Chiment *et al.*, 1986). However, GSH concentration changes are not specific for SO_2-induced stress. They indicate that some plant stress has occurred, but not which one (Hughes, 1988).

MBB adults also preferred to feed on O_3-treated soybean foliage. Foliage that had been exposed to the highest O_3 dose was preferred (Endress and Post, 1985). Similar results were obtained for soybeans exposed to O_3 in open-top chambers (Chappelka *et al.*, 1988).

15.2.3 Yield effects

Most work with air pollution–insect interactions has not progressed beyond the insect response stage. There are few reports that relate possible air pollution–plant–insect interactions on crop yields. Several examples, however, are summarized in Table 15.1.

Peas were grown to maturity in solar domes at Lancaster University (Warrington *et al.*, 1987). When peas were grown in nonfiltered ambient air, with pea aphids, there was a 42% reduction in pea seed yield. When ambient air was supplemented with 0·45 ppm SO_2, aphid populations increased 1·8 times compared to ambient air alone, and pea seed yield was reduced 52%. When the ambient air plus SO_2 was used with no aphids there were no effects on pea seed yield.

In a field study, caged soybean plants were exposed to SO_2 and six pairs of newly-emerged adult MBB were introduced into each cage. Adult MBB caused 45% foliar damage on control plants and 75% on plants exposed to SO_2. Female MBBs that fed on SO_2-treated leaves grew larger and were more fecund. After correction for field variation, no significant differences were found for the average number or weight of pods per plant between SO_2-treated and control plants. Mean stem biomass, however, was less for SO_2-treated plants than for controls, indicating less growth for SO_2-treated plants (Hughes *et al.*, 1983).

Bark beetles of various kinds are known to invade stress-weakened trees. This hastens tree decline and death and reduces timber yields. In the San Bernardino Mountain forests, east of Los Angeles, ponderosa pines (*Pinus ponderosa*) weakened by O_3 stress became more susceptible to invasion by pine beetles (*Dendroctonus brevicomis*) (Cobb *et al.*, 1968; Miller, 1983). Similar results were noted for O_3-weakened eastern white pines (*Pinus strobus*), in the Blue Ridge Mountains of Virginia (Skelly *et al.*, 1983).

Based on results from the few available studies, it is clear that very little is known about the possible effects of air pollution–plant–insect interactions on plant yields.

TABLE 15.1

Examples of air pollution–plant–insect interactions and plant yields

Plant	Pollutant	System	Insect	Effects	Reference
Peas	SO$_2$	Solar domes	Pea aphids	52% reduction in pea seed	Warrington *et al.* (1987)
	Ambient air			42% reduction in pea seed	
Soybeans	SO$_2$	Caged plants	Mexican bean beetles	No significant differences in average no. or wt of pods/plants	Hughes *et al.* (1983)
Ponderosa pine Eastern white pine	Ambient O$_3$	Natural stands	Bark beetles	Accelerated tree decline	Cobb *et al.* (1968); Miller (1983); Skelly *et al.* (1983)

15.3 INTERACTIONS BETWEEN PLANTS AND PATHOGENS

As for insects, there is an abundance of anecdotal accounts of real or alleged increases or decreases in plant disease incidence in relation to air pollution sources or along pollutant gradients. Unlike the situation with insects, it is well established that air pollutants can directly or indirectly affect the course of development of plant diseases (Heagle, 1973, 1982; Hughes and Laurence, 1984; Huttunen, 1984; Laurence, 1981; Laurence *et al.*, 1983; Manning, 1975; Smith, 1981).

15.3.1 Experimental results

Bacteria—Alfalfa (*Medicago sativa*) leaves exposed to O_3 for 4 h at 0·20 ppm, were less susceptible to *Xanthomonas alfalfae* than leaves not exposed to O_3. Conversely, alfalfa leaves with *X. alfalfae* infections were less sensitive to O_3 (Howell and Graham, 1977). Lesion numbers for bacterial infection of soybean leaves by *Pseudomonas glycinea* were reduced by O_3, except for inoculations made 2 days after exposure to O_3 (Laurence and Wood, 1978). Dry bean plants with infections caused by *Xanthomonas phaseoli* were less susceptible to O_3 injury in the field. Ozone injury did not protect plants from infection by *X. phaseoli* (Temple and Bisessar, 1979). Sulfur dioxide suppressed expansion of lesions caused by *Corynebacterium nebraskense* on maize (*Zea mays*) and *X. phaseoli* var. *sojensis* on soybean leaves (Laurence and Aluisio, 1981). Similar results for pre- and post-inoculation exposures with SO_2 were found for *X. phaseoli* on red kidney bean leaves, with an extended latent period between inoculation and symptom expression (Laurence and Reynolds, 1982, 1986). SO_2 had no effect on the resident phase of the pathogen.

The effects of SO_2 on the epidemiology of common blight of red kidney bean was investigated in the field, utilizing an open-air, chamberless fumigation system, with a gradient of 0 to 1·00 ppm SO_2, applied for 3 periods, 2 to 3 times per week through the growing season. Bean plants were spray-inoculated with *Xanthomonas campestris* p.v. *phaseoli* (syn. *X. phaseoli*). Over time, SO_2 exposures resulted in significant decreases in the rate of lesion appearances on leaves (Reynolds *et al.*, 1987).

These few examples suggest that air pollutants, especially SO_2, inhibit bacterial diseases, decrease lesion size, and often increase latent periods.

Fungi—There are a great many fungal pathogens of crop plants. This is one reason why most of the examples of interactions between air pollutants and pathogens involve fungi. Because of their diversity and different modes of parasitism, fungal diseases have been reported to be either enhanced, inhibited, or not affected by air pollutants.

Obligate parasitic fungi grow and survive only on specific host plants. Host penetration is usually not extensive, and mycelium and spores are usually exposed on plant surfaces. The potential for direct pollutant effects on growth and reproduction of these fungi is very great (Heagle, 1975).

Heagle (1970) made the first report of the effects of O_3 on a rust disease under defined experimental conditions. Crown rust-differential varieties of oats (*Avena sativa*) were inoculated with *Puccinia coronata* and exposed to O_3 at 0·10 ppm for 10 days. Ozone significantly reduced the growth of uredia of *P. coronata*. Sporulation of the wheat stem rust fungus *P. coronata* f. sp. *avenae* was inhibited by O_3, but not spore germination or infection (Heagle and Key, 1973*a*). Wheat (*Triticum aestivum*) leaves infected by *Puccinia graminis* f. sp. *tritici* exhibited less O_3 injury than noninfected leaves (Heagle and Key, 1973*b*). Pretreatment of young wheat plants with 0·087 ppm O_3 for 3 days or 0·107 ppm O_3 for 7 days, followed by inoculation with urediospores of *Puccinia recondita* f. sp. *tritici* (brown rust), resulted in a great reduction in the number of rust pustules on O_3-treated plants compared to those grown in carbon-filtered air and then inoculated (Dohmen, 1987). Ozone inhibited the size of primary uredia of *Uromyces phaseoli* on bean leaves, but there were more of them and often there were secondary uredia at their margins. This would result in a considerable increase in inoculum (Resh and Runeckles, 1973).

Sulfur dioxide, at concentrations less than 0·20 ppm, decreased the incidence and severity of bean rust (*U. phaseoli*) and the size and percentage germination of urediospores (Weinstein *et al.*, 1975). *P. graminis* on wheat was reduced by SO_2, but was cultivar dependent. Sulfur dioxide, at 0·10 ppm for 100 h, starting 2 days after inoculation, reduced rust pustules on a cultivar with horizontal resistance to rust race 15-B, but did not inhibit rust on a cultivar with vertical resistance (Laurence *et al.*, 1979). Barley rust (*Puccinia hordei*) was reduced on winter barley by SO_2 from an open-air fumigation system in England (McLeod, 1988).

Ozone at concentrations of 0·05, 0·10, and 0·15 ppm, did not affect germination of conidia of *Erysiphe graminis* f. sp. *hordei*. When sporulating colonies on barley were exposed to O_3, the percentage infection subsequently caused by exposed conidia was significantly reduced. Ozone exposures during spore incubation also significantly reduced infections. Ozone exposures that caused chlorosis of barley leaves significantly increased colony and spore mass length (Heagle and Strickland, 1972). Ozone had little effect on *Microsphaera alni*, the fungus that causes powdery mildew of lilac (*Syringa vulgaris*). Mature conidia were resistant to O_3, and spore germination and infection were not affected. Sulfur dioxide however, greatly decreased infection of lilac leaves by *M. alni*. Spore germination, penetration, and hyphal production were all inhibited (Hibben and Taylor, 1975).

Heagle (1977) exposed corn leaves to O_3 for 6 days at 0·06, 0·12, and 0·18 ppm and then inoculated them with *Helminthosporium maydis*. He found that sporulation was increased on leaves exposed to 0·06 and 0·12 ppm, but not to 0·18 ppm. When exposures occurred 6 days after inoculation, increasing O_3 concentrations decreased sporulation. Exposure of corn plants to 0·15 ppm SO_2 for 14 h d^{-1} for 8 days before inoculation decreased lesion numbers caused by *H. maydis* (Laurence *et al.*, 1979). Sulfur dioxide at concentrations less than 0·20 ppm, had no effect on *Alternaria solani* (early blight) on tomato leaves (Weinstein *et al.*, 1975). SO_2 increased the number of lesions on Scots pine

(*Pinus sylvestris*) needles caused by *Scirrhia acicola*, when applied 5 days after inoculation. Similar results were obtained with $SO_2 + O_3$, but needle injury was greater than additive (Weidensaul and Darling, 1979).

Fehrmann *et al.* (1986) reported that foliar pathogens of cereals have become more prevalent in central Europe during the past two decades. They investigated the effects of preinfection exposures of several cultivars of wheat and barley to O_3, SO_2, and $O_3 + SO_2$ on infection of wheat by *Ascochyta* (*Didymella exitialis*) and *Gerlachia nivalis* (*Fusarium nivale*) and barley by *Drechslera teres* (*Helminthosporium teres*) and *D. sorokiniana*. For *G. nivalis* and *Ascochyta*, less so for *Drechslera sorokiniana*, O_3 promoted disease development. Sulfur dioxide was often suppressive, with $SO_2 + O_3$ sometimes promoting disease. Cultivars varied considerably in their responses to pollutants and pathogens.

In a unique experiment, the effects of SO_2 on foliar pathogens of winter barley and winter wheat were determined in the field in southern England, using an open-air fumigation system. Plants were exposed to four concentrations of SO_2 (0·01–0·06 ppm, arithmetic mean concentrations) from emergence in autumn to maturity the following summer. Normal agricultural practices were used. Barley was sprayed in November with the fungicide triademefon to prevent powdery mildew (*Eryspihe graminis* f. sp. *hordei*). Wheat was sprayed in the fall with the fungicide propiconazole to prevent infections by *Septoria tritici*. For barley, SO_2 at 0·023, 0·038, and 0·058 ppm inhibited the rust fungus *P. hordei*. This allowed eyespot incidence (*Pseudocercosporella herpotrichoides*) to increase at the 0·023 and 0·038 ppm SO_2 concentrations. Foliar infections, caused by *Rhynchosporium secalis* also decreased with increasing SO_2 concentrations. Sulfur dioxide decreased *P. herpotrichoides* (eyespot) and *Fusarium* (brown foot rot of winter wheat). Powdery mildew (*Erysiphe graminis* f. sp. *tritici*) infestations were higher, rather than lower, at all SO_2 concentrations (McLeod, 1988; McLeod and Roberts, 1987). In the same experiments, the presence and absence of surface-growing black mould fungi (*Cladosporium herbarum* and *Alternaria alternata*), which discolor grain and reduce its quality and economic value, were monitored by McLeod *et al.* (1986). Incidence was reduced in the open-field fumigation system at 0·023 ppm SO_2 and eliminated at 0·058 ppm SO_2.

The antioxidant EDU, and the fungicide Du-Ter, were used separately, and in combination, to demonstrate interactions between ambient O_3 and early blight (*Alternaria solani*) on potatoes (*Solanum tuberosum*) in the field in Ontario. Adaxial O_3 lesions on potato leaves were rapidly colonized by *A. solani,* under conditions conducive to rapid disease development. EDU did not affect *A. solani* in culture tests (Holley *et al.*, 1985). Bisessar (1982) also found that O_3 affected incidence of *A. solani* on potato leaves.

In the air pollutant–plant–pathogen literature, interactions between O_3–plant–*Botrytis* have received the most attention. *Botrytis cinerea* and related species are common necrotrophic pathogens of above-ground plant parts. *Botrytis* has been found in the phylloplane mycoflora of more than 35 plant genera

(Dickinson, 1976). It is a common leaf surface inhabitant, and it is logical to expect that it would invade air pollution-weakened or damaged tissues (Manning, 1976).

In 1968, Norland potatoes in a field in Paxton, Massachusetts, were injured by O_3. Ozone-injured potato foliage was also extensively invaded by *B. cinerea*. As a result, early senescence occurred, and yields were reduced. The syndrome was reproduced under experimental conditions, and a new disease complex of potato was described (Manning *et al.*, 1969). Ozone-injured geranium (*Pelargonium hortorum*) leaves were also shown to be more susceptible to *B. cinerea* (Manning *et al.*, 1970). Infections appeared to begin in O_3-caused necrotic lesions. Naturally occurring *B. cinerea* was consistently isolated from washed pinto bean leaf discs with visible O_3 injury (Manning, 1976).

Ozone at 0·15 ppm decreased sporulation of *B. cinerea*. Treated spores did not germinate well and were less able to invade geranium leaves. At 0·30 ppm, spore germination was poor, and spores were unable to infect geranium leaves. Both of these indicate that O_3 directly affected sporulation and ability of spores to germinate and infect leaves (Krause and Weidensaul, 1978*a,b*).

Certain onion (*Allium cepae*) cultivars are susceptible to injury by O_3 and to *B. cinerea*, *B. allii*, and *B. squamosa*. Wukasch and Hofstra (1977*a, b*) used open-top chambers to study the effect of ambient O_3 on onion leaf dieback, caused by added inoculum of *B. squamosa* and naturally occurring *Botrytis* spp. Charcoal filtration resulted in a six-fold reduction in leaf dieback, with reductions in *Botrytis* infections. In a related experiment with onions in open-top chambers, the antioxidant EDU and/or a fungicide were used to reduce O_3 injury, *Botrytis* spp., or both and to demonstrate an interaction between O_3 and *Botrytis* in onions. Application of EDU reduced O_3 injury and *Botrytis* infections. Although nontoxic to *Botrytis in vitro*, EDU appeared to slightly reduce *Botrytis* infections on onion leaves in the field. Despite this, an interaction between O_3 injury and increased *Botrytis* infections was demonstrated.

Increased foliar leaching occurred on onion leaves exposed to O_3 for 4 h. When conidia of *B. cinerea* were placed in dew on older O_3-treated onion leaves, significantly more Botrytis lesions occurred per cubic centimeter than in controls, indicating that increased exudation is also a factor in enhancing *Botrytis* infections (Rist and Lorbeer, 1984*a*). In related work, it was found that O_3 increases infections by *B. cinerea,* but not by *B. squamosa.* Ozone had no detectable effect on expansion of pre-established lesions caused by *B. cinerea* or *B. squamosa* (Rist and Lorbeer, 1984*b*).

Recent work by Tiedemann and Fehrmann (personal communication), indicates that a 6-day preinfection SO_2 exposure can result in a three- to four-fold increase in necrotic leaf spots on grape (*Vitis vinifera*) leaves caused by *B. cinerea*. A similar, but less pronounced effect occurred with O_3. Ozone and SO_2 effects were additive (Fehrmann *et al.*, 1986).

Due to the technical difficulties in working with roots and root diseases or a lack of appreciation of their importance, little is known about the effects of air pollutants on root diseases. Roots are very important components of air pollution-mediated decline problems, but they have not been extensively investigated.

The effects of O_3 on the rhizoplane mycoflora of pinto bean roots was determined in 1971 by Manning *et al.* Pinto bean plants were exposed to O_3 at 0·10 to 0·15 ppm for 28 days. More fungal colonies were consistently isolated from roots and hypocotyls of plants exposed to O_3 than from those grown in carbon-filtered air. Differences in the successional root mycoflora surface were quantitative rather than qualitative. None of the fungi isolated was pathogenic to pinto bean plants grown in carbon-filtered air. Ozone caused accelerated plant senescence and accelerated the rate of colonization of normal root decay fungi.

Ozone delayed the growth of young tomato plants inoculated with the vascular wilt pathogen *Fusarium oxysporum* f. sp. *lycopersici* and delayed the onset of vascular wilt disease by several weeks (Manning, 1978; Manning and Vardaro, 1976). Ozone had no effects, however, on the progress of wilt of young cabbage plants, caused by *F. oxysporum* f. sp. *conglutinans* (Manning *et al.*, 1971).

Unlike wilt pathogens, cortical decay fungi cause slow nondramatic plant declines. *Pyrenochaeta lycopersici* causes a slow decline disease (brown root rot) of tomato. Tomato plants were exposed to O_3 for 60 days (0·08–0·10 ppm, $7 h^{-1} 5 d^{-1} wk^{-1}$), with and without *P. lycopersici*. Ozone significantly affected dry weights of tomato roots. Root dry weight differences were also significant when comparisons were made between plants with brown root rot exposed to O_3 and plants with brown root rot grown in carbon-filtered air (Manning and Vardaro, 1974). Alfalfa seedlings, grown in steamed or non–steamed field soil, containing *Fusarium avenaceum*, *F. oxysporum*, and *F. solani*, were exposed to 0·06 to 0·08 ppm O_3 for 13 weeks. Both nonsteamed soil and O_3 reduced alfalfa growth, with soil plus O_3 effects greater than additive (Cooley and Manning, 1986). Ozone did not affect severity of hypocotyl cortical decay of soybeans caused by *Fusarium oxysporum*. Plants infected with *F. oxysporum* and exposed to O_3 exhibited greater reductions in relative growth rate and net assimilation rate, as well as more intense foliar O_3 symptoms than did noninoculated plants also exposed to O_3 (Damicone *et al.*, 1987).

A classic example of the effects of O_3 on a forest ecosystem is the decline of Jeffrey and ponderosa pines in the San Bernardino Mountains east of the Los Angeles basin. This decline had been extensively and meticulously documented (Miller, 1983).

Heterobasidion annosum (*Fomes annosus*) commonly invades roots of declining or stressed trees. Effects of O_3 on *H. annosum* and infection of O_3-affected pines by *H. annosum* have been determined (James, 1977). *H. annosum* more readily invaded freshly cut stumps of O_3-injured trees when inoculations were made (James *et al.*, 1980*a*). When roots of ponderosa pines, severely injured by O_3, were inoculated with *H. annosum* in the field, roots of these trees became infected more often than did roots of healthy trees. Exposure of potted inoculated Jeffrey and ponderosa pine seedlings to O_3

resulted in more infection in those exposed to O_3 than those not exposed. Ozone dose and seedling injury directly affected colonization of host tissue by *H. annosum* (James *et al.*, 1980*b*). In laboratory fumigation studies, O_3 decreased growth and reproduction of *H. annosum* and colonization of wooden discs (James *et al.*, 1982). These effects were not thought to be significant under natural conditions.

In a related study on the effects of O_3 on Eastern white pine in the Blue Ridge Mountains of Virginia, it was observed that trees weakened by O_3 were more subject to root disease caused by *Verticicladiella procera* (Skelly *et al.*, 1983).

Vesicular arbuscular endomycorrhizal fungi and ectomycorrhizal fungi are usually not considered to be root pathogens, but they are intimately associated with plant roots and play important roles in water and nutrient uptake and may help defend against invasion by root pathogens. The extent and condition of plant mycorrhizae are an excellent indicator of plant health. Many stress factors, including air pollutants, can affect mycorrhizal incidence and function.

Soybeans infected with *Glomus geosporum* in open-top chambers were less sensitive to the adverse effects of O_3. *G. geosporum* produced 40% fewer chlamydospores on roots of plants exposed to 0·079 ppm O_3 than on roots of those exposed to 0·025 ppm O_3 (Brewer and Heagle, 1983). Endomycorrhizae were reduced as was chlamydospore production by *Glomus fasciculatus* on roots of citrange seedlings exposed to O_3 (McCool *et al.*, 1979). Similar results were obtained for tomato seedlings and *G. fasciculatus* exposed to 0·15 and 0·30 ppm O_3, with photoassimilate partitioning reduced by O_3 treatments (McCool and Menge, 1983).

Loblolly pine seedlings, with ectomycorrhizae caused by *Pisolithus tinctorius*, were not adversely affected by O_3, SO_2, or $O_3 + SO_2$ (Mahoney *et al.*, 1985). Ozone caused significant decreases in ectomycorrhizae on roots of white birch and white pine seedlings (Keane and Manning, 1987).

Nematodes—There are few reports on the effects of air pollution on plants in relation to effects on plant parasitic nematodes.

Sulfur dioxide had no adverse effects on four soybean nematodes (*Belonolaimus longicaudatus*, *Heterodera glycines*, *Paratrichodorus minor* and *Pratylenchus penetrans*). Ozone and $O_3 + SO_2$, however, reduced reproduction in *H. glycines* and *P. minor*. Reproduction of the foliar nematode *Aphelenchoides fragariae* on begonia leaves was severely reduced when leaves were damaged by O_3 or $O_3 + SO_2$ (Weber *et al.*, 1979). Young tomato plants (1- to 3-weeks-old) infected with *P. penetrans* were more sensitive to mixtures of O_3 and SO_2. The interaction of 0·20 ppm O_3 and 0·80 ppm SO_2 (which suppressed shoot growth) enhanced reproduction of the nematode on tomato roots. All other $O_3 + SO_2$ interactions did not affect nematode reproduction (Shew *et al.*, 1982).

Bisessar and Palmer (1984) used the antioxidant EDU to determine the effects of ambient O_3 on incidence of northern root-knot nematodes (*Meloidogyne hapla*) on field-grown tobacco (*Nicotiana tabacum*). Application of EDU reduced O_3 injury and also reduced the incidence of nematode galls on tobacco roots.

Viruses—Virus infections in plants have a pronounced effect on how plants respond to air pollutants. Virus-infected plants are usually less affected by air pollutants than are virus-free plants. This is described as "protection" from air pollutant injury. As most virus infections are debilitating to plants, however, the value of this "protection" is questionable.

Tobacco plants (*Nicotiana sylvestris*) infected with tobacco mosaic virus (TMV) were reported to be resistant to O_3 at 0·30 ppm during winter months. During the summer months, however, virus-infected and virus-free tobacco were both equally susceptible to O_3 injury. Mosaic-type viruses like TMV caused less severe symptoms in summer, and a higher O_3 concentration (0·40 ppm) was required to cause injury to tobacco (Brennan and Leone, 1969). Partial protection of pinto bean leaves from O_3 injury was provided when plants were inoculated 4 to 6 days prior to exposure to O_3 (0·25 ppm for 4 h) with bean common mosaic, tobacco mosaic, alfalfa mosaic, tomato ring-spot, and tobacco ring-spot viruses (Davis and Smith, 1974, 1976). Inoculation of tobacco leaves with tobacco etch virus 9 days prior to exposure to O_3 protected them from O_3 injury (Moyer and Smith, 1975). Field-grown tobacco infected with TMV had 60% less O_3 injury than noninfected tobacco (Bisessar and Temple, 1977). The translocatable nature of virus protection against O_3 injury was demonstrated by Vargo *et al.*, (1978). Noninoculated bean primary leaves were partially protected from O_3 injury if their companion primary leaves had been inoculated with TMV 8 to 10 days before exposure to O_3. Tobacco plants with systemic tobacco streak virus, however, were more susceptible to O_3 injury rather than less susceptible (Reinert and Gooding, 1978). Subacute doses of SO_2 (0·10 or 0·20 ppm) for 5 to 10 days increased the titre of southern bean mosaic virus in bean and maize dwarf mosaic virus (MDMV) in maize. Infection by MDMV and symptom expression were also intensified by SO_2 (Laurence *et al.*, 1981).

Disease epidemiology—With the exception of the work of Laurence and Reynolds (1982, 1986) and Reynolds *et al.* (1987) on the basic work required to investigate the epidemiology of common blight of the bean, no one has systematically studied the effects of air pollutants on plant disease epidemics.

In a theoretical sense, air pollutants can increase, decrease, or not affect the course of development of a disease epidemic (change in disease over time in a population). Madden and Campbell (1987) have considered some of the possible effects of air pollutants on epidemics. for example, initial disease level (y_0) and the apparent infection rate (r, rate of disease increase) could be altered by air pollutants. Given the right data, it would be possible to model the effects of air pollutants on plant disease epidemics.

15.3.2 Yield effects

There are only a few reports of air pollution–plant–pathogen interactions and the effects of these interactions on plant yields (Table 15.2). Very few experiments are designed to determine these interactions.

In an open-air fumigation system, SO_2 inhibited development of common blight of bean, but also inhibited the yield of noninfected control plants. Yield

TABLE 15.2

Examples of air pollution–plant–pathogen interactions and plant yields

Plant	Pollutant	System	Pathogen	Effects	References
Kidney bean	SO_2	Open-air fumigation	*Xanthomonas campestris* pv. *phaseoli*	Reduction in yield of noninfected plants. No reduction in yield of infected plants	Reynolds *et al.* (1987)
Onion	Ambient O_3	Open-top chambers EDU + fungicide	*Botrytis cinerea* and *B. squamosa*	Onion bulb yield increased 28%. Onion bulb yield increased 39% by EDU	Wukasch & Hofstra (1977a, b)
Potatoes	Ambient O_3	Field plots EDU + fungicides	*Alternaria solani*	EDU + fungicide DU-TER increased tuber fresh weight and specific gravity	Holley *et al.* (1985)
Soybeans	O_3	Open-top chambers	*Glomus geosporum*	Pod yields for mycorrhizal plants reduced by 25%; yields for nonmycorrhizal reduced by 48%	Brewer and Heagle (1983)
Ponderosa pine	Ambient O_3	Natural stands	*Heterobasidion annosum*	Increased root rot in O_3-injured trees and accelerated decline	James *et al.* (1980b)
Eastern white pine	Ambient O_3	Natural stands	*Verticicladiella procera*	Increased root rot in O_3-injured trees and accelerated decline	Skelly *et al.* (1983)
Tomato	O_3 SO_2	Chambers	*Pratylenchus*	Nematodes did not affect fruit yields	Shew *et al.* (1982)

of infected plants, however, was not decreased. This may have been due to the late appearance of symptoms in the growing season (Reynolds *et al.*, 1987).

Two field experiments were conducted to determine the effects of O_3 and *Botrytis* spp. on onion bulb yields. Carbon-filtered air in open-top chambers reduced onion leaf dieback caused by *Botrytis squamosa* and *Botrytis* spp. six-fold and increased onion bulb yields by 28%. In a second experiment, use of the antioxidant EDU, with or without a fungicide, reduced O_3 injury and *Botrytis* infections and resulted in a 39% increase in onion bulb yields (Wukasch and Hofstra, 1977*a*, *b*).

An interaction between O_3 and early blight (*Alternaria solani*) on three potato cultivars was demonstrated using EDU and a fungicide. *A. solani* colonized adaxial O_3 lesions on potato leaves. Significant increases in tuber fresh weights and specific gravity for all three cultivars were greater than additive for the combination of fungicide and EDU, when compared to increases for either single treatment (Holley *et al.*, 1985).

Soybeans infected with the mycorrhizal fungus *Glomus geosporum* were less affected by O_3 in open-top chambers. Pod yields for mycorrhizal plants were reduced by 25%, while yields for nonmycorrhizal plants were reduced by 48% (Brewer and Heagle, 1983).

Ponderosa pine in California with O_3 injury were more susceptible to root rot caused by *H. annosum* (*Fomes annosus*) (James *et al.*, 1980*b*). Eastern white pine in Virginia that are weakened by O_3 are more susceptible to *Verticicladiella procera* (Skelly *et al.*, 1983). Both root pathogens hasten tree decline and death and reduce timber yields.

Infection of tomato roots by *Pratylenchus penetrans* and exposure of the plants to O_3, SO_2, and various combinations of O_3 and SO_2 did not affect tomato fruit yields

5.4. PESTICIDE USAGE IN YIELD STUDIES

Virtually all yield studies are conducted on the basis of a single interaction between the plant and one air pollutant. All other factors are held as constant as possible. For initial studies, this may be essential to determine the effects of the pollutant alone. Eventually, however, more factors need to be included so that the total response of the plant to the pollutant can be determined (Heck, 1982). A more holistic approach needs to evolve.

Investigators have been cautioned that plant pests or pathogens can affect the responses of plants to air pollutants and that pesticides can also alter responses. In a standard book on air pollution research methods, it is stated that "Disease and insect infestations should be avoided, but if they occur, care should be exercised in selecting a pesticide" (Leone and Brennan, 1979). Malathion, for example, contains sulfur and would not be a good choice to manage insects in a long-term study on the effects of SO_2 on plants.

Some examples of pesticide usage in some recent open-top chamber and open-field studies on the effects of O_3 or SO_2 on major crop plants are given in

TABLE 15.3

Examples of pesticides used in open-top and open-field studies on air pollution effects on crops

Crop	Pollutant	System	Parameter measured	Pesticide	Purpose	Reference
Winter wheat	O_3	Open-top chambers	Photosynthesis and yield	Karathane	Powdery mildew	Amundson et al. (1987)
Winter wheat	O_3	Open-top chambers	Yield	Malathion	Aphids	Heagle et al. (1979)
Winter wheat	SO_2	Open-air system	Foliar diseases	Propiconazole	Septoria tritici	McLeod (1988)
Winter barley	SO_2	Open-air system	Foliar diseases	Tridemefon	Powdery mildew	McLeod (1988)
Soybeans	O_3	Open-top chambers	Yield	Permethrin cyhexatin	Insects and mites	Heagle et al. (1986)
Soybeans	O_3	Open-top chambers	Yield	Kalthane	Spider mites	Amundson (1983)
Lettuce	O_3	Open-top chambers	Yield	Orthene, thuricide and sevin—every week	Lepidoptera larvae	Temple et al. (1986)
Alfalfa	O_3	Concentration gradient	Dose-response function	Diazinon—every week	General insects	Oshima et al. (1976)

Table 15.3. These reports were chosen randomly, but they are typical examples of pesticide usage. Many investigators do not provide information on pesticides used, but mention only that "normal cultural practices were followed". It is unusual to find a report that mentions that no pesticides were used, because they were not needed (Kress and Miller, 1985). Wheat or soybean growers may not routinely spray their crops for disease or insect management and diazinon is probably not applied to alfalfa once a week. It is, however, likely that orthene, thuricide, and sevin would be applied to lettuce on a weekly basis.

Pesticide usage in open-top chambers suggests that insect infestations and disease incidence in the chambers are probably unusual and relate to the effects of the chamber. The insect or disease problem must be eliminated or the single variable experiment cannot be carried out. As a result, little is learned about the effects of the insect or pathogen.

15.5 CONCLUSIONS

There are many reports in the literature regarding air pollution–plant–insect or pathogen interactions. Many of these are anecdotal or unverified accounts of cause and effect relationships. There are a few examples where interactions have clearly been established in relation to crop yields.

In the past few years, entomologists have taken a new interest in examining air pollution interactions with insects, especially aphids. These experiments need to be continued and expanded to include effects on crop yields.

With few exceptions, most of the reports of air pollution–pathogen interactions are from single experiments. There are few studies on effects on disease epidemiology and crop yields.

Experiments need to be designed to investigate the long-term effects of air pollution–plant–insect and/or pathogen interactions in relation to crop yields. Otherwise, we will not determine the true nature of the total effects of air pollution on plants and crop yields.

ACKNOWLEDGEMENTS

The authors gratefully acknowledge personal communications and copies of new pending literature from:

J. N. B. Bell, Imperial College, Silwood Park, UK.
A. R. McLeod, Central Electricity Generating Board, Leatherhead, UK.
G. P. Dohmen, Gesellschaft für Strahlen- und Umweltforschung, Neuherberg, FRG.
P. R. Hughes, Boyce Thompson Institute, Ithaca, N.Y., USA.
S. Warrington, Lancaster University, UK.

REFERENCES

Alstad, D. N., G. F. Edmunds Jr, and L. H. Weinstein. (1982). Effects of air pollutants on insect populations. *Annu. Rev. Entomol.,* **27**, 369–84.

Amundson, R. G. (1983). Yield reduction of soybean due to exposure to sulfur dioxide and nitrogen dioxide in combination. *J. Environ. Qual.,* **12**, 454–9.

Amundson, R. G., R. J. Kohut, A. W. Schoettle, R. M. Raba, and P. B. Reich. (1987). Correlative reductions in whole-plant photosynthesis and yield of winter wheat caused by ozone. *Phytopathology,* **77**, 75–9.

Bisessar, S. (1982). Effect of ozone, antioxidant protection and early blight on potato in the field. *J. Am. Soc. Hortic. Sci.,* **107**, 597–9.

Bisessar, S. and P. J. Temple. (1977). Reduced ozone injury on virus-infected tobacco in the field. *Plant Dis. Rep.,* **61**, 961–3.

Bisessar, S. and K. T. Palmer. (1984). Ozone, antioxidant spray and *Meloidogyne hapla* effects on tobacco. *Atmos. Environ.,* **18**, 1025–7.

Braun, S. and W. Flückiger. (1985). Increased population of the aphid *Aphis pomi* at a motorway. Part 3—the effect of exhaust gases. *Environ. Pollut. (Series A),* **39**, 183–92.

Brennan, E. and I. A. Leone. (1969). Suppression of ozone toxicity symptoms in virus-infected tobacco. *Phytopathology,* **59**, 263–6.

Brewer, P. F. and A. S. Heagle. (1983). Interactions between *Glomus geosporum* and exposure of soybeans to ozone or simulated acid rain in the field. *Phytopathology,* **73**, 1035–40.

Chappelka, A. H., M. E. Kraemer, T. Mebrahtu, M. Rangappa, and P. S. Benepal. (1988). Effects of ozone on soybean resistance to the Mexican bean beetle (*Epilachna varivestis*). *Environ. Exp. Bot.,* **28**, 53–60.

Chiment, J. J., R. Alscher, and P. R. Hughes. (1986). Glutathione as an indicator of SO_2-induced stress in soybean. *Environ. Exp. Bot.,* **26**, 147–52.

Cobb, F. W., Jr, D. L. Wood, R. W. Stark, and J. R. Parmeter, Jr. (1968). Photochemical oxidant injury and bark beetle (Coleoptera: Scolytidae) infestation of ponderosa pine. IV. Theory on the relationships between oxidant injury and bark beetle infestation. *Hilgardia,* **34**, 141–52.

Cooley, D. R. and W. J. Manning. (1986). Ozone and root and crown rot: interactive effects on growth of alfalfa. *Phytopathology,* **76**, 651.

Damicone, J. P., W. J. Manning, S. J. Herbert, and W. A. Feder. (1987). Growth and disease response of soybeans from early maturity groups to ozone and *Fusarium oxysporum*. *Environ. Pollut.,* **48**, 117–30.

Davis, D. D. and S. H. Smith. (1974). Reduction of ozone sensitivity of pinto bean by bean common mosaic virus. *Phytopathology,* **64**, 383–5.

Davis, D. D. and S. H. Smith. (1976). Reduction of ozone sensitivity of pinto bean by virus-induced local lesions. *Plant Disease Rep.,* **60**, 31–4.

Dickinson, C. H. (1976). Fungi on the aerial surfaces of higher plants. In *Microbiology of aerial plant surfaces,* ed. by C. H. Dickinson and T. F. Preece, 293–324. London, Academic Press.

Dohmen, G. P. (1985). Secondary effects of air pollution: enhanced aphid growth. *Environ. Pollut. Ser. A* **39**, 227–34.

Dohmen, G. P. (1987). Secondary effects of air pollution: ozone decreases brown rust disease potential in wheat. *Environ. Pollut.,* **43**, 189–94.

Dohmen, G. P., S. McNeill, and J. N. B. Bell. (1984). Air pollution increases *Aphis fabae* peat potential. *Nature,* **5946**, 52–3.

Endress, A. G. and S. L. Post. (1985). Altered feeding preference of Mexican bean beetle *Epilachna varivestis* for ozonated soybean foliage. *Environ. Pollut.,* Ser. A **39**, 9–16.

Fehrmann, H., A. von Tiedemann, and P. Fabian. (1986). Predisposition of wheat and barley to fungal leaf attack by preinoculative treatment with ozone and sulphur dioxide, *Z. Pflanzenkrankheiten Pflanzenschutz*, **93**, 313–18.

Flückiger, W. and S. Braun. (1986). Effect of air pollutants on insects and host plant/insect relationships. *Proc. Conf. How are the effects of air pollutants on agricultural crops influenced by the interaction with other limiting factors*, Brussels, ECE. 79–91.

Heagle, A. S. (1970). Effect of low-level ozone fumigations on crown rust of oats. *Phytopathology*, **60**, 252–4.

Heagle, A. S. (1973). Interactions between air pollutants and plant parasites. *Annu. Rev. Phytopathol.*, **11**, 305–88.

Heagle, A. S. (1975). Response of three obligate parasites to ozone. *Environ. Pollut.*, **9**, 91–5.

Heagle, A. S. (1977). Effect of ozone on parasitism of corn by *Helminthosporium maydis. Phytopathology*, **67**, 616–18.

Heagle, A. S. (1982). Interactions between air pollutants and parasitic plant diseases. *Effects of gaseous air pollution in agriculture and horticulture*, ed. by M. H. Unsworth and D. P. Ormrod, 333–48. London, Butterworths Scientific.

Heagle, A. S. and A. Strickland. (1972). Reactions of *Erysiphe graminis* f. sp. *hordei* to low levels of ozone. *Phytopathology*, **62**, 1144–8.

Heagle, A. S. and L. W. Key. (1973a). Effect of ozone on the wheat stem rust fungus. *Phytopathology*, **63**, 397–400.

Heagle, A. S. and L. W. Key. (1973b). Effect of *Puccinia graminis* f. sp. *tritici* on ozone injury in wheat leaves. *Phytopathology*, **63**, 609–13.

Heagle, A. S., S. Spencer, and M. B. Letchworth. (1979). Yield response of winter wheat to chronic doses of ozone. *Can. J. Bot.*, **57**, 1999–2005.

Heagle, A. S., V. M. Lesser, J. O. Rawlings, W. W. Heck, and R. B. Philbeck. (1986). Response of soybeans to chronic doses of ozone applied as constant or proportional additions to ambient air. *Phytopathology*, **76**, 51–6.

Heck, W. W. (1982). Future directions in air pollution research. In *Effects of gaseous air pollution in agriculture and horticulture*, ed. by M. H. Unsworth and D. P. Ormrod, 411–35. London, Butterworths Scientific.

Hibben, C. R. and M. P. Taylor. (1975). Ozone and sulphur dioxide effects on the lilac powdery mildew fungus. *Environ. Pollut.*, **9**, 107–14.

Holley, J. D., G. Hofstra, and R. Hall. (1985). Effect of reducing oxidant injury and early blight on fresh weight and tuber density of potato. *Phytopathology*, **75**, 529–32.

Howell, R. K. and J. H. Graham. (1977). Interaction of ozone and bacterial leaf spot of alfalfa. *Plant Dis. Rep.*, **61**, 565–7.

Hughes, P. R. (1988). Insect populations on host plants subjected to air pollution. In *Plant stress–insect interactions*, ed. by E. A. Heinrichs. New York, John Wiley (in press).

Hughes, P. R. and J. A. Laurence. (1984). Relationship of biochemical effects of air pollutants on plants to environmental problems: insect and microbial interactions. In *Gaseous air pollutants and plant metabolism*, ed. by M. K. Koziol and F. R. Whatley, 361–77. London, Butterworths Scientific.

Hughes, P. R., J. E. Potter, and L. H. Weinstein. (1981). Effects of air pollutants on plant–insect interactions: reactions of the Mexican bean beetle to SO_2-fumigated pinto beans. *Environ. Entomol.*, **10**, 741–4.

Hughes, P. R., J. E. Potter, and L. H. Weinstein. (1982). Effects of air pollution on plant–insect interactions: increased susceptibility of greenhouse-grown soybeans to the Mexican bean beetle after plant exposure to SO_2. *Environ. Entomol.*, **11**, 173–6.

Hughes, P. R., A. I. Dickie, and M. A. Penton. (1983). Increased success of the Mexican bean beetle on field-grown soybeans exposed to sulfur dioxide. *J. Environ. Qual.*, **12**, 565–8.

Hughes, P. R., J. J. Chiment, and A. I. Dickie. (1985). Effect of pollutant dose on the response of Mexican bean beetle (Coleoptera: Coccinellidae) to SO$_2$-induced changes in soybean. *Environ. Entomol.*, **14**, 718–21.

Huttunen, S. (1984). Interactions of disease and other stress factors with atmospheric pollution. In *Air pollution and plant life*, ed. by M. Treshow, 321–56. New York, John Wiley.

James, R. L. (1977). *The effects of photochemical air pollution on the epidemiology of Fomes annosus*. PhD Dissertation, University of California, Berkeley.

James, R. L., F. W. Cobb Jr, W. W. Wilcox, and D. L. Rowney. (1980*a*). Effects of photochemical oxidant injury of ponderosa and Jeffrey pines on susceptibility of sapwood and freshly cut stumps to *Fomes annosus*. *Phytopathology*, **70**, 704–8.

James, R. L., F. W. Cobb, Jr, P. R. Miller, and J. R. Parameter, Jr. (1980*b*). Effects of oxidant air pollution on susceptibility of pine roots to *Fomes annosus*, *Phytopathology*, **70**, 560–3.

James, R. L., F. W. Cobb, and J. R. Parmeter Jr. (1982). Effects of ozone on sporulation, spore germination, and growth of *Fomes annosus*. *Phytopathology*, **72**, 1205–8.

Jeffords, M. R. and A. G. Endress. (1984). Possible role of ozone in tree defoliation by the gypsy moth (Lepidoptera:Lymantriidae). *Environ. Entomol.*, **13**: 1249–52.

Keane, K. D. and W. J. Manning. (1987). Effects of ozone and simulated acid rain and ozone and sulfur dioxide on mycorrhizal formation in paper birch and white pine. In *Acid rain: scientific and technical advances*, ed. by R. Perry, *et al.*, 608–13. London, Selper Ltd.

Krause, C. R. and T. C. Weidensaul. (1978*a*). Effects of ozone on sporulation, germination, and pathogenicity of *Botrytis cinerea*. *Phytopathology*, **68**, 195–8.

Krause, C. R. and T. C. Weidensaul. (1978*b*). Ultrastructural effects of ozone on the host–parasite relationship of *Botrytis cinerea* and *Pelargonium hortorum*. *Phytopathology*, **68**, 301–7.

Kress, L. W. and J. E. Miller. (1985). Impact of ozone on field-corn yield. *Can. J. Bot.*, **63**, 2408–15.

Laurence, J. A. (1981). Effects of air pollutants on plant–pathogen interactions. *Z. Pflanzenkrankheiten Pflanzenschutz*, **87**, 156–72.

Laurence, J. A. and F. A. Wood. (1978). Effects of ozone on infection of soybean by *Pseudomonas glycinea*. *Phytopathology*, **68**, 441–5.

Laurence, J. A. and L. H. Weinstein. (1981). Effects of air pollutants on plant productivity. *Annu. Rev. Phytopathol.*, **19**, 257–71.

Laurence, J. A. and A. L. Aluisio. (1981). Effects of sulfur dioxide on expansion of lesions caused by *Corynebacterium nebraskense* in maize and *Xanthomonas phaseoli* var. *sojensis* in soybean. *Phytopathology*, **71**, 445–8.

Laurence, J. A. and K. L. Reynolds. (1982). Effects of concentration of sulfur dioxide and other characteristics of exposure on the development of lesions caused by *Xanthomonas phaseoli* in red kidney bean. *Phytopathology*, **72**, 1243–6.

Laurence, J. A. and K. L. Reynolds. (1986). The joint action of hydrogen fluoride and sulfur dioxide on the development of common blight of red kidney bean. *Phytopathology*, **76**, 514–17.

Laurence, J. A., L. H. Weinstein, D. H. McCune, and A. L. Aluisio. (1979). Effects of sulfur dioxide on southern corn leaf blight of maize and stem rust of wheat. *Plant Dis. Rep.*, **63**, 975–8.

Laurence, J. A., A. L. Aluisio, L. H. Weinstein, and D. C. McCune. (1981). Effects of sulfur dioxide on southern bean mosaic and maize dwarf mosaic. *Environ. Pollut.*, **24**, 185–91.

Laurence, J. A., P. R. Hughes, L. H. Weinstein, G. T. Geballe, and W. H. Smith. (1983). Impact of air pollution on plant–pest interactions: implications of current research and strategies for future studies. Ecosystems Research Center Report No. 20. Ithaca, New York, Cornell University.

Leone, I. A. and E. Brennan. (1979). Plant growth and care. In *Methodology for the assessment of air pollution effects on vegetation.* ed. by W. W. Heck, S. V. Krupa and S. N. Linzen, 5-1–5-14. Pittsburgh, PA, Air Pollution Control Association.

Madden, L. V. and C. L. Campbell. (1987). Potential effects of air pollutants on epidemics of plant diseases. *Agric. Ecosys. Environ.,* **18,** 251–62.

Mahoney, M. J., B. I. Chevone, J. M. Shelly, and L. D. Moore. (1985). Influence of mycorrhizae on the growth of loblolly pine seedlings exposed to ozone and sulfur dioxide. *Phytopathology,* **75,** 679–82.

Manning, W. J. (1975). Interactions between air pollutants and fungal, bacterial and viral plant pathogens. *Environ. Pollut.,* **9,** 87–90.

Manning, W. J. (1976). The influence of ozone on plant surface microfloras. *Microbiology of aerial plant surfaces,* ed. by C. H. Dickinson and T. H. Preece, 159–72. London, Academic Press.

Manning, W. J. (1978). Chronic foliar ozone injury: effects on plant root development and possible consequences. *Calif. Air Environ.,* **7,** 3–4.

Manning, W. J. and P. M. Vardaro. (1974). Ozone and *Pyrenochaeta lycopersici*: effects on growth and development of tomato plants. *Phytopathology,* **64,** 582.

Manning, W. J. and P. M. Vardaro. (1976). Ozone and *Fusarium*: effects on the growth and development of a wilt-susceptible tomato and a wilt-resistant tomato. *Proc. Am. Phytopathol. Soc.,* **3,** 227.

Manning, W. J., W. A. Feder, P. M. Papia, and I. Perkins. (1971). Effect of low levels of ozone on growth and susceptibility of cabbage plants to *Fusarium oxysporum* f. sp. *conglutinans. Plant Dis. Rep.,* **55,** 47–9.

Manning, W. J., W. A. Feder, I. Perkins, and M. Glickman. (1969). Ozone injury and infection of potato leaves by *Botrytis cinerea. Plant Dis. Rep.,* **53,** 691–3.

Manning, W. J., W. A. Feder, and I. Perkins. (1970). Ozone injury increases infection of geranium leaves by *Botrytis cinerea. Phytopathology,* **60,** 669–70.

Manning, W. J., W. A. Feder, P. M. Papia, and I. Perkins. (1971). Influence of foliar ozone injury on root development and root surface fungi of pinto bean plants. *Environ. Pollut.,* **1,** 305–12.

McCool, P. M. and J. A. Menge. (1983). Influences of ozone on carbon partitioning in tomato: potential role of carbon flow in regulation of the mycorrhizal symbiosis under conditions of stress. *New Phytol.,* **94,** 241–7.

McCool, P. M., J. A. Menge, and O. C. Taylor. (1979). Effects of ozone and HCl gas on the development of the mycorrhizal fungus. *Glomus fasciculatus* and growth of 'Troyer' citrange. *J. Am. Soc. Hortic. Sci.,* **104,** 151–4.

McLeod, A. R. (1988). Effects of open-air fumigation with sulphur dioxide on the occurrence of fungal pathogens in winter cereals. *Phytopathology,* **78,** 88–94.

McLeod, A. R. and T. M. Roberts. (1987). The importance of the total ecosystem response to air pollution. Preprint. 87-36.1. Pittsburg, PA, Air Pollution Control Association.

Mcleod, A. R., N. Magan, and M. V. Proctor. (1986). The effect of atmospheric sulphur dioxide on phylloplane fungi of cereals. Central Electricity Generating Board Report TPRD/L/3070,R86, UK.

McNary, T. J., D. G. Milchunas, J. W. Leetham, W. K. Lauenroth, and J. L. Dodd. (1981). Effect of controlled low levels of SO_2 on grasshopper densities on a northern mixed-grass prairie. *J. Econ. Entomol.,* **74,** 91–3.

McNeill, S., M. Amino-Kanu, G. Houlden, J. M. Bullock, S. Citrone, and J. N. B. Bell. (1987). The interaction between air pollution and sucking insects. *Acid rain: scientific and technical advances,* ed. by R. Perry, *et al.,* 602–7. London, Selper Ltd.

Miller, P. R. (1983). Ozone effects in the San Bernardino national forest. In *Air pollution and the productivity of the forest.* ed. by D. D. Davis, A. A. Millen, and L. Dochinger. 161–97. Washington, D.C., Izaak Walton League of America.

Moyer, J. W. and S. H. Smith. (1975). Oxidant injury reduction on tobacco induced by tobacco etch virus infection. *Environ. Pollut.,* **9,** 103–6.

Oshima, R. J., M. P. Poe, P. K. Braegelmann, D. W. Baldwin, and V. Van Way. (1976). Ozone dosage-crop loss function for alfalfa: a standardized method for assessing crop losses from air pollutants. *J. Air Pollut. Control Assoc.*, **26**, 861–5.

Przybylski, Z. (1979). The effects of automobile exhaust gases on the arthropods of cultivated plants, meadows and orchards. *Environ. Pollut., Ser. A*, **19**, 157–61.

Reinert, R. A. and G. V. Gooding, Jr. (1978). Effect of ozone and tobacco streak virus alone and in combination on *Nicotiana tabacum*. *Phytopathology*, **68**, 15–17.

Resh, H. M. and V. C. Runeckles. (1973). Effects of ozone on bean rust, *Uromyces phaseoli*. *Can. J. Bot.*, **51**, 725–7.

Reynolds, K. L., M. Zanelli, and J. A. Laurence. (1987). Effects of sulfur dioxide exposure on the development of common blight in field-grown red kidney beans. *Phytopathology*, **77**, 331–14.

Rist, D. L. and J. W Lorbeer. (1984*a*). Moderate dosages of ozone enhance infection of onion leaves by *Botrytis cinerea*, but not by *B. squamosa*. *Phytopathology*, **74**, 761–7.

Rist, D. L. and J. W. Lorbeer. (1984*b*). Ozone-induced leaching of onion leaves in relation to lesion production by *Botrytis cinerea*. *Phytopathology*, **74**, 1217–20.

Shew, B. B., R. A. Reinert, and K. R. Barker. (1982). Response of tomatoes to ozone, sulfur dioxide, and infection by *Pratylenchus penetrans*. *Phytopathology*, **72**, 822–6.

Skelly, J. M., Y.-S. Yang, B. I. Chevone, S. J. Long, J. E. Nellessen, and W. E. Winner. (1983). Ozone concentrations and their influence on forest species in the Blue Ridge Mountains of Virginia. In *Air pollution and the productivity of the forest*. ed. by D. D. Davis, A. A. Miller, and L. Dochinger. 143–59. Washington, D.C., Izaak Walton League of America.

Smith, W. H. (1981). *Air pollution and forests*. New York, Springer-Verlag.

Temple, P. J. and S. Bisessar. (1979). Response of white bean to bacterial blight, ozone and antioxidant protection in the field. *Phytopathology*, **69**, 101–3.

Temple, P. J., O. C. Taylor, and L. F. Benoit. (1986). Yield response of head lettuce (*Lactuca sativa* L.) to ozone. *Environ. Exp. Bot.*, **26**, 53–8.

Trumble, J. T., J. D. Hare, R. C. Musselman, and P. M. McCool. (1987). Ozone-induced changes in host–plant suitability: interactions of *Keiferia lycopersicella* and *Lycopersicon esculentum*. *J. Chem. Ecol.*, **13**, 203–18.

Vargo, R. H., E. J. Pell, and S. H. Smith. (1978). Induced resistance to ozone injury of soybean by tobacco ringspot virus. *Phytopathology*, **68**, 715–9.

Warrington, S. (1987). Relationship between SO_2 dose and growth of the pea aphid, *Acyrthosiphon pisum*, on peas. *Environ. Pollut.*, **43**, 155–62.

Warrington, S., T. A. Mansfield, and J. B. Whittaker. (1987). Effect of SO_2 on the reproduction of pea aphids, *Acyrthosiphon pisum*, and the impact of SO_2 and aphids on the growth and yield of peas. *Environ. Pollut.*, **48**, 285–94.

Weber, D. E., R. A. Reinert, and K. R. Barker. (1979). Ozone and sulfur dioxide effects on reproduction and host–parasite relationships of selected plant-parasitic nematodes. *Phytopathology*, **69**, 624–8.

Weidensaul, T. C. and S. L. Darling. (1979). Effects of ozone and sulfur dioxide on the host–pathogen relationship of Scotch pine and *Scirrhia acicola*. *Phytopathology*, **69**, 939–41.

Weinstein, L. H., D. C. McCune, A. L. Aluisio, and P. Van Leuken. (1975). The effect of sulphur dioxide on the incidence and severity of bean rust and early blight of tomato. *Environ. Pollut.*, **9**, 145–55.

Wukasch, R. T. and G. Hofstra. (1977*a*). Ozone and *Botrytis* spp. interaction in onion leaf dieback: field studies. *J. Am. Soc. Hortic. Sci.*, **102**, 543–6.

Wukasch, R. T. and G. Hofstra. (1977*b*). Ozone and *Botrytis* interactions in onion leaf dieback: open top chamber studies. *Phytopathology*, **67**, 1080–4.

Session VI

STATISTICAL AND SIMULATED
MODELING APPROACHES

(Victor C. Runeckles, *Chairman*; Eric M. Preston, *Co-Chairman*)

16

STATISTICAL APPROACHES TO ASSESSING CROP LOSSES

JOHN O. RAWLINGS, VIRGINIA M. LESSER, and
KAREN A. DASSEL

North Carolina State University, Raleigh, North Carolina, USA

16.1 INTRODUCTION

The National Crop Loss Assessment Network (NCLAN) was developed in 1980 to coordinate research on the impact of gaseous pollutants on agricultural crops. From the beginning the program had the dual objectives of providing information for an economic assessment of the impact of gaseous pollution (i.e., ozone, sulfur dioxide, nitrogen dioxide) on the agricultural industry, and of developing an understanding of the effects of pollutants (primarily ozone) on agricultural crops (Heck *et al.*, 1984*a, b*). Over the seven-year period 1980–86, the program investigated 14 different species at six test sites across the country in a total of 43 separate studies.

The NCLAN data library and data synthesis project was started in 1986 with the objectives of providing an archive for all NCLAN data and of doing a coordinated analysis of all yield–response data. The archiving of the data is complete, as is the basic data analysis, that of characterizing the yield dose–response relationships for the economic assessment. The explanation of the methods of analysis and the summary report of the dose–response relationships are being reported by Lesser and Rawlings.

The purpose of this presentation is to discuss some of the statistical methodologies used by NCLAN, both in design of experiments and in data analysis. In the design of field experiments, there seldom are "hard" data that pinpoint which experimental design will be optimum for a given field in a given year, nor is there sufficient evidence on the response that will be observed to define the optimum treatment design. The statistician must usually rely on general principles developed from experience and theory, and on the information the researcher can provide about his experience with experiments of this type in this environment. Subjective judgements are involved at many stages in the process, and there is plenty of room for alternative opinions. We will discuss our rationale for the choices we made and, where we can anticipate questions about alternatives, discuss their relative merits as we see them.

16.2 EXPERIMENTAL METHODS

16.2.1 Controlled experimentation versus observational studies

NCLAN early in the program adopted the use of open-topped chambers (Heck *et al.*, 1984*a*, *b*) in controlled field experimentation to assess the impact of ozone. In these studies, the ozone treatment levels and other treatment factors (moisture stress levels, cultivars, or sulfur dioxide pollution levels) are imposed on the crop in controlled, replicated studies. Significant treatment effects in controlled studies imply causal relationships. (Nonsignificant effects do *not* imply *no* causal relationship.) *Observational studies* are an alternative approach to controlled studies. In observational studies, the relationships among levels of the treatment factor (ozone), other environmental and management variables, and crop production, are observed as they occur in nature across the country. From these relationships, some measure of the link between ambient levels of ozone and yield is deduced. In its simplest form, this is a large multiple regression problem, the partial regression coefficient for "ozone level" being the measure of its impact on yield.

The appeal of this approach is that large amounts of data are available (USDA and meteorological data bases), and the data "cover" the country. There are situations in which this approach would be preferred, such as in exploratory investigations of a phenomenon to develop hypotheses on possible causal relationships. However, there are lethal problems with observational studies if the objective is to infer causality. And, if the results are to be used to set or modify air pollution standards, there is an implicit, if not explicit, causal inference. Even without the "causality" problem, such data usually present very difficult problems in sorting out the effect of a particular variable. The variables in a natural system do not operate independently, and there is no opportunity with observational studies to use randomization to disrupt the natural correlations. (Random selection is used to ensure that sampling sites and samples within sites are representative, but this does not remove correlations among variables.) There are two major problems.

(1) The multicollinearity of the explanatory variables that is to be expected in observational data makes it impossible to clearly assign effects to individual variables in the analysis. The effect being attributed to one variable is in reality a somewhat arbitrary piece of the joint effect of all variables with which it is highly correlated. For further discussion on the effects of collinearity in observational studies, refer to the report by the American Statistical Association (ASA Coordinating Committee, 1985) on the review of EPA-funded research on acid deposition effects on

forests and crops, and the book by W. G. Cochran (1983), *Planning and analysis of observational studies.*

(2) There will always be a few variables on which, for one reason or another, data are not available. Since these "overlooked" variables are still correlated with the measured variables, one can never be certain that an observed significant effect of a variable in the analysis is not the reflection of a correlated variable not included in the analysis, a "lurking" variable.

16.2.2 Dose–response approach versus a two-treatment NF, CF design

Early informal discussions on assessing the ozone effects on crop plants involved the relative merits of the dose–response studies being run by NCLAN versus studies in which all effort is concentrated on the ozone dose range of immediate interest. In particular, the proposal suggested devoting the available chambers to only the charcoal-filtered (CF) and nonfiltered (NF) treatments. There are merits to this proposal: (1) The ozone treatment levels are created only by filtering ambient air (CF) so there is no need for ozone-dispensing equipment. This makes the experimental procedure simpler and less expensive. (2) With the simpler system, it is realistic to expect that the pollution studies could be expanded to better sample the national environments. (3) Research effort is being concentrated on the dose region of immediate interest. The limitation of this approach is that one obtains information on only the dose interval between charcoal-filtered air and ambient air.

The use of only the two ozone levels generated by the CF and NF treatments might be sufficient for an economic assessment of the impact of *current* levels of ozone pollution, but NCLAN was equally interested in the more general question of plant response to ozone, and interest was certainly not restricted to the CF–NF interval. Further, while it may sound inconsistent at first, a dose–response approach, with appropriate choice of treatments, can provide *more* precision on the estimated ozone effect for a given interval than a simple means comparison, even when all effort has been devoted just to that specific interval. In addition, the dose–response approach provides information on effects of ozone over the entire experimental dose range. If one identifies a specific interval of primary interest, the standard error of the estimated relative yield loss (RYL) from ozone over that interval will be smaller from a properly designed response-curve study than from a study in which the same number of experimental units is devoted to the two endpoints of the interval.

We can illustrate this with the following comparison of experimental designs. Assume a constant error variance and a constant number (N) of chambers for all experiments. The experimental designs to be compared allocate the N chambers equally to $t = 2$, 3, 4, or 5 ozone treatment levels. In each case, the lowest two treatment levels are assumed to be the endpoints of the interval of primary interest, say $O_3 = 0.025$ and 0.055 ppm (0.025 and $0.055 \, \mu l \, l^{-1}$) to match (approximately) the O_3 levels in the CF and NF treatments. The $t = 2$ experiment is referred to as the CF–NF experiment, and consists of assigning half the chambers to each of the two treatments at the ends of the interval of interest. For reference, the variance of the estimated relative yield loss, Var(RYL), for this design is Var(RYL) = 0·0017 if the coefficient of variation (CV) is 10%, $N = 24$, and the true relative yield loss is zero. Var(RYL) decreases slightly as the true relative yield loss increases from zero.

In those experiments with three or more treatment levels, one has two choices on how RYL is estimated. Either RYL can be estimated by using only the CF and NF treatment means, or a response model can be fit and RYL estimated for the CF–NF interval from the response equation. The former ignores the relevant information contained in the other treatment means, and the latter, of course, requires that some model be adopted.

The comparisons of the designs are expressed in terms of the relative efficiencies of the two methods for estimation of RYL (Table 16.1). The relative efficiency (RE) of method A to method B is defined as the ratio of the variance of the estimate for method B to that for method A. All efficiencies are relative to the efficiency of the estimate obtained from the CF–NF means comparison from the two-point design. The left half of Table 16.1 gives the number of distinct design points (number of treatments) and the dose levels of these treatments. The "Means Est." column of Table 16.1 shows the relative efficiencies of the estimates of RYL when the estimates are based only on the CF and NF treatment means, but where research effort has been equally distributed among all treatments. The decrease in the relative efficiencies directly reflects the decrease in number of chambers devoted just to these two treatments. Clearly, if one is solely interested in the CF–NF interval and RYL is to be estimated only from the NF and CF means, one would not recommend additional treatments.

The story is different, however, if RYL is estimated from an appropriate response equation. The "linear" column (Table 16.1) shows the relative efficiencies when the true response to ozone is linear and a linear model has been fit. Notice that one additional treatment of 0·085 ppm (167 μg m^{-3}) outside the interval of interest more than doubles the relative efficiency. The

TABLE 16.1

Relative efficiency of estimates of RYL from mean estimation and dose–response estimation for various designs relative to mean estimation from a two-point design. All comparisons are based on an equal number of experimental units for the design and assume the fitted model is the true model

Treatment design		Means Est	Response models		
No. pts	Levels of X		Linear	Quad.	Cubic
2	(0·025, 0·055)	1·00	1·00	—	—
3	(0·025, 0·055, 0·085)	0·66	2·69	0·66	—
4	(0·025, 0·055, 0·085, 0·115)	0·50	5·17	0·83	0·50
4	(0·025, 0·055, 0·100, 0·150)	0·50	9·76	1·18	0·50
5	(0·025, 0·055, 0·085, 0·115, 0·15)	0·40	8·65	1·16	0·50
5	(0·025, 0·055, 0·100, 0·150, 0·20)	0·40	17·36	1·83	0·61
5	(0·025, 0·055, 0·100, 0·200, 0·30)	0·40	44·44	3·69	1·00

relative efficiency continues to increase as the upper dose level is increased. This is a familiar phenomenon with linear models, but it is not as well known for other models. The "quad." column (Table 16.1) shows the results if the true and fitted response models are quadratic. Three design points are required to estimate the model parameters, but with only three points the response equation estimate of RYL is the same as the means estimate. With four design points and the upper dose level somewhere between 0·12 and 0·15 ppm, relative efficiencies exceed 1·0. The improvement in efficiency is not nearly as marked as it was for the linear model, but this is to be expected; more effort is being devoted to the estimation of parameters. The last column (Table 16.1) shows the results if the true and fitted response models are a cubic polynomial. The last five-point design shows that, again, the precision of the two-point design can be matched and if the levels were increased further the response curve approach with a cubic polynomial would provide higher precision on the CF–NF comparison.

The point of this comparison is that response surface methodology with a properly designed experiment will provide more precise estimation of the relative yield loss for any specific dose interval than will be provided by a two-point design concentrating on that specific interval. At no additional cost, one has information on all possible intervals within the design space. Of course, the dose–response approach requires that the model be representative of the true relationship, and therefore it is imperative that the models used be as realistic and as representative of the data as possible. The relative efficiencies given assume there has been no mis-specification of the model.

16.2.3 Using dose levels above ambient

The previous discussion was simply to show that response curve methodology applied to some dose regions was more efficient than devoting all effort to the specific interval of immediate interest. It was apparent in that illustration that the increased efficiency came from allowing the treatment levels to extend outside the specific interval. This section will discuss more specifically the use of dose levels well above ambient levels.

The optimum choice of number of design points and placement of these points depends on the response model being assumed, and the use to be made of the fitted response equation. Dassel (1987) has studied the optimum experimental design for a dose–response study where the Weibull response model is appropriate. Discussion of experimental design strategy where a Weibull-type response is assumed is presented by Dassel and Rawlings (1988). Dassel (1987) shows that, as with the polynomial models, there is a choice of design points that gives greater precision for the estimation of RYL than is obtained from a means estimation based on two treatments. As with the polynomial models, the placement of these points becomes important.

The basic form of the Weibull dose–response function used in the NCLAN analyses is

$$Y_i = \alpha e^{-(X_i/\omega)^\lambda}$$

where X_i is some appropriate measure of ozone dose and α, ω, and λ are parameters controlling the function. Arguments for the use of this function and interpretations of the parameters will be given later. The only point needed for the present discussion is to note that, when dose $X_i = \omega$, yield Y_i is reduced to $\alpha e^{-1} = 0.37$, or 37% of the yield at zero dose. The Weibull design that is optimum for estimation of parameters places the highest dose level well above the $X = \omega$ level; that is, the highest dose is above the point where 63% of the yield has been lost. For estimation of RYL, the optimum placement of the maximum dose depends on the estimation interval. Figure 16.1 shows the effect of the maximum dose level on Var(RYL) for two prediction intervals. These comparisons are for four- and eight-point designs in which the lowest dose is fixed at the level where yield will be decreased to 98% of the original yield, the maximum ozone dose is chosen such that yield will be decreased to the indicated value on the abscissa, and the remaining dose levels are equally spaced between these two levels (Equal-X allocation). Each curve shows Var(RYL)/σ^2 for one of two prediction intervals denoted by the percentage yield remaining at the beginning and end of each interval; PY(90, 70) and PY(80, 60) denote the intervals where yield decreases from 90% to 70% and 80% to 60% respectively.

The horizontal dotted lines in Fig. 16.1 show Var(RYL)/σ^2 for an experiment in which the same amount of effort is devoted to just the two end points of each estimation interval and RYL is estimated from the two treatment means. The figure shows that for every prediction interval, there are appropriate choices of design points where Var(RYL) from the Weibull response model will be smaller than that from the two-point design, and that by judicious choice of the upper dose level, high precision can be attained for all intervals. Other work has shown that the best choice of maximum dose to

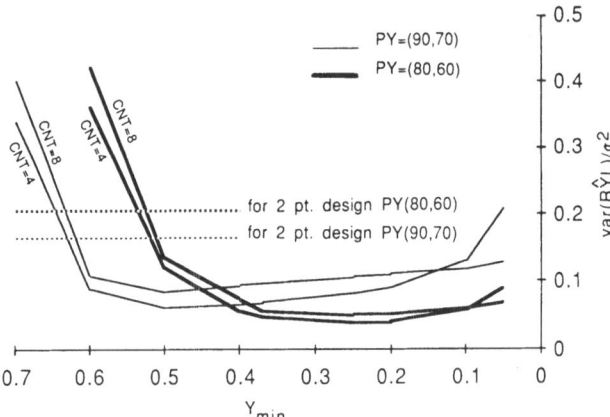

Fig. 16.1. Variance of estimated relative yield loss, Var(RYL)/σ^2, as a function of the remaining yield at the highest dose, Y_{min}, for estimation intervals PY = (90, 70) and (80, 60) for CNT = 4 and 8 distinct dose levels. The minimum dose is such that the remaining yield is $Y_{max} = 0.98$, and the dose levels are allocated between the two limits with the equal-X strategy. There are $N = 24$ experimental units in both cases. The horizontal dotted lines give Var(RYL)/σ^2 for estimation from two-point designs.

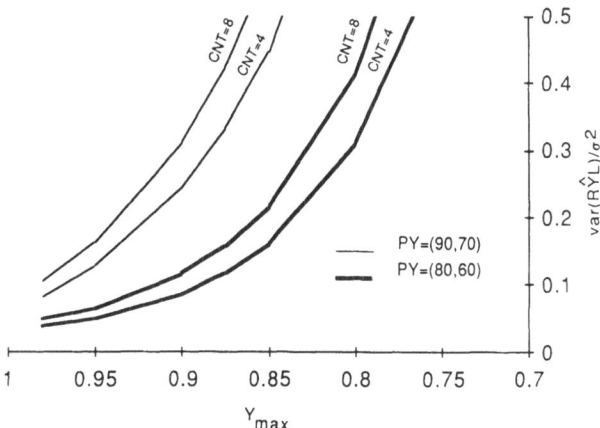

Fig. 16.2. Variance of estimated relative yield loss, Var(RYL)/σ^2, as a function of the remaining yield at the lowest dose, Y_{max}, for estimation intervals PY = (90, 70) and (80, 60) for CNT = 4 and 8 distinct dose levels. The maximum dose is such that the remaining yield is $Y_{min} = 0.25$ and the dose levels are allocated between the two limits with the equal-X strategy. There are $N = 24$ experimental units in both cases.

give good precision on all estimation intervals is not very different from the optimum for the estimation of parameters for the Weibull model. The theoretical results say that, if the Weibull response model is appropriate, the highest dose level should be above $X = \omega$ which in all species studied by NCLAN is well above any ambient levels of ozone.

The effect of changing the level of the lowest dose is shown in Fig. 16.2. For this illustration, the highest dose is fixed at the level that gives a 75% yield loss (or $Y_{max} = 0.25$). There is no flexibility in the placement of the lowest dose without incurring loss of precision. The lowest dose must be close to zero. This investigation of designs for the Weibull response model suggests that the dose levels should cover the region from very near zero dose to above $X = \omega$. With ozone pollution, these limits extend beyond the region of primary interest. This phenomenon of improving the precision of information within the primary region of interest by using observations outside the region is not unique with the polynomial and Weibull response models. It is a general principle that will apply to all continuous-response curves.

This effect of the placement of the highest dose level is illustrated using the NCLAN data for peanut, *Arachis hypogaea*, cv. NC 6 (Heagle *et al.*, 1983; Heck *et al.*, 1984b). (Citations associated with NCLAN studies are to publications that describe the study and the findings of the principal investigators. The statistical results from those data reported in this publication are the responsibility of the authors.) The peanut study at Raleigh in 1980, was a randomized complete block design using six treatment levels of ozone, CF, NF, CA15, CA45, CA75, and CA105, with the highest dose level being 0·123 ppm. An experiment with a maximum dose level of 0·076 ppm was simulated by reanalyzing these data with the two highest treatments deleted. The results from the analysis were adjusted to reflect the same error variance,

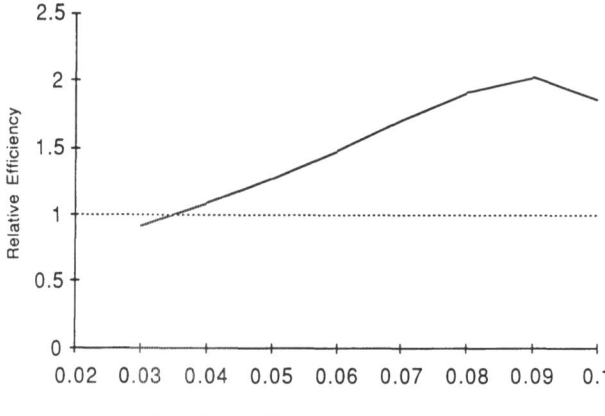

Fig. 16.3. Relative efficiency of estimates of RYL of the original NCLAN peanut study with highest dose level at $X = 0.123$ ppm to estimates from a simulated study with maximum dose level at $X = 0.076$ ppm. The estimation interval is from 0.025 ppm to the indicated dose.

degrees of freedom, and total number of experimental units as in the original experiment. Figure 16.3 shows the relative efficiencies (RE) of the RYL estimates of the original design to the simulated design,

$$RE = \frac{\text{Var(RYL) for the simulated design}}{\text{Var(RYL) for the original design}}$$

The lower limit for each estimation interval for RYL is $O_3 = 0.025$; the upper limit is the indicated dose in Fig. 16.3. The relative efficiencies go from $RE = 0.92$, for an estimation interval of $O_3 = 0.025$ to $O_3 = 0.03$, to $RE = 2.03$, for an estimation interval of $O_3 = 0.025$ to $O_3 = 0.09$. Except for the first interval, the higher dose level in the original experiment, 0.123 ppm versus 0.076 ppm, gave greater precision for the estimated yield losses than did the simulated experiment. Expressed in another way, the confidence interval estimate of the relative yield loss was as much as 42% longer in the simulated experiment with the lower maximum dose.

As we have illustrated, there are good statistical reasons for using the dose–response approach with dose levels well above the region of immediate interest. Two concerns about the use of relatively high dose levels have been raised which we would like to address briefly. The first is the concern that the response at the upper dose levels would dominate the fitting in such a way that the response in the lower dose range would not be well represented. The upper tail would 'wag the dog', so to speak. This is an appropriate concern, but it is handled by taking care that the response models are realistic and reflect the behavior of the data. Any consistent lack-of-fit of the model to sections of the dose range would be a flag that the model is inadequate. This is evident, for example, when a polynomial response model is fit to an observed response that has a plateau. The fitted equation will almost always overshoot the plateau before it turns back.

The second concern expressed with respect to using higher than ambient dose levels is that the high levels may trigger a biological process different from that causing the yield reductions at ambient levels, and as a consequence the response equations using higher dose levels would not be reliable for prediction at the low-dose levels. It is certainly reasonable to expect more than one biological process to be involved in the plant response to ozone. Even at ambient levels there may be more than one process involved; ozone may be triggering the stomates to close and it may be killing cells. This creates no problem in model fitting as long as the response to increasing dose can be expected to be continuous and reasonably smooth. Sharp discontinuities in the response would make modeling more difficult.

With respect to the NCLAN ozone data, none have shown discontinuities that suggest any abrupt change in the process. All observed responses follow a typical degradation curve approaching zero as the dose gets very high. Modeling the response has been very successful with relatively simple models, and there has been no suggestion from the model fitting that predictions in the low-dose region have been systematically biased by having higher than ambient levels of ozone. Some evidence is given in the discussion on choice of models.

16.3 CHOICE OF MODELS

One of the more controversial areas of the NCLAN analysis seems to have been the choice of model to characterize the relationship between yield and ozone. There are undoubtedly many mathematical functions that will adequately describe the relationship, particularly if one is concerned with only one or a few of the data sets at a time. The early NCLAN analyses characterized the responses with polynomial equations of the appropriate degrees. Polynomials were used because (1) they are the old 'standby' with which everyone is familiar, and (2) they are linear in the parameters and simple to fit with linear least squares. Polynomial models cannot be considered biologically realistic response curves in most situations. They have served very usefully as approximations over limited regions of the independent variable.

However, as NCLAN data accumulated, it became evident that polynomial models were not completely satisfactory. Some of the species showed a very plateau-like response before ozone seemed to have any impact. A quadratic model did not fit the plateau well, even though it may not have shown significant lack of fit. This is illustrated in Fig. 16.4. In this particular case, the observed treatment means also show a higher yield at ambient than at CF levels of ozone, but this is the exception rather than the rule. Whether the increases in yield in the near ambient levels of ozone observed in this and a few other NCLAN studies is real or a result of random variation will be discussed later. The point to be noted from Fig. 16.4 is that the inherent curvature of the quadratic polynomial will not allow it to represent a plateau-type response as does the Weibull. The linear response models were not always satisfactory in that they sometimes missed small plateaus that might

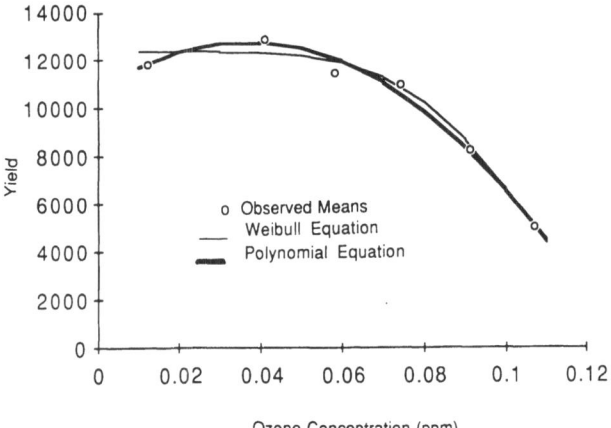

Fig. 16.4. Quadratic polynomial and Weibull models fit to the 1981 Argonne corn study (Kress and Miller, 1985*a*), hybrid PAG-397, showing "peaking" of the quadratic response function at near ambient levels of ozone.

be present. Finally, polynomial models were not going to be very convenient for comparing responses over experiments. Polynomial equations of differing degree are difficult to compare; the coefficients change meaning as additional terms are added to the equation, so that the coefficients from a linear response in soybeans at Raleigh are not comparable to the coefficients of a quadratic response in soybeans at Beltsville. These differences in degree of polynomial also come into play when one starts testing homogeneity of responses over experiments.

All of these features led to a search for a nonlinear model that might represent the yield responses to ozone more consistently and more simply. The criteria below were used.

(1) Foremost is that within the limits of experimental error the model must adequately represent the observed responses.
(2) The model must be flexible enough (with choice of parameters) to represent the range of observed responses over species. This was for simplicity, for ease of combining information over studies, and for testing homogeneity of responses.
(3) The model should have a biologically realistic form. (This is not saying that it must be derived from biological principles.) The reasons for this criterion were, in part, simple aesthetics, but more importantly, a realistic model reduces the potential for introducing biases in fitted values due to the model not being completely correct. Again, the quadratic polynomial fit to the plateau-type response is an example where a model of incorrect form, but which shows no significant lack of fit, might produce biased estimates.
(4) The parameters of the model should be easily interpreted.

The NCLAN data had shown a range of response varying from an exponential decay type response, to a nearly linear response, to a very marked

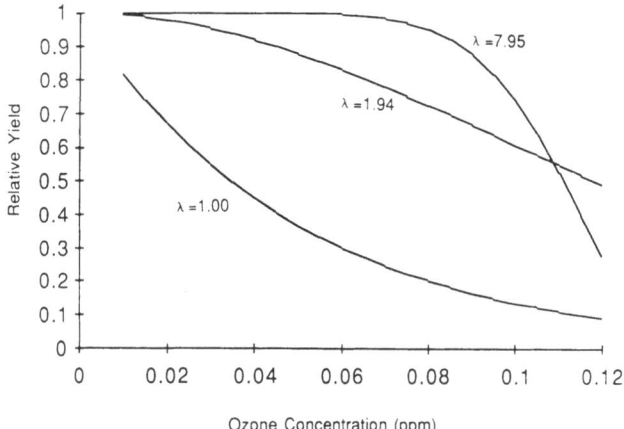

Fig. 16.5. Response curves obtained from the Argonne (Kress and Miller, 1985*b*) and Boyce Thompson (Kohut *et al.*, 1987) wheat studies showing the range of types of responses observed in the NCLAN data.

plateau before ozone had an impact (Fig. 16.5). A variation of the logistic growth model was considered. The logistic is a three-parameter S-shaped growth curve, and with appropriate choice of parameters, it might cover the range of responses. However, the parameters of the logistic are difficult to interpret and there is a strict symmetry about the midpoint of the curve that somewhat limits its flexibility. The Mitscherlich growth model is inadequate in that it accommodates only the exponential decay-type response. The probit transformation to utilize the normal distribution function as a response model also imposes a symmetry of response and closes the door to use of other transformations to improve the distribution of the residuals. (There is no closed form of the normal distribution function, so it cannot be used directly as a model.)

The model finally settled on was a variation of the Weibul distribution function,

$$Y_i = \alpha e^{-(X_i/\omega)^\lambda}$$

This is a very flexible function with three parameters, all of which have simple interactions. The parameter α is the theoretical yield at zero ozone; ω is the dose at which ozone has reduced yield to $\alpha e^{-1} = 0.37\alpha$, or 37% of the yield at zero dose; λ is the parameter that controls the shape of the curve (Fig. 16.6). If $\lambda = 1$, the Weibull response model reduces to the exponential decay function, a very common function representing many phenomena. If $\lambda \doteq 1.3$, the Weibull curve is very nearly a straight line for dose levels below ω. The larger values of λ give a very pronounced plateau before the impact of ozone is realized. See Rawlings and Cure (1985) for more discussion on the Weibull model.

The Weibull function has been used extensively in reliability studies to model failure of systems. More recently, the Weibull function has been used in biology to model plant disease progression (Pennypacker *et al.*, 1980). In

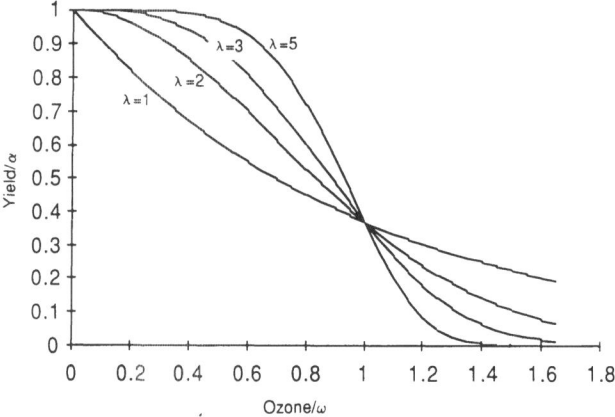

Fig. 16.6. Illustration of the flexibility of the Weibull response model with choices of λ.

reliability studies, the parameter λ has a physical interpretation as the number of independent components of the system that must fail before the system fails. Or, alternatively, one can think of λ as the number of 'hits' the system must experience before it is 'killed'. One could propose an obvious analog in the plant–ozone interaction as a justification for the Weibull model. Could the parameter λ relate to the number of 'hits' a cell must take from ozone molecules before it is inactivated? Clearly, large values of the parameter λ identify species that appear more 'resistant' to the effects of ozone. The reason for the higher resistance would be the subject of further research. Does the species have a better repair mechanism? Or does it have a greater capacity to inactivate the ozone before it causes damage, or to filter out the ozone by closing stomates?

The Box–Tidwell model has been used as a model to characterize yield response to ozone (Kopp *et al.*, 1984). The form of this model is

$$Y_i = \alpha + \beta X_i^{\gamma} + \varepsilon_i$$

This function (and its name) comes from a method proposed by Box and Tidwell (1962) for finding a power transformation on X that would 'straighten' a one-bend relationship. For $\gamma = 1$, the Box–Tidwell relationship gives the linear equation (Fig. 16.7). For $\gamma > 1$, the rate of loss ($\beta < 0$) of yield increases with dose, giving a response curve that is concave downward. Larger positive values of γ give an increasingly pronounced plateau before the response goes toward negative infinity at an increasing rate. For $0 < \gamma < 1$, the rate of loss of yield decreases with dose, giving a response curve that is concave upward.

From purely empirical considerations of adequacy of model fit, there is little reason to choose the Box–Tidwell over the Weibull or vice versa for many NCLAN data sets. Since the Box–Tidwell model includes the linear response as a special case, it clearly will be an adequate model for the cases where the observed response is essentially linear. For the concave upward curves, the Box–Tidwell model tends to concentrate the losses somewhat more heavily in

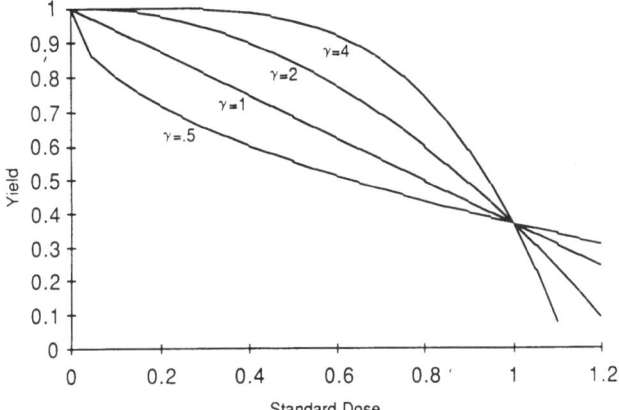

Fig. 16.7. Illustration of the shapes of the Box–Tidwell model with choices of the power parameter γ.

the low dose region than does a similarly shaped Weibull. However, it is doubtful that the differences could be detected in any one data set on a purely empirical basis. The Box–Tidwell function can accommodate a plateau-type response (by choosing γ large) as long as the response beyond the plateau is limited. It cannot, however, accommodate a plateau-type response where the loss is sufficient to begin to show the asymptotic approach to zero yield (i.e., an S-shaped response), since it has only one direction of curvature.

Finally, the Box–Tidwell function does not provide for the asymptotic approach to zero yield for high doses as one would expect for responses to a toxic substance; regardless of the value of γ, the function will eventually give negative yields ($\beta < 0$). In this sense, the function is not biologically realistic, even though its flexibility in the lower dose range would allow it to accommodate most NCLAN data sets. There are a few NCLAN data sets that appear to show the change in direction of curvature of the response as it starts to approach the zero asymptote. A function such as the Box–Tidwell that does not reflect realistic biological behavior in the upper dose levels can lead to biased estimates of losses when high dose levels are used. It is for this reason that we have used 'realistic biological response' as a criterion in choice of models.

Our experience with the Weibull function as a dose–response model in fitting the NCLAN data has been very good. The only problem occurred in a few individual studies where the observed response to ozone was so small relative to experimental error that the nonlinear program could not converge on a solution. (In most of these cases, ozone effects were not significant in either the analysis of treatment differences or in fitting the linear response.) The analysis of the fitted Weibull models and their residuals has not revealed any identifiable systematic biases. Usually, the fitted polynomial and Weibull response curves were very similar. In those cases where they differed appreciably, the Weibull more often than not appeared to give the more reasonable representation of the data.

One of our original concerns about using the Weibull model was that it does not permit yield to increase with increasing dose; it is a monotonically decreasing function. (This is also true for the Box–Tidwell model and all classical 'growth' models when used as loss functions.) When NCLAN first discussed whether the basic form of the Weibull was a realistic biological form for representing the impact of ozone, the possibility was discussed that, because of selection and adaptation or some deleterious effect of charcoal filtering, true yield might be higher at NF levels of ozone than at the CF level.

Many individual studies gave observations with higher yields for the NF treatment than for the CF treatment, and in some the NF treatment mean was higher than CF treatment mean. These types of results raised the question of whether there might be an increase in yield with increasing O_3 in CF to NF range. However, none of the individual studies had sufficient power to label any of these increases as statistically significant. To obtain greater power to address this question, the CF–NF comparisons from all studies over all species were combined. The summary of all NCLAN data does *not* support the hypothesis of increasing yield with increasing O_3.

Table 16.2 summarizes a sign comparison of the CF and NF treatments over all NCLAN studies. The chi-squares given are for the null hypothesis of equal true yield for NF and CF. The comparisons were made between paired experimental units receiving the NF and CF treatments within blocks and at

TABLE 16.2

Sign comparison of CF and NF treatments over all NCLAN studies, and the chi-square test of $H_0: p = 0.5$

Species	No. times CF > NF	No. times CF < NF	Chi-square $H_0: p = 0.5$
Alfalfa	15	5	4·04
Barley	3	5	0·13
Corn	16	18	0·03
Cotton	17	6	4·34
Forage	12	12	0·04
Kidney bean	6	2	1·13
Lettuce	0	4	2·25
Peanut	4	0	2·25
Sorghum	3	0	1·33
Soybean	67	29	14·26[a]
Tobacco	1	1	0·0
Tomato	8	6	0·07
Turnip	9	7	0·06
Wheat	22	14	1·36
Total	183	109	18·75[a]

[a] Computed chi-square exceeds the critical value for $p = 0.01$. "Total chi-square" is the 1 df test of the agreement of the totals over species with the null hypothesis of equal probability.

the same level of all other treatment factors involved. Thus, several paired comparisons were made in each experiment. Only three of the 14 species, corn, lettuce, and barley, showed more than half the cases with NF having the higher yield. None of the chi-squares for these three cases showed a significant departure from the null hypothesis of equal yields. (The critical value for a two-tailed test with $\alpha = 0.05$ is $\chi^2 = 3.841$.) Corn and lettuce are two species that showed a strong plateau, very little effect of ozone in the low doses, so one would expect essentially equal distribution of the counts as is observed. And the barley study showed no significant effect of ozone. One species, soybean, showed a significant departure from the null hypothesis but this was in the direction of lower yield for the NF treatment. Overall, 63% of 292 comparisons showed CF > NF. This is a significant departure from equal yield and in favor of lower yields for NF. Further, the chi-square test of heterogeneity among the species is not significant. Thus, for each species and over all species, these comparisons do not support a conclusion of a higher yield at ambient air levels of ozone than at CF levels.

The sign comparison, which failed to support the alternative hypothesis that yield under ambient conditions might be higher than yield under CF air, ignored the magnitude of the differences, some of which were quite large. Of the seven species with the most data, the largest relative yield gain from the CF treatment mean to the NF treatment mean was 20% for cotton (*Gossypium hirsutum* L., cultivar Acala SJ-2) under moisture stress conditions in the California study in 1981 (Temple *et al.*, 1985; Heck *et al.*, 1984*b*). The second largest mean gain was 15% for wheat (*Triticum aestivum*, L., cultivar Abe) grown at Argonne in 1983 (Kress and Miller, 1985*b*). However, in spite of the fact that these were the largest of the treatment mean differences observed in that direction, neither of these mean differences is significantly different from zero by the standard *t*-test. Thus, neither the sign comparison nor mean comparisons support a conclusion that true yield increases from CF to NF. Thus, a monotonically decreasing response model like the Weibull would appear to be consistent with the data.

The final comparison to be made relative to choice of models is of the estimates of relative yield losses obtained from the fitted polynomial and Weibull models. This is to address the concern that the use of the Weibull model might bias the estimated yield losses in the low-dose range of the curve. We have already noted that inspection of the plots of the fitted equations and the residuals sometimes indicated an inadequacy of the polynomial models, even though the lack-of-fit was not significant, but that no systematic lack-of-fit from the Weibull model was detected. (Recall that the Weibull model was chosen for its flexibility in providing the types of responses being observed in the NCLAN studies.)

In order to compare possible biases of the two models in the low dose region, the estimated yield losses were compared for the lowest dose interval in each study, that between the CF and NF treatments. For each CF and NF treatment, the fitted polynomial and Weibull equations are used to estimate the mean yields at the observed levels of ozone, and then the relative yield losses between the CF and NF levels of ozone are estimated. For comparison, the estimated relative yield losses from the CF and NF treatment means are included, although these have very low precision. The individual studies for the seven species on which the most information is available were used: alfalfa, corn, cotton, forage, sorghum, soybean, and wheat. There were a total of 58 cases. Of these, there were no estimates from the Weibull model for four cases because of lack of convergence of the nonlinear program. These cases were deleted for the comparisons.

The simple comparison of the mean relative yield losses (Table 16.3) for the polynomial and Weibull models shows that the Weibull, on the average, is giving slightly lower estimates of loss than is the polynomial. Thus the Weibull model does not appear to be overestimating losses compared to the polynomial model. Relative to the average losses estimated from the CF–NF means, both the polynomial and Weibull models appear to be overestimating the yield losses in the low-dose range, 2·4% for the Weibull and 3·3% for the polynomial.

Whether this is to be regarded as a serious bias depends on one's conclusion from the earlier discussion on the likelihood of an increase in yield from CF to NF. Both the CF–NF estimates and the polynomial estimates, unlike those from the Weibull (or any monotonically decreasing response model), allow for increasing yield (negative losses) in this interval. Consequently, one might expect the estimated losses from the Weibull to be slightly larger. If one adopts the attitude that the true yields cannot increase with increasing O_3 so that the true losses cannot be negative, then the best estimation procedure for any model would set to zero all negative estimates of loss. Following this strategy, the constraint inherent in the Weibull model would be imposed on the CF–NF and polynomial estimates to yield only zero or positive estimates of loss. For comparison, the bottom portion of Table 16.3 gives the means when this constraint is imposed; all negative estimates of loss have been set to zero. The difference between the means for the CF–NF and Weibull methods disappears; the mean loss from the polynomial model increases to give 1·3% greater loss than the CF–NF and Weibull methods. This suggests that the Weibull model is giving better estimated losses in the low dose range than is

TABLE 16.3

Mean RYL for CF–NF interval over 54 comparisons using the CF–NF means estimation and the polynomial and Weibull response equations

	CF–NF means	*Polynomial*	*Weibull*
Mean RYL	6·51	9·79	8·92
Std Dev.	13·21	7·62	6·77
If negative estimates set to zero:			
Mean RYL	8·92	10·20	8·92
Std Dev.	10·45	6·89	6·77

the polynomial model, and that the original difference in estimated loss between the Weibull and the CF–NF methods resulted from the constraint (presumed to be biological) built into the Weibull model.

Two points need to be made with respect to models that impose such constraints on the estimates. First, if the constraint properly reflects the biology, negative estimates of yield loss are unrealistic, and setting such estimates to zero will always give better estimates of loss (closer to the true loss) for any one regression than will the negative estimates. However, such estimators will tend to be biased, and averaging the estimates over several regressions, as was done in Table 16.3, will tend to compound the bias. For this reason, it is better to refit the model to the pooled data (which has been done in the analysis of the NCLAN data) than to average the losses from individual regressions. One would expect the average losses (Table 16.3) to overstate the difference between the CF–NF means estimates and the Weibull model estimates, compared to the losses obtained from fitting response models to pooled data. Second, searching for and using a biological realistic model is one way of incorporating into the analysis prior information (auxiliary information) on the behavior of the biological system. This use of auxiliary information, in this case in the form of the Weibull response model, will increase the precision of the estimates. This is reflected in part by the smaller standard deviation among the Weibull estimates in Table 16.3.

The behavior of these estimated losses is more easily seen by plotting the estimates from one method against those from another. Figure 16.8 shows RYL estimated from the CF–NF means plotted against the corresponding estimates from the Weibull equations. The most notable features of the plot are the negative estimates and the large dispersion of the 'means' estimates (vertical dispersion). The same noisy behavior of the means estimates is shown in Fig. 16.9 plotted against the estimates obtained from the polynomial model. The polynomial model also gives five negative estimates (one is hidden). These

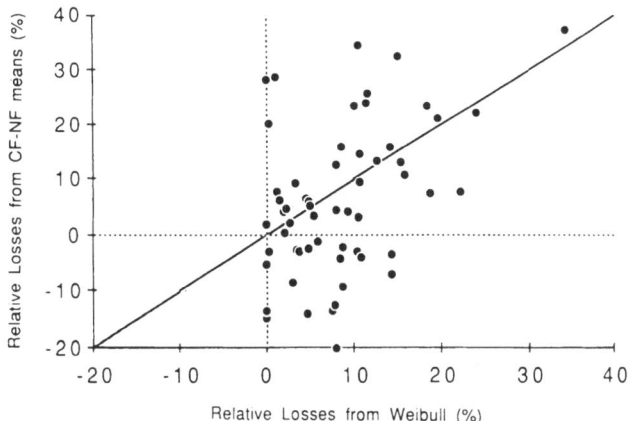

Fig. 16.8. The relative yield losses estimated from the CF, NF treatment means compared to the estimates obtained from the Weibull response equations. Data are from the NCLAN experiments on seven species. The diagonal line is the line of equal estimates for the two methods.

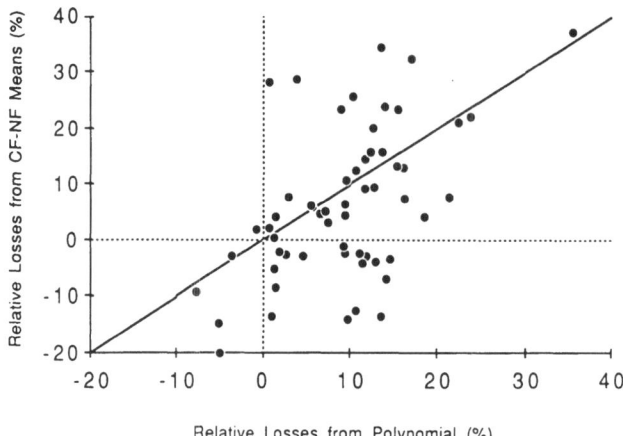

Fig. 16.9. The relative yield losses estimated from the CF, NF treatment means compared to the estimates obtained from the polynomial response equations. Data are from the NCLAN experiments on seven species. The diagonal line is the line of equal estimates for the two methods.

negative estimates for the polynomial model occur when a quadratic polynomial is fit to a plateau with the result that the fitted line rises above the plateau.

Figure 16.10 shows the polynomial estimates of RYL plotted against the Weibull estimates. The negative estimates from the polynomial are again evident. Except for the negative estimates of loss for the polynomial, the agreement between the two models is reasonable, although there is some tendency for the polynomial estimated losses to be slightly larger than the Weibull estimates when the losses are less than about 10%. The largest discrepancy between the polynomial estimate and the Weibull estimate was

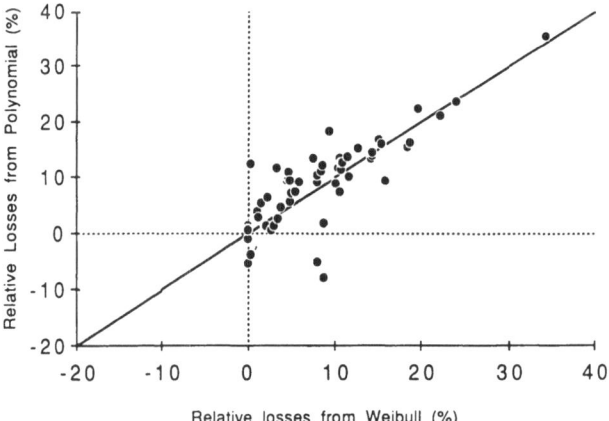

Fig. 16.10. The relative yield losses estimated from the polynomial response equations compared to the Weibull response equations. Data are from the NCLAN experiments on seven species. The diagonal line is the line of equal estimates for the two methods.

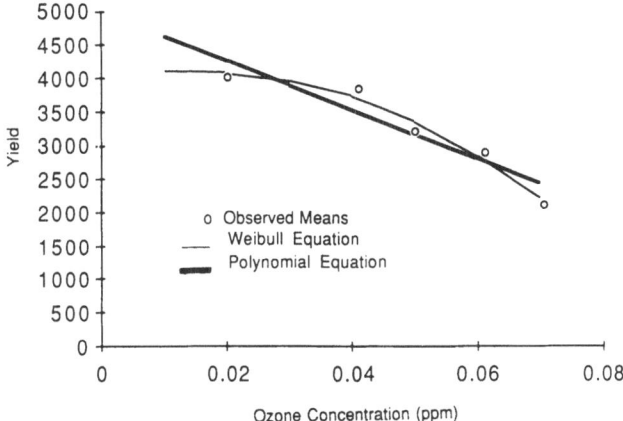

Fig. 16.11. Polynomial and Weibull response curves from Raleigh, 1985, well watered cotton study (Heagle *et al.*, 1988).

from soybeans (*Glycine max*, L., Merr.), Argonne 1983, Corsoy cultivar, where the polynomial response gave a 12·6% RYL and the Weibull gave less than 0·5% loss (Kress *et al.*, 1986). This is an example where the linear polynomial did not take into account the plateau, whereas the Weibull model did. The second largest discrepancy is from the cotton study in Raleigh, 1985, under well watered conditions, which shows an 18% loss from the polynomial and a 9% loss from the Weibull (Heagle *et al.*, 1988). Again, the Weibull model allowed for curvature that was not great enough to require a quadratic term in the polynomial model (Fig. 16.11).

In summary, NCLAN has chosen to use controlled experimentation with dose–response methodology to investigate the effects of ozone on crop production. The controlled studies have shown a direct causal effect of ozone on crop losses and provided dose–response information over ozone levels from ~0·02 ppm to well above current ambient levels. In all but six studies, the effect of ozone was statistically significant. Polynomial and Weibull response models were used to characterize the dose–response relationships. Our experience with the two families of response models has led to increasing support for the Weibull model for these data. Its realistic, flexible form has produced satisfactory fits to the data in all cases where there was a noticeable response to ozone. The simple interpretations of the Weibull parameters have made for easy comparison of responses over studies. The fitted model provides a simple procedure for estimating relative yield losses, relative to any specified level of ozone, for all levels of ozone within the experimental dose range. Measures of precision can be placed on these estimated losses. While we do not pretend that the Weibull function is the only function that would serve the purpose, we have seen no inadequacies in the Weibull model that would dictate a change in the model.

16.4 DATA ANALYSIS METHODOLOGY

The remainder of this discussion deals with the statistical methods adopted for the combined analysis of NCLAN data to produce the final dose–response relationships. There are two distinct steps to the analysis of the data for each species. The first step involves a check on the validity of the least squares assumptions, and of whether a transformation of the yield variable and/or weighted least squares would be helpful in the analysis. The second step uses transformations and weighted least squares as dictated by the results of the residuals analysis in step one to fit response models to each experiment, and then reduce the number of response equations to the minimum number that would adequately represent the species.

Even though the NCLAN program was a coordinated research program, the individual experiments differed considerably. The individual experimental designs were determined by what the principal investigator thought best for the local situation. The number of levels of ozone, and the presence and number of levels of other treatment factors, such as cultivars, SO_2, and moisture stress, differed depending upon the resources and the objectives adopted for the individual study. The realized ozone treatment levels differed, even from block to block, because ozone was being added to ambient levels. These differences in experimental design make it impossible to combine the results in the conventional combined analysis of variance. Consequently, all combined analyses used a general linear models approach with ozone levels (and sulfur dioxide levels, if present) treated as continuous response variables. All other experimental factors, site/year, block, moisture stress, and cultivar, were treated as class variables.

16.4.1 Analysis of residuals

Experimental error variances were expected to differ from experiment to experiment, particularly when different experimental sites were involved. It was anticipated that part of the heterogeneity in variances could be attributed to different mean yield levels and the commonly observed association of mean and variance. Often in such cases, an appropriate power transformation of the variable will remove a major part of the heterogeneity of variances. Otherwise, weighted least squares would be required to account for the heterogeneity. Likewise, a power transformation is often useful in improving the normality of the residuals.

To remove the effect of plot size on the behavior of the residuals, the experimental data were reconstructed so that, so far as possible, the basic experimental units were the same size in the different studies. This often meant that the data for one experiment were subdivided into data sets for the north half and the south half of the plot, or into data sets for each cultivar, for example. The appropriate analysis of variance was then run on each "standardized" data set, and the analysis of variance residuals from all data sets for the species were combined for examination. For these analyses, the ozone treatment levels were treated as class variables in order to eliminate the effect that choice of response model might have on the behavior of the residuals.

Three basic procedures were used to determine a combination of transformation of variable and weighted least squares to make the behavior of the residuals reasonably satisfactory.

(1) Bartlett's chi-square test of homogeneity was used to assess heterogeneity of error variances across experiments.

(2) Normal plots of the residuals and plots of the residuals against the estimated mean values from the analysis of variance, and against the level of ozone were used to detect nonnormality, patterns of dispersion, and outliers. (Observations corresponding to large residuals were checked for accuracy. No errors were found and no observations were dropped from the analyses.)

(3) The Box–Cox method (Box and Cox, 1964) was used to identify the power transformation (λ) that would most nearly satisfy the analysis of variance model, normality, and homogeneity of variance. In many cases, the Box–Cox analysis was used with both weighted and unweighted least squares to remove the "pressure" on the method to try to correct site-to-site heterogeneity of variances.

In many cases, the plots of the residuals from the unweighted least squares analysis of variance showed marked departures from expectations. The plots against the estimated mean showed increasing dispersion with increasing mean (Fig. 16.12), and a distinct step in the normal plot (Fig. 16.13). In nearly all cases the 'step' in the normal plot was attributable to heterogeneous site-to-site variances, and was eliminated when weighted least squares was used. (Note: Inspection of the residuals from weighted least squares must be on the weighted scale. The residuals output in PROC GLM (SAS Institute Inc., 1985), for example, are on the original scale and must be reweighted for

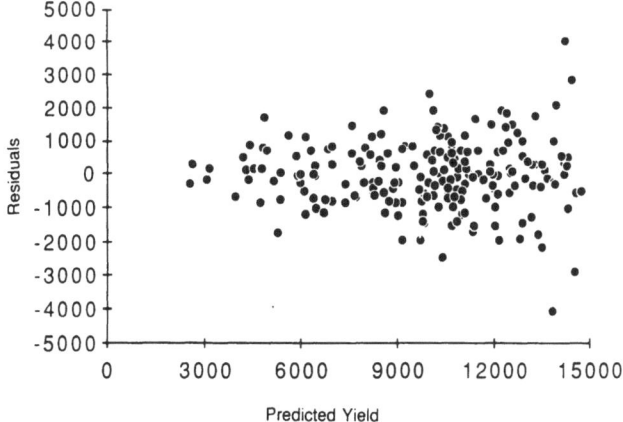

Fig. 16.12. Residuals versus predicted yields from the Argonne corn studies, 1981 (Kress and Miller, 1985a) and 1985 (Kress *et al.*, unpublished), showing increasing dispersion with increasing predicted yield.

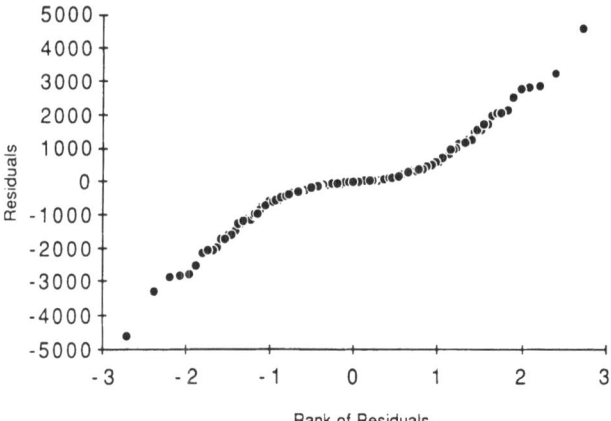

Fig. 16.13. Normal plot for the residuals from the Argonne corn studies, 1981 (Kress and Miller, 1985a) and 1985 (Kress *et al.*, unpublished), showing the S-shaped, nonnormal behavior.

inspection.) The net result of the analysis of the residuals was that ordinary least squares analysis on the original scale was used for seven species (barley, forage, lettuce, sorghum, peanut, tomato, and tobacco), and weighted least squares analysis on the original scale was used for all other species except two, corn and soybeans. In the cases of corn and soybeans, the Box–Cox analysis clearly pointed toward a square root transformation in addition to weighted least squares, as shown in Fig. 16.14 for corn. For illustration, the residuals for corn after weighted least squares on the square root transformed data are shown in Fig. 16.15, residuals versus fitted values, and Fig. 16.16, normal plot of residuals.

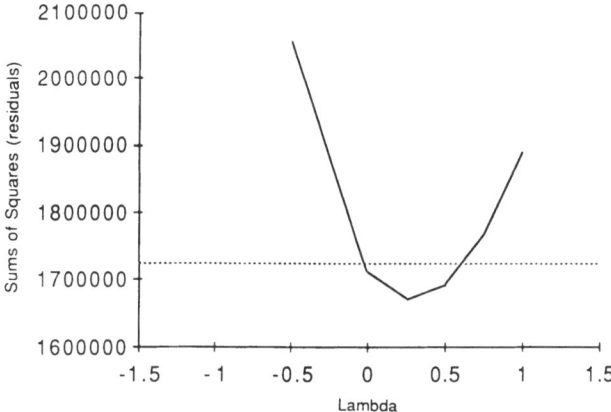

Fig. 16.14. Box–Cox plot for the Argonne corn data (Kress and Miller, 1985a) suggesting a square root transformation. The value of lambda at the minimum residual sum of squares is the suggested power transformation. Vertical lines dropped from the horizontal dotted line provide an approximate 95% confidence interval estimate of the power transformation; both $\lambda = 0$ and $\lambda = 0\cdot5$ are included in the interval.

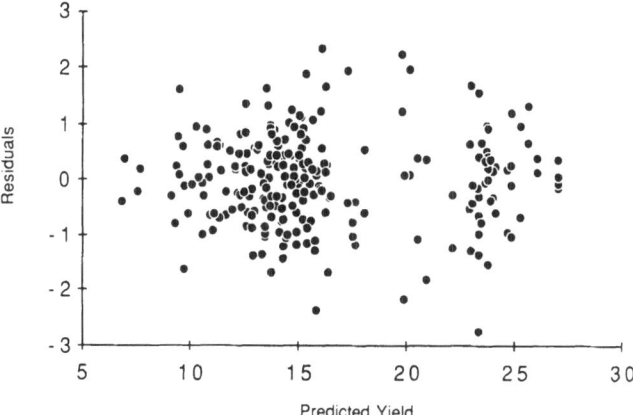

Fig. 16.15. Residuals from the Argonne corn data (Kress and Miller, 1985*a*) obtained after the square root transformation.

16.4.2 Estimation of dose–response equations

Transformation of yield and weighted least squares were used for each species as indicated by the previous analyses. The response models used accounted for the effects of ozone with either polynomial models of the appropriate degree or the Weibull model. Both models were used on all species (except barley, for which the response was not sufficient for the Weibull to converge). Sulfur dioxide effects were accounted for with a polynomial response. (A bivariate Weibull model to account jointly for the effects of ozone and sulfur dioxide was tried, but the yield response to sulfur dioxide was never sufficient to define the Weibull curve.) Effects of all other factors (blocks, cultivars, years, etc.) were incorporated by extending the

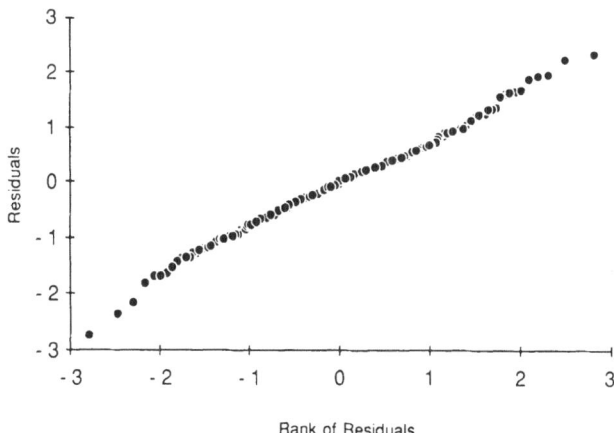

Fig. 16.16. Normal plot for the residuals from the Argonne corn data (Kress and Miller, 1985*a*) after the square root transformation.

alpha-term of the model (Rawlings and Cure, 1985). The analysis started with the individual experiments. Homogeneity of response to ozone was tested over levels of any other treatment factors included in the experiment, cultivars, moisture stress, and sulfur dioxide. When the responses to ozone were homogeneous, a common ozone response was imposed on all levels of the factor(s).

The second stage of developing the response equations was to group the responses from the individual experiments into homogeneous subgroups for the species. The subgrouping was based on tests of homogeneity of response over experiments. In some cases, a particular experimental response could be logically placed in more than one homogeneous subgroup. In these cases, preference was given to matching it with other studies from the same site. The treatment factors not common among the experiments were handled as follows in the test of homogeneity over experiments. Cultivar differences between experiments were treated as part of the experiment-to-experiment difference unless the same cultivar was used in all studies. Thus, cultivar differences, as well as any environmental differences, could be one of the contributing factors to heterogeneity in ozone response between studies. All studies were regarded as 'well watered' unless drought stress was a treatment factor and the data were coded with indicator variables as $M = 1$ or $M = 0$, accordingly. Sulfur dioxide levels were assigned the value of zero in all studies in which SO_2 was not a treatment factor. This is a realistic value for the ambient SO_2 levels at the test sites. These analyses led to what we have called the homogeneous subset models, or the Category II models.

The third and final step in building the response models was to compute the common response to ozone for each species, the Category III models. In all but three cases, this involved combining over nonhomogeneous responses, and consequently, these Category III models should be regarded at this stage as a crude average of the species behavior. They are not to be interpreted as the appropriate response for the species in all situations.

16.4.3 Measures of confidence in estimated losses

The measures of precision of the estimated relative yield losses presented for the final dose–response equations have assumed that the study-to-study effects are fixed effects. This is unrealistic if the statement on losses is to be interpreted as representative of the geographical region. A major component of the differences among studies must be due to random environmental effects, and a more realistic measure of confidence would treat study-to-study effects as random. For the balanced case, we know how to incorporate the estimates of treatment by environment interaction components of variance into the measures of precision. For the unbalanced case, and particularly with nonlinear models, the procedures are not as well defined and computer programs do not exist (to our knowledge) to do this. We are currently working on the methodology and plan to incorporate it into the MIXMOD program of F. Giesbrecht (Giesbrecht, 1983, 1984). This work is progressing and is expected to provide appropriate measures of precision for the regional inferences.

One caveat must be given. As in all cases where inferences over random environments is intended, an underlying assumption is that the environments sampled by the experiments are representative of the population of environments of interest. This assumption is always difficult to satisfy in agricultural work. The research cannot be conducted on random sites, and certainly future environments cannot be sampled by present studies. The best one can do is hope the bias from the inadequate sampling is not too serious. It is clear in the case of the NCLAN studies that the sampling of environments for regional and national inferences is extremely limited. The best set of data is for soybeans, in which several years and four test sites are involved. Although a single site cannot represent the environments of an entire region, such as the Midwest soybean region, nevertheless, the relative consistency of the responses over the test sites and years leads one to be relatively confident in the general results that soybean losses from ozone are substantial, approximately 18% between the ozone levels of 0·025 and 0·05 ppm.

The cotton data, on the other hand, involving tests in North Carolina and California, differ markedly in their responses. Even the 2 years of data in California at the same test site give very different responses. Until the environmental factors that have caused these differences can be identified and accounted for in the response equations, one cannot be very confident in the projection of cotton losses for the region. It is this degree of certainty or uncertainty that we hope to quantify from the continuing research.

16.5 WHAT MIGHT HAVE BEEN DONE DIFFERENTLY

From the statistician's point of view, what would we like to have seen done differently? And what additional information would be helpful in using the data already generated? When one begins to critique the NCLAN program and its methodology, it is important that two points be kept in mind. First, from the beginning, the program had the dual objectives of obtaining an assessment of the impact of gaseous pollutants on agricultural crops and developing an understanding of the mechanisms of damage caused by the pollutants. These dual objectives are not compatible when it comes to developing an optimal strategy for the conduct of the studies. Second, much of the NCLAN program and methodology was evolutionary in nature so that protocol and secondary objectives shifted as information was developed. For example, the mechanics of controlled proportional dispensing were not fully developed when NCLAN began so that 'constant addition' dispensing was initially used. The initial protocol of using a 7-h/day exposure was later modified to using a 12-h/day exposure on the basis of NCLAN experimental results that indicated an effect from elevated levels of ozone during the last 5 h of the daily 12-h period.

As already implied, the most serious limitation in the NCLAN data stems from the inadequate sampling of environmental conditions, an inadequate number of test sites. It would have greatly strengthened the results if resources would have permitted an adequate environmental sampling for the regional and national inferences that are implied in an economic assessment. This is an easy criticism to level at almost any agricultural study that has the objective of making an inference over time and space. The problem of adequately sampling the random environmental conditions is almost never satisfactorily met. It is doubly difficult with the very broad NCLAN objective of providing a national assessment for all of agriculture. The limited time and resources meant that some subjective judgements had to be made as to which species were to be studied and the amount of effort to be devoted to each. The primary criteria for these decisions were the relative economic importance of the species and their relative sensitivities to ozone. The latter was based on information from the NCLAN studies and, as a result, was subject to change as the program developed.

It would have simplified the final analyses if the NCLAN designs had all been standardized to include the same plot sizes, experimental designs, treatment factors, and levels of these factors. This, however, would have been impractical in some instances (cultivars are adapted only to certain regions, and ambient levels of ozone are not constant over space and time), and would have removed the flexibility to obtain physiological data and to follow important leads as they arose. Nevertheless, it is important that experimental methods and designs be kept as nearly constant as practical to facilitate combining information over studies.

A great deal has been learned from the coordinated analysis of the NCLAN data. It would have been productive in terms of feedback to the entire NCLAN program if the coordinated analysis had been started much earlier.

Considerably more information is needed for the development of realistic dose–response models that incorporate the interactive effects of important environmental factors. NCLAN was able to obtain limited information on the interactive effects of moisture stress and SO_2, but better information is needed on these factors and the multitude of other environmental factors that might be affecting the response to O_3. For the purpose of identifying the important interacting factors on which research might concentrate, more extensive use should be made of factorial experiments and, in particular, fractional factorials.

Finally, more research is needed on the definition of appropriate dose metrics that take into account stage of development of the plant, the rate of physiological activity of the plant, and relative weighting of different concentrations of ozone. It may be that more progress could be made by recognizing the multidimensional nature of the exposure regime, and adopting the use of multiple measures of dose in the dose–response equations. For example, perhaps a mean measure such as the 12-h dose metric used by NCLAN plus a measure of dispersion that allows for the additional impact of peaks is necessary. It is quite likely that differences in effective ozone dose, however dose is to be defined, are a factor in the observed site-to-site differences in ozone response.

ACKNOWLEDGEMENTS

Journal Series Paper No. 11370 of the North Carolina Agricultural Research Service, Raleigh, NC 27695-7601. Research partly supported by Interagency Agreement between the USEPA and the USDA: Interagency Agreement No. DW 12931347, and Specific Cooperative Agreement No. 58-43YK-6-0041 between the USDA and the NC Agricultural Research Service. The authors acknowledge the contributions of the principal investigators of NCLAN: A. S. Heagle, H. E. Heggestad, R. J. Kohut, L. W. Kress, and P. J. Temple, in the use of their data in this report. The use of trade names in this publication does not imply endorsement by the USDA or the North Carolina Agricultural Research Service of the products named, nor criticism of similar ones not mentioned.

REFERENCES

ASA Coordinating Committee (J. O. Rawlings, Chairman, E. Landau, and G. P. Patel). (1985). ASA reviews of EPA-funded acid precipitation research. *Am. Statistician* **39**, 243–59.

Box, G. E. P. and D. R. Cox. (1964). An analysis of transformations. *J. R. Statistical Soc., Series B,* **26**, 211–43.

Box, G. E. P. and P. W. Tidwell. (1962). Transformation of the independent variables. *Technometries,* **4**, 531–50.

Cochran, W. G. (1983). *Planning and analysis of observational studies.* New York, John Wiley.

Dassel, K. A. (1987). *Experimental design for the Weibull function as a dose response model.* PhD dissertation, Institute of Statistics Mimeograph Series No. 1910T, Raleigh, North Carolina State University.

Dassel, K. A. and J. O. Rawlings. (1988). Experimental design strategy for the Weibull dose–response model. *Environ. Poll.,* **53**, 333–49.

Giesbrecht, F. G. (1983). An efficient procedure for computing MINQUE of variance components and generalized least squares estimates of fixed effects. *Comm. Statistics, Theory, Methods,* **12**(18), 2169–77.

Giesbrecht, F. G. (1984). *MIXMOD, a SAS procedure for analysing mixed models.* Institute of Statistics Mimeograph Series No. 1659. Raleigh, North Carolina State University.

Heagle, A. S., M. B. Letchworth, and C. Mitchell. (1983). Injury and yield responses of peanuts to chronic doses of ozone in open-top chambers. *Phytopathology,* **73**, 551–5.

Heagle, A. S., J. E. Miller, W. W. Heck, and R. P. Patterson. (1988). Injury and yield response of cotton to chronic doses of ozone and soil moisture deficit. *J. Environ. Qual.* (in press).

Heck, W. W., W. W. Cure, J. O. Rawlings, L. J. Zaragosa, A. S. Heagle, H. E. Heggestad, R. J. Kohut, L. W. Kress, and P. J. Temple. (1984a). Assessing impacts of ozone on agricultural crops: I. Overview. *J. Air Pollut. Control Assoc.,* **34**, 729–35.

Heck, W. W., W. W. Cure, J. O. Rawlings, L. J. Zaragosa, A. S. Heagle, H. E. Heggestad, R. J. Kohut, L. W. Kress, and P. J. Temple. (1984b). Assessing impacts of ozone on agricultural crops: II. Crop yield functions and alternative exposure statistics. *J. Air Pollut. Control Assoc.,* **34**, 810–17.

Kohut, R. J., R. G. Amundson, L. Colavito, L. van Leuken, and P. King. (1987). Effects of ozone and sulfur dioxide on the yield of winter wheat. *Phytopathology,* **77,** 71–4.

Kopp, R. J., W. J. Vaughan, and M. Hazilla. (1984). Agricultural sector benefits analysis for ozone: Methods evaluation and demonstration. US Environmental Protection Agency, Office of Air Quality Planning and Standards, EPA report no. EPA-450/5-84-003, Research Triangle Park, NC. Available from: NPIS, Springfield, VA; PB85-119477/XAB.

Kress, L. W. and J. E. Miller. (1985a). Impact of ozone on field corn yield. *Can. J. Bot.,* **63,** 2408–15.

Kress, L. W. and J. E. Miller. (1985b). Impact of ozone on winter wheat yield. *Environ. Exp. Botany,* **25,** 211–28.

Kress, L. W., J. E. Miller, H. J. Smith and J. O. Rawlings. (1986). Impact of ozone and sulphur dioxide on soybean yield. *Environ. Pollut., Ser. A,* **41,** 105–23.

Pennypacker, S. P., H. D. Knoble, L. E. Antle, and L. V. Madden. (1980). A flexible model for studying plant disease progression. *Phytopathology,* **70,** 232–5.

Rawlings, J. O. and W. W. Cure. (1985). The Weibull function as a dose–response model to describe ozone effects on crop yields. *Crop Sci.* **25,** 807–14.

SAS Institute Inc. (1985). *SAS User's Guide: Statistics,* Version 5 Edition. Cary, NC, SAS Institute Inc.

Temple, P. J., O. C. Taylor, and L. F. Benoit. (1985). Cotton yield responses to ozone as mediated by soil moisture and evapotranspiration. *J. Environ. Qual.* **14,** 55–60.

17

ASSESSING THE MECHANISMS OF CROP LOSS FROM AIR POLLUTANTS WITH PROCESS MODELS

R. J. LUXMOORE

Oak Ridge National Laboratory, Oak Ridge, Tennessee, USA

17.1 INTRODUCTION

The modeling of plant physiological processes with digital computers is entering its third decade. Continuing advances in computing technology have greatly reduced the cost of simulation and have greatly expanded the scope of problems that may be investigated. It also appears that these advances are continuing at an accelerating rate with the development and use of supercomputers, parallel processors, and optically based machines. These are exciting and challenging opportunities that require some vision as to what can be realistically undertaken in the near term and be prepared for in the longer term. This paper considers the advantages and disadvantages of process modeling of physiological responses of crops to air pollutants and considers the role of process modeling in a crop loss assessment program such as could have been developed in the National Crop Loss Assessment Network (NCLAN) in the United States. A brief review of plant models is given, and a concept of how process models can contribute to crop loss assessment on a site-specific and regional basis is outlined. As such the paper has a more forward-looking focus and less of a retrospective analysis of air pollutant effects modeling.

A process model is defined as a mathematical simplification of the physical, chemical, and physiological mechanisms determining the flows, storages, and conversions of carbon, water, and chemicals in soil–plant–atmosphere systems. Models vary considerably in complexity, and most do not account for all aspects of the above definition even though considerable gains in knowledge have been made in the last decade.

17.2 PROCESS MODELING

17.2.1 Disadvantages of process models

There are no disadvantages, in a scientific sense, in using process models in assessment programs. However, in an environment of limited research funds, the advantages and disadvantages are sometimes viewed in comparative terms with the advantages and disadvantages of experimental research and alternative modeling methods. It is readily acknowledged that there are limitations in the use of process models, which relate directly to the simplifications of the real world adopted in model formulation. As model complexity increases, the number of parameters increases, and this bears on the experimental approaches required to provide suitable data for modeling. Conversely, it is also acknowledged that experimental methods have limitations in addressing biological issues. The use of open-top chambers for air pollutant exposure to plants is a relevant example. The main advantages of process models come when they are used as an integral part of research with experimental studies. In this way the limitations of modeling and experimental methods can be mutually understood thereby facilitating the extrapolation of results for assessment needs. Process modeling approaches were proposed for NCLAN (Kercher *et al.*, 1982), but their use has been on a restricted basis rather than being an integral part of the experimental research. The decreasing cost of simulation methods is a significant factor in future considerations.

17.2.2 Use of process modeling

Process modeling is relatively meager in scope in air pollution effects research in comparison with the modeling activities in other areas of environmental biology. Crop-pollutant models have often had a specific mechanistic focus on pollutant uptake using empirical relationships to account for other plant mechanisms or environmental effects. The reluctance to invest in comprehensive whole-plant physiological models contrasts with the general acceptance of short-term experiments generating empirical data of limited scope and applicability.

The Carbon Dioxide Research program operated by the US Department of Energy (DOE) has a major commitment to process models for the investigation of CO_2 enrichment affects on vegetation (Dahlman, 1985). Similarly, the National Forest Response Program of the US Department of Agriculture (USDA)-Forest Service and US Environmental Protection Agency has a commitment to integrated experimental and process modeling investigations of forest decline. The USDA-Agricultural Research Service has recently formed a Model and Data Base Coordination Laboratory that seeks to promote the use of mechanistic plant models in research and extension. These are encouraging developments, and the reticence of the last decade seems to be giving way to an earnest effort at linking process modeling and experimental research. In addition, the new initiative in the United States by the National Science Foundation (NSF), USDA, and DOE to support centers of excellence in the plant sciences (Crawford, 1987) should spur the synthesis of basic plant physiological knowledge in well-considered long(er) term research.

17.2.3 Mechanistic, phenomenological, and regression approaches

It is useful to distinguish between three types of crop response models (Table 17.1, Fig. 17.1). Regression relationships discussed in part by Rawlings and Dassel (1988) are the simplest and have the characteristic of "instantaneous simulation". As soon as the input conditions are specified, the final result is completely defined by the regression relationship. The exposure–response regression relationships used in crop loss assessment can have significant problems in defining exposure (Forster, 1983), and the recent comparative summary prepared by Reich (1987) on plant response to O_3 shows that dose defined in terms of actual plant uptake provides a much more consistent estimator of plant response than values based on exposure (atmospheric concentration × time). Crops, hardwoods, and conifers displayed similar declines in photosynthesis and growth in response to total O_3 uptake calculated as the product of O_3 exposure and the mean leaf diffusive conductance for each species (Reich, 1987).

Regression relationships are also used in mechanistic and phenomenological modeling (Fig. 17.1) to define attributes of the soil–plant system of interest before the time sequence of calculations begins. Thus, the regression relationships do not completely predetermine the simulation outcome. Phenomenological models are less mechanistic than mechanistic codes (Table 17.1), but the latter often contain some phenomenological attributes (e.g., crop phenology). Both mechanistic and phenomenological types are considered to be process models in this paper since both aspects (mechanisms, phenomena) are usually included in each model type with differing degrees of emphasis.

Several reviews of crop models have been published providing some useful

TABLE 17.1
Seven comparisons between three types of models

Mechanistic models	Phenomenological models	Regression relationships
Defines mechanisms	Represents mechanisms	Represents the system
Synthesis of available knowledge	Synthesis of selected knowledge	Synthesis of experimental data
Requires many component characteristics	Requires several component correlations	Requires few, perhaps one, correlation
Empirical soil and plant characteristics	Empirically lumped parameters	Completely empirical
Useful at sites with extensive databases	Adaptable to many sites	Applicable to sites of determination
Provides robustness of understanding	Provides essence of system	Minimal understanding
Valuable for analysis of chronic impacts in complex systems	Useful compromise approach	Usually the best predictor but restricted use

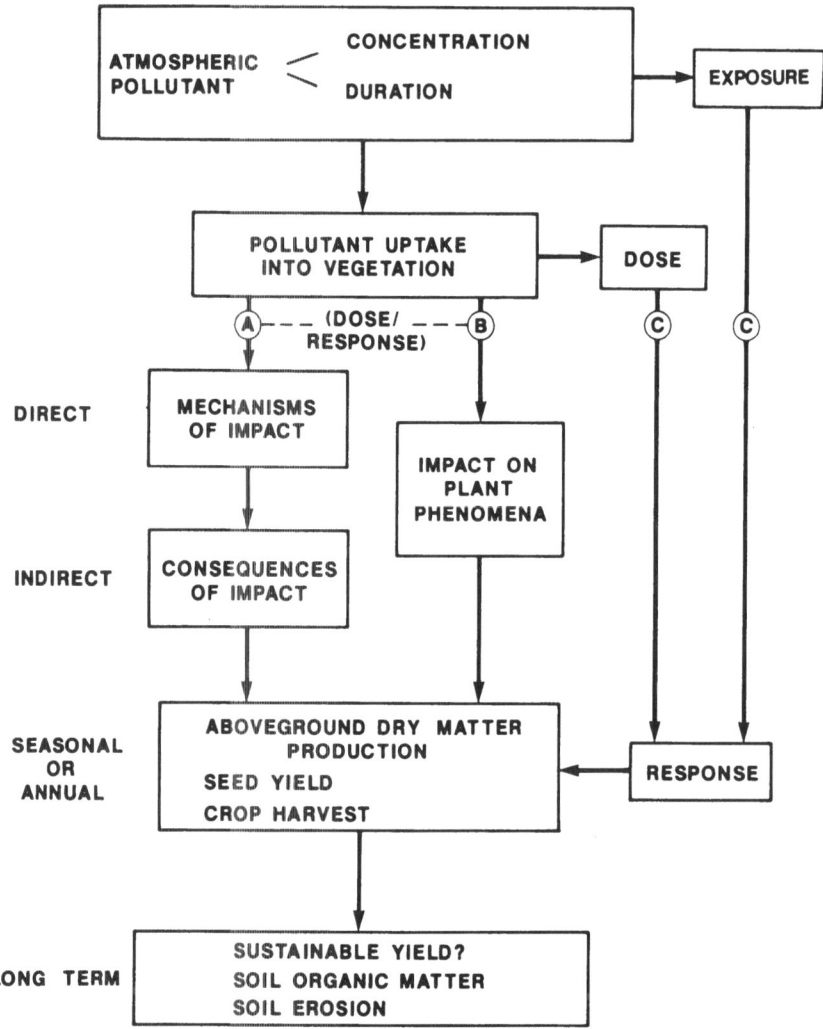

Fig. 17.1. Components of mechanistic (A), phenomenological (B), and empirical (C) models using exposure or dose–response relationships at differing levels.

evaluations for air pollution applications. Rupp *et al.* (1979) compiled a bibliography of 200 modeling reports and commented on the utility of the models for assessing impacts of energy technologies, including air pollution effects on agricultural crops. A wide range of models were evaluated (regression, phenomenological, and mechanistic), and many were viewed as suitable for air pollution assessment through incorporation of dose–response relationships. Relatively few specific air pollutant–crop models were identified at that time.

Several mechanistic models of air pollutant effects on crops, grasslands, and forests have been reported in recent years, and Krupa and Kickert (1987) have summarized their evaluations of these model developments. Four of the models were for crops, two for grasslands, and the remainder (nine) were for woody vegetation. There is a surprising scarcity of mechanistic crop-pollutant response models. The most consistent effort has been with gaseous pollutant uptake modeling. A brief review of models of pollutant uptake and crop effects is given in the next two sections dealing with component processes and whole-plant modeling. The discussions also include references to models that could be usefully adapted to air pollutant effects modeling.

17.3 COMPONENT PLANT PHYSIOLOGICAL PROCESSES

Many gains in understanding of plant physiological processes and responses to environmental factors have come from process modeling. The Penman–Monteith equation (Monteith, 1965) has been highly successful in predicting plant transpiration for a wide range of soil–plant weather systems (Sharma, 1984). Detailed models of physiological processes have been developed in recent years. Parkhurst (1986) reported investigations of the influence of internal leaf structure on steady-state CO_2 diffusion from stomata to a chemical reaction sink in palisade cells using finite element methods. The model showed that stomata behave as point sources of CO_2 from the viewpoint of leaf cells, and large variations in CO_2 concentration can occur in mesophyll cells. Optimal rates of photosynthesis were associated with very large substomatal cavities. Leaves with such cavities may also support high rates of gaseous pollutant uptake.

A model for translocation of photosynthate through phloem based on the Münch–Horwitz theory (Goeschl and Magnuson, 1986) has been responsible for generating several ideas on the links between translocation and plant water relations, which were initially untestable due to the lack of suitable experimental systems. This modeling research spurred the initiative to develop the ^{11}C tracer technique for plant systems in The United States. An increase in phloem loading rate was predicted by the model to increase photosynthate concentration in sieve tubes and increase translocation speed (Fig. 17.2) if phloem unloading was not limiting. These forecasts were confirmed experimentally (Magnuson *et al.*, 1986) with the ^{11}C technique. Converse effects would be expected with reduced phloem loading rates as may occur under air pollutant stress. The combination of low photosynthate concentration in phloem and low translocation speed could be significant mechanisms contributing to reduced root growth of plants exhibiting a decline in photosynthesis with gaseous pollutant exposure.

Recent reviews of plant modeling given in the two volumes edited by Wisiol and Hesketh (1987) summarize many other aspects of component process models. Conceptual frameworks of pollutant impact on plant physiological processes (Tingey and Taylor, 1982; Garsed, 1985) provide an integration of knowledge and hypothesis, which are the springboards for mathematical modeling, but few have taken the plunge. Modeling is a continuing iterative process as new knowledge is gained. For example, the roles of stress ethylene

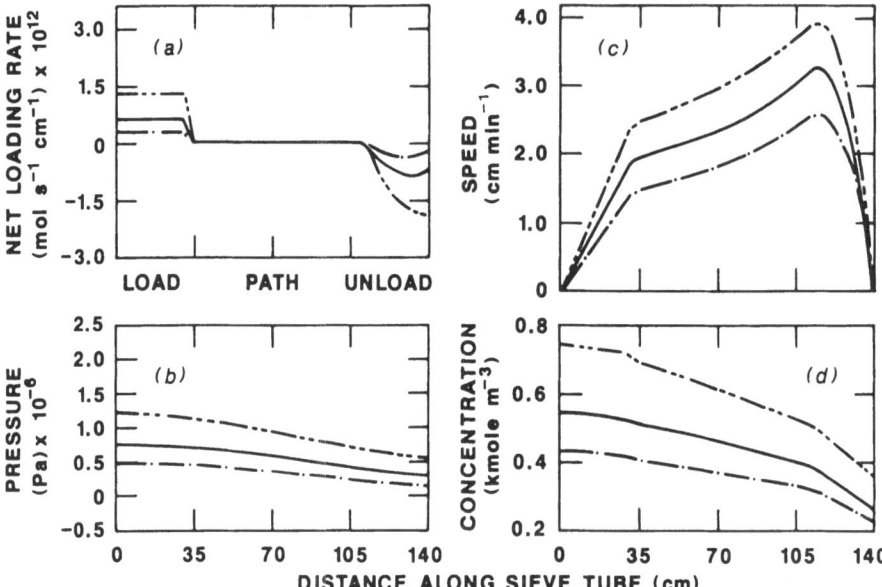

Fig. 17.2. Phloem solute concentration, pressure gradient, and translocation speed decrease with decrease in phloem loading rate (from Goeschl and Magnuson, 1986), a case that may occur with air pollutant stress. Loading rate: $-\cdot-\cdot-$, 1×10^{-11} mole s^{-1}; ——, 2×10^{-11} mole s^{-1}; $----$, 4×10^{-11} mole s^{-1}.

(Mehlhorn and Wellburn, 1987), endogenous growth (Mansfield and Davies, 1985), and stress regulators (Gollan et al., 1986) will eventually be included in mechanistic models as a means of determining the significance of these phenomena in a whole plant context. Further aspects of two component processes, pollutant uptake and photosynthesis, are considered in the next section.

17.3.1 Modeling pollutant uptake

Gaastra (1959) laid the foundation for the mathematical description of gas exchange processes of leaves with an Ohm's law type equation in which the uptake of carbon dioxide or loss of water vapor from leaves was given as the concentration gradient divided by pathway resistances including those due to the leaf boundary layer and stomata. This approach was adopted by Bennett *et al.* (1973) for gaseous pollutant uptake. The authors rearranged terms of the equation to obtain an expression for the average internal pollutant concentration C_i as follows:

$$C_i = C_a - RQ_a \tag{1}$$

where C_a is the average external pollutant concentration occurring during pollutant uptake (Q_a) and R is the pathway resistance term. O'Dell *et al.* (1977) followed the same approach for leaves and showed a linear relationship between SO_2 uptake and external concentration for this highly soluble gas with an assumed C_i of zero. Pollutant uptake into a plant canopy has been represented by the terms in eqn (1) in which the foliage is represented by a single plane in what has been called the "big-leaf" model (Sinclair *et al.*, 1976).

Shreffler (1976) used eddy diffusion relationships involving crop height, zero-plane displacement, roughness, and leaf area density to predict the flux of a gaseous pollutant to a surface with defined pollutant concentration. The method provided satisfactory comparisons with published results on thorium deposition to an artificial grasslike surface (7·5 cm height). Coughenour (1981) accounted for these meteorological transport processes in SO_2 deposition into prairie grasslands. Wind profile relationships for the canopy were used to estimate aerodynamic resistances; however, Coughenour suggested that these detailed equations were unnecessary since SO_2 deposition was insensitive to large variations ($\times 0.5$, $\times 2.0$) in wind speed.

Layered canopy models (Waggoner, 1975) accounting for within-canopy turbulent exchange processes with a resistance analogy approach often show the transport resistance within the canopy to be very small compared to the stomatal resistance, and the "big-leaf" and layered canopy approaches can give similar results for gas-exchange predictions (Sinclair *et al.*, 1976). Recently, Baldocchi (1988) has shown that a layered canopy model provides an improved estimator for SO_2 deposition in a deciduous forest over that obtained for the "big-leaf" model.

All of the resistances in the uptake of pollutant gases into foliage vary with environmental or biological factors (Unsworth, 1981), and these have been reasonably well characterized with empirical functions. The most important controlling resistance in the system is that due to stomata. There is not a complete mechanistic understanding of stomatal action at the present time, but there has been a large amount of data collected on stomatal behavior, which can provide a reasonable representation of stomatal dynamics using a combination of mechanistic and empirical functions (Zeiger *et al.*, 1987). Continuing advances in understanding of pollutant uptake by foliage will come from layered canopy models that account for the sun and shade leaves within each layer and the clumpiness (nonrandom) of leaf display. The nonlinear effects of irradiance on stomata and photosynthesis are important factors determining pollutant uptake and effects. The direct adsorption of some pollutant gases (HNO_3 vapor) onto foliar surfaces is being increasingly quantified (Taylor *et al.* 1988) and the significance of the cuticular pathway of pollutant uptake needs to be evaluated in relationship to uptake through stomata. Process modeling should be a useful tool in this research.

17.3.2 Modeling photosynthesis

There are four main components of photosynthesis in C_3 plants with an additional component for C_4 plants due to the spatial separation of Calvin cycle reactions within the vascular sheath tissues (Fig. 17.3) from the initial biochemical uptake of CO_2 in the mesophyll. The biochemical model of photosynthesis developed by Farquhar and von Craemmerer (1982) links steps 2 and 3 of Fig. 17.3 and contrasts with many empirical photosynthesis models that are based on step 4. Farquhar and von Craemmerer (1982) identified up to

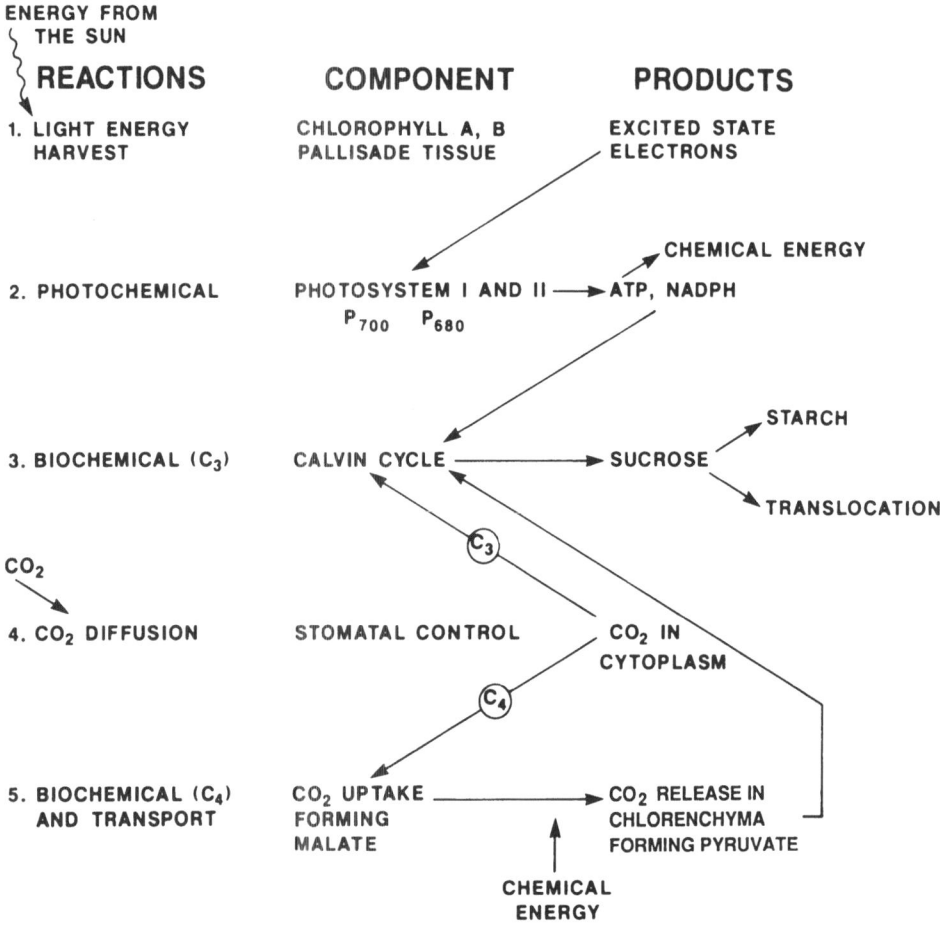

Fig. 17.3. The main attributes of photosynthesis included a range of components that may be disrupted by pollutants.

five potentially rate-limiting processes in the photochemical and biochemical components of photosynthesis and defined mathematical relationships for these processes in their model. The incorporation of pollutant effects into this model is a worthy challenge. Similarly the challenge of linking models like the internal leaf transport model (Parkhurst, 1986), the photosynthesis model (Farquhar and von Craemmerer, 1982) and the phloem transport model (Goeschl and Magnuson, 1986) is a large, but potentially useful step in the synthesis of understanding of whole-leaf physiological processes. The linkage of these three models would also be valuable for integrated investigations of foliar–pollutant interactions.

Recently, a transient-state model of photosynthesis has been used in investigations of cycling photosynthetic rates (Giersch, 1986), a phenomenon that has been recognized for some 30 years and also known to occur for stomatal aperture (Barrs, 1971). Characterization of the responses of shade leaves to sunflecks and the need to differentiate between the photosynthetic behavior of sun and shade foliage are challenges that can be met with multilayer canopy models. These levels of detail are useful for investigation of modes of action of air pollutants on leaf processes in a canopy context. The modeling of modes of pollutant action on the component processes of photosynthesis has not been undertaken to any significant extent; however, the modeling of stomatal and net assimilation responses has been undertaken.

Schut (1985) investigated the effects of short-term O_3 exposure on leaf physiology with a series of six models using published data on O_3 effects on photosynthesis and transpiration. By assuming that O_3 concentration within the leaf is zero and that the leaf maintains a constant internal CO_2 concentration, O_3 uptake, CO_2 uptake, and stomatal conductance were proportional. Photosynthesis was represented by a CO_2 gradient/resistance equation, and inhibition due to O_3 was made proportional to O_3 uptake. The consequences of a threshold O_3 uptake due to the effects of the absorptive capacity of cell walls and deactivation of O_3 by enzymes or chemical reaction was evaluated with model 2 (Fig. 17.4(a)). Below the threshold rate of O_3

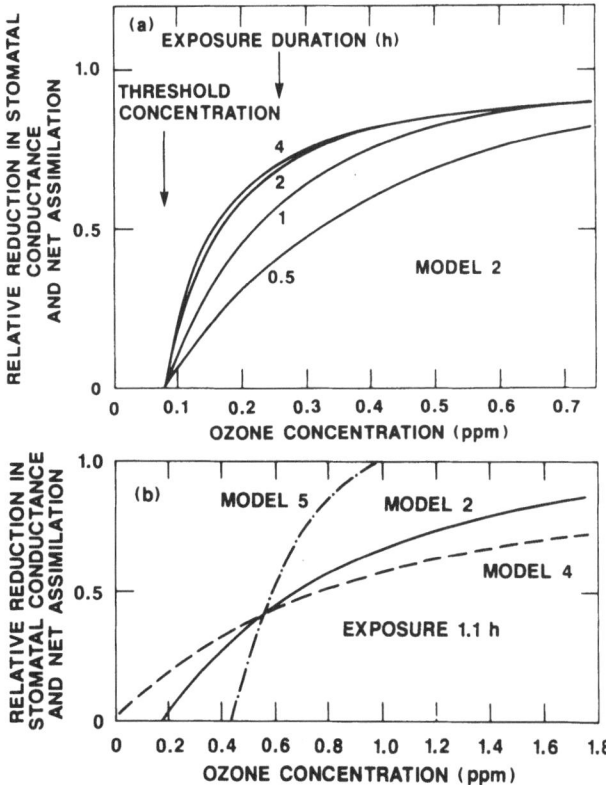

Fig. 17.4. Relative reduction in stomatal conductance and net assimilation simulated for various O_3 concentrations: (a) for Model 2 with a threshold O_3 effect and exposure durations up to 4 h, and (b) comparison of three models at 1·1-h exposure duration, where Model 2 includes a threshold O_3 effect, Model 4 has repair of O_3 damage increasing with increase in damage, and Model 5 has repair of O_3 damage dependent on net assimilation (adapted from Schut, 1985).

uptake, no disruption of photosynthesis was represented. An associated threshold O_3 concentration in the atmosphere was calculated, and values in the range from 0·040–0·11 ppm ($\approx$80–220 μg O_3 m^{-3}) were estimated for two herbaceous species. Reductions in stomatal conductance and net assimilation rate were predicted with higher O_3 concentrations (Fig. 17.4(a),(b)), and these reductions increased with exposure duration.

An alternative model, including repair of O_3 impacts proportional to photosynthesis rate in excess of a base rate (model 5, Fig. 17.4(b)), provided a favorable representation of reversible and irreversible O_3 injury. This was not the case for model 4 based on the assumption of repair being proportional to O_3 uptake (Schut, 1985). Ozone levels that caused photosynthesis to decrease below a base rate resulted in increased injury instead of repair. Repair has been shown to be slow in relation to partial stomatal closure responses. The closure responses were inferred to result from CO_2 buildup within the leaf associated with inhibition of photosynthesis. Simulation of visible leaf injury at high O_3 exposure was investigated by Schut (1985) with a normal distribution of susceptible sites over the leaf surface. The predicted fraction of injured leaf surfaces tended to be higher than observed responses for *Vicia faba* in one comparison with experimental data; nevertheless, the approach is worthy of further investigation.

17.4 MODELS USING WHOLE-PLANT APPROACHES

This section describes several alternative modeling approaches that have been or could be usefully adapted to the needs of crop loss research and assessment. They all involve whole-plant physiological processes but to differing levels of detail.

17.4.1 Solar conversion

One of the simplest statements of plant growth is given in terms of the product of two components, solar conversion efficiency and absorbed light energy [see eqn (2)].

Dry weight gain = Solar conversion efficiency • Absorbed light
$$
\text{(g plant}^{-1}) \qquad\qquad \text{(g MJ}^{-1}) \qquad\qquad \text{(MJ plant}^{-1}) \qquad (2)
$$

Monteith (1977) showed linear relationships between dry weight gain and absorbed radiation for a number of crop plants in Britain and found an average solar conversion efficiency of 1·4 g MJ^{-1} of absorbed solar energy for the crops. Air pollution impacts on vegetative growth could be investigated using this relationship and attributed to the efficiency term, which incorporates pollutant effects on net photosynthesis and carbon allocation, or to the absorbed energy term which incorporates pollutant effects on leaf area, canopy geometry, leaf reflectance, and leaf transmittance of light. It is noted that water stress and nutrition effects are also included in both of these terms.

Determination of the specific effects of pollutants on crop yield would require empirical data on the crop harvest index (crop yield/above ground dry weight at the harvest time) responses to air pollutant dose. The combination of equation (2) and harvest index values could provide a useful summary for regional assessment of crop loss due to pollutants. Simulation results from more complex plant physiological models could also be summarized in terms of pollutant effects on solar conversion efficiency, absorbed light, and crop harvest index.

17.4.2 Growth analysis

Growth analysis methods have also been used to evaluate air pollutant effects. Lieth (1982) introduced pollutant stress functions into the growth rate equations of Richards (1959) to account for one or more short-term stress effects on plant growth. In an application of the method to snap bean *(Phaseolus vulgaris* L., cv Bush Blue Lake 290) growth and O_3 stress, three parameters were quantified: one for the reduction in growth rate immediately following exposure, another to define the rate of recovery from stress, and the third for the final growth rate relative to the prestress rate. A single 3-h O_3 exposure to 0, 15, 30, 45, or 60 nl l^{-1} was given to 15-day-old bean plants, and the parameter values were fitted to experimental data from subsequent harvests. The model results showed that a decline in growth rate with increase in O_3 exposure was offset by an increased rate of recovery at the higher O_3 exposure levels. This method of modeling provides a mathematical description of stress effects without identification of any mode of action. Extrapolation of parameters from an experiment with one O_3 exposure to a multiple exposure experiment was not successful.

17.4.3 The GROW1 model

The growth model report of Kercher (1977) outlines the concepts of a general crop model (GROW1) of photosynthesis, carbon allocation, respiration, and plant growth in relation to gaseous sulfur forms. Several of the component algorithms were considered preliminary and have been extended in an application to sugar beets (Kercher and King, 1985). Air pollutant effects were incorporated into the photosynthesis equation, allowing for a stimulation at low pollutant concentration (a nutrient effect) and linearly decreasing photosynthesis above a threshold pollutant concentration (S_0). The nutrient effect is avoided with a zero threshold concentration. Simulation of SO_2 effects showed the expected results of increased growth below S_0 and decreased growth above S_0. The evaluation of pollutant × abiotic stress interactions could be usefully attempted with this type of model.

17.4.4 Grassland model

Coughenour (1981) evaluated a resistance network model for SO_2 deposition and uptake into a grassland–soil system. Some arbitrary resistance values were selected due to limited experimental characterization. This resulted in some uncertainty in the relative significance of the high resistance-large surface area pathway through cuticular absorption and leaching. Reasonable agreements between some experimental observations of sulfur concentration of dead *Agropyron smithii* shoots and simulation results were obtained through adjustment of a surface deposition resistance term. The resulting value for this resistance, $30 \, s \, cm^{-1}$, compared reasonably with the value of $36 \, s \, cm^{-1}$ for *Vicia faba* leaves obtained by Black and Unsworth (1979). Use of the surface resistance value for dead shoots in live shoot calculations overestimated the sulfur concentration of live shoots, and a mesophyll resistance of $10 \, s \, cm^{-1}$ was invoked to reduce SO_2 uptake via the stomatal pathway. This very high mesophyll resistance value for SO_2 uptake is unlikely; a value of zero is generally used for this highly soluble gas. The upshot is that the deposition resistance to live shoots is higher than for dead shoots. Comparisons of simulations with parameter adjustments provide a sensitivity analysis which can lead to new insights or hypotheses for further experimental investigation. The simulations of sulfur movement through the grassland indicated that the greatest responses would be increases in soil sulfate concentration and in the sulfur content of dead material and of labile soil organic matter. The overall effect of $0 \cdot 1 \, \mu l^{-1} \, SO_2$ exposure for 336 days was a 380% increase in the total sulfur increment in the grassland system over the control case. The longer term impacts of sulfur deposition have been investigated with the more comprehensive grassland model of Heasley *et al.* (1984).

17.4.5 Prairie-ruminant ecosystem model

The SAGE model of Heasley *et al.* (1981) has also been applied to investigations of SO_2 impacts on a northern mixed prairie grassland (Heasley *et al.*, 1984). SAGE accounts for the flows of C, N, and S in prairie vegetation, soil, and ruminant grazers with a basic time step of 1 day, with some processes being simulated within a day. Photosynthesis is represented by CO_2 uptake through a leaf diffusion resistance network to a carboxylation sink involving Michaelis–Menten kinetics.

Sulfur dioxide deposition is simulated with aerodynamic and boundary layer resistances for each of six canopy layers and for the litter-soil surface determined from wind profile calculations. Equations for a resistance analog representation of flow within a canopy are used to predict the SO_2 uptake by each leaf layer, and the calculations include the leaf area and stomatal resistance associated with each layer of foliage. Within the leaf, SO_2 is converted to sulfite and then to sulfate, and these conversion rates determine the incorporation of atmospheric sulfur into the leaf sulfur pool. During simulation when SO_2 enters foliage at a greater rate than the rate of sulfate incorporation, sulfite builds up and senescence of leaf tissue increases. The effect of SO_2 on stomata is also included, represented by an increase in conductance below a threshold concentration and a decrease at higher SO_2 levels. In addition, sulfur levels can positively or negatively influence photosynthesis through the algorithms determining the carboxylation enzyme levels.

This model offers considerable opportunity for evaluating mechanisms of SO_2 uptake, effects on leaf physiology, and the propagation of these impacts through the prairie grassland ecosystem. Increasing SO_2 exposure caused a consistent but small decline in dry matter production. However, in a 30-year simulation, dry matter production was similar for cases with no SO_2 or with 50 μl^{-1}. The main long-term effect on vegetation for 50 μl^{-1} SO_2 exposure was a reduction in the labile carbon pool associated with a higher respiration rate. There were no effects on the dry matter production, a very small decrease in soil organic matter, and an increase in cattle production with SO_2 exposure. The latter effect was attributed to a larger nitrogen retention by cattle with the enhanced sulfur diet. Such combinations of experimental and modeling investigations illustrate the most valuable role of process models in combining details into a consistent framework as a basis for long-term extrapolation and testing of concepts.

17.4.6 Phenomenological soybean model

King (1987) adopted a phenomenological approach in his model of soybean (*Glycine max*) responses to O_3 exposure and soil water stress. The basis of his approach involved modification of the transpiration model of Hanks (1974) to account for O_3 effects on seasonal crop water use. This represents one means whereby knowledge of soil and plant phenomena can be used as the basis for crop loss assessment through incorporation of exposure–response relationships obtained under controlled experimental conditions. Hourly O_3 levels above a threshold value are summed to give a daily exposure, and these are summed in the calculation of a cumulative O_3 exposure function. Water stress and long-term O_3 exposure reduce current O_3 uptake through stomatal closure, and these effects are included in the cumulative O_3 exposure function. The current value of this cumulative function is used to account for O_3 effects on current daily transpiration in addition to the direct effects of current soil water availability. Daily transpiration is summed over the growing season. Dry matter production is finally calculated as the product of seasonal water use and the transpiration ratio (dry matter production per unit water use).

There are five input parameters that characterize the soil water and O_3 effects on transpiration (Table 17.2). Insufficient data were available from one crop to estimate all parameters, and, therefore, data from sunflower and pinto bean were used in the soybean application. This is standard operating procedure in modeling research that is not integrated with experimental research. Fortunately, the model behavior is not very sensitive to the parameters from nonsoybean sources. The most sensitive parameter was B (Table 17.2), which concerned the available soil water effects on transpiration. This identifies the importance of root density and the volume of root exploration, and highlights the need for soil water characteristics in crop loss assessment. Sensitivity analyses of soil hydraulic properties and their influence on transpiration have shown that large changes (four orders of magnitude) in hydraulic conductivity can have only minor effects on transpiration (Luxmoore *et al.*, 1976), whereas variation in the soil water retention characteristics (relationship between water content and soil water tension) can have a significant influence on evapotranspiration (ET). Grassland vegetation on a

TABLE 17.2

Five parameters from the King (1987) model of crop loss, parameter values,
and sensitivity to water stress and ozone dose

---

(1) Definitions

B $\dfrac{\text{Soil water content readily available to plant}}{\text{Maximum available soil water}}$

k Sensitivity factor for transpiration to O_3 dose

Th Threshold concentration for O_3 effects on transpiration

β Exponent factor for protection from current O_3 due to water
stress and prior O_3 dose

τ Exponent factor for the relative effect of O_3 on transpiration ratio and
seasonal transpiration

(2) Values

Parameter	Value	Data source
B	0·5	Soil profile data
k	$0{\cdot}00571\,\mu l^{-1}$	Soybean (open-top chamber)
Th	$0{\cdot}01\,\mu l\,l^{-1}$	Soybean (open-top chamber)
β	1·36	Sunflower in controlled environment
τ	0·38	Pinto bean grown in saline solutions

(3) Importance

Parameter	Water stress	Ozone exposure
B	c	b
k	c	b
Th	a	a
β	a	a
τ	a	a

---

[a] Low sensitivity.
[b] Medium sensitivity.
[c] High sensitivity.

clay soil had 20% greater ET than the same vegetation on a silt loam soil in the
simulations of Luxmoore and Sharma (1984).

The relative soybean dry matter production was shown to decrease linearly
with increase in mean O_3 concentration above a threshold level of $0{\cdot}01\,\mu l\,l^{-1}$
(Fig. 17.5). In a water stress scenario, dry matter production was reduced 18%
without O_3 exposure, and the sensitivity of crop loss to increasing O_3
concentration was reduced under water stress conditions as shown by a higher
threshold concentration and reduced increment of crop loss per incremental
increase in O_3 concentration (Fig. 17.5). These aspects were built into the
model. The utility of the model is in the flexibility of application to a wide
range of soil and weather conditions.

King's model does not account for O_3 effects on grain yield. Although high
dry matter production can be correlated with high grain yield, this is not
always the case in areas with limited rainfall. Denmead and Shaw (1960)
showed that water stress during flowering of corn can dramatically decrease
yield. Although nitrogen fertilization can increase dry matter production, leaf
area and soil water use early in the growing season, reduced water availability
during the grain formation, and filling stages may depress final yield. It is
conceivable with some weather patterns that early-season O_3 damage to

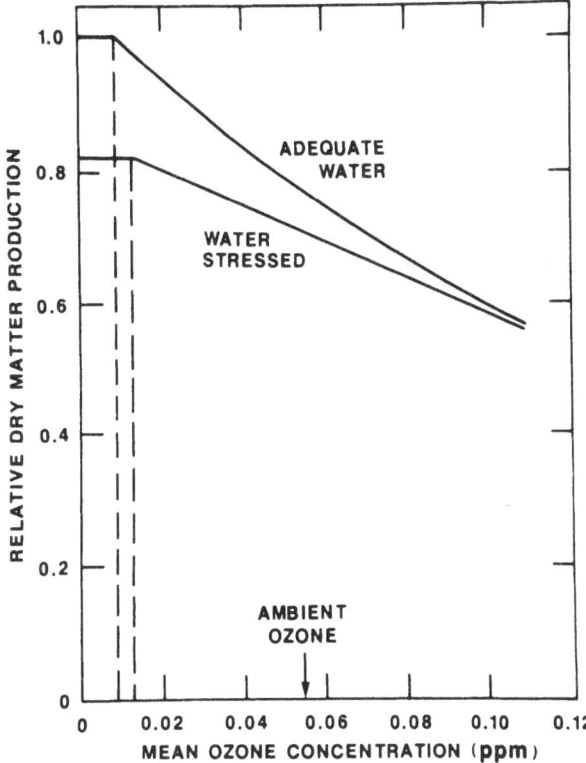

Fig. 17.5. Relative loss in soybean dry matter production with increase in O_3 concentration for two water supply situations (from King, 1987). *Water stressed* represents 1980 conditions for southern Illinois and *adequate* represents a repeat simulation with soil water fixed at field capacity.

vegetative growth of a crop could result in soil water being conserved, allowing more favorable soil water status during flowering and a greater grain yield than for a crop without O_3 stress. Mechanistic whole-plant models are capable of simulating these aspects of crop loss assessment.

17.4.7 The ECCES model

The ECCES model developed at the Riso National Laboratory in Denmark is designed to estimate environmental impacts from energy production. The code contains submodels for atmospheric dispersion and deposition, soil chemistry, and pollutant uptake by selected crops (Brodersen *et al.*, 1986). The model is concerned with particulate and aerosol pollutants and does not consider gaseous pollutants at the present stage of development.

17.5 MODELING CROP LOSS FROM AIR POLLUTANTS

Process models can support regional assessment type programs such as NCLAN in the United States in both analysis and synthesis. An example of the former is the study by Schut (1985) that evaluates the effect of short-term O_3 exposure on photosynthesis. An example of the latter is the effort of King (1987) with his phenomenological modeling of soil water deficits and O_3 effects on soybean growth.

The real value of process modeling for assessment purposes may come from the adaptation of the comprehensive crop models that have been under development during the last decade. Incorporation of pollutant uptake and physiological effects into an existing comprehensive model is quite feasible if the models have some favorable attributes. The specifications recommended for mechanistic modeling of air pollution effects are

(1) a water-budget model with explicit stomatal conductance simulation for sunlit and shaded foliage;
(2) an hourly time step to represent diurnal physiological processes in relation to pollutant exposure dynamics during the day; and
(3) a coupled photosynthesis model that has source–sink relationships and allows for feedback effects on photosynthesis.

King (1987) took two notable steps in his modeling developments. He adapted an existing model and used a water-budget model as the basis for the modeling approach. Both of these steps are recommended in any further crop loss modeling developments at the process level. Water budget models are well developed with sound foundations in physics and physiology (Jarvis, 1981) and are inherently self correcting. It is suggested that more insights will be gained by adapting existing crop models than by developing new pollutant-effects codes with limited physiology.

17.5.1 In support of complexity

Complex simulation models that require a large number of parameters are typically a collection of algorithms that require complex bookkeeping and cross-checking (mass balance and error tests) of calculations. Some outputs, such as crop yield, are not sensitive to all the complexity all of the time; nevertheless, computation with all the complexity provides robustness for the occasions when the computations are not useless, being invoked as appropriate by the computer and not being omitted for the sake of simplification or to reduce central processing unit (CPU) costs. Simplification of complex codes will be developed eventually, and there will be robustness in these simplifications if they are based on a thorough working knowledge of the system complexities and their behavior. The abbreviation of this procedure by premature simplification needs to be avoided.

17.5.2 Adaptation of crop models for site-specific assessment

A very large investment has been made in the mechanistic modeling of soybean growth and yield with the development of the GLYCIM (Acock *et al.*, 1985) model. The adaptation of this model to compute pollutant uptake into foliage and effects on leaf processes is a reasonably straightforward modification. The mechanistic propagation of the pollutant impact through whole-plant physiological processes (within phenological constraints that have been carefully determined in experiments conducted in association with model development) offers a rich resource for gaining insights into mechanisms of crop loss due to pollutant insult.

A number of crop-specific and general crop growth models have been developed, and reviews by Reynolds and Acock (1985), Joyce and Kickert (1987) and King and DeAngelis (1987) summarize and compare many potentially useful codes that could be adapted for investigation of pollutant impacts on crops. Some examples of candidate models are given in Table 17.3 with some comments on their attributes. Many of these models have been tested against experimental data (validated); however, their adaptation to air pollutant impacts on crops would require appropriate testing. The information to develop mechanistic algorithms of pollutant impact can be obtained from

TABLE 17.3

Examples of process models of plant growth and yield that could be adapted to crop loss assessment

Crop	Model	Comment	Reference
Alfalfa	SIMED	Hourly carbon gain and growth simulator; tested against field data	Holt *et al.* (1975)
Cotton	GOSSYM	Well-tested model now used in an expert system for crop management	Baker *et al.* (1983)
Grassland	SAGE	Already developed for SO_2 impacts	Heasley *et al.* (1981)
Maize	CORNF	Daily growth, yield and crop phenology simulator	Stapper and Arkin (1980)
Potato		Similar to sugar beet model	Ng and Loomis (1984)
Soybean	GLYCIM	Extensive development and testing for CO_2 enrichment research	Acock *et al.* (1985)
Soybean		Adaptation of existing models for O_3 effects.	Kercher *et al.* (1982)
Sugar beet	BEETGRO	Model is concerned with carbon gain and allocation; water and nutrients are not limiting	Hayes (1984)
Wheat	TAMW	Daily growth, phenology and water use	Maas and Arkin (1980)
General	BACROS	Well-developed code largely tested with maize data	de Wit *et al.* (1978)
General	GROW1	Crop model for effects of gaseous forms of sulfur	Kercher (1977)
General	UTM	Coupled set of transport models including pollutants; adaptable to crops	Luxmoore (1988)

published sources as well as from ongoing research into mechanisms as called for by Mansfield and McCune (1988). The overall testing and validation of mechanistic crop loss models needs to be coordinated with field experimental studies.

17.5.3 Data sources

Experimental stations and monitoring networks operated by universities and by state and federal agencies are useful sources of meteorological, soil, and agronomic data for process model applications, and these stations cover a very wide range of locations within the United States (Fig. 17.6). Also, the air quality monitoring network in the United States provides gaseous pollutant concentration data over an extensive area (Barchet, 1987). The selection of sites with extensive data sets from these sources could form nodal sites for site-specific applications within a crop loss assessment region.

The typical approach in process model application to new situations involves the assemblage of local data, such as topographic maps; meteorological records from the nearest one, two, or three monitoring stations; and county survey reports on land use, soil types, and vegetation distributions. Discussions with local extension agents can also provide valuable insights, including local crop phenology and productivity of soils. All of these data sources are useful in choosing input parameter values for crop simulation that account for the influences of local spatial variability in soil physical and chemical properties.

Spatial variability in soil hydraulic properties of a field can be represented by a frequency distribution of scaling factors (usually lognormal) that provides a convenient, simplified approach for representing soil and is well suited to stochastic modeling. In scaling, a factor of 1 is assigned to a reference soil with known hydraulic properties, and the properties of finer textured soils (factors <1) and coarser textured soils (factors >1) can be derived from the reference properties. In some studies the frequency distribution of scaling factors has been obtained from infiltration surveys (Sharma and Luxmoore, 1979).

17.5.4 Stochastic methods

Each variable in the input data stream of a simulator comes from some frequency distribution with a mean, variance, and some characteristic shape that represents the heterogeneity of the variable over some defined region. The influence of this variation on output variables (sensitivity analysis) can be determined by two main stochastic methods. The simple Monte Carlo approach involves the random selection of input values from each frequency distribution to generate a full set of input values, which yields one set of output values. The process is repeated many times to generate frequency distributions from the output variables. Mean values for the outputs with unbiased estimates of the confidence intervals can be obtained, but very large numbers of simulations (1000 +) are usually required to get useful confidence intervals for process models with many variables.

An alternative method called Latin hypercube sampling (Iman and Conover, 1980) has been developed. This method requires fewer simulation runs and produces unbiased confidence intervals for the output variables, which are smaller than those obtained by the simple Monte Carlo method for the same number of simulation runs. Iman and Conover (1980) and Gardner *et al.* (1983) describe the method and outline an application of Latin hypercube sampling. The frequency distributions of inputs are first stratified into equal probability classes, and all class intervals (differing lengths for nonuniform distributions) are used in the simulation without replacement (Fig. 17.7). Thus, once a particular class has been used, it is not reused until all other class intervals of the frequency distribution have been selected, and in this way the complete distribution is accounted for in the results. Correlations between input variables can be represented (e.g., fine soil texture, high soil nitrogen) if needed; otherwise, values are selected randomly from their respective distributions. Any artificial correlations between random inputs or unexpected relationships between correlated variables obtained in the sampling process are identified and eliminated so that statistical relationships between inputs and model response can be maintained with a low number of simulation runs. Computer programs that implement the Latin hypercube approach typically save the output variables in relationship to the input variables, allowing alternative input frequency distributions to be evaluated without the need for repeated simulation. The Latin hypercube sampling method is well suited to the use of process models in crop loss assessment, if computer costs are not excessive.

The problem of computer costs can be significant in process modeling, particularly when stochastic methods are used. The best solution these days is the purchase of a CPU to avoid CPU user charges. The computer can usually be linked to a printer and graphics plotter associated with the mainframe computer at a research institution, avoiding hardware costs for these items. For example, significant progress has been made recently with the Unified Transport Model (UTM, Luxmoore, 1988) at Oak Ridge National Laboratory. An annual simulation of the coupled component models for soil–plant water relations, plant growth, nutrient and pollutant uptake and effects, and soil chemistry usually takes 4 min on the mainframe computer at significant cost ($60). On standby (usually overnight) the cost decreases by half. The code is now operational on a purchased minicomputer, taking 27 min for an annual simulation without CPU cost. Even with the longer computation time, the actual job turnaround time is greatly reduced due to the elimination of the time required in the mainframe job queue. For the cost of 450 mainframe simulations, we can currently upgrade to a faster dedicated minicomputer, allowing the UTM to execute in 8 min per annual simulation. In a full day of execution, it would be possible to make 100 annual simulations, and in 2 days the approximately 200 runs needed for stochastic modeling could be obtained for a site. A year of effort could provide a vast amount of mechanistic modeling for 100 or so locations throughout the dryland cropping areas and the basis for regional assessment. This example illustrates the feasibility of scaling up process models in space, and the same approach can be applied in scaling up over long time scales.

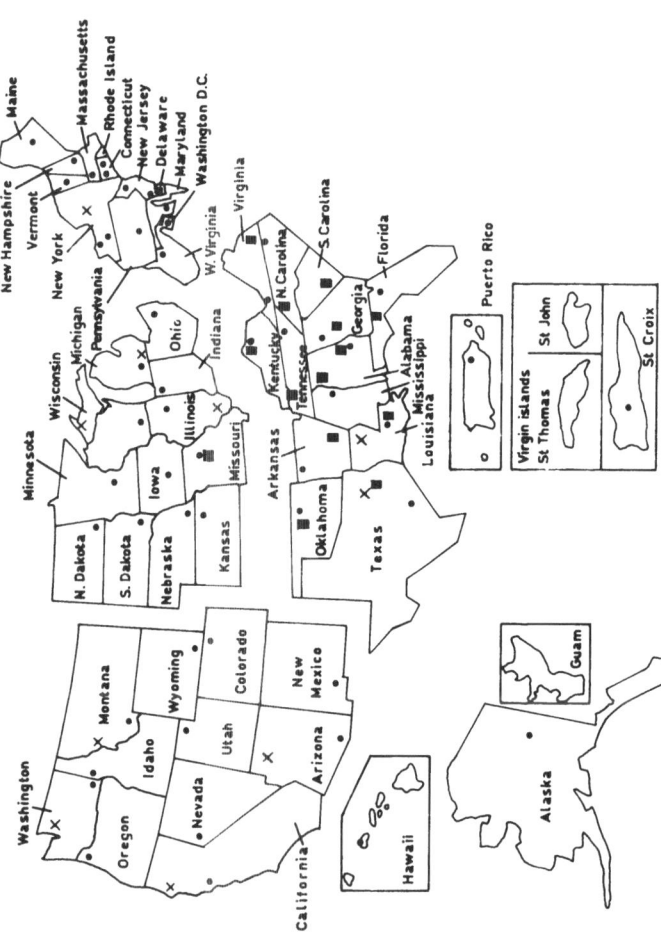

Fig. 17.6. Locations of many state and federal forestry and agricultural experiment stations in the United States. (a) ●, State Agricultural Experiment Stations; ×, Cooperating Forestry Schools; ■ Colleges of 1890 (including Tuskegee University).

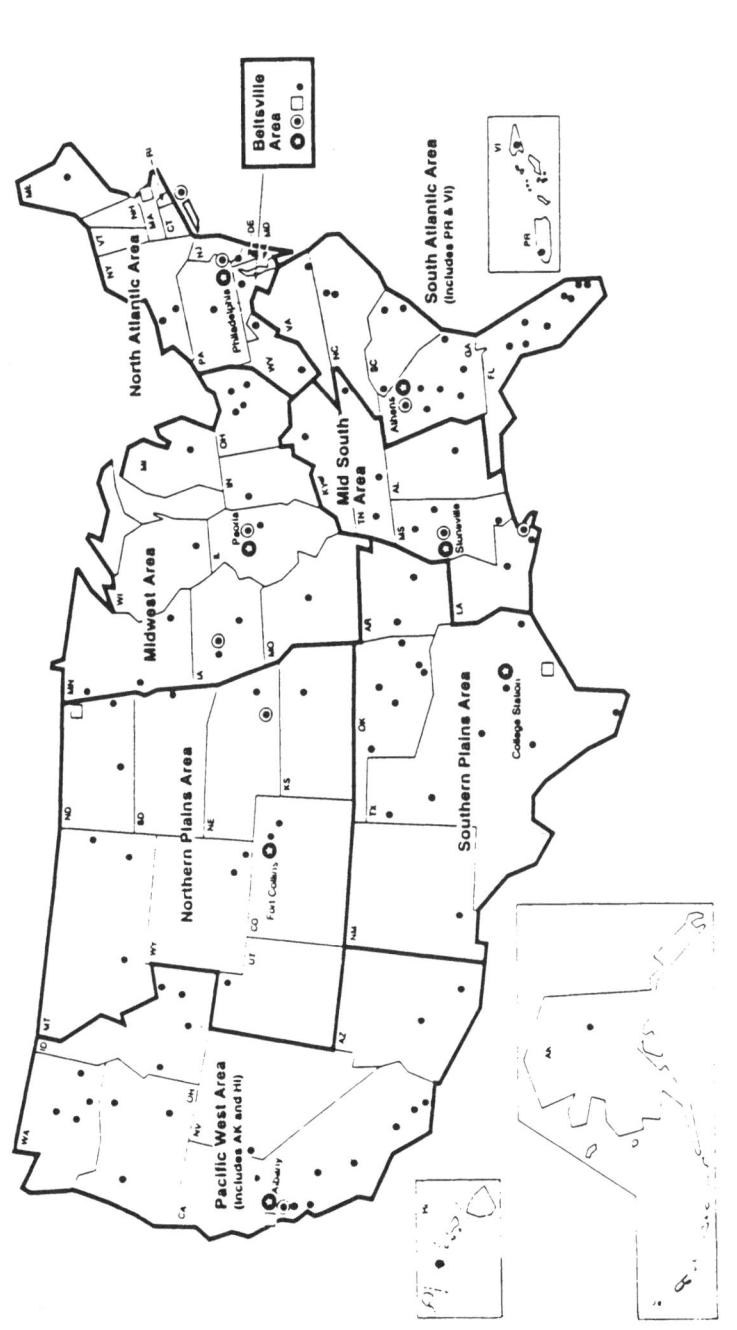

Fig. 17.6.—*contd.* (b) ⊙, Area Headquarters; ●, Research Centers; ⊙, Research Locations of the US Department of Agriculture–Agricultural Research Service.

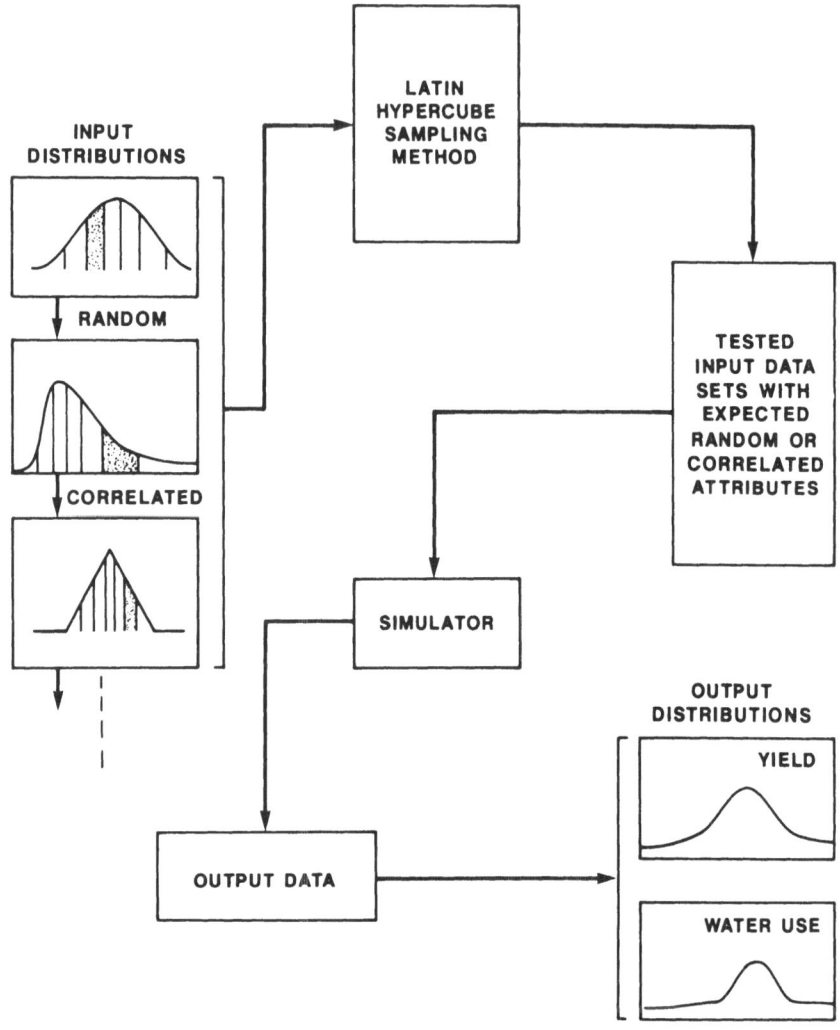

Fig. 17.7. Steps in stochastic modeling using the Latin hypercube sampling method with a crop simulator.

17.5.5 Regional assessment

Four features facilitate the use of process models in regional assessment: (1) the computer capability is affordable, (2) several mechanistic crop models are available that can be adapted for pollutant uptake and effects, (3) detailed data sets can be derived from monitoring and experimental station sources, and (4) the Latin hypercube technique for stochastic evaluation of soil–plant–weather combinations is developed. This approach involves considerable effort in data management, and it would be advantageous if it were developed with cooperation between all agencies with similar interests. For example, the USDA (crop production), EPA (pollutant impacts), and DOE (CO_2 enrichment) could use the same site-specific databases for a region and stochastic analysis methods to assess regional crop yields from different perspectives.

Site-specific modeling of crop response for the areas surrounding agricultural experiment stations could form the node sites for modeling the regional responses to air pollutants. Pattern of regional-scale spatial and temporal variability could be evaluated using geostatistics (Journal and Huijbregts, 1978) by applying semivariogram analysis to the node sites (30 or more needed) within a region. This procedure identifies the space and time scales with correlation such as may occur in areas where neighboring sites have similar soils and weather patterns. In such cases, a second geostatistical procedure, kriging, can be used to provide interpolated variable values for intermediate locations with sparse input data sets. Kriging also provides unbiased confidence intervals, a necessary component of crop loss assessment. In regions with no spatial correlation between neighboring sites, the regional response is given by the mean of the frequency distribution of nodal site responses.

King (1986) used Latin hypercube sampling methods to estimate the net CO_2 exchange of the 64°N to 90°N latitude. This region is dominated by tundra and coniferous forest vegetation, and daily time step models of net primary production for these two vegetation types were implemented with frequency distributions of climate variables representative of this global zone. The simulated net CO_2 flux for one specific site differed from the mean flux from 200 simulations representative of the region. There is no support for the concept of a "representative" site being representative of a region in the King (1986) investigations. His research is supportive of the use of site-specific models with frequency distributions for input variables representative of a region for the generation of regional responses by the Latin hypercube sampling method.

The main requirements in using the Latin hypercube sampling approach for regional assessment are that (1) the simulator represents all relevant phenomena at the largest time and space scales under consideration, (2) the model algorithms be applicable to the full range of variable values at the largest scale of time and space, and (3) the data distributions for the input variables be representative of the region and time scales under investigation.

The main advantages of the Latin hypercube sampling method are as follows: (1) the means of the output distributions provide the best estimates of the expected responses for the region and the distributions can be used to calculate unbiased confidence intervals for the mean response, (2) the method greatly reduces the number of simulations required relative to the simple Monte Carlo approach, and (3) the output distributions are in a form that can be used as input distributions in a higher hierarchical scale model or an economic model. The method is conceptually and statistically sound and it offers one means of addressing large-scale issues from a mechanistic basis.

It will take some time to implement this approach to regional crop loss assessment, and it will take some statesmanship to develop the efficiency of a cooperative agency program. Significant gains in insight of plant response to biotic and abiotic factors from comprehensive and well-planned "number crunching" will provide the robustness needed to develop simplifications for the broad-scale modeling needs of regional assessment. It is anticipated that with continuing gains in computational power and with robust simplification developments, regional assessment of crop loss could become a sophisticated science making full use of the available knowledge, data, and computer resources.

17.6 SUMMARY AND CONCLUSIONS

Process models have assisted crop loss assessment to some extent, through *analysis* of mechanisms of pollutant uptake and impact on crops, and through *synthesis* of data on crop response to pollutants for predictive purposes. However, a review of pollutant–crop models shows that the use of process modeling of pollutant effects on crop growth and yield has not been well developed, and comments on nonpollutant process models dealing with specific plant physiological mechanisms and alternative approaches of simulating whole plant behavior have been included. The adaptation of existing nonpollutant models for crop loss purposes through the incorporation of appropriate pollutant dose–physiological response algorithms is suggested to be a useful consideration in future initiatives.

One approach to applying process models in site-specific, and regional crop loss assessment is outlined. The method is facilitated by the availability of low-cost computation options, available plant growth models that can be adapted to pollutant impact studies, extensive data sets available from monitoring and experimental stations in agricultural areas, and the advances in stochastic computer methods (Latin hypercube sampling) that can account for regional variability effects of weather, soil, and plant characteristics. This approach to crop loss assessment can be applied on a site-specific or regional basis with the advantage of requiring fewer simulations than with the traditional Monte Carlo method.

In regions with sparse input data sets for process models, a nodal approach may be considered. The areas within a region with detailed data sets (e.g., experimental stations) may be simulated as the nodal sites. If there is no spatial correlation between adjacent nodes as determined by geostatistical procedures, then the regional response is obtained from the mean of the frequency distribution of nodal responses. In regions with similar crop loss responses at neighboring nodal sites, geostatistical methods (kriging) can provide isopleth responses for the region with confidence intervals. Latin hypercube sampling and geostatistical methods both allow the mean regional response to be estimated with unbiased confidence intervals. The application of detailed process models in crop loss assessment is a worthy and feasible challenge in any future assessment initiative. The full use of available knowledge, data, and computer resources required for such an effort will provide the means for regional assessment to become a sophisticated science with a sound mechanistic basis.

ACKNOWLEDGMENTS

The assistance of Randy A. Hoffman and Nancy S. Dailey with the search and compiling of reference materials is gratefully acknowledged. Research sponsored in part by the Ecological Research Division, Office of Health and Environmental Research, and in part by the Office of Planning and Environment, US Department of Energy, under contract DE-AC05-840R21400 with Martin Marietta Energy Systems, Inc. Publication No. 3039, Environmental Sciences Division, Oak Ridge National Laboratory, P.O. Box 2008, Oak Ridge, Tennessee 37831-6038.

REFERENCES

Acock, B., V. R. Reddy, F. D. Whisler, D. N. Baker, J. M. McKinnon, H. F. Hodges, and K. J. Boote. (1985). *The soybean crop simulator GLYCIM: Model documentation 1982.* US Department of Agriculture, Washington, DC, Report PB85171163-AS.

Baker, D. N., J. R. Lambert, and J. M. McKinnon. (1983). *GOSSYM: A simulator of cotton crop growth and yield.* Technical Bull. 1089, Clemson, SC. South Carolina Agricultural Experiment Station.

Baldocchi, D. (1988). A multi-layer model for estimating sulfur dioxide deposition to a deciduous oak forest canopy. *Atmos. Environ.*, **22**, 869–84.

Barchet, W. R. (1987). Acidic deposition and its gaseous precursors. In *NAPAP interim assessment.* Vol. III: Atmospheric processes and deposition. Chapter 5, 1–116. Washington, DC, National Acid Precipitation Assessment Program.

Barrs, H. D. (1971). Cyclic variations in stomatal aperture, transpiration, and leaf water potential under constant environmental conditions. *Annu. Rev. Plant Physiol.*, **22**, 223–35.

Bennett, J. H., A. C. Hill, and D. M. Gates. (1973). A model for gaseous pollutant sorption by leaves. *J. Air Pollut. Control Assoc.*, **23**, 957–62.

Black, V. J. and M. H. Unsworth. (1979). Resistance analysis of sulfur dioxide fluxes to *Vicia faba*. *Nature*, **282**, 68–9.

Brodersen, K., P. B. Montensen, and T. Petersen. (1986). *ECCES—A model for calculation of environmental consequences from energy systems predicting ion concentrations and acidification effects in terrestrial ecosystems.* Roskilde, Denmark, RISO-M-2615 Riso National Laboratory.

Coughenour, M. B. (1981). Relationship of SO_2 dry deposition to a grassland sulfur cycle. *Ecol. Model.*, **13**, 1–16.

Crawford, M. (1987). Plant science grant program nears approval. *Science*, **236**, 1620.

Dahlman, R. C. (1985). Modeling needs for predicting responses to CO_2 enrichment: Plants, communities and ecosystems. *Ecol. Model.*, **29**, 77–107.

Denmead, O. T. and R. H. Shaw. (1960). The effect of soil moisture stress at different stages of growth on the development and yield of corn. *Agron. J.*, **52**, 272–4.

de Wit *et al.* (1978). *Simulation of assimilation, respiration and transpiration of crops.* New York, John Wiley.

Farquhar, G. D. and S. von Craemmerer. (1982). Modelling of photosynthetic response to environmental conditions. In *Physiological plant ecology II. Water relations and carbon assimilation*, ed. by O. L. Lange, P. S. Nobel, C. B. Osmond and H. Ziegler, 549–88. Encycl. Plant Physiol. New Ser., Vol. 12B. Berlin, Springer-Verlag.

Forster, B. A. (1983). Pollution variability and the shape of the dose–response curve. *J. Air Pollut. Control Assoc.*, **33**, 774–5.

Gaastra, P. (1959). Photosynthesis of crop plants as influenced by light, carbon dioxide, temperature and stomatal diffusion resistance. *Meded. Landbouwhogeschool Wageningen*, **59**, 1–68.

Gardner, R. H., B. Röjder, and U. Berström. (1983). *PRISM: A systematic method for determining the effect of parameter uncertainties on model predictions*. Studsvik Energiteknik AB report NW-83/555. Nyköping, Sweden.

Garsed, S. G. (1985). SO$_2$ uptake and transport. In *Sulfur dioxide and vegetation: Physiology, ecology, and policy issues*, ed. by W. E. Winner, H. A. Mooney, and R. A. Goldstein, 75–95. Stanford, CA, Stanford University Press.

Giersch, C. (1986). Oscillatory response of photosynthesis in leaves to environmental perturbations: A mathematical model. *Arch. Biochem. Biophys.*, **245**, 263–70.

Goeschl, J. D. and C. E. Magnuson. (1986). Experimental tests of the Münch–Horwitz theory of phloem transport: Effects of loading rates. *Plant Cell Environ.*, **9**, 95–102.

Gollan, T., J. B. Passioura, and R. Munns. (1986). Soil water status effects the stomatal conductance of fully turgid wheat and sunflower leaves. *Aust. J. Plant Physiol.*, **13**, 459–64.

Hanks, R. J. (1974). A model for predicting plant growth as influenced by evaporation and soil water. *Agron. J.*, **66**, 660–5.

Hayes, J. T. (1984). Application of a crop growth simulation model of potential sugar beet production to California. *Publications In Climatology* Vol. 37, Elmer, New Jersey, Thornthwaite Associates, Laboratory of Climatology.

Heasley, J. E., W. K. Lauenroth, and J. L. Dodd. (1981). Systems analysis of potential air pollution impacts on grassland ecosystems. In *Energy and ecological modelling*, ed. by W. J. Mitsch, R. W. Bosserman, and J. M. Klopatek, 347–59. Amsterdam, Elsevier.

Heasley, J. E., W. K. Lauenroth, and T. P. York. (1984). Simulation of SO$_2$ impacts. In *The effects of SO$_2$ on a grassland. A case study in the Northern Great Plains of the United States*, ed. by W. K. Lauenroth and E. M. Preston, 161–84. New York, Springer-Verlag.

Holt, D. A., R. J. Bula, G. E. Miles, M. M. Schreiber, and R. M. Peart. (1975). *Environmental physiology, modeling and simulation of alfalfa growth: I. Conceptual development of SIMED*. Report 907, Agricultural Experimental Station, West Lafayette, IN, Purdue University.

Iman, R. L. and W. J. Conover. (1980). Small sample sensitivity analysis techniques for computer models, with an application to risk assessment. *Commun. Statist.-Theor. Meth.*, **A9**, 1749–1842.

Jarvis, P. G. (1981). Stomatal conductance, gaseous exchange and transpiration. In *Plants and their atmospheric environment*, ed. by J. Grace, E. D. Ford, and P. G. Jarvis, 175–204. London, Blackwell.

Journal, A. G. and C. J. Huijbregts. (1978). *Mining geostatistics*. New York, Academic Press.

Joyce, L. A. and R. N. Kickert. (1987). Applied plant growth models for grazing lands, forests and crops. In *Plant growth modeling for resource management*, ed. by K. Wisiol and J. D. Hesketh, Vol. 1, 17–55. Boca Raton, FL, CRC Press.

Kercher, J. R. (1977). GROW1: A crop growth model for assessing impacts of gaseous pollutants from geothermal technologies. Livermore, CA, Lawrence Livermore National Laboratory, UCRL-52247.

Kercher, J. R. and D. A. King. (1985). Modeling effects of SO$_2$ on the productivity and growth of plants. In *Sulfur dioxide and vegetation: physiology, ecology, and policy issues*, ed. by W. E. Winner, H. A. Mooney, and R. A. Goldstein, 357–72. Stanford, CA, Stanford University Press.

Kercher, J. R., D. A. King, and G. E. Bingham. (1982). Approaches for modeling crop-pollutant interactions in the NCLAN program. Livermore, CA, Lawrence Livermore National Laboratory, UCRL-86898.

King, A. W. (1986). *The seasonal exchange of carbon dioxide between the atmosphere and the terrestrial biosphere: Extrapolation from site-specific models to regional models.* PhD thesis, Knoxville, TN, The University of Tennessee.

King, A. W. and D. L. DeAngelis. (1987). Information for seasonal models of carbon fluxes in agroecosystems. Oak Ridge, TN, Oak Ridge National Laboratory, ORNL/TM-9935.

King, D. A. (1987). A model for predicting the influence of moisture stress on crop losses caused by ozone. *Ecol. Model.,* **35,** 29–44.

Krupa, S. and R. N. Kickert. (1987). An analysis of numerical models of air pollutant exposure and vegetation response. *Environ. Pollut.,* **44,** 127–58.

Lieth, J. H. (1982). *Light interception, growth dynamics, and dry matter partitioning in a phytotron-grown snap bean (Phaseolus vulgaris L.) crop: A modeling analysis with reference to air pollution effects.* PhD thesis, Raleigh, NC, North Carolina State University.

Luxmoore, R. J. (1988). Modeling chemical transport, uptake and effects in the soil–plant–litter system. In *Analysis of biogeochemical cycling processes in Walker Branch Watershed,* ed. by D. W. Johnson and R. I. Van Hook, 351–84. Heidelberg, Springer-Verlag.

Luxmoore, R. J. and M. L. Sharma. (1984). Evapotranspiration and soil heterogeneity. *Agric. Water Managemt,* **8,** 279–89.

Luxmoore, R. J., J. L. Stolzy, and J. T. Holdeman. (1976). Some sensitivity analyses of an hourly soil–plant–water relations model. Oak Ridge, TN, Oak Ridge National Laboratory, ORNL/TM-5343.

Maas, S. J. and G. F. Arkin. (1980). TAMW: A wheat growth and development simulation model. Temple, TX; Texas Agricultural Experiment Station, Report 80-3.

Magnuson, C. E., J. D. Goeschl, and Y. Fares. (1986). Experimental tests of the Münch–Horwitz theory of phloem transport: effects of loading rates. *Plant Cell Environ.,* **9,** 103–9.

Mansfield, T. A. and W. J. Davies. (1985). Mechanisms of leaf control of gas exchange. *BioScience,* **35,** 158–64.

Mansfield, T. A. and D. C. McCune. (1988). Problems of crop loss assessment when there is exposure to two or more gaseous pollutants. In *Assessment of crop loss from air pollutants.* Proceedings from the International Conference, Raleigh, NC, USA, ed. by W. W. Heck, O. C. Taylor, and D. T. Tingey, 317–44. London, Elsevier Applied Science.

Mehlhorn, H. and A. R. Wellburn. (1987). Stress ethylene formation determines plant sensitivity to ozone. *Nature,* **327,** 417–18.

Monteith, J. L. (1965). Evaporation and environment. In *The state and movement of water in living organisms,* ed. by C. E. Fogg. *Soc. Exp. Biol. Symp.,* **19,** 205–34.

Monteith, J. L. (1977). Climate and efficiency of crop production in Britain. *Phil. Trans. R. Soc. London,* Ser. B, **281,** 277–94.

Ng, N. and R. S. Loomis. (1984). *Simulation of growth and yield of the potato crop,* Simulation Monograph. Wageningen, Pudoc.

O'Dell, R. A., M. Taheri, and R. L. Kabel. (1977). A model for uptake of pollutants by vegetation. *J. Air Pollut. Control Assoc.,* **27,** 1104–9.

Parkhurst, D. F. (1986). Internal leaf structure: a three-dimensional perspective. In *On the economy of plant form and function,* ed. by T. J. Givnish, 215–49. Cambridge, Cambridge University Press.

Rawlings, J. O. and H. Dassel. (1988). Statistical approaches to assessing crop losses. In *Assessment of crop loss from air pollutants.* Proceedings of the International Conference, Raleigh, NC, USA, ed. by W. W. Heck, O. C. Taylor and D. T. Tingey, 389–416 . London, Elsevier Applied Science.

Reich, P. B. (1987). Quantifying plant response to ozone: A unifying theory. *Tree Physiol.*, **3**, 63–91.

Reynolds, J. F. and B. Acock. (1985). Predicting the response of plants to increasing carbon dioxide: A critique of plant growth models. *Ecol. Model.*, **29**, 107–29.

Richards, F. J. (1959). A flexible growth function for experimental use. *J. Exp. Bot.*, **10**, 290–300.

Rupp, E. M., R. J. Luxmoore, and D. C. Parzyck. (1979). Energy technology impacts on agriculture with a bibliography of models for impact assessment on crop ecosystems. Oak Ridge, TN, Oak Ridge National Laboratory, ORNL/TM-6694.

Schut, H. E. (1985). Models for the physiological effects of short O_3 exposures on plants. *Ecol. Model.*, **30**, 175–207.

Sharma, M. L. (ed.). (1984). *Evapotranspiration from plant communities.* New York, Elsevier Applied Science.

Sharma, M. L. and R. J. Luxmoore. (1979). Soil spatial variability and its consequences on simulated water balance. *Water Resour. Res.*, **15**, 1567–73.

Shreffler, J. H. (1976). A model for the transfer of gaseous pollutants to a vegetational surface. *J. Appl. Meteorol.*, **15**, 744–6.

Sinclair, T. R., C. E. Murphy, and K. R. Knoerr. (1976). Development and evaluation of simplified models for simulating canopy photosynthesis and transpiration. *J. Appl. Ecol.*, **13**, 813–30.

Stapper, M. and G. F. Arkin. (1980). *CORNF: A dynamic growth and development model for maize (Zea mays L.)* Temple, TX, Texas Agricultural Experiment Station, Report 80-2.

Taylor, G. E., P. J. Hanson, and D. D. Baldocchi. (1988). Pollutant deposition to individual leaves and plant canopies: Sites of regulation and relationship to injury. In *Assessment of crop loss from air pollutants.* Proceedings of the International Conference, Raleigh, North Carolina, USA, ed. by W. W. Heck, O. C. Taylor, and D. T. Tingey, 227–57. London, Elsevier Applied Science.

Tingey, D. T. and G. E. Taylor. (1982). Variation in plant response to ozone: A conceptual model of physiological events. In *Effects of gaseous air pollution in agriculture and horticulture,* ed. by M. H. Unsworth, 113–38. London, Butterworths Scientific.

Unsworth, M. H. (1981). The exchange of carbon dioxide and air pollutants between vegetation and the atmosphere. In *Plants and their atmospheric environment,* ed. by J. Grace, E. D. Ford, P. G. Jarvis, 111–38. London, Blackwell.

Waggoner, P. E. (1975). Micrometeorological models, Vol. 1. In *Vegetation and the atmosphere,* ed. by J. L. Monteith, 205–28. London, Academic Press.

Wisiol, K. and J. Hesketh (eds). (1987). *Plant growth models for resource management. Vol. I and II.* Boca Raton, FL, CRC Press.

Zeiger, E., G. D. Farquhar, and I. R. Cowan (eds). (1987). *Stomatal function.* Stanford, CA, Stanford University Press.

18

REGIONAL/NATIONAL CROP LOSS ASSESSMENT MODELING APPROACHES

SHARON K. LEDUC and CLARENCE M. SAKAMOTO

National Oceanic and Atmospheric Administration, University of Missouri, Columbia, Missouri, USA

18.1 INTRODUCTION

Historically, weather has been the factor most often considered when assessing crop losses at the regional or national levels. More recently other factors such as air pollution and soil loss have also been considered. Crop loss can be measured as change in crop yield or production relative to a reference base. The most common reference compares the current or future crop yield with a base period (years). Maximum yield, an estimated potential considering current technologies, and maximum observed yield, derived from data for earlier years or inferred from crop insurance data, represent two more references. Crop loss also directly relates to land use, including not only the area under production but also the type of crops grown. A gradual shift of crop type and area may indirectly result from nonenvironmental (government policy, price) and/or environmental (air pollution) effects. For example, CO_2-induced climate change is expected to cause a shift in where crops are grown (Parry, 1985).

An ideal system for assessing crop loss begins with data collection and ends with user information, a stage not adequately addressed by most researchers. This paper proposes combining several modelling methods currently available into such a system. Discussion of the strengths and weaknesses of the models, hypotheses on the effect of potential improvements, and suggestions for developments follow.

18.2 CURRENT MODELING METHODS

Many researchers have identified direct relationships between air pollution and crop yield. Ozone and byproducts damage plant cell membranes,

445

impairing photosynthesis and reducing crop yield (e.g., Amundson *et al.*, 1986). Ryan *et al.* (1981) report that concentrations of O_3 and sulfur cause physiological damage to vegetation and that experiments carried out under field conditions indicate that the yield for sensitive species may be reduced by 50% or more. Benedict (1979) estimates that 70% of damage to vegetation by air pollutants in the US results from O_3 concentrations. Experimental data from open-top chambers indicate significant crop losses induced by ambient O_3 (Heck *et al.*, 1983).

Crop loss is also affected indirectly by air pollution through climate change, even more difficult to quantify than the direct effects of air pollution on crops. Decker and Achutuni (1987) reviewed studies on changes in climate due to CO_2 concentration and the subsequent impact on crop production. Bland and Benedict (1979) conducted laboratory studies on the possible synergistic or antagonistic effects resulting from the interaction of CO_2 and oxidant mixtures. These interactions need to be adequately evaluated or quantified for assessments outside the laboratory. The evaluation of laboratory results for field use requires observational data. The sample and the plan for obtaining that sample are very important. This statistical problem is closely related to the problem of experimental design (Fienberg and Tanur, 1986).

The following examples use weather rather than air pollution as the primary impact on crop loss. However, the methods used may have potential for assessing the impact of pollutants.

18.2.1 Regression models from surface observations

Variations of regression models with crop yield as the dependent variable have been used for many years and situations (Mearns, 1987). The statistical techniques may vary, but they are similar in that they estimate the crop yield (or loss) from explanatory variables, predominantly standard weather observations. The variability associated with an estimate derived from a regression model is a function of the standard error of the regression and how different the new observation is from the observations used to estimate the regression (deviation from the mean). The Large Area Crop Inventory Experiment (LACIE) used these models to estimate yield accurately enough to satisfy the requirement that estimated production be within 10% of the reported value 90% of the time (LeDuc *et al.*, 1979; Hill *et al.*, 1980).

Regression models need at least five observations per predictor variable. Linear relationships are usually estimated and the predictor variables transformed if the relationship is nonlinear. The transformation is an approximation which may not be valid for all possible values of the variables. If sufficient data are not available to develop a regression relationship for a particular situation, a regression model developed for a similar situation can be used, i.e., an analog model. However, new data are still needed to establish the standard error for the application of the existing model in a new situation.

Predictor variables for regression models are generally aggregated and complex, e.g., total precipitation for a large area for a defined time period. Using available predictor variables which are already aggregated over time and/or space simplifies the computations and data input requirements. However, a good areal estimate of precipitation is often difficult to measure with existing data sources because weather stations are often not located in agricultural areas.

Regression techniques generally provide good estimates when conditions are average with respect to the data used to develop the regression models. However, when unusual conditions, or outliers, occur, the regression models may provide unreliable estimates. An example is the 1970 yield for corn in the US that was drastically reduced by corn blight on T-strain hybrids. This situation could not have been accurately indicated by the predictor variables considered in the model.

A regression model for crop yield in the Tahoua administrative region of Niger in Africa (Fig. 18.1) is used as an example. The model predicts Y, millet yield in t ha^{-1}, based on X, the cumulative precipitation received in July and August in mm, or

$$Y = 0 \cdot 026 + 0 \cdot 00242X$$

The fit of the model and estimate for 1986 are shown in Fig. 18.2.

Feyerherm and Paulsen (1981a, b) developed a more complex but physically

Fig. 18.1. Location of Millet Area in Tahoua, Niger.

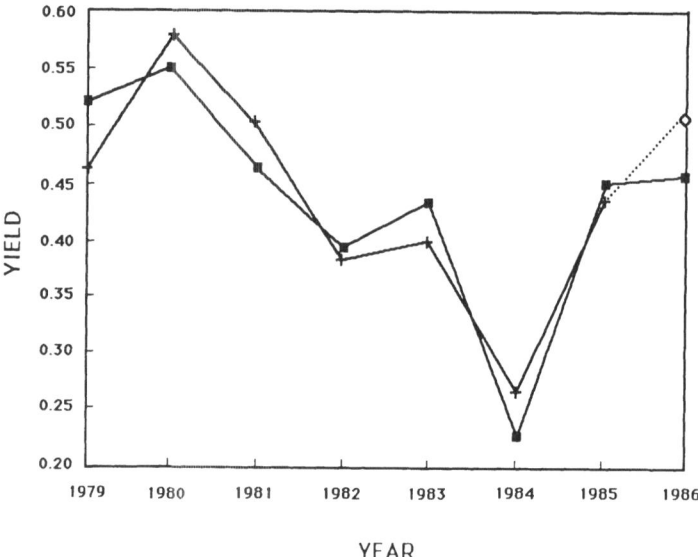

YEAR

Fig. 18.2. Millet yield (t ha^{-1}) for Tahoua, Niger (■), estimate from regression on precipitation (+) and 1986 forecast from regression (◇).

appropriate regression model. The objective was a model which could be adapted to many locations with only minor adjustments. Their model for winter wheat yield uses a simulated crop calendar and soil moisture budget. The planting data are based on observed data or are estimated by a 'starter' model. The stages of crop development, from emergence through ripening, are simulated using daily weather values. Average temperatures and cumulative precipitation are used over stages of development (Cotter, 1984).

The model is based on a weather index (WX) of the form

$$WX = 72 \cdot 6 + ET + XPR + TEMP$$

where ET are evapotranspiration effects, XPR are excessive precipitation effects and TEMP are temperature effects. The average WX for a state is the predictor variable, along with a linear function of year. Yields are adjusted for rust loss and technology. The technology factor is a function of differential yielding ability and nitrogen fertilizer. Bootstrap test results for Ohio are shown in Fig. 18.3. The root mean square error was 3·7 bu acre^{-1}, due largely to an overprediction of 25·9 bu acre^{-1} in 1973.

The regression method using surface observations is limited by data constraints and the time and effort necessary to develop the models.

18.2.2 Regression models from remotely sensed data

Remotely sensed data, in particular satellite data, offer another input to assess crop conditions (Barnett and Thompson, 1983; Wiegand and Richardson, 1984; van Dijk, 1986). Aircraft and some ground-based data also qualify as being remotely sensed, but satellite observations offer frequent and less

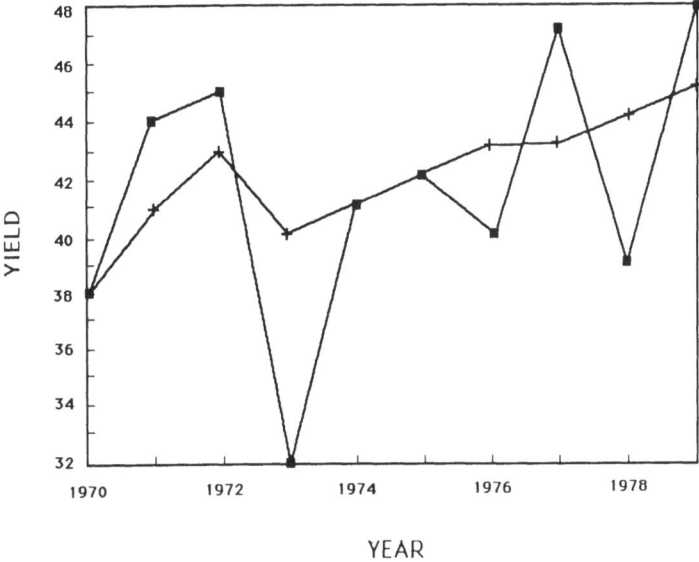

Fig. 18.3. Winter wheat yield (bu acre^{-1}) for Ohio (■) and estimate from regression on average weather index (+).

costly monitoring with homogeneous spatial representation. Satellite data can also be incorporated with observations and information from the ground (AISC/CIAM, 1986).

Satellite data vary in cost depending on resolution, total area, frequency, and number of channel sensors utilized in the analysis. A comparison of four satellite remote sensing systems by Hutchinson (National Research Council, 1987) shows the variations in the cost per unit area ($ km^{-2}), ranked according to resolution: SPOT 1 (the French System Probatoire d'Observation de la Terre), 0·41; LANDSAT's Thematic Mapper (TM), 0·10; LANDSAT's Multispectral Scanner (MSS), 0·02; and Advanced Very High Resolution Radiometer (AVHRR), from the polar orbiting satellite of the National Oceanic and Atmospheric Administration (NOAA), 0·0003. The frequency of data collection varies respectively: from 5 to 20 days; 16 days; 16 days; 12 h.

The degree to which crop yield and crop loss can be determined from satellite data has not been established. However, much research indicates that it is possible to estimate yield from satellite (Barnett and Thompson, 1983; Hatfield, 1983; Wiegand and Richardson, 1984; van Dijk, 1986). The precision and sources of variability for that estimation need to be established and improved.

An experiment to estimate regional crop yields from satellite data in the Sahel and Horn regions in Africa, conducted during 1985 and 1986, illustrates the approach. Evaluation of this experiment is not complete at this time. Data from two channels of AVHRR on the NOAA-9 polar orbiting satellite were used. The resolution of the basic satellite data was about 4 km on the ground. The daily data were filtered temporally, then smoothed both temporally and

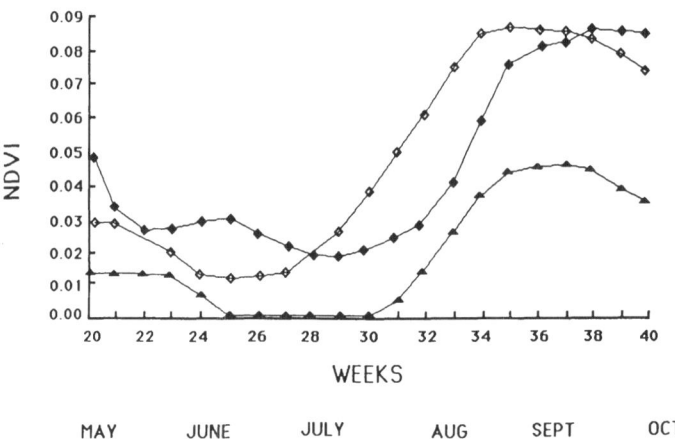

Fig. 18.4. Time series of normalized difference vegetation index (NDVI) average for area (5·0–6·0°E, 14·5–15·0°N) in Tahoua, Niger during 1985 (▲), 1986 (◇), and 1987 (◆).

spatially to provide year to year comparisons of satellite indications throughout the growing season for a specified area on the ground (Fig. 18.4). The predictor variable was the value of the satellite indication, i.e., the normalized difference vegetation index (NDVI), during the early reproductive period of the crop (van Dijk, 1986). The NDVI is the difference between the near infrared reflectance and the visible reflectance divided by the sum of the two reflectances. The NDVI was found to have a strong correlation with green biomass on the ground (Tucker *et al.*, 1985). The condition of the crop and the amount of biomass at the start of reproduction gives an indication of potential crop yield. If the crop is stressed, yield can be expected to be reduced.

The NDVI and other functions of reflected radiation measure green biomass on the ground and are positively correlated with leaf area index (LAI) of corn and other crop canopies (Gardner *et al.*, 1985; Hatfield *et al.*, 1985). However, other factors that may lead to crop damage and potential crop loss, such as insects, disease, pollution, or weather phenomena, may be present but not detectable when reproduction starts. A high NDVI or LAI is not a sufficient indicator of high yield.

The yield indication from satellite is based on an estimate from the regression of yield on NDVI at the early reproductive stage. Because the yield data are available for large areas (regional or national levels), these large aggregated areas are assessed. However, the predictor variable NDVI can be disaggregated to a pixel or point location. The regression relation, because it is linear, can be applied to each pixel and the yield estimates at the pixels aggregated to derive an estimate for the area. When used this way, however, it is difficult to establish a confidence interval for the yield estimate for the region because of the correlation between the pixels. Consequently the pixels are usually aggregated to a single value of NDVI for a region. From this a confidence interval can be established for the yield estimate for the region.

In the case of crop loss due to acid deposition the use of remotely sensed measurements offers unique opportunities to investigate the causes of stress.

Research on detecting spruce forest damage from acid deposition through use
of the remotely sensed Landsat Thematic Mapper has been very encouraging
(NASA, 1987). A current study in West Germany attempts to detect forest
stress and damage using spectral data (Thomas, 1987). If the techniques are
successful they will be applied to forest areas in the United Kingdom. Some
pollutants may acidify soil water and affect uptake of water while others cause
changes in forest canopy pigmentation. Detecting these changes using remote
sensing is in the research stage. However, if the capability can be established
for forests, the method should be evaluated for application to field crops.

18.2.3 Plant process (growth) models

Although considerable work on plant process models was done in recent
years (Wisiol, 1987), few approaches addressed regional assessment. A great
deal of expertise and quantified knowledge is incorporated into these models
and a great deal is known regarding their use in resource management. This
type of model may be quite complex. The large number of feedback
mechanisms and interactions used and the quality control required on the
inputs make operation of the models difficult. One problem is that although
specialists can identify model weaknesses they are rarely able to improve them
because models so far have been constructed to simulate the system, *not* to
explain *how* the system was simulated. Hopefully the current efforts among
researchers to cooperate in the standardization and documentation of the
modules or basic components of plant process models will encourage exten-
sions and improvements (Acock, 1987).

Plant process models generally apply to a very small area, such as a field or
plot. For this reason their main use is for assessment at the farm level. Hodges
et al. (1987) examined the utility of these models for large area assessment for
the US Corn Belt, a first attempt at regional/national assessment. Varietal
coefficients for corn (maize) were derived for each of 51 locations. Estimated
dates when 50% planting was completed were used. A silt loam soil was
assumed for all locations. Daily weather data from principal stations and
climate outlooks were obtained from the National Weather Service. The
method of aggregating the yield estimates at the 51 locations into an estimate
for the US Corn Belt was based on the 1981 Agricultural Census figures for
acreage of corn harvested. The yield in a county was assumed to be the same
as the yield at the station nearest the center of the county. The model
estimated a yield for both irrigated and non-irrigated corn, weighted according
to 1982 acreages. The final estimate of corn production from the model varied
from 92 to 101% of the official production figures of USDA during the 1982 to
1985 period. A production forecast was available six times during the growing
season (Fig. 18.5).

One advantage of this system is that forecasts can be made within the
growing season based on climate outlooks or chosen scenarios. One disadvantage for real-time application is that the sample of locations where the plant
process models can provide point forecasts are weather stations at airports
rather than in representative agricultural areas.

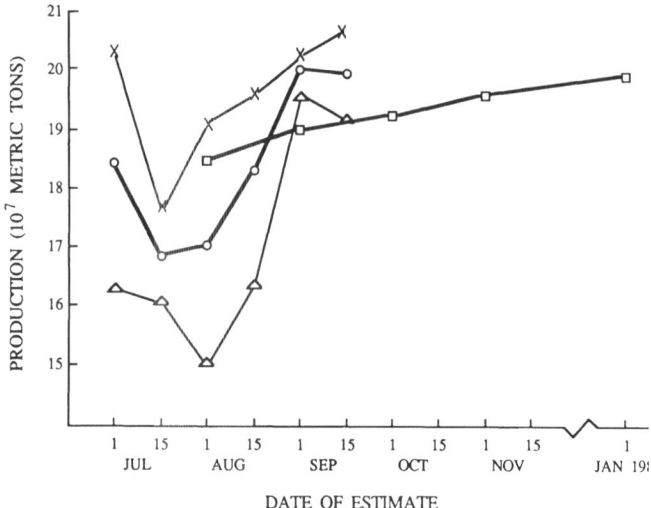

Fig. 18.5. US corn production (10^7 t) USDA estimate (□), plant process model forecast based on similar years: highest forecast (×), mean or average forecast (○), lowest forecast (△). (From Hodges *et al.* (1987).)

18.2.4 Objective surveys

Surveys for crop yield are usually based on random samples. Data from these samples can be used in predictive models, e.g., for corn the counts of ears and plants as well as average kernal row length (from destructive sampling) can be used as predictor variables in equations (specific for each maturity group) for gross yield (Steele, 1987). The random samples are drawn from a population defined either by classification of cropped area or by a list of persons who farm. Keeping the inventory of either one of these populations current is not as difficult as establishing the initial inventory of the populations, but it does require continued effort and resources. Taking objective measurements requires extensive training, detailed project design, and methods to control and check the quality of the results. The appropriate theory and its application require constant monitoring. The strength of the system lies in the control of the estimates and errors.

18.2.5 Combination of models

The assessment of regional or national crop yield should incorporate all available information. If several models exist, they should all be considered. Several models available to estimate crop yield for regions in the Sahel will serve as examples. A regression model using precipitation as the predictor variable and a regression model which used NDVI as the predictor variable were combined. An estimate derived from a regression equation with the usual assumptions that the errors are normally and independently distributed has a *t*-distribution.

The estimate of millet yield for the Tahoua region of Niger using the 1986 rainfall reports was 0.51 t ha^{-1}. The uncertainty in the estimate is indicated by

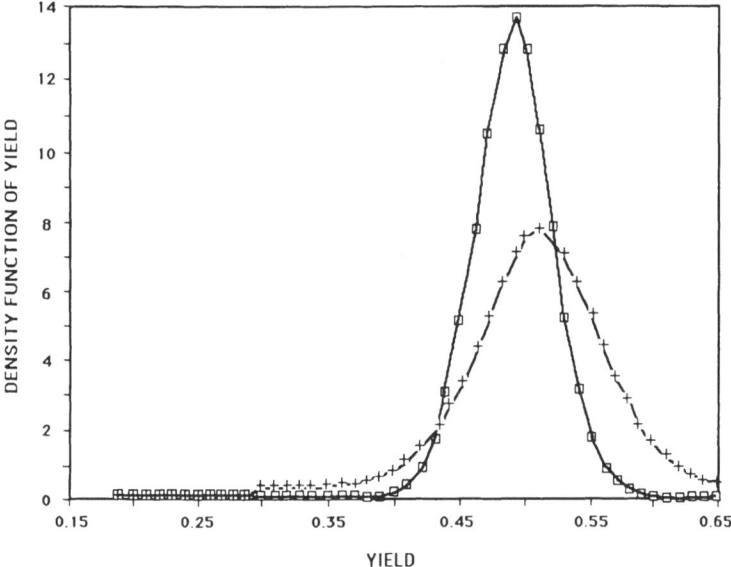

Fig. 18.6. Density function of millet yield forecast ($t\,ha^{-1}$) for Tahoua, Niger in 1986 from regression on precipitation (+) and regression on NDVI (□). (Area under each curve equals one.)

the probability density function of the t-distribution. An independent estimate of $0\cdot49\,t\,ha^{-1}$ was obtained using a regression model with NDVI as the predictor. This estimate also has uncertainty as seen in the probability density function (Fig. 18.6). In this particular example the estimates were very close, not the case for all regions.

The two estimates, in particular the probability density functions, were weighted with respect to the R^2 of the regression equations, i.e., the amount of explained variation in yield. Other factors should probably be considered as well, for example, the range of observations used in the developmental data set. One might wish to increase the weight placed on the estimate derived from precipitation data, because the observational data set using precipitation as the predictor variable contained more years and consequently a wider range of conditions.

The probability density function resulting from the combination of the precipitation based estimate and the NDVI based estimate (Fig. 18.7) has a mode and median at $0\cdot49\,t\,ha^{-1}$, with a 50% chance that the yield is between $0\cdot47$ and $0\cdot51\,t\,ha^{-1}$. In fact, the final yield statistic issued by the Government of Niger for millet in the Tahoua region in 1986 was $0\cdot464\,t\,ha^{-1}$, just below the 50% confidence interval.

This method of combining models, i.e., using the probability density functions of the estimates from the different models to produce a single probability density function, is not necessarily the best. It does, however, provide information on the estimate of the current yield, especially with regard to the uncertainty associated with the estimate.

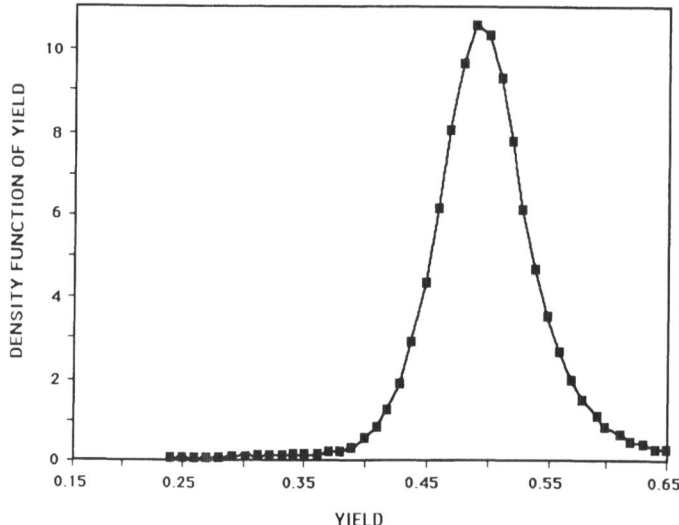

Fig. 18.7. Density function millet yield forecast ($t\,ha^{-1}$) for Tahoua, Niger in 1986, derived from two density functions of Fig. 18.6. (Area under curve equals one.)

18.3 PROPOSED SYSTEM

Descriptions of hypothetical systems using combinations of the above modeling techniques follow.

18.3.1 Combining three models

Independent estimates from models of the plant process (or growth) can be used along with the yield estimates from the precipitation-based regression model and the NDVI based regression model previously discussed. Assume, for this hypothetical case, that a plant process model has been adapted for use at seven locations in the Tahoua region of Niger (Fig. 18.8). The variability of model estimates of yield for the seven locations from the yield for the region of Tahoua for one or more years is used to estimate the spatial variability of yield. To get an estimate for Tahoua, a single weighted estimate of yield is derived from the estimates at the seven locations which are aggregated based on reported acreage for those years. (This is the same approach used by Hodges *et al.* (1987).) The regression between these weighted estimates from the process model and the reported yields establishes the distribution of the 1986 estimate.

To get an estimate for 1986, the data for each of these locations is used in the plant process model adapted for that location, providing a single yield estimate for each location. The weighted average of these seven yield estimates is based on the area upon which millet was grown in 1986. This estimate has a *t*-distribution based on the regression relation between yields and weighted estimates for previous years. Integration of the plant process (PP) model with the previously discussed regression models requires determining a weighting of

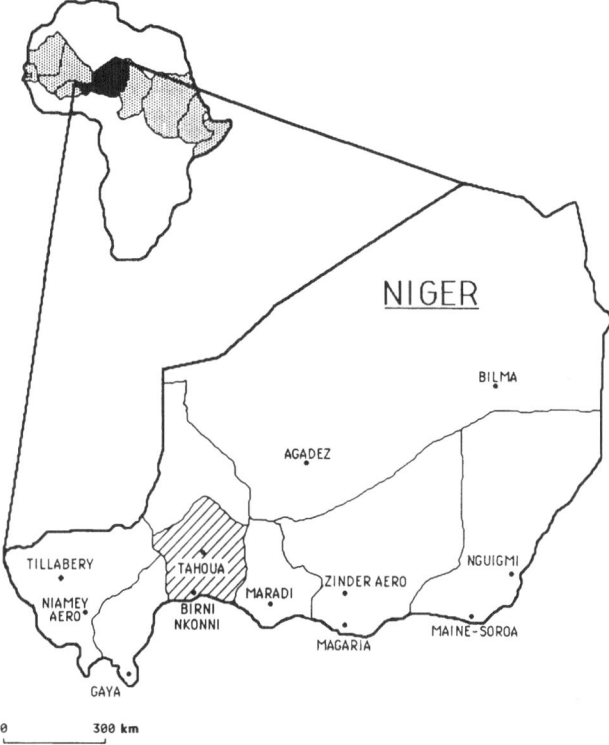

Fig. 18.8. Locations in Tahoua, Niger for estimates from plant process model, i.e., stations with daily meteorological database.

this probability function in addition to the weightings of the two probability density functions from the other two regression estimates. The amount of variation explained, i.e., R^2 adjusted for degrees of freedom, is a reasonable first choice for the weights.

This hypothetical application is based on existing databases in the Tahoua region consisting of seven subregions with crop area and production data. Daily meteorological data are also available for at least one station in each subregion (Fig. 18.8). The agricultural statistics and meteorological data for the Tahoua region start in 1960. The daily meteorological data are used in the plant process model at each of the seven locations (j) and provide a yield estimate (Y_{ij}) for each year (i). The crop areas for each subregion (a_{ij}) are used to aggregate yield at each station to estimate the yield (Y_i^*) for the Tahoua region for year i:

$$Y_i^* = \sum_j Y_{ij} a_{ij} \Big/ \sum_j a_{ij}$$

where the sum is over the j locations, for each of the years 1960–1985. The regression of regional yield (Y) on plant process generated yield (Y^*) provides a method to forecast the 1986 yield and the distribution of that forecast. Assume the explained variation or R^2 from the above sample of (Y_i, Y_i^*)

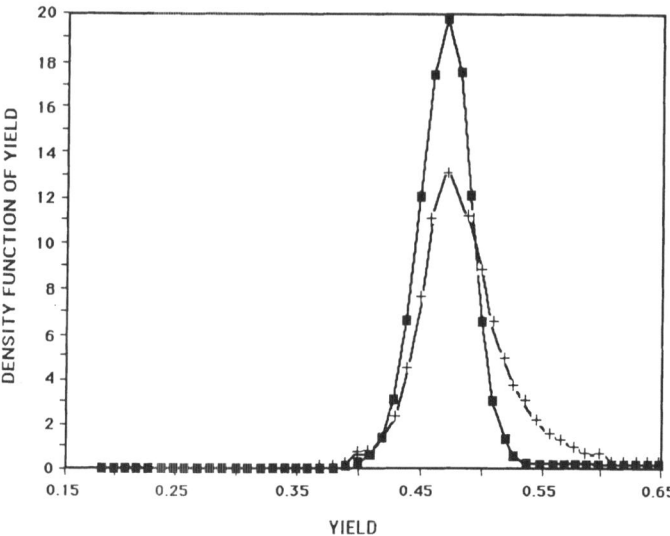

Fig. 18.9. Density function of millet yield forecast (t ha^{-1}) for Tahoua, Niger in 1986, from hypothetical density function from plant process model (■) and density function derived from (■) and the density function of Fig. 18.7 (+). (Area under each curve equals one.)

$i = 1960, \ 1961, \ldots, 1985$ is 0.75 and that Y^*_{1986} is $0.47\,t\,ha^{-1}$ and the hypothetical t-distribution for this forecast is as shown in Fig. 18.9. To combine these forecasts, the three probability density functions are combined based on the R^2 values from the three regressions:

$$\bar{f}(Y) = \frac{R^2_{pcp}f_{pcp}(Y) + R^2_{NDVI}f_{NDVI}(Y) + R^2_{PP}f_{PP}(Y)}{R^2_{pcp} + R^2_{NDVI} + R^2_{PP}}$$

where the subscript indicates the predictor variable (pcp is precipitation, NDVI is normalized difference vegetation index, and PP is plant process model), R^2 is explained variance and $f(Y)$ is the appropriate t-distribution for each forecast of yield. The resulting density function f (ALL) on Fig. 18.9 has a median value 0.48 and the 50% confidence interval is from 0.46 to $0.50\,t\,ha^{-1}$ (compare with Fig. 18.7). The plant process model changes the yield estimate, but not the variability associated with it.

18.3.2 Including objective surveys

The system can be further enhanced by including objective surveys, if they are available. Well designed surveys, carried out with quality controls, will carry more weight in a system approach. The probability density function of these estimates can also be incorporated since a normal distribution is appropriate and standard deviations can be estimated. The weight assigned to this probability density function needs to be established. This estimate is incorporated to define the resulting probability density function from the three independent yield estimates. Predictions from objective survey methods might replace or at least dominate the other methods at times since they have the capability to include other factors such as pests and disease.

18.4 GEOGRAPHIC INFORMATION SYSTEMS

Creating a single system for estimating crop loss from a variety of modeling approaches requires the capability to manage and manipulate vast quantities of data, apply analysis methods, and display the results. Once the important factors in assessing crop loss are geographically defined, a computer based Geographic Information System (GIS) can be used to sample, model and analyze a specific area, aggregating or disaggregating data to the appropriate scale.

Before a GIS can be used effectively, the surveys must be well designed for generalization to larger populations. Decisions must be made regarding what data to collect and at what level to store the data, which is difficult because researchers always want access to *all* the data. However, for assessment purposes, maintaining data at the lowest level of aggregation, i.e., raw data, can quickly lead to an unmanageable situation. In addition, the privacy consideration limits availability of certain raw data, e.g., agricultural survey data. Flexibility to manipulate the data is required. To plan for flexibility one needs to understand relationships between the various types of data. The better these relationships are understood the better the system design. Research needs are application specific (Niemann, 1987).

Perhaps a scenario would help to demonstrate how a GIS might be used. Because counties would probably be the lowest level of aggregation to be maintained, geographic boundaries for counties are the first input into the system. Statistics regarding the percentage of area cropped to specific crops for each county could be included. Locations, i.e., by county, of pollutants, measured concentrations, and time of measurements are entered into the database. Combining these data and applying relationships between them, one can see where impacts might be expected, for what crops, and perhaps quantify the impact. Areas of possible antagonistic or synergistic effects can be identified and other areas for taking samples determined from the location of known pollution sources. Including satellite data on a regular basis allows classification and specification of the variability of vegetation within a county by tracking: atmospheric parameters; determining land use; quantifying spatial and temporal variability of vegetation; and classifying homogeneous areas. Since time of pollution measurements can be compared to satellite derived vegetation indices before and after episodes of short term air pollution, crop loss in dry matter could be inferred during that period. Taking the system one step further, models could be used to quantify the crop loss. Dose–response models which include both moisture stress and O_3 could be used to estimate the dry matter yield (King, 1987). Process models then could be used to infer crop loss in grain yield from dry matter changes. Model estimates could be integrated to provide estimates of crop loss for counties. These county estimates could be aggregated to regional or national levels for the assessment of crop losses due to air pollutants.

The USDA/National Agricultural Statistics Service recognizes the importance of a Geographic Information System and is currently testing a prototype system to aid in the construction of an area frame used in the random sampling design (Carney *et al.*, 1987). The potential benefits include: better control over the homogeneity of the primary sampling units within the strata; more flexibility in stratification strategies; and incorporation of automated remote sensing techniques.

18.5 SUMMARY AND CONCLUSIONS

The assessment of crop loss at the national/regional level should be approached in stages. Existing capabilities should be utilized, but the link with new research is necessary to improve them and also to support the research. A GIS incorporates all available data and methodologies into one system.

Selection or modification of an existing GIS for this application should be done now. A limited area should be selected for initial application, perhaps southern California where large variations in O_3 effects can be examined. This could be accomplished at the county level. This is similar to the large area production forecasting system (Holt, 1988) which uses a weather factor for each county. For that system the factor is derived from a process model. This proposed system would have a weather/O_3 factor derived from the dose–response and process models similar to that proposed by King and Nelson (1987). The GIS should take existing data and models and provide the final assessment. The GIS should also be flexible enough to accommodate future data and models and have an interactive capability to quality control data and model results. The use of the GIS will clarify research needs if used appropriately. The large structured databases needed for the GIS should also be developed now. The final steps are identifying models which can be used now in the assessment process and incorporating them into a single system.

A sampling plan for observational data should be developed for the initial application. This will be necessary to establish confidence in the assessment. This effort should be a cooperative venture between EPA, NOAA, and USDA. As Fleagle (1987) notes, "EPA has statutory responsibility for dealing with air-quality programs, but its research capabilities are distinctly limited, and it has not taken the leadership role in a broad inter-agency approach to problems of air quality". Contrary to this, NCLAN was considered a success. A formalized documentation and implementation of the results are needed.

ACKNOWLEDGEMENTS

The authors would like to thank V. C. Runeckles, E. Preston, and D. A. King for their reviews and comments, R. Terry for significant editorial support, J. Trujillo for the graphics, M. Grissum for typing and G. Johnson for comments on GIS. This work was sponsored under NOAA grant (NA87AA-H-RA076) to the Cooperative Institute for Applied Meteorology.

REFERENCES

Acock, B. (1987). *Electronic journal of agricultural systems software and standard generic modular structure for plant simulations.* Proposals. Beltsville, MD USDA/ARS/BARC.

AISC/CIAM. (1986). *Briefing package: methodology for NOAA/NESDIS/AISC climate impact assessment program in Africa.* National Academy of Science.

Amundson, R. G., R. M. Raba, A. W. Schoettle, and P. B. Reich. (1986). Response of soybean to low concentrations of ozone: II. Effects on growth, biomass allocation, and flowering. *J. Environ. Qual.,* **15,** 161–7.

Barnett, T. L. and D. R. Thompson. (1983). Large-area relation of Landsat MSS and NOAA-6 AVHRR spectral data to wheat yields. *Remote Sens. Environ.,* **13,** 277–90.

Benedict, H. M. (1979). Economic impact of air pollutants on plants in the United States. Final report for Coordinating Research Council, Menlo Park, CA, SRI International.

Bland, M. K. and H. M. Benedict. (1979). *Air pollutant effects on vegetation.* Menlo Park, CA, SRI International.

Carney, B., M. L. Holko, J. Nealon, and J. Cotter. (1987). *Digital area sampling frame development.* Research Proposal. Washington, DC, USDA/NASS.

Cotter, J. F. (1984). *Evaluation of the Feyerherm '82 winter wheat model for estimating yields in Indiana, Kansas, Montana and Ohio.* SRS Staff Report No. AGES841023. Washington, DC, Statistical Research Division, Statistical Reporting Service, US Dept. of Agriculture.

Decker, W. L. and R. Achutuni. (1987). *A review of national and international activities on modeling the effects of increased CO_2 concentrations on the simulation of regional crop production.* College of Agriculture, University of Missouri-Columbia.

Feyerherm, A. M. and G. M. Paulsen. (1981a). Development of wheat yield prediction model. *Agron. J.,* **73,** 277–82.

Feyerherm, A. M. and G. M. Paulsen. (1981b). An analysis of temporal and regional variation in wheat yields. *Agron. J.,* **73,** 836–7.

Fienberg, S. E. and J. M. Tanur. (1986). A long and honorable tradition: Intertwining concepts and constructs in experimental design and sample surveys. *Proceedings of the conference on survey research methods in agriculture,* Leesburg, VA. American Statistical Association, Alexandria, VA.

Fleagle, R. G. (1987). The case for a new NOAA charter. *Bull. Am. Meteor. Soc.,* **68**(11), 1417–23.

Gardner, B. R., B. L. Blad, D. R. Thompson, and K. E. Henderson. (1985). Evaluation and interpretation of thematic mapper ratios in equations for estimating corn growth parameters. *Remote Sens. Environ.,* **18,** 225–34.

Hatfield, J. L. (1983). Remote sensing estimations of potential and actual crop yield. *Remote Sens. Environ.,* **13,** 301–11.

Hatfield, J. L., E. T. Kanemasu, G. Asrar, R. D. Jackson, P. J. Pinter, Jr, R. J. Reginato, and S. B. Idso. (1985). Leaf-area estimates from spectral measurements over various planting dates of wheat. *Int. J. Remote Sensing,* **6,** 167–75.

Heck, W. W., R. M. Adams, W. W. Cure, A. S. Heagle, H. E. Heggestad, R. J. Kohut, L. W. Kress, J. O. Rawlings, and O. C. Taylor. (1983). A reassessment of crop loss from ozone. *Environ. Sci. Technol.,* **17,** 572A–81A.

Hill, J. D., N. D. Strommen, C. M. Sakamoto, and S. K. LeDuc. (1980). LACIE—an application of meteorology for United States and foreign wheat assessment. *J. Appl. Meteor.,* **19**(1), 22–34.

Hodges, T., D. Botner, C. Sakamoto, and J. Hays Haug. (1987). Using the CERES-Maize model to estimate production for the US cornbelt. *Agric. Meteor.,* **40,** 293–303.

Holt, D. A. (1988). Crop assessment. In *Assessment of crop loss from air pollutants*. Proceedings of the international conference, Raleigh, N.C., USA, ed. by W. W. Heck, O. C. Taylor, and D. T. Tingey, 9–26. London, Elsevier Applied Science.

King, D. A. (1987). A model for predicting the influence of moisture stress on crop losses caused by ozone. *Ecological Modelling*, **35**, 29–44.

King, D. A. and W. L. Nelson. (1987). Assessing the impacts of soil moisture stress on regional soybean yield and its sensitivity to ozone. *Agric., Ecosystems Environ.*, **20**, 23–35.

LeDuc, S. K., C. M. Sakamoto, N. D. Strommen, and L. T. Steyaert. (1979). Some problems associated with using climate–crop–yield models in an operational system: An overview. *Int. J. Biometeorol.* **24**(2), 104–14.

Mearns, L. (1987). *Empirical crop-climate models: An historical perspective*. Boulder, CO, National Center for Atmospheric Research.

NASA. (1987). *Earth science and applications division, the program and plans for FY 1987–1988–1989*. National Aeronautics and Space Administration.

National Research Council, Board on Sciences and Technology for International Development. (1987). *Final report: Panel on the NOAA climate impact assessment program for Africa*. Washington, DC, National Research Council.

Niemann, B., Jr. (1987). *Usefulness of GIS for natural resource planning, management and monitoring: an evaluation of eleven applications with research needs identified*. Presentation at IGIS '87. Washington, DC, Association of American Geographers.

Parry, M. L. (ed.). (1985). The sensitivity of natural ecosystems and agriculture to climatic change. *Clim. Change*, **7**, 1–3.

Ryan, John W., E. Loehman, W. Lee, E. Trundsen, M. Bland, R. Groen, F. Ludwig, T. Eger, S. Eigst, D. Conley and R. Cummings. (1981). *An estimate of the non-health benefits of meeting the secondary national ambient air quality standards*. SRI International Project No. 2094 under Contract 27-AQ-8060 for National Commission on Air Quality, Washington, DC.

Steele, R. J. (1987). *Corn objective yield operational vs non-invasive maturity category determinations*. SRB-87-04. Washington, DC, USDA/NASS.

Thomas, R. (ed.). (1987). NRSC projects for the department of the environment: Acid rain and natural vegetation. Newsletter No. 9. Farnborough, Hants, UK. National Remote Sensing Centre.

Tucker, C. J., C. L. Vanpraet, M. J. Sharman, and G. Van Ittersum. (1985). Satellite remote sensing of total herbaceous biomass production in the Senegalese Sahel: 1980–1984. *Remote Sens. Environ.*, **17**, 233.

van Dijk, A. (1986). *A crop condition and crop yield estimation method based on NOAA/AVHRR satellite data*. PhD dissertation, University of Missouri-Columbia.

Wiegand, C. L. and A. J. Richardson. (1984). Leaf area, light interception and yield estimates from spectral component analysis. *Agron. J.*, **76**, 543–8.

Wisiol, K. (ed.). (1987). *Plant modeling for resource management, Vol. I: Current models and methods* and *Vol. II: Quantifying plant processes*. Boca Raton, FL, CRC Press.

Session VII

ECONOMIC CONSIDERATIONS
AND POLICY IMPLICATIONS

(O. Clifton Taylor, *Chairman*; Patricia M. Irving, *Co-Chairman*)

19

MODEL REQUIREMENTS FOR ECONOMIC EVALUATIONS OF POLLUTION IMPACTS UPON AGRICULTURE

RICHARD M. ADAMS

Oregon State University, Corvallis, Oregon, USA

and

THOMAS D. CROCKER

University of Wyoming, Laramie, Wyoming, USA

19.1 INTRODUCTION

Everything we say in this chapter unconditionally presumes that man is the measure of all things. Whatever a man does must be the best thing for him to do, given his knowledge of the moment—otherwise he would not do it. Thus, preferences are assumed to be revealed by self-governed behavior. More broadly, the economic concept of value is based on and is derived from individual human preferences. This is not as restrictive as it may first appear; while this perspective does not grant the natural environment equal status with human beings, we include the preferences of those persons who believe that it ought to be so granted, along with their probable high valuations of environmental assets, such as air quality.

Contrary to common usage, neither we nor other economists view 'economics' and 'money' as synonymous. For example, human behavior and the health or esthetic effects of a pollutant on that behavior are directly economic. No laws governing economic activity are innate in the material objects of ordinary cognition. Economic propositions relate to subjective desires, which motivate individuals to alter the facts they conceive about their environments. Objects do not become pertinent to economic analysis until someone believes that they can be used for some purpose, subjective or otherwise. Thus, the effects of a pollutant on vegetation are economic only insofar as that vegetation contributes to human health and happiness.

19.2 THE ECONOMIC ASSESSMENT PROBLEM

The preceding section conveys the stance of economics with respect to the basis of values; it fails to state the units in which values are to be measured, or the context that bestows meaning on these values. Assume, for example, that a person derives purely esthetic satisfaction from an area of lush vegetation. If there is a local decline in vegetation lushness, the person will likely believe that he has been made worse off. However, if there are other worldly things capable of providing him with satisfaction, then additional provision of these other things may cause him to feel as well off as he would without the decline in vegetation lushness. Finally, if these other things can be secured by the expenditure of income, or of time that can be used to earn income, then there is some additional income that, given the lushness decline, would make the individual feel no worse off. The unit, therefore, in which economists measure value is money, stated in terms of income. Implicit in the acceptance of this unit is the presumption that, even if the thing being valued cannot be secured in the marketplace, there are collections of other things from which a person can receive equal satisfaction. These other things, which have market prices attached, then serve as vehicles to infer the values of assets and services for which no directly observable monetary value exists.

To say that value can be inferred from monetary price is not to say that monetary price is value. In Fig. 19.1, MU depicts a representative consumer's marginal (incremental) utility from his demand function for an agricultural crop, X, during a year. The MU curve depicts the additional satisfaction the consumer will obtain from the purchase and use of an additional unit of the product. The observable prices of other commodities that provide the consumer with equal satisfaction determine the positioning of the MU function with respect to the vertical axis. Thus, demand is synonymous with the consumer's maximum willingness-to-pay. The difference between this maximum and what the consumer in fact has to pay is his consumer (buyer) surplus.

Let MC_0 represent the marginal cost of supplying an additional unit of the product. It is a supply function because it represents the minimum additional rewards (price) growers must receive so they will not shift their resources to other activities. The observable earnings their resources could receive in these

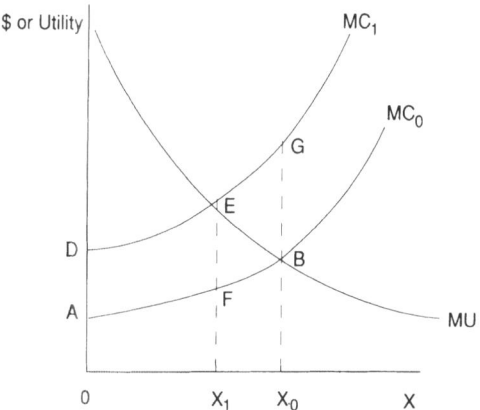

Fig. 19.1. The influence of pollution.

alternative activities determine the minimum reward in the current activity and the positioning of MC_0 with respect to the vertical axis. Thus, supply is synonymous with the grower's minimun necessary reward. His surplus (quasi-rent) is the difference between what he actually receives (price) and this minimum necessary reward. Economic efficiency implies the maximization of the sum of consumer and grower surpluses.

The shift of the supply function from MC_0 to MC_1 simulates the economic effect of an increase in environmental pollution that makes it more costly for the grower to produce any given level of output. The positive vertical intercepts of MC_0 and MC_1 imply that the grower bears some costs whether he produces anything or not; for example, the grower might have to pay the rent on leased land, even if he does not plant a crop.

Given the cost functions for producing various quantities of output, the grower in Fig. 19.1 plans an output level X_0 under the original pollution level and X_1 under the new level. Under the original level, the total cost of meeting demand is the sum of the incremental costs of attaining X_0. This is given by the area $OABX_0$. The corresponding area with the new and higher pollution level is $ODEX_1$. The difference between these two areas is the change in the producer's total costs because of the pollution increase. Since $OAFX_1$ is common to the two areas, the change in total costs is given by the difference between ADEF and X_1FBX_0. ADEF is simply the change in cost necessary to achieve output level X_1, and ADGB is the change in cost necessary to maintain the original output X_0. Note that the only circumstances in which ADGB would be a correct measure of the total economic value of the pollution increase is if the MU function were disregarded. That is, if and only if the demand for the crop was completely unresponsive to changes in crop price (the minimum reward the grower must receive), such that the demand function exactly coincided with the GBX_0 line, would a measure of this sort actually measure the total economic value of the pollution change. Otherwise, the area ADGB neglects that change in value to the consumer expressed in the movement along the MU function as the marginal cost function shifts upward in response to the pollution increase. If the output demanded is responsive to variations in that crop's price, the ADGB measure does not capture the value of the sum of the excess of the utility of the output no longer produced over the sum of the costs of producing this output.

Simple comparisons of these areas in Fig. 19.1 demonstrate that it is possible to observe zero, or quite small changes in consumer expenditures (grower gross revenues), and still experience substantial changes in economic value. Only the equivalence or the near-equivalence of ADEF and X_1FBX_0 is required. By its very nature, an expenditure measure is a unit money outlay multiplied by the number of units. Given a pollution increase and its consequent effects upon the cost of maintaining a given crop output level, the consumer may adjust the price he is willing to pay for an additional output unit, as well as the actual number of units that he purchases. If the pollution increases production costs, and thus crop price, the number of units purchased will deviate farther from the consumer's satiation level. Thus, the value the

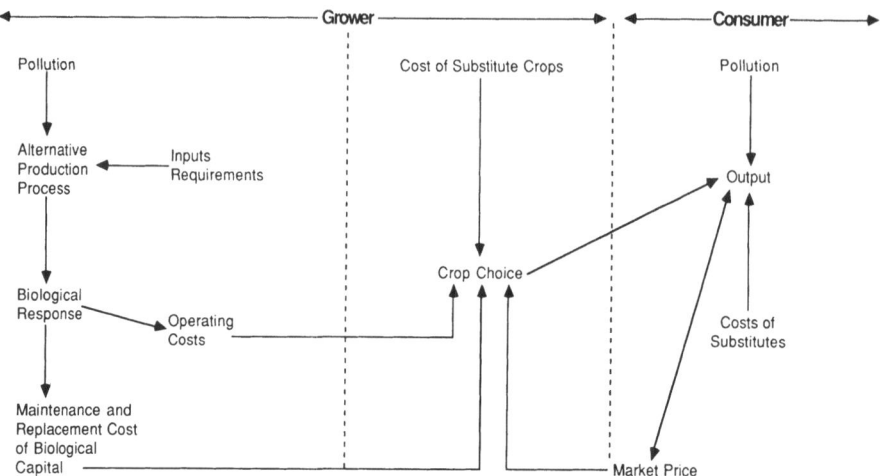

Fig. 19.2. The sources of economic consequences.

consumer attaches to a movement toward this level will increase. According to whether the proportionate reduction in quantity purchased is greater than or less than the proportionate change in his willingness-to-pay, observed expenditures on the output may increase or decrease, respectively.

Figure 19.2 traces the set of factors that drive the relationships depicted in Fig. 19.1. Assume that a grower is trying to decide what kind and how much of a crop to produce. Some production processes for these crops are susceptible to pollution damages. For a given expected pollution level, the portion of the figure lying to the left of the left dotted line represents the problem of determining the least costly way of producing a given crop. It answers the question: 'Given the alternative ways I have to produce any particular level of output for this crop, which ways allow me to employ the minimum combinations of those costly inputs necessary for production?'. Variations in pollution levels will alter these minimum necessary combinations. Identification of these combinations for each pollution level enables the grower to maximize the output of the crop that he obtains from a given expenditure on inputs. However, the solution to this problem cannot tell him which of the alternative crops to produce; it can only show the least costly plan for producing a particular crop, and the manner in which this plan will vary as pollution varies.

The portion of Fig. 19.2 lying between the dotted lines depicts the problem of choosing among the alternative crops, given prior identification of the least costly means with which to produce each crop. In this portion, the grower is allowed to adapt by substituting crops that are more or less prone to quantity and/or quality reductions from expected pollution levels. The problem with this middle portion can be contrasted with the problem with the first portion, where the grower could only adapt by manipulating the time and/or the process used to produce a particular product. Basically, the grower has a set of alternative processes defined by biological and physical science knowledge of the laws of nature, as well as by institutional constraints (the laws of man). He estimates the least cost process for each of these known and allowable processes for each crop, and he then selects the combination of crops that he expects to generate his maximum rewards for given output prices and pollution levels.

The consumer resides in the right-hand portion of Fig. 19.2. There are two routes whereby pollution damages can influence his behavior, thus altering the satisfaction he obtains from the crop in question. First, a pollution increase that is registered in crop output reductions will increase grower costs, as shown in Fig. 19.1. Consequently, the minimum price the grower must receive in order to be willing to commit himself to supplying any given quantity of the crop must increase. Figure 19.1 displays both the grower and the consumer value consequences of this. However, the figure does not account for pollution impacts upon the quality of the crop output. Quality reductions will reduce the consumer's maximum willingness-to-pay for any particular quantity of the crop. In Fig. 19.1, this quality reduction would be pictured by a leftward, downward shift in the consumer's demand function, MU. Calculation of the value consequences would then involve a more complicated but nevertheless similar (to Fig. 19.1) comparison of areas under original and new grower's supply and consumer's demand functions.

19.3 EMPIRICAL IMPLEMENTATION

Figures 19.1 and 19.2 are strictly heuristic devices; actual quantitative assessment requires acknowledgement of the trade-off between completeness and analytical and empirical tractability. A number of analytically defensible and empirically tractable techniques are available to develop quantitative representations of the relationships in Fig. 19.2 and of the appropriate areas in Fig. 19.1.

For example, biological and economic relations can be integrated in a sound economic model by treating the economic problem as a mathematical programming problem. While the approach often requires extensive economic, biological, and physical data to represent the agricultural environment with tolerable accuracy, as well as some means of solving the problem such as an appropriate computer algorithm, it has been widely used because of its ability to project the economic value consequences of as yet unrealized environmental effects. Alternatively, economic models based on actual grower and consumer behavior can be used to reflect pollution-induced changes in yields and in economic values. Policy makers do not have to base decisions on incomplete or incorrect methods that capture only a portion or entirely misrepresent the Fig. 19.2 sources of change in economic value consequences. Just *et al.* (1982) provide a correct and detailed, but not highly technical, review of measurement methods.

Because of their ability to register the economic consequences of prospective as well as realized environmental states, mathematical programming methods have been widely employed in assessments of the economic value consequences of pollution-induced damages to agriculture. Adams *et al.* (1982, 1984) apply price-endogenous versions of this method. The basic outlines of their workings are readily grasped. They are representative of all mathematical programming formulations.

Applied price-endogenous, mathematical programming formulations have both a grower or farm-level component and a more aggregate or sector component. Specifically, the grower level is portrayed by production 'activities', which represent the alternatives available with which to produce various crops. Especially in Adams *et al.* (1984) the production side is modelled in considerable detail (12 annual crops plus hay and beef, pork, and milk production), input use, and environmental and other fixed constraints. Such detail is needed to adequately model potential producer mitigative behavior in the face of environmental change. As noted by Crocker (1982), failure to account for these adaptive opportunities tends to overstate potential environmental damages and understate benefits. In the aggregate, these production activities determine the total supply of the commodities in the model. The supply of each commodity is then used in combination with demand relationships for each crop or livestock product to determine the market price, following the logic of Figs 19.1 and 19.2.

The solution procedure underlying the model solves for the set of market prices that maximizes an 'objective function'. The objective function is simply a mathematical expression that takes on economic significance when its variables are defined to represent prices and quantities of commodities. Graphically, it is analogous to the area in Fig. 19.1 above each crop's supply curve and below its demand curve, to the left of the intersection point, or the sum of consumers' and producers' surpluses. The economic surplus values that the objective function can take are constrained by other relationships, such as total farm land available or total processing capacity. By specifying the variables in the model to conform to real economic value as defined by supporting historical data, the model solution approximates actual conditions for a given time period.

Since the demand functions and aggregate production responses in the model are specified at the national level, the model's solution provides objective function values at the national level. By then imposing alternative O_3 pollution levels on the model as manifested in yields predicted by dose–response functions, changes in crop production and prices and ultimately the objective function may be measured. Comparisons of changes in the model objective function value between these hypothetical O_3 situations and current ambient O_3 pollution levels provide economic estimates of the benefits or costs of these alternative levels of O_3 pollution. This kind of information can then be used by policy makers to assess the economic efficiency of changing current O_3 pollution standards. Comparisons of changes in consumers' and producers' surplus proportions can also be used to evaluate distributive (equity) consequences.

Figure 19.3 presents a flow chart of the model workings. The left-hand side shows the essential link between meteorological, biological, and economic data

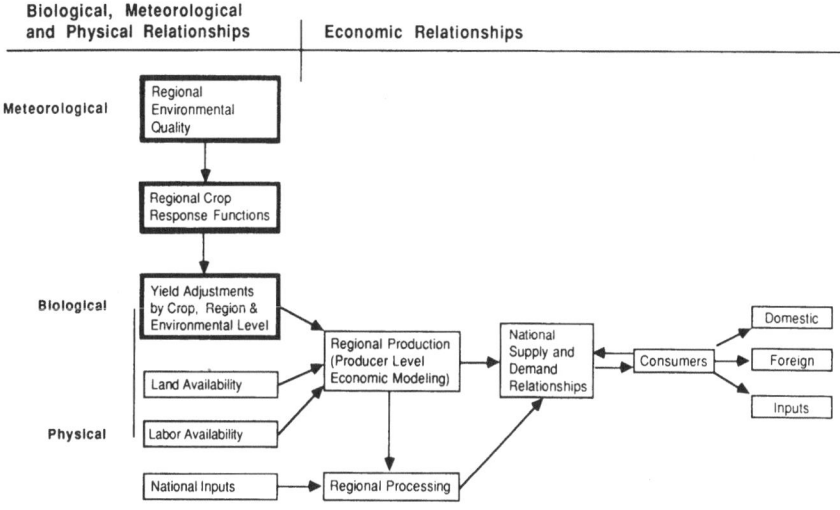

Fig. 19.3. Flow chart of a generalized mathematical programming model of agriculture.

needs. The key inputs for the model are given in the bold-outlined blocks, which deal with environmental quality, crop dose–response, and yield adjustment data. These three input boxes then serve to drive the economic analysis portion of the model, represented in the right-hand side of Fig. 19.3. The focus of this discussion is on the inputs in the three bold-outlined blocks. The data on regional environmental quality serve two purposes. First, when using actual ambient conditions, the model can be solved for a given year for which results are already available. For example, 1983 was used as a base for validation of Adams *et al.* (1988). This base solution serves as the benchmark against which to compare hypothetical environmental conditions associated with proposed standards or regulations. Second, if alternative standards or assumptions on environmental quality are to be considered, the implications of these conditions in each geographical region considered in the model must be defined. The projected regional changes in pollution are then fed into the dose–response block of the figure. Ideally, this box contains regional dose–response functions for each major crop grown in that region, adjusted for the effects of other stresses. By assuming responses to be somewhat homogeneous across regions, a smaller number of response functions may be acceptable. These functions are typically estimated using response data generated from experiments on the environmental stress or stresses in question, as in the NCLAN program.

By adjusting the dose parameters in each response function, alternative yield changes relative to ambient conditions are estimated. The changes in dose would be determined by the regional environmental quality block. The projected regional yields arising from the interplay of dose and response functions become the driving force in the economic analysis. These data are needed for the third block. For tractability, these yield changes should be measured as relative changes; that is, the dose–response functions project proportionate or percentage yields, not actual yields. The use of proportionate adjustments also allows for potential extrapolation of site-specific data to regional response, if the proportionate responses are statistically homogeneous (Rawlings and Cure, 1985). These proportionate yield adjustments are then used to modify the base yields in the economic model. It is this resultant differences in yields between base or actual yields and those under the hypothetical environmental situation which is at the heart of the economic analysis.

Depending on the scope and extent of the environmental assessment, a number of yield scenarios may be required. In each case, the differences in yields trigger changes in the economic model which ultimately translate into changes in the model solution or objective function values, i.e., the measure of economic surplus. Comparison of the differences in objective function values, as measured in dollars, is then an estimate of the benefits or costs of each environmental scenario. Thus, the interface between the meteorological, biological, and economic analyses is essentially captured in the linkages in the three boxes described in Fig. 19.3.

19.4 SUMMARY

The models, procedures, and data needs outlined in this paper are meant to be illustrative of techniques available for assessing the agricultural benefits of controlling air pollution or other environmental stresses. The treatment here has been general, defining major conceptual concerns and data needs for performing benefits analyses on alternative environmental situations.

As more biological and meteorological data have become available, numerous assessments of economic benefits to agriculture from air pollution control have been performed. Properly structured estimates can provide decision-makers with the compromises associated with environmental regulation. How such benefit estimates are ultimately used in setting policy is outside the purview of the economist. However, for the benefit estimates to be meaningful, the economic assessment must capture the principal mechanisms involved in the agricultural economy, including supply and demand phenomena. Failure to account adequately for both the physical or biological and economic aspects of the problem will result in possibly misleading and even contradictory assessment estimates.

REFERENCES

Adams, R. M., T. D. Crocker, and N. Thanavibulchai. (1982). An economic assessment of air pollution damages to selected annual crops in Southern California. *J. Environ. Econ. Managemt,* **9,** 42–58.

Adams, R. M., S. A. Hamilton, and B. A. McCarl. (1984). *The economic effects of ozone on agriculture.* EPA-600/3-84-090. Corvallis, OR, Environmental Protection Laboratory, US Environmental Protection Agency.

Adams, R. M., J. D. Glyer, and B. A. McCarl. (1988). The NCLAN economic assessment: approach, findings and implications. In *Assessment of crop loss from air pollutants*. Proceedings of the international conference, Raleigh, NC, USA, ed. by W. W. Heck, D. T. Tingey, and O. C. Taylor, 473–504. London, Elsevier Applied Science.

Crocker, T. D. (1982). Pollution damage to managed ecosystems: economic assessments. In *Effects of air pollution on farm commodities*, ed. by J. S. Jacobson and A. A. Miller, 103–24. Arlington, VA, Izaak Walton League of America.

Just, R. E., D. L. Hueth, and A. Schmitz. (1982). *Applied welfare economics and public policy*, Englewood Cliffs, NJ, Prentice-Hall, Inc.

Rawlings, J. D. and W. W. Cure. (1985). The Weibull function as a dose–response model to describe ozone effects in crop yields. *Crop Sci.*, **25**, 807–14.

20

THE NCLAN ECONOMIC ASSESSMENT: APPROACH, FINDINGS AND IMPLICATIONS

RICHARD M. ADAMS, J. DAVID GLYER

Oregon State University, Corvallis, Oregon, USA

and

BRUCE A. McCARL

Texas A & M University, College Station, Texas, USA

20.1 INTRODUCTION

The role of food and fiber in human welfare make agriculture an important sector in most economies. It is thus not surprising that concerns about pollution or other adverse environmental changes often focus on agricultural effects. In the United States, potential adverse effects of air pollutants such as ozone (O_3) on agriculture and other vegetation are one motivation for regulating such pollutants. In turn, this regulatory interest leads plant scientists, economists, and policymakers to expend considerable effort on measuring and interpreting the economic and other consequences of pollution on agriculture. Given the recent interest regarding the effects of air pollutants on vegetation and the regulatory importance of economic information on such effects, it is instructive to examine the current state of knowledge about the economic effects of O_3 on agriculture.

In this paper, we draw upon the recent experience of the National Crop Loss Assessment Network (NCLAN) program to explore the current understanding of the economic effects of O_3 on agriculture. The specific objectives of the paper are to (1) review what has been learned to date from recent economic assessments of O_3 effects on agriculture and (2) present the results of the latest and final NCLAN economic assessment, an assessment that builds on much of this previous research. This publicly funded research can also be viewed as prototypical of the approaches for addressing other environmental stresses on agriculture, such as global climate change. Some of the qualitative findings from these NCLAN assessments can reduce the initial information required to understand the agricultural implications of future environmental changes.

The next section of this paper presents a review of empirical studies that draw upon the NCLAN data and procedures. This review serves to develop generalizable findings from these multiple studies. It also serves to provide a perspective on the economic estimates generated in the present study. The following section then describes the specific methodology, procedures, and assumptions used in the current NCLAN economic assessment. The final section discusses the results and implications from this most recent NCLAN assessment effort and suggests areas for future research. Throughout, we adopt an expository treatment, avoiding detailed mathematical and statistical formulations. More detailed treatment can be found in past and current NCLAN manuscripts (e.g., Adams *et al.*, 1984*a*).

20.2 RESULTS FROM PREVIOUS ASSESSMENTS OF O₃ USING NCLAN DATA

Any assessment of the economic consequences of pollution on consumers and producers requires three kinds of information: (1) the differential changes that pollution control causes in each person's production and consumption opportunities, (2) the responses of input and output market prices to these changes, and (3) the input and output changes that those affected can make to minimize losses or maximize gains from changes in production and consumption opportunities and in the prices of these opportunities. Plant science studies of dose–response functions, combined with aerometric data on O_3 pollution levels are the primary source of information for the first requirement. Evaluation of the latter two requirements represents the economics portion of a benefits assessment. If pollution control causes substantial changes in outputs, price changes can occur which, in turn, lead to further market-induced output changes. Moreover, even if prices are constant, plant science information still fails to provide accurate indications of output changes when individuals can alter production practices and the types of outputs produced. Thus, accurate information on the economic consequences for agriculture of pollution can be achieved only if the reciprocal relations between physical and biological changes and the responses of individuals and institutions are explicitly recognized.

The above discussion summarizes the biological and economic problems embedded in economic assessments of O_3. It also suggests the need for an integrated approach for implementing such protocols for policy analysis. This section reviews specific assessments that provide empirical evidence of the economic consequences of O_3 pollution on US agriculture. Table 20.1 portrays the conditions specified and the empirical results obtained for these selected studies. With the exception of the duality study of Mjelde *et al.* (1984), the remaining studies cited in the table employed simulation, particularly mathematical programming approaches.

Caution should be exercised in trying to compare numerical estimates across these studies, given that crops, response information, and assumed economic and environmental conditions differ considerably. However, even though

TABLE 20.1

Studies of the economic effects of O_3 on agriculture based on NCLAN data

Study	Region	Ambient[a] concentration	Model features				Crops	Results (1980 US dollars)		
			Prices changes	Crop substitutions	Input substitutions	Quality changes		Consumer benefits	Producer benefits	Total benefits
Howitt et al. (1984)	California	Universal reduction to 0·04 ppm	Yes	Yes	Yes	No	18 crops	$17 × 10^6$	$28 × 10^6$	$45 × 10^6$
Adams and McCarl (1985)	Corn Belt	Reduction in NAAQS[b] from 0·12 ppm to 0·08 ppm	Yes	Yes	Yes	No	Corn, soybeans, wheat	$2 079 × 10^6$	$-1 411 × 10^6$	$668 × 10^6$
Mjelde et al., (1984); also Garcia et al., (1986)	Illinois	10% increase from 0·0465 ppm	No	Yes	Yes	No	Corn and soybeans	None	$226 × 10^6$	$226 × 10^6$
Adams and Crocker (1984)	US	Universal reduction from 0·053 ppm to 0·04 ppm	Yes	Yes	Yes	No	Corn, soybeans, cotton	Not reported	Not reported	$2 220 × 10^6$
Adams et al., (1984b)	US	Universal reduction from 0·048 ppm to 0·04 ppm	Yes	No	No	No	Corn, soybeans, cotton, wheat	Not reported	Not reported	$2 400 × 10^6$
Adams et al. (1986)	US	25% reduction from 1980 level for each state	Yes	Yes	Yes	No	Corn, soybeans, cotton, wheat, sorghum, barley	$1 160 × 10^6$	$550 × 10^6$	$1 770 × 10^6$
Kopp et al., (1985)	US	Universal reduction from 0·053 ppm to 0·04 ppm	Yes	Yes	Yes	No	Corn, soybeans, wheat, cotton, peanuts	Not reported	Not reported	$1 300 × 10^6$
Shortle et al., (1988)	US	Universal reduction from 0·053 ppm to 0·04 ppm	Yes	No	No	Yes	Soybeans	$880 × 10^6$	$-90 × 10^6$	$790 × 10^6$

[a] 7-h growing season geometric mean. Given a log-normal distribution of air pollution events, a 7-hour seasonal O_3 level of 0·04 ppm is approximately equal to an hourly standard of 0·08 ppm, not to be exceeded more than once a year (Heck et al., 1982).
[b] Averaging time of 1 h; not to be exceeded more than once a year.

numerical estimates and the conditions under which they were derived may differ, the NCLAN-based studies exhibit common patterns of behavioral responses and sensitivities to imposed conditions and to data accuracy and precision. These commonalities both serve to summarize the major findings arising from prior NCLAN research and can serve to restrict the dimensions of their analytical and empirical problems. Most of these commonalities follow directly from economic theory; others are more subtle and arise from the characteristics of the agricultural setting. Together, they serve as a summary of current understanding of the economic consequences of O_3. They also offer a set of priorities for future assessments of environmental change, including the final NCLAN economic analysis.

20.2.1 Increasing air pollution causes losses in the total economic surplus from the production and consumption of agricultural outputs to increase at an increasing rate

The estimates of the following studies display this pattern: Howitt *et al.* (1984); Adams and Crocker (1982); Adams *et al.*, (1984*b*); Kopp *et al.* (1985); Adams and McCarl (1985); Adams *et al.* (1984*a*); Adams *et al.* (1986); Shortle *et al.* (1988). No thoroughly consistent pattern emerges as to the absolute magnitude of this rate of change in total surplus.

20.2.2 Growers can gain from increases in air pollution

Adams and McCarl (1985) found that substantial increases in O_3 increase the quasi-rents of Corn Belt growers. Shortle *et al.*, (1988) also estimated that Corn Belt growers benefit from O_3 increases. A sufficient condition for these findings is that the pollution-induced percentage reduction in output quantity for a given crop be less than the percentage increase in market price that the supply reduction causes. When secondary effects on livestock producers are included in national assessments, the net effect of increases in O_3 on all producers is a loss in welfare (Adams *et al.*, 1984*a*). Even here, however, producers in some regions benefit while others lose. Pre-NCLAN studies generally conclude that growers always lose from increased air pollution. However, those studies typically disregard market price changes and therefore predict producer losses with increased pollution. Such studies cannot be taken seriously as evidence supporting the hypothesis of universal producer losses.

20.2.3 Losses to consumers are a significant portion of the total losses that air pollution causes agriculture

Of the studies that accounted for air-pollution-induced changes in the prices of agricultural outputs, consumer surplus losses as a percentage of total losses range from 50% (Adams *et al.*, 1986) to 100% (Adams and McCarl, 1985). Given that producers can sometimes benefit from air pollution increases, methods that disregard consumer impacts can understate total losses in surplus and grossly misstate the distribution of these welfare effects.

20.2.4 As air pollution changes, percentage changes in total economic surplus are less than the percentage changes in biological yields that triggered these losses

Any study that accounts for price effects of crop and input substitutions will obtain this result. The result is a direct outgrowth of microeconomic theory. Consider a deterioration in air quality. From the grower's perspective, the decrement in air quality is met by substitution of crops, cultivars, input combinations, and even growing locations. Consumers will respond analogously to any price increase; that is, they will substitute away from the good that has had an increase in relative price. Depending upon price elasticities, percentage gains in surplus could also exceed percentage changes in yields. Thus, the substitutions allow producers and consumers to attenuate the losses they would otherwise suffer from air quality declines.

20.2.5 Air pollution changes can cause growers of pollution-tolerant crops to experience production changes greater than those initially suggested by the pollution increase

This phenomenon is explicitly noted by Howitt *et al.* (1984), where growers of pollution-tolerant crops such as broccoli, cantaloupe, carrots, and sugar beets were also estimated to experience production declines along with growers of sensitive crops as air pollution increased. Several studies (e.g., Brown and Smith, 1984; Kopp *et al.*, 1985) of the corn–soybean–wheat economy in the US Midwest have produced the same finding. This counterintuitive result occurs because of a "crowding out" phenomenon. If, after a pollution increase, the pollution-sensitive crops continue to have higher returns per acre than the pollution-tolerant crops, and if inputs such as land and water are scarce, growers of the pollution-intolerant crops will substitute land, water, and similar inputs for air quality. The more profitable intolerant crops now require more land and water to produce the market equilibrium quantities.

20.2.6 Changes in air pollution affect the productivity of and aggregate demand for factors of production

Several studies based on NCLAN data have demonstrated that changes in air pollution will change the demand for specific inputs. Mjelde *et al.* (1984) estimate that a 10% increase in O_3 in Illinois results in a 4% decline in the demand for variable inputs such as labor, water, and fertilizer. Howitt *et al.* (1984), Garcia *et al.* (1986), Kopp *et al.* (1985), and Brown and Smith (1984) have all shown that increased O_3 increases the demand for land inputs.

These input demand effects result from two interdependent processes. First, changes in air pollution may change the marginal productivity of a given input. For example, if changes in air pollution are viewed as a neutral technological change, reductions in air pollution increase the productivity of all inputs proportionally. The productivity of any input in the agricultural production process is thus increased, increasing the grower's willingness to pay for the input. The second process that results in increased demand for *some* inputs refers to the "crowding out" phenomenon noted above. That is, some crops and regions receive a greater relative increase in yields from reductions in air pollution. Consequently, an expansion of individual crop acreage or total regional crop production may occur. This will result in an increase in aggregate input use in these more favored crops and regions. Conversely, crops and regions that do not realize net relative productivity gains from pollution control may reduce individual or total crop acreage, thus causing a reduction in aggregate input use. The exact nature of these input demand changes will be a function of the input mixes used for each crop or region and the relative effects of pollution changes within and across regions.

20.2.7 The contribution of additional biological information to the accuracy and precision of economic assessments declines as the level of information increases

How much plant response information is needed in order to perform credible economic assessments? Several NCLAN-related studies have attempted to provide guidance on this issue: Adams et al., (1984b); Howitt et al. (1984); Adams and McCarl (1985); Kopp et al. (1985); and Adams and Crocker (1982). Typically, these studies have focused on how the economic estimates change as the researcher changes assumptions about the nature of the underlying plant response data, such as functional form of the response model, the number of observations on which to estimate such models, and the number of cultivars over which to estimate "representative" response models. The results have generally demonstrated that the gains in economic precision (as measured by changes in the confidence intervals of the estimates) tend to decline as more information is obtained on each issue. These findings, however, do not mean that additional information on particular model parameters is never warranted. Rather, the costs of acquiring that information need to be weighed against the consequences of being "wrong" with respect to the use of the model predictions.

20.2.8 Changes in air pollution have differential effects on the comparative advantage of agricultural production regions

Several NCLAN-based studies have focused on the consequences of air pollution across large geographical areas, such as the United States, e.g., Adams and Crocker (1982); Kopp et al. (1985); and Adams et al. (1986). To correctly represent aggregate supply response across a large geographical region, economists typically define the large regions as being composed of a collection of distinct subregions representing different characteristics in terms of crop production alternatives, resource and environmental conditions, and other factors within each subregion. A useful feature of such spatial equilibrium models is that the aggregate economic consequences of the policy alternatives as well as the effects of each policy on each subregion are measured. The NCLAN literature demonstrates that the economic effects can vary sharply across regions. For example, studies by Adams et al. (1986) and Kopp et al. (1985) indicate that reductions in air pollution result in gains to producers in areas simultaneously characterized by high ambient pollution levels and crop mixes dominated by species sensitive to air pollution. In such areas, the percentage gain in yields is greater than the overall reduction in agricultural prices at the national level, thus resulting in an increase in regional net income. This leads to expanded crop acreage. Conversely, areas with the opposite set of characteristics experienced reductions in revenues and acreages of some crops and hence a loss of "market share" for these crops. Thus, while these analyses all found net gains to society from reductions in air pollution, some subregions gained at the expense of others.

20.2.9 Air pollution has economic transboundary effects that alter international trade flows with attendant gains and losses to exporters and importers

The economic gains from free and open trade are well recognized. International trade in agricultural commodities has profound effects on the welfare of consumers and growers in both exporting and importing countries. A few of the recent NCLAN-based studies of US agriculture incorporate a trade component into the model, e.g., Adams *et al.* (1986); Shortle *et al.* (1988). Since much of the increase in agricultural output generally moves into the export market, the results of changes in air pollution have been to alter the supply of export commodities. The net effect is that foreign consumers of these commodities capture some of the consumer gains while foreign producers lose. Indeed, Adams *et al.* (1986) estimated that of the total consumer gains from ambient O_3 reductions, 60% accrued to non-US consumers. Conversely, increases in air pollution imply a reduction in the welfare of importing countries. As a result, national environmental policies can readily have transboundary *economic* implications even in the absence of a pollution transboundary phenomenon.

20.3 THE CURRENT NCLAN ASSESSMENT: METHODOLOGY

The overall objective of this paper is to provide an updated and more complete estimate of the economic effects of O_3 on US agriculture. As noted above, performing such an analysis involves a multistage process in which biological data (e.g., yield changes for selected crops under O_3 scenarios) are used to alter the yield values in a quantitative model of the US agricultural economy. The subsequent changes recorded in certain economic dimensions of the model provide a measure of the economic consequences of those changes in biological data.

This section discusses the structure and assumptions of the economic model used in this NCLAN analysis. Conceptually, the economic model is the same as that used in Adams *et al.* (1984a). The features of this model, including a technical appendix, are described in detail therein.

The primary differences between the model reported in Adams *et al.* (1984a) and the current version of the model, as noted subsequently, include (1) an update of the model to reflect economic, agronomic, and environmental conditions through 1983; (2) slightly greater spatial resolution (from 55 to 63 contiguous production regions) in the United States; (3) generalization of the model to allow characterization of either individual or multi-year "base" periods (e.g., 1983 or 1981–1983); (4) reestimation and expansion of livestock components of the model; (5) adjustment of demand and supply elasticities to reflect changes in national and world markets; (6) greater disaggregation of O_3 levels; (7) more specific scenarios concerning changes in O_3, including effects of a proposed seasonal standard; (8) more fully modeled adjustments for moisture stress; and (9) the addition of hay, alfalfa, and rice to the crops modeled in the analysis.

The methodology contains a linking of micro (farm level) and sector models, as suggested by McCarl (1982) and others. The sector model component is represented by a price-endogenous mathematical programming model of the agricultural sector, i.e., an activity analysis spatial equilibrium model (Taka-yama and Judge, 1971). Such sector models have been used extensively by agricultural economists to simulate the effect of alternative agricultural policies or technological change (Duloy and Norton, 1973; Heady and Srivistava, 1975). Among the various analytical techniques available to formulate policy models, mathematical programming has proven to be a particularly useful tool given its ability to predict consequences of as yet unrealized policies.

20.3.1 The farm model

Multiple activities (within crops) are used to generate the set of primary agricultural commodities that interface with the sector models based on historical crop mixes. These primary commodities are listed in Table 20.2. In addition to crops, the analysis includes livestock products. An endogenous livestock component is needed, given that livestock are the principal "con-sumers" of some important commodities, such as corn and soybeans.

The acreage relationships observed over time are used to develop a mix of

TABLE 20.2

Primary commodities included in the economic model

Field crop commodities	Livestock commodities
Cotton	Milk
Corn	Culled dairy cows
Soybeans	Culled dairy calves
Wheat	Culled beef cows
Sorghum	Live heifers
Oats	Live calves
Barley	Nonfed beef available for slaughter
Rice	Fed beef available for slaughter
Sugar cane	Calves available for slaughter
Sugar beets	Feeder pigs
Silage	Hogs available for slaughter
Hay	Poultry

crop activities from the United States Department of Agriculture's (USDA) Farm Enterprise Data System (FEDS) budgets for these regions. This procedure is intended to provide an economically and technically realistic portrayal of producers' behavioral responses.

An alternative to accounting explicitly for these relationships would have been to simply use the FEDS budgets and a given annual crop mix to define crop alternatives (activities) for each representative farm. While analytically simpler than trying to account for crop mix-yield changes, this was not done because it ignores potentially important microeconomic information. Specifically, since expected profit is assumed to be a primary factor in the producer's planting decisions (acreage and crop mix) and since these decisions can affect yields, it is important to account for these responses in the modeling of farm-level production.

20.3.2 The sector model

The producer-level responses generated through historical data interface with the macro component of the sector model to obtain a measure of social benefits. Consistent aggregation is achieved by building on the above micro conditions, which allows the aggregate and micro processes to be linked. The macro model features constant elasticity of demand relationships for the outputs (commodities) of the micro models. The elasticities vary with end use and across domestic and export markets. Given the long-run nature of the model, export demands assume major importance. Export demand elasticities for corn, sorghum, wheat, soybeans, soybean oil, and soybean meal are derived from the USDA Policy Analysis Group. The export elasticities vary from -0.18 for cotton, to -0.82 and -0.80 for soybeans (whole) and sorghum, respectively. Assuming supply and demand functions that are integrable and independent of sector activity, first-order conditions are then achieved in the macro model specification. The objective function of this specification is

$$\text{Maximize} \quad Q = \sum_i g_i(Z_i) - \sum_j e_j(X_j) - \sum_m C_m Y_m \quad (1)$$

where Q is the sum of ordinary consumers' and producers' surplus and the integrals are evaluated from zero to Z_i^*, the amount of the ith commodity produced and sold to consumers; and from zero to X_j^*, the amount of the jth factor used. The parameters are as follows:

$g_i(Z_i)$ is the area under the demand function for the ith products;
$e_j(X_j)$ is the area under the supply function for the jth factor; and
C_m is the miscellaneous cost of production, subject to a set of technical and behavioral constraints.

Given the micro and macro structure of a model, the sector model solution then simulates a long-run, perfectly competitive equilibrium. The full empirical detail of the objective function is provided in Appendix A of Adams *et al.* (1984*a*).

Following Samuelson (1952), the objective function (Q) may be interpreted as a measure of ordinary consumers' and producers' surplus (quasi-rents) or net social benefit. Analytically, this is defined as the area between the demand and supply curves to the left of their intersection. The demand functions are specified at the national level, as are aggregate production responses. Thus, the solution of the sectoral model provides objective function values at the national level.

The linking of the detailed producer behavioral model with a macro model measured in consumers' and producers' surplus provides a useful policy model. Specifically, changes in economic surplus arising from different model assumptions can simulate the economic effects of policy options. Justification for the use of economic surplus in policy analysis is well documented in the literature (Willig, 1976; Just *et al.*, 1982) and is particularly relevant to agricultural uses in which aggregate distributional consequences are of concern. By imposing alternative ambient O_3 assumptions on farm-level behavior, as manifested in yield changes predicted by NCLAN response data, changes in production and consumption and economic surplus may be measured. Changes in consumers' and producers' surplus between the alternative environmental states and current ambient concentrations indicate the benefits for these alternative O_3 levels.

20.3.3 Solution procedure and data summary

The sector model is solved under a mix of demand curves (constant elasticity, stepped demand, each displaying a range of elasticities) using the MINOS software package (Murtaugh and Saunders, 1977). A schematic of the model structure (in a two-region example) is presented in Fig. 1 of Adams *et al.* (1984*a*). The basic data used in implementing the model are USDA FEDS budgets. The United States is disaggregated into 10 regions consisting of the 48 states. In addition, the Corn Belt (Iowa, Indiana, Ohio, Illinois, Missouri), California, and Texas are disaggregated into 22 subregions, resulting in a total of 63 production regions. The primary crop coverage (commodities) is presented in Table 20.2. Table 20.3 summarizes the secondary commodities arising from these primary items. In its current version, the FEDS budgets used in the analysis were updated to 1983, using 1983 yields, acreage, and

TABLE 20.3
Secondary commodities in the economic model

Field crop commodities	Livestock commodities
Soybean meal	Veal
Soybean oil	Nonfed beef
Poultry feed	Fed beef
Feed grains	Pork
Protein supplement dairy feed	Milk products
High protein swine feed	Low protein swine feed

prices. The modified budgets include transportation costs, chemicals, machinery, fuel and repairs, and interest.

Labor and land availability depend on endogenous prices. The miscellaneous production costs were altered following the procedures outlined in Fajardo *et al.* (1981). (These procedures involve calculating miscellaneous costs so that they equal the difference between the value of production and the cost of the endogenously priced inputs, given that this is a long-run equilibrium model.) Finally, demand levels for products and the supply prices are drawn from 1983 agricultural statistics. As discussed earlier, elasticity values are from USDA and vary according to end use, for both domestic and export markets (R. House, personal communication).

One feature of the sector model of interest here is the export component. Specifically, since many US commodities enter world trade, any economic model of US agriculture must contain a representation of world demand for these commodities. Eight primary commodities in the model enter world trade: cotton, corn, soybeans, wheat, sorghum, rice, barley, and oats. To reflect their respective demand situations, constant elasticity of export demand functions are assumed. For purposes of this analysis, the assumption need only hold approximately and locally because we are dealing with only modest changes in prices and quantities. As a part of the model solution, equilibrium prices, quantities, and consumer surplus for these eight export commodities are derived. This presents a measure of the effects of imposed O_3 levels (in the United States) on not only domestic producers and consumers, but also foreign consumers (and indirectly, foreign producers). As a result, the transboundary effects of environmental policies in one country (the United States) can be observed in an aggregate sense by noting corresponding changes in foreign consumption and welfare.

20.4 PROCEDURES, DATA, AND ASSUMPTIONS IN THE O_3 ANALYSES

Data and assumptions from several disciplines must be integrated to model the effects of tropospheric O_3 on US agriculture. The integration must be consistent with the analysis performed by each of the participating research teams. This includes the dose–response data generated in the crop science experiments, the statistical analysis of those data, the calculation of regional O_3 levels, and prediction of O_3 exposures experienced by crops in the field. Finally, these data were combined with producer and consumer responses via the economic model. In the model, consumers adjust consumption patterns as yield changes affect market prices and producers respond to both yield changes and endogenous price changes. This process allows the economic actors wide latitude in consumption and in crop patterns which, in turn, has a major impact on the net effects precipitated by shifts in ambient O_3 levels.

Each of the pieces that underlies the economic analysis of O_3 effects are discussed in this section. Specific details are contained in a report to the US

EPA. The components, to be discussed in the following order, are (1) the dose–response functions, (2) atmospheric O_3 levels, growing seasons, and "effective" O_3 doses received by different crops in different states, (3) treatment of the effects of moisture stress, and (4) scenarios for changes in ambient O_3 concentrations.

20.4.1 Exposure–response functions

Exposure–response experiments have been carried out on cultivars of 14 distinct crops by NCLAN scientists over the past 6 years. The economic analysis focuses on the eight major field crops (alfalfa, barley, corn, cotton, hay (a clover-fescue mix), sorghum, soybeans, and wheat) in this set. In addition, data from an experiment conducted for the California Air Resources Board (Katz *et al.*, 1985) were used to obtain yield effects for rice.

The experiments involved fumigating crops in field chambers 7 or 12 h per day at different O_3 concentrations. The O_3 levels were measured and transformed into seasonal average exposures for each experimental treatment. The functional correspondence between seasonal doses and yields was then estimated using the Weibull function (Heck *et al.*, 1983). The flexible, nonlinear Weibull function is biologically reasonable and captures aspects of plant response to O_3 that linear fits do not, with important economic consequences. As noted earlier, other nonlinear functional forms, such as the Box–Tidwell family have qualitatively similar implications as the Weibull. The form of the Weibull function is

$$y = \alpha \exp\left(\frac{-x}{\sigma}\right)^c \qquad (2)$$

The estimated values used in this analysis are reported in Heagle *et al.* (1988), and the functions for the nine crops with yield adjustments are graphed in Fig. 20.1(a) and (b). The three parameters of the Weibull each have an interpretation. First, α is not an absolute yield scale parameter that does not have any effect in this analysis because interest here is on relative changes in yields at different O_3 levels. These relative changes are used to adjust historically determined yields for each of the 63 regions in the model. The σ parameter, measured in parts per million of O_3, scales plant sensitivity to O_3. Curvature of the function is determined by c. A value of 1·34 results in an almost linear response until the yield has declined to about 37% of maximum, after which it declines more slowly. Higher values of c imply downward curvature until yield has been reduced to about 37% of maximum; then the curvature reverses as yield declines further. For all nine crops in the model that have significant O_3 effects, the curvature (over the relevant range of O_3 levels) is concave downward (i.e., $c > 1·34$). Economic implications of this curvature are discussed in the analysis of results.

Statistical fits were obtained from field experiments for each cultivar and site for each crop, as well as combinations across crop cultivar experiments. For the economic analysis, only the aggregate of all sites and cultivars of a crop (e.g., soybeans) was used. Selection of an aggregate or pooled response is

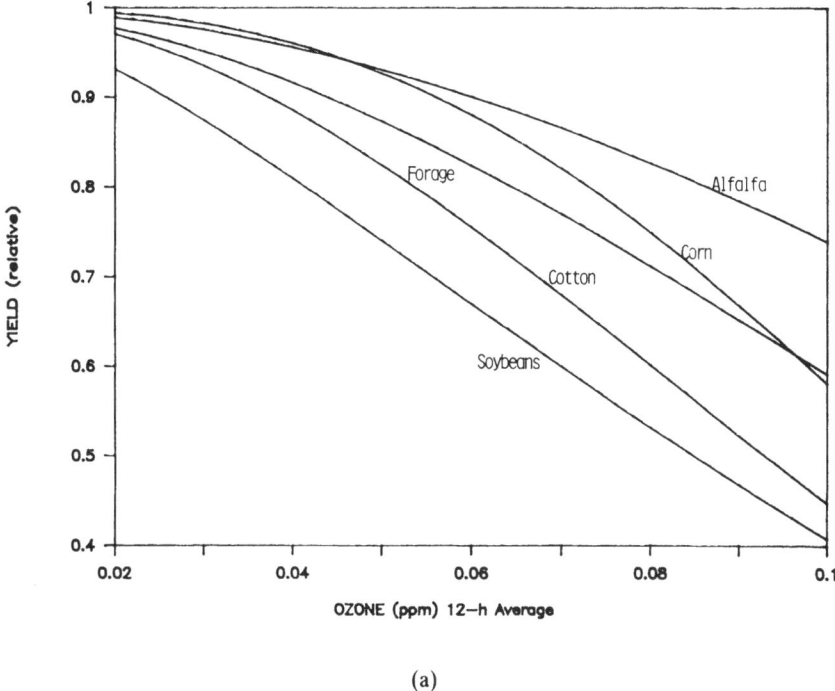

(a)

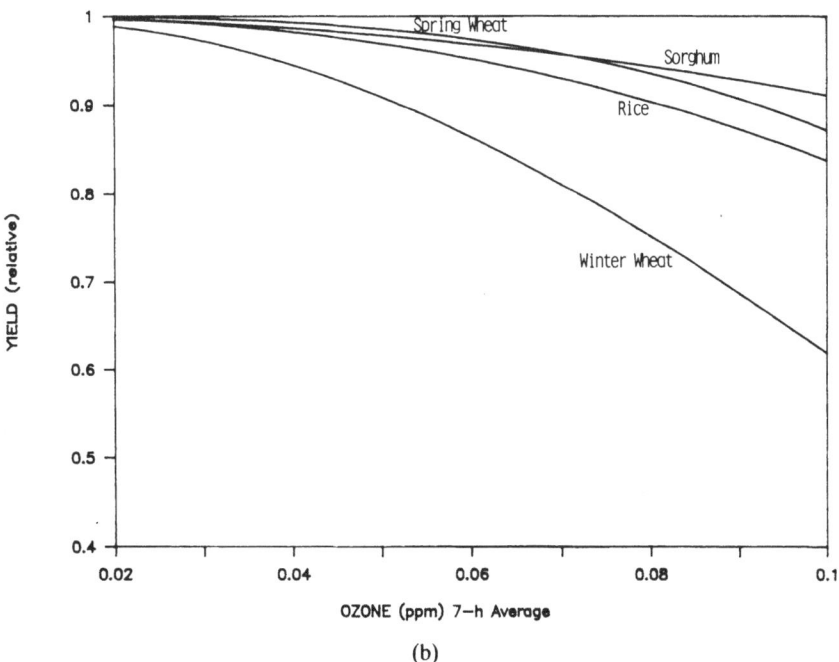

(b)

Fig. 20.1. Example of Weibull dose–response functions for each crop used in the economic assessment. Barley did not exhibit a statistically or economically significant response to O_3.
(a) After 12 h; (b) after 7 h.

motivated by several factors. First, the model focuses on aggregate economic analysis. While individual cultivars common to a particular region might show a different experimental result than the pooled response, on average, the combined analysis is a more balanced representation of national-level response. Further, since the pooled sample size is much larger than that for individual cultivars, the analysis is less prone to spurious correlation than using a variety of individual results for different regions. This also avoids the problem of deciding which cultivar to use when multiple cultivars are grown in a given region. Finally, confidence bounds are available on the combined fits, facilitating an analysis of the sensitivity of the economic results to the precision in the response parameters. This type of analysis is not feasible with a large collection of individual cultivar equations. It should be noted that the use of pooled response is inherently conservative (understates yield effects) because slightly larger yield effects would result from a given reduction in O_3 if producers shifted cultivars because of differential response to O_3. (The approach to measuring yield effects of O_3 by using net dry weight is also modestly conservative because it ignores quality aspects of yield. While no systematic analysis of yield quality has been instituted in the NCLAN program, it seems likely that crop quality may decline with yield. Shortle *et al.* (1988) extend the analysis of yield effects by measuring changes in both oil and protein content of soybeans, but this is the extent of quantitative analysis of the issue.)

20.4.2 O_3 data and assumptions

The NCLAN crop experiments characterize O_3 exposures in the form of a seasonal average of daily 7- or 12-h periods with the highest O_3 readings. Hence, use of the NCLAN response functions requires information on O_3 levels in agricultural areas consistent with the exposure measure used in this experiment. In previous assessments (e.g., Adams *et al.*, 1986), a spatial interpolation of O_3 readings stored in EPA's Storage and Retrieval of Aerometric Data system (SAROAD) served to provide rural ambient O_3 estimates.

Using an updated set of EPA SAROAD values, Lefohn *et al.* (1987) calculated monthly averages on both a "maximum 7-h" and a "maximum 12-h" basis for a 5-year period. Kriging was used to spatially interpolate the SAROAD values. Tests of the validity of the methodology (using sites not included in the original data set) indicate good predictive ability, especially for areas characterized by a high density of monitoring stations and where topography and atmospheric conditions lead to homogeneity of O_3 levels in the spatial dimension. In areas where these conditions do not hold, especially the west, there is greater uncertainty. Fortunately, except for California, which does have a large number of monitoring sites, the agricultural output in such areas is modest.

The kriging process for generating state averages of O_3 was also used to construct confidence bounds on the estimates. These bounds are a function of the number and spatial distribution of the monitoring stations and of the heterogeneity of the measurements. The bounds form the basis for a test of the aggregate sensitivity of the model to the accuracy of the overall level of the actual ambient O_3 levels. This is done by adding (subtracting) the confidence bounds from the base level to obtain high (low) estimates of O_3 levels.

The 7- and 12-h seasonal means correspond to the protocols used in the crop experiments and to the O_3 metrics used in estimating the exposure–response functions. The seasonal measure is also similar to a standard proposed by the Office of Air Quality Planning and Standards (OAQPS, 1986). This standard is based on an 8-h daily average over a 3-month period and is discussed below in developing the scenarios that were' used to simulate changes in ambient O_3 levels.

The estimated rural seasonal O_3 levels for each year serve to define the actual base or benchmark level for use in the economic analysis. Since the model is intended to represent equilibrium behavior using recent data, the average 1981 to 1983 O_3 levels are used as the ambient O_3. These years represent a range of O_3 levels, with 1983 a higher O_3 year, and 1981 and 1982 somewhat lower, relative to the average, with the net effect that the average is slightly below the average for 1980 to 1984, and approximately 10% lower than 1980, the year used as a base in the 1984 preliminary assessment (Adams *et al.*, 1984*a*). This base O_3 level is then altered to develop changes in O_3 for use in the response function, as described below.

20.4.2.1 Growing windows

To approximate the exposure of O_3 to which crops are exposed, it is necessary to model the growing season for each crop and region. Growing seasons will vary by crop and by state. Further, there is a distribution of growing seasons within a state because producers have flexibility with planting (and hence harvesting) dates. Finally, there are periods in the plant phenology when plants are more sensitive to O_3, e.g., during flowering of soybeans. This results in a range of plant yield sensitivity between the planting and harvesting dates.

Several other features of plant growth and the experimental protocols were considered in the construction of the crop O_3 exposure. The amount of solar radiation (and temperature levels) affects the growth rates of plants, implying that the middle of the season should receive greater weight than either the beginning or ending periods. Also, in most experiments, fumigations with O_3 began several weeks after emergence. The O_3 levels experienced by the plants during this early growth period were not included in the calculation of the O_3 levels used to generate the dose–response functions. If they had been, the same yield reductions would have been observed at lower calculated doses, indicating greater sensitivity to changes in O_3 (a steeper response function). This results in overly conservative calculations of yield changes. By narrowing the seasonal window, the tails of the growing season are given less weight. Since O_3 levels are generally higher in the core of the growing season, this results in slightly higher calculation of seasonal averages. The implication of these features is that greater weight be given to the core of the growing season. The specifics of this procedure are outlined below.

The growing season was initially defined using the USDA publication 'Usual

Planting and Harvesting Dates for U.S. Field Crops' (USDA, 1984c), which contains information on the range of planting and harvesting dates by crop and state. These dates were also checked by several NCLAN scientists for reasonableness. Next, estimates of lags from planting to emergence and from the point at which yield is fully determined to the typical harvest dates were obtained for each crop (Doorenbos *et al.*, 1979). This information determines the beginning and end points for the growing season (t_b and t_e), and implicitly defines the midpoint ($t_m = (t_b + t_e)/2$). A core window (usually 90 days) was defined around these end points [t_1, t_2] and received full weight. The period between t_b and t_1 (between t_2 and t_e) received proportionally increasing (decreasing) weight. The weighting function, $w(t)$ is defined as

$$
\begin{aligned}
w(t) &= 0 & t \leq t_b \\
&= (t - t_b)/(t_1 - t_b) & t_b < t_1 \\
&= 1 & t_1 \leq t \leq t_2 \\
&= (t_e - t)/(t_2 - t_e) & t_2 < t < t_e \\
&= 0 & t \geq t_e
\end{aligned}
\tag{3}
$$

The average is then calculated as

$$
\frac{\sum w(t)Z(t)}{\sum w(t)}
\tag{4}
$$

where $Z(t)$ is the daily O_3 level, and t is the Julian date.

The above procedure explicitly accounts for the distribution of planting dates and growth rates for each crop in concert with the experimental details that give rise to the dose–response relations. By then adjusting these dates and weights on a state-by-state basis, it is believed that the estimated O_3 exposures provide a good approximation to real-world conditions.

20.4.3 Moisture stress

An important issue in the O_3 analysis is the interaction with moisture stress, because moisture stress is one of the primary factors affecting yields of field crops in the United States. Large areas of the Midwest and South depend entirely on naturally occurring moisture for production. Inadequate levels can reduce yields substantially, as evidenced by droughts in the Corn Belt. A number of NCLAN experiments were aimed at testing potential interactions between O_3 and moisture stress. While explicit interactions were not statistically significant, such effects are not easy to detect.

To address this relationship, King (1988) modeled drought effects with a plant process model, concentrating on the major crops (corn, soybeans, cotton, wheat, and forage) in the Corn Belt and adjacent areas. This simulation process projects diminished sensitivity to O_3 when yields have already been reduced by moisture stress. Using disaggregated data provided by Control Data Corporation (King, 1988), alterations in the yield adjustments due to O_3 are obtained.

The net effect of these adjustments is to lower modeled sensitivity of yields to O_3. However, because the exposure–response relationships incorporate experiments with a mix of moisture effects, areas that are not stressed (areas that received adequate rainfall in the 1981 to 1983 base period, or that were irrigated, as in much of the West), result in slightly increased O_3 sensitivity.

King (1988) provides the details of the procedure and adjustments for five major crops for 1980 to 1983. The results for 1981 to 1983 were combined in this analysis by averaging the effects across the 3 years using production weights. The years 1981, 1982, and 1983 involved, respectively, modest, low, and substantial drought stress. This situation results in greater than average moisture stress, but the extent is diminished by using production weights because 1983, the drought year, had reduced acreage due to the government payment-in-kind (PIK) program.

20.4.4 Scenarios for changes in O_3

To simulate the welfare effects of changes in O_3 levels requires development of specific assumptions concerning alternative O_3 levels. Previous assessments use different assumptions in developing O_3 changes. For example, Adams *et al.* (1984*a*) hypothesize proportional changes in O_3 levels from ambient in a specific year (e.g., 25% from 1980 levels). Benefits arising from such proportional changes are then calculated. Adams *et al.* (1984*a*), Kopp *et al.* (1985), and Howitt and Goodman (1988) specify seasonal averages similar to the 7-h seasonal average discussed above. Benefits are then calculated for changes from ambient to an array of specified seasonal levels.

In this analysis, both proportional changes and changes linked to seasonal levels are evaluated. The latter case is meant to be suggestive of a possible seasonal standard. Specifically, scenarios are developed to approximate a standard proposed by the US EPA (OAQPS, 1986). The proportional changes are calculated to provide a comparison with the results arrived at by Adams *et al.* (1984*a*).

The standard proposed by OAQPS is a seasonal average incorporating a 3-month average of daily 8-h averages. The 8-h average is the average of the eight (consecutive) 1-h periods in a day with the highest average, whether from 8 am to 4 pm or 11 am to 7 pm, etc. As such, the hours averaged may shift from day to day. The 3-month average represents the 3 consecutive months with the highest average. OAQPS selected this 3-month period for the standard because it corresponds to the core of the growing season.

Another feature of a proposed seasonal standard involves the stochastic nature of the weather and air pollution (and human elements that cause it). Because of pollution variability, current National Ambient Air Quality Standards (NAAQS) for O_3 use the highest 1-h average on the second highest pollution day. To allow for this variability in evaluating a seasonal secondary NAAQS, scenarios are needed that reflect different probabilities that the seasonal standard will be violated in a given year. (From a regulatory perspective, this stochastic aspect is important, for if individual regions aim exactly at a standard, the distribution of air pollution events implies the standard will be exceeded in about half of the years.)

The way in which these probabilities are captured here is to start with accurately kriged O_3 information, 1980 through 1984. The seasonal average across this time period is then taken as the current long-term average. (Analysis of the Corn Belt indicates that there is no time trend to the averages for individual EPA monitoring sites. Similarly, McCurdy (1987) finds no indication of a time trend over the period 1979–83. Thus there is little reason to believe that most agricultural areas would evidence such a trend.) Next, a sample standard deviation is calculated to reflect the intertemporal variability of each region. This standard deviation is taken to be representative of actual variability. Given the mean and the standard deviation, it is straightforward to construct an O_3 level, X_i, to which a region must actually aim so that 95% of the time the standard will not be exceeded; i.e.,

$$X_i = S - k \cdot SD_i \qquad (5)$$

where S is the standard (for instance, 0·05 ppm O_3 using the 3-month, 7-h average), SD_i is the simple standard deviation for state i, and k is the number of standard deviations needed for a given confidence level. For a sample of 5 years (1980–1984), a 95% probability level implies $k = 2 \cdot 132$. (This formulation assumes that the O_3 variability is approximately normally distributed. An analysis of SAROAD O_3 data for 1980 through 1984 in the Corn Belt indicates that this is approximately correct for individual sites using the analysis of residuals provided in SHAZAM (White, 1978). Aggregation to state averages should strengthen this approximation.)

From a regulatory perspective, if the state's long-term mean level $\bar{Z}_i$ is sufficiently low (below X_i, not just below the standard, S), there is no adjustment taken, and state ozone levels remain unchanged. If the long-term mean exceeds X_i, then define the adjustment ratio, r_i, as

$$r_i = X_i / \bar{Z}_i \qquad (6)$$

The ratio is used to multiply *all* monthly O_3 levels in the state (region) i. The smaller the ratio, the greater the O_3 reduction necessary to meet the standard. Both the level of the standard and the probability of exceedance determine the degree of control of the overall standard.

The economic implications of this seasonal standard are potentially quite different than for an O_3 scenario that merely adjusts O_3 levels by a given proportion. These implications relate primarily to distributional consequences. For example, producers in high pollution areas will gain because their yields will likely increase substantially more than prices decline because not all other areas experience equivalent increases in yields. Thus, prices will decrease but not as much as when all areas experience the same reduction in O_3. The losers are producers in areas that are already relatively pollution free. For them, yields are little affected, but prices decline. This is the same situation facing foreign producers.

From a regional, air quality management perspective, this may be more equitable because reduced O_3 in a polluted area is a benefit to agricultural producers, which counters the cost that must be borne by that region to reduce the pollution. "Clean" areas do not have to limit pollution-generating activities, but agricultural producers in such regions must bear the price-depressing effects of increased yields. The consumer benefits, however, are dispersed throughout domestic and foreign markets.

20.5 RESULTS AND IMPLICATIONS

Section 20.4 presented the key features of the economic model used in this assessment. As discussed, the NCLAN economic model includes many more features of producer and consumer behavior, features that reflect the complex set of interactions that underlie economic markets. It also integrates across the diverse components of the US agricultural sector to obtain measures of aggregate and regional economic activity, including measures of producer and consumer welfare. As with any model, however, complexity does not guarantee predictive ability. Thus, it is important to establish that this model is a reasonable approximation of the agricultural sector over the period of interest. To validate the model, we test the endogenous prices and output for the base years 1981 through 1983. Successful validation provides one indication that the model is appropriate for evaluating the effects of O_3 on agriculture.

Table 20.4 provides a comparison of the actual average prices and the quantities produced with those determined by the model solution at 1981 to 1983 average ambient O_3 and moisture stress conditions. As is evident, the prices for all commodities match reasonably well, whereas the quantities generally understate actual levels by 5 to 10%. Overall then, model prices and quantities for both crop and livestock commodities appear to capture the relative magnitudes of equilibrium prices and quantities observed in the 1981 through 1983 period.

20.5.1 Comparison with the previous NCLAN assessment

Section 20.2 of this paper contains a review of previous economic assessments based on NCLAN data. The results of Adams *et al.* (1984a), referred to as AHM, are of particular interest here because they are a component of the overall NCLAN program and are based on an earlier version of the economic model. In addition, while the AHM results were described as "preliminary", the results have appeared in the published literature on O_3 pollution. Thus, to place past and current results in perspective, we compared the AHM results with those obtained in this current analysis. It must be stressed that the present results reflect different exposure–response estimates and aerometric and economic conditions. Specifically, the current assessment is based on an average of 1981 through 1983 conditions, including foreign and domestic demand, technological conditions and levels of O_3 and moisture stress.

By comparison, AHM results were based on conditions in 1980. Generally, 1980 had higher levels of O_3 and moisture stress than the 1981 through 1983

TABLE 20.4
Model prices and quantities vs actual: 1981–1983

Commodity	Prices ($ per unit)		Quantities (million)	
	Model	Actual	Model	Actual
Cotton	284·97	281·90	10·24	11·79
Corn	2·68	2·68	6 452·28	6 839·00
Soybeans	5·73	5·65	2 018·02	1 915·00
Wheat	3·56	3·50	2 260·78	2 419·00
Sorghum	2·53	2·50	662·00	730·00
Rice	8·13	8·01	120·00	145·00
Barley	2·23	2·20	425·18	498·00
Oats	1·55	1·67	503·04	526·00
Silage	21·76	n.a.	45·92	n.a.
Hay	62·08	65·76	76·82	82·00
Milk	13·35	13·65	1 283·61	1 359·80
Pork	169·54	165·90	139·38	151·70
Fed beef	232·27	239·70	135·81	156·25
Nonfed beef	131·44	145·15	87·60	78·37

Prices for all crops are dollars per bushel, except for cotton ($ per 480-lb bale), rice ($ per hundredweight), and silage and hay ($ per ton). Meat prices are $ per cwt and are average retail prices for finished meat products.
Sources: USDA, ESA, Statistical Bulletin No. 715, Washington, DC. USDA, *Agricultural Statistics, 1984.* Washington, DC.

period. The models also differ in the extent of crop coverage, in that rice, and alfalfa and other hays have been incorporated. Further, the current assessment utilizes more complete adjustments for moisture stress and reflects increased precision in the calculations of seasonal O_3 exposures and in the exposure–response estimates for each crop. Thus, it is expected that the economic consequences predicted in the current assessment present a more accurate measure of economic effects and differs somewhat from those of AHM.

As with earlier analyses, an item of primary interest in the model solutions are the *changes* noted in the solutions of each O_3 scenario when compared with the base solution. These changes represent the economic effects of the respective O_3 change. Table 20.5 contains the economic surplus values for the base case solution and the changes in economic surplus from that base for 10, 25, and 40% changes in O_3 from 1981 through 1983 ambient levels. In addition, summary totals of aggregate national economic surplus are provided for the AHM analyses, along with a breakdown of associated producer and consumer effects for both the current and AHM assessments.

For consistency with the current analysis, the AHM results are drawn from the moisture stress case reported as Analysis V. The 1980 figures from AHM have been inflated by the Consumer Price Index to 1982 dollars to further facilitate comparison. This results in a 17% increase from those given in AHM,

TABLE 20.5

Comparison of current economic surplus estimates with 1984 assessment, in 1982 dollars

Ozone assumption	Producers' surplus	Consumers' surplus	Total surplus
	LEVELS ($ millions)		
Base[a]	21 220	139 676	160 896
	CHANGES ($ millions)		
Current			
+25%	−690	−1 363	−2 053
−10%	286	522	808
−25%	572	1 318	1 890
−40%	683	2 097	2 780
1984 Model[b]			
+25%	−650	−1 515	−2 165
−10%	403	296	699
−25%	754	1 074	1 828
−40%	1 047	1 590	2 637

[a] The base levels are from the current assessment, which is in 1982 dollars, and reflect an average of 1981–1983 O_3, production, and moisture levels. Both assessments include adjustments for moisture stress (Analysis V of Adams *et al.*, 1984a). The corresponding base values for AHM in 1982 dollars are 30 443; 134 503; and 165 087 million for producers, consumers, and total surplus, respectively.
[b] The 1984 assessment used 1980 as a base. The dollar totals reported here have been equated by using the CPI indexes for 1980 and 1982.

Analysis V. The agricultural sector was economically stronger in 1980 than in 1981 to 1983, as measured by exports and debt-to-equity ratios. For this reason and the others noted above, the two assessments are not strictly comparable.

As is evident from Table 20.5, in aggregate terms the results across the two assessments are quite similar. For example, when O_3 is reduced uniformly by 25% across all regions, the estimated economic benefits are 1890 million for the current assessment and 1828 million for AHM (in 1982 dollars). These values are approximately 1·9% of total annual agricultural revenues for 1981 through 1983. The results for the 10 and 40% O_3 changes reveal the same close correspondence across the current and the AHM assessment (e.g., $2780 million versus $2637 million for the current and AHM assessments, respectively, at 40% O_3 reductions). The overall similarities across the assessments in the face of different exposure–response estimates and aerometric and economic conditions is attributable to countervailing influences of those differences. Specifically, increased response sensitivity across crops and expanded crop coverage in the current assessment tends to be offset by somewhat lower exports, crop prices, and O_3 levels in the 1981 through 1983 period.

20.5.2 Policy analyses

To facilitate comparison with earlier studies, one set of O_3 change scenarios repeated the constant proportional O_3 changes used in AHM, as reported in Table 20.5. However, this study differs from previous published analyses in focusing on potential *seasonal* standards (developed in Section 20.3) as part of the set of O_3 scenario evaluations. In addition to this simulated seasonal standard, the following discussions will continue to present changes that result for a uniform 25% reduction in ambient O_3 levels. The seasonal standard analysis is intended to represent O_3 conditions that would prevail if states exceed a given seasonal level in only 5% of the years. By including the probabilistic aspects of the standard, the analysis developed here captures both average levels and year-to-year variability in O_3. For the purposes of this discussion, we focus on a seasonal standard of 0·050 ppm (196 μg m^{-3}). As discussed below, the economic consequences of this standard are close to the net aggregate benefits achieved with a proportional 25% reduction in O_3. This standard also falls midway in the range of potential seasonal standards (0·04–0·06 ppm) suggested by OAQPS (1986). The 0·050 ppm standard is sufficiently high that some states never exceed the standard, while others would have to affect substantial reductions in pollution to be in compliance.

Changes in economic welfare are analyzed for both the 25% reduction and the 0·050 ppm (95% compliance) seasonal standard by comparing changes in producers' and consumers' (both domestic and foreign components) surplus, as reported in Table 20.6. In addition to these central cases, several variations are also presented in Table 20.6 to allow assessment of the sensitivity of these estimates to the extent and nature of O_3 reductions. For percentage changes,

TABLE 20.6

Changes in economic surplus arising from alternative O_3 scenarios: constant percentage reductions and seasonal standards[a]

Ozone assumption	Total surplus ($ millions)	Producers' surplus ($ millions)	Consumers' surplus ($ millions)	Domestic consumers' surplus ($ millions)	Export surplus ($ millions)
−10%	808	286	522	313	209
−25%	1 890	572	1 318	738	580
−40%	2 780	683	2 097	1 127	970
60/95[b]	612	414	198	158	40
50/95	1 674	769	905	531	374
40/95	2 645	818	1 827	985	847
50/50	853	631	222	186	136
50/90	1 465	738	727	473	254
50/99	2 117	843	1 274	698	576

[a] Changes in economic surplus are the changes from the base values reported in Table 20.5.
[b] These numbers refer to scenarios for changes in O_3 levels. 60/95 is a 0·060 ppm seasonal standard with 95% probability of not exceeding the standard, as explained in the text.

these are the 10 and 40% reductions used in AHM. For the seasonal standard, both the level of the standard and the probability of compliance are varied. Table 20.6 reports results of standards of 0·04 and 0·06 ppm with 95% compliance, as well as 0·05 ppm with 90 and 99% compliance.

The aggregate benefits of a 25% reduction and a 0·05 ppm standard with 95% compliance (50/95) are similar, $1890 and $1674 million, respectively. In aggregate terms, producers gain more with the seasonal standard than with the constant 25% rollback (46 versus 30% of total benefits). This difference is due to the regions (and hence the crops) that are most affected. The inclusion of an endogenous livestock sector also allows regional producers to capture a greater share of those respective yield increases, which points out the importance of having a fully integrated model. The distribution of consumer gains is stable across the two scenarios, with domestic consumers obtaining more than half (56–59%) of the benefits. This reflects the diminished impact of the export sector vis-à-vis the AHM assessment, where two-thirds of additional consumers' surplus accrued to foreign consumers.

At the regional level, the distribution of the producer gains differs markedly between the seasonal standard and the constant percentage reduction. These regional implications are presented in Table 20.7. (For comparison, the

TABLE 20.7
Producers' surplus, by region, in millions of 1982 dollars[a]

| | Baseline surplus | 25% reduction in O_3 | | 50/95 Ozone[b] | | 1984[c] |
		Change in surplus	Percentage change	Change in surplus	Percentage change	Percentage change
North east	126	25	19·76	39	30·95	4·36
Lake	1 858	36	1·94	−15	−0·81	1·04
Corn Belt	5 836	278	4·76	475	8·14	3·18
N. Plains	3 096	34	1·10	3	0·10	0
Appalachia	1 226	102	8·06	120	9·48	2·87
Southeast	675	37	5·48	48	7·11	4·58
Delta	963	28	2·91	42	4·36	7·42
S. Plains	1 629	−17	−1·04	−17	−1·04	1·30
Mountain	2 384	44	1·85	20	0·84	2·47
Pacific	1 214	28	2·31	73	6·01	21·03
US						
Total	19 047	584	3·07	779	4·09	3·04
Foreign	2 173	−12	−0·55	−10	−0·46	0
Total	21 220	572	2·70	769	3·62	3·04

[a] Totals may not add due to rounding.
[b] 50/95 is a 0·050 ppm 7-h seasonal O_3 standard with a 95% probability of compliance.
[c] 1984 is the 25% reduction from the AHM assessment. However, these changes come from a model without adjustment for moisture stress. Moisture stress would reduce the percentage changes listed here by a fifth, on average.

distribution of producers' surplus from AHM is also presented. Major shifts are apparent.) As the numbers in the table suggest, producers in higher O_3 areas, especially those with greater variability, gain most with the 50/95 standard. The Northeast, Corn Belt, Appalachia, and Southeast regions all gain substantially. These gains come at the expense of lower O_3 regions of the Lake states, Northern and Southern Plains, and Mountain areas. This is due to relatively low O_3 levels in these regions (and hence little yield improvement), although crop prices decline by the same amount across all regions. The regional aggregation masks more extreme differences within some regions. For instance, the Pacific region is composed of California (two regions), Oregon, and Washington where the O_3 reductions are 35, 22, 0, and 0%, respectively. California producers gain substantially while those in Washington and Oregon suffer from the decline in product prices.

As noted before, the 0·05 ppm standard is midway in the range of seasonal standards discussed by OAQPS. With 95% compliance, seasonal standards of 0·04 ppm and 0·06 ppm bracket the 50/95 scenario, with gains of $612 and $2645 million, respectively. Likewise, the compliance parameter can be varied. Such variations dramatically affect the economic benefits. Referring back to Table 20.6, for 90 and 99% compliance with a 0·05 ppm standard (50/90, 50/99), respective increases in economic surpluses of $1465 and $2117 million result. A 0·05 ppm standard with 50% compliance results in benefits of only $853 million, half that of the 50/95 scenario. In all cases, the gains of modest O_3 reduction accrue mainly to producers (two-thirds for the 60/95 standard), while greater reductions consistently favor consumers. Consumer benefits favor the domestic sector, a reversal from the 1984 AHM assessment, which reflects changes in international markets.

An implication of these preliminary evaluations of seasonal standards is that the degree of compliance is an important element in setting a standard. Ignoring year-to-year variability in implementing a seasonal standard would be similar to choosing a standard with only a 50% chance of actually attaining that seasonal level. Further, from a regulatory perspective, it is probably unreasonable to assess the impact of a standard by assuming that all regions exactly equal the proscribed level.

20.5.3 Impact of O_3 standards on US farm programs

Economists, policymakers, and others have expressed considerable interest in the cost and effectiveness of government intervention in agriculture. Despite continuing controversy, the electorate, through its public officials has consistently intervened on behalf of the agricultural sector. Thus, the agricultural industry typically responds to a mix of market and institutional signals. Since agriculture is also affected by other government policies, some of which are interactive, its status as a revenue-supported industry may have implications when examining effects of such interactive policies. For example, McGartland (1987) has questioned past estimates of benefits from O_3 reductions because farm program provisions and resulting surplus production are not fully modeled in these assessments.

One reason such analyses are lacking is the difficulties attendant to modeling interactions between O_3 environmental policies and US Farm Program provisions. For example, farm programs vary from year to year, depending on economic and environmental conditions worldwide. Where the interest is in measuring long-term economic implications of O_3 on agriculture, it is questionable as to whether short-term Farm Program provisions can or should be included in the economic analysis.

Conceptually, changes in ambient O_3 levels also change the production possibilities of producers, thus acting like a technological change. At the macro level, O_3 effects will not be neutral because (1) input supplies, O_3 levels, and crop mixes differ across regions; (2) crops are differentially sensitive to O_3; and (3) the response functions are nonlinear. Thus, it is difficult to model effects of program provisions in the aggregate.

Recognizing these difficulties, we attempt to account for major features of the 1985 Farm Program by incorporating the target price response at the producer (micro) level. Although the provisions of the program are conditional, such a preliminary inquiry into overall effects can suggest the magnitude of bias, if any, associated with exclusion of Farm Bill provisions in environmental policy analyses.

There are several practical difficulties involved in imposing the 1985 Farm Program provisions on the current economic model. For example, the economic model is tuned to economic, technological, and environmental conditions for 1981 through 1983. (The economic information for the 1985 crop year has just been published on the consistent basis needed for this analysis. Resetting the model with the new data is an extremely complicated task. Since 1985 was a year of considerable disequilibrium due to the extremely high value of the dollar in exchange markets, we decided not to use 1985 for the full analysis. Also, the detailed crop budgets are more suited to the 1981 through 1983 period.) The 1985 Farm Bill provisions, however, are specific to 1985 conditions. (Since the modeling process is cast in terms of equilibrium conditions, the target prices of the 1985 Farm Bill themselves should really be considered endogenous. In addition, current regulations base deficiency payments on an average of historic, most current, yields. Thus, the effect of changes in future yields will be confined to provisions of loan payments. We ignore this aspect of the Farm Bill to maintain our focus on long-run equilibrium.) Ozone and moisture stress data for 1985 are not currently available. Thus, to evaluate the effects of O_3-induced supply increases in this analysis requires that the 1985 Farm Bill be integrated with 1981 through 1983 economic conditions.

The procedure we use here to obtain an impression of the effects of O_3 reductions on Farm Program payments consists of three general steps. In the first stage, the target prices are introduced for crops covered in the 1985 Farm Program (cotton, corn, wheat, rice, barley, sorghum, and oats). Producers are presumed to use the target prices as their expected market prices, hence the model is solved under this initial set of prices. The quantities produced under these price expectations are then used in the next stage. Specifically, stage two fixes the quantities of program crops in the model at the levels determined by producer response to the target prices. Market clearing prices for these new quantities result from the stage two solution. The difference between the target price and the market clearing price, times the quantity produced, under the program, is equal to the deficiency payment for each program crop.

The final stage repeats the above steps, but uses the yield adjustments occasioned by changes in ambient O_3 levels. Comparison of the farm program solutions, including deficiency payments, with and without the O_3 adjustment provides an estimate of the possible impact of the 1985 Farm Program on the benefits from changes in O_3 levels.

The results of this procedure are reported in Table 20.8. As the comparative values indicate, the inclusion of the target price provisions reduce the benefits of O_3 control reported in the "without" analyses (e.g., Tables 20.5 and 20.6) between 10 and 12%. The modifications to the benefit estimates are more modest than suggested by McGartland (1987), which is a result of a relatively inelastic supply for the affected crops at the margin. In general, the more inelastic the supply, the smaller the welfare changes (and hence, possibly distortions) of movements along the supply curve. Also, as noted by Segerson (1987), it is inappropriate to ascribe this reduction in benefits from O_3 control to the cost of that environmental control program. Rather, it is the increased cost of the Farm Program under the new "technological" conditions.

There are several shortcomings in this analysis. The Farm Program requires acreage set-asides as a condition of participation. This provision has not been directly incorporated in the analysis. The impact is to overstate output (for both ambient and reduced O_3 solutions). This overstates the size of the deficiency "price" and the quantity to which it applies. Second, there are both loan and payment limit provisions. These factors combine to limit participation to between 50 and 90% of available acreage, depending on the crop. Participation is a farm-level decision, which cannot be fully modeled with the representative farms because it depends on farm size, farm-specific historical yields, and the relative productivity of marginal land that can be set aside. However, a gross approximation can be made by interpolating between full participation and no participation, i.e., no farm program. The participation

TABLE 20.8

Evaluation of the interaction between O_3 reduction and 1985 Farm Bill provisions

Ozone scenario	Change in economic surplus ($ millions)	Change in estimates ($ millions)
25%		
With 1985 Farm Bill	1 692	
Without 1985 Farm Bill	1 890	198
50/95		
With 1985 Farm Bill	1 473	
Without 1985 Farm Bill	1 674	201

rate is taken to be the weighted average of the individual crop programs where the weights are crop values (in the undistorted model). Since the effective participation rate in 1985 was about two-thirds, the net effect of O_3 reductions on program payments is less than reported in Table 20.6, amounting to approximately $136 million for both scenarios.

20.5.4 Sensitivity analysis
While the current assessment represents an improvement in plant science and aerometric data over AHM and other studies in the literature, there is still some uncertainty remaining in these elements. The imprecision in the exposure–response estimates is addressed by calculating 95% confidence intervals around the respective crop response curves. These are used to provide bounds on the estimated benefits. Because the errors across equations are not completely correlated, using 95% confidence intervals for each crop response results in an *aggregate* confidence level much higher than 95%. The magnitude of the benefits change using these modified response functions, but qualitatively the results are similar to those reported in Table 20.6. The sensitivity results are presented in Table 20.9. For the two O_3 scenarios, the upper (lower) bounds are 20 to 25% higher (lower) than our benchmark estimates. Given that these bounds are obtained from an extreme assumption regarding the correlation of errors across crop response functions, they provide a relatively narrow range of estimated benefits.

Uncertainty in the aerometric data is addressed by using the standard errors on the kriged O_3 levels for each state and each month (and year). Typically, these standard errors are of the order of 3 to 8% of the O_3 levels. If these errors are independent, then aggregate economic benefits will be virtually unaffected because the errors cancel out by the law of large numbers when averaging across the 4 to 6 months of the growing season, the 3 years (1981–1983), and the 63 production regions in the model. Temporally, it is expected that the errors will be completely uncorrelated. However, because a network of adjacent monitoring sites is used in the kriging process there is some correlation spatially. To approximate the effects of these errors we assume that, on average, every four states have completely correlated errors but these groupings are mutually uncorrelated. These groupings are combined

TABLE 20.9

Sensitivity of economic effects to dose–response and aerometric data uncertainties

Ozone assumption	Benefits ($ millions)				
	Benchmark estimate[a]	Dose–response bounds		Ozone data bounds	
		Lower	Upper	Lower	Upper
25% reduction	1 890	1 444	2 356	1 854	1 927
50/95 standard	1 674	1 335	2 031	1 605	1 745

[a] The benchmark estimate includes the baseline O_3 data and dose–response estimates.

with the uncorrelated errors in the time domain to yield an overall 95% confidence interval for our benefit estimates. (The 95% confidence intervals use the base O_3 levels $\pm 1 \cdot 96 \cdot$SE. The standard error (SE) is approximately the individual standard errors times k where $k = 1/\sqrt{n_1 \times n_2}$, and $n_1 =$ weighted average number of time periods for each state (13 1/2) and n_2 is the number of independent regions (63/4). Thus $k = 0 \cdot 0686$.)

As a result of the relatively small magnitude of the kriging errors themselves, and the independence of those errors, the interval estimates due to the uncertainty in the O_3 levels are quite narrow (Table 20.9). For the 25% reduction they are $\pm \$36$ million (2%), while for the 50/95 standard they amount to $\pm \$70$ million (4%). The effect of uncertainty in the case of a standard will be larger than for comparable percentage rollbacks because the starting point in the standard is changed while the ending point is unaffected. (For example, consider a region with an O_3 level of $0 \cdot 060$ ppm (and a standard deviation over time of $0 \cdot 0025$ ppm). A 25% reduction and a 50/95 standard both imply an average level of $0 \cdot 045$ ppm. However, if the regional O_3 level were $0 \cdot 064$ ppm, a 25% reduction now implies a level of $0 \cdot 048$ ppm after control while the standard still implies $0 \cdot 045$ ppm.)

It was noted earlier that the marginal value of additional biological response (such as greater precision of the exposure–response functions) in defining economic effects declines relatively quickly. The sensitivity analysis confirms this general theme. The reason for this is that the aggregate economic effects are summed over a large number of crops and growing regions, averaging out by the law of large numbers.

20.6 CONCLUSIONS

Research performed in the NCLAN program over the past seven years has greatly enhanced the understanding of the effects of O_3 on major agronomic crops. The availability of NCLAN plant science and aerometric data, coupled with the US EPA's need for regulatory impact analyses, motivated a series of assessments of the economic effects of O_3 on US agriculture. This chapter first has reviewed economic findings and approaches from previous assessments and then provides an overview of the assumptions, procedures, and results from the latest in this series of NCLAN economic assessments.

The long-term nature of the NCLAN program provides a somewhat unique "Bayesian" experiment on the value of continued biological and economic research as a means for improving regulatory information. For example, the review of previous economic studies using NCLAN data found a set of common themes (priors) concerning the qualitative economic effects of O_3 on agriculture. In general, the quantitative results of this latest assessment confirmed these earlier common themes. The two exceptions are the observations that producers may gain from O_3 increases and that pollution-tolerant crops will experience greater changes than suggested by the pollution change. In this analysis, producers experienced welfare losses in the *aggregate* from increases in pollution, but producers in certain regions and producers of specific agricultural products did experience gains. Also, the "crowding out" phenomenon noted in some earlier studies generally did not occur, in part because land and other resource constraints were not binding in most regions. Overall, findings of this latest assessment demonstrate a high degree of stability in the qualitative economic information provided by the NCLAN program, supporting the use of this information in future assessments.

Results from the current assessment also confirm earlier quantitative findings; namely, that O_3 pollution imposes substantial economic costs on society. Specifically, increases in the yields of eight NCLAN crops associated with a 25% reduction in 1981–1983 ambient O_3 levels result in a benefit of approximately $1900 million (in 1982 dollars). Rollbacks of O_3 by 40% reveal net benefits of almost $3000 million. These values amount to approximately 1·9 and 2·8%, respectively, of total annual agricultural revenues during this period. Including some Farm Program provisions in the analysis does not appreciably affect the magnitude of these benefit estimates. While the general range of benefits parallel earlier NCLAN assessments, the current values are generated from more accurate exposure–response and aerometric data, and greater spatial resolution and crop coverage than those earlier studies. The current estimates are strengthened by these refinements.

One unique feature of this analysis is the definition of policy analyses aimed at evaluating possible seasonal standards for vegetation, in contrast to the current NAAQS hourly measures. The analysis of a range of seasonal standards indicates that the magnitudes of potential benefits are similar to some recorded with the constant percentage rollbacks, but the distributional consequences are substantially different. For example, a seasonal standard of 0·050 ppm with 95% compliance across each region would produce approximately $1700 million in benefits annually (in 1982 dollars) which is about the same amount as that for the 25% rollback. However, the regional implications are quite different, with those areas with the greatest air quality improvement realizing the greatest gain. Areas already in or near compliance may actually lose due to a decline in national crop prices from increased supply. In addition to suggesting the potential efficiency gains from a seasonal standard, this type of analysis continues to support the need for economic models that adequately capture effects across the many facets that make up economic markets.

The analysis reported here is based on what we believe are the best available biological and meteorological data. The economic model also provides a reasonable representation of the key economic dimensions of agriculture. However, neither the natural and physical science data nor the economic model are free of potential errors. Some of the same limitations noted in earlier assessments apply here including (1) continued extrapolation from a limited set of crops and cultivar responses to represent regional or national response; (2) uncertainty about appropriate exposure measures; (3) relatively few data points on response surfaces at below ambient levels; (4) measurement and interpolation errors in developing rural O_3 estimates; (5) uncertainties in linking proposed standards to observed O_3 levels in rural areas; (6) potential errors in export and other demand elasticities used in the economic model; and (7) potential errors arising from applying a long-run economic model to short-run or disequilibrium situations. These (and perhaps other) limitations must be noted. In a Bayesian perspective, however, information from previous NCLAN studies suggest that the increases in precision gained from elimination of similar uncertainties are relatively minor.

Finally, the NCLAN program was successful in developing a scientific consensus on a set of biological and economic issues that are important in performing policy analyses. Most of these issues could not have been effectively addressed without the willing cooperation of researchers within each discipline. Reasons for this success go beyond the individual researchers (although the social dynamics of the NCLAN participants seemed fairly harmonious) to the management structure of the program. A management committee structure that included representatives of each discipline, coupled with direct EPA participation to ensure consistency with the regulatory goals of the program, created a decision framework that allowed for disciplinary give and take, within the constraints of meeting a well-defined objective. Also, the long-term nature of the program facilitated periodic re-evaluations of critical data needs, with a corresponding reallocation of resources to meet newly identified needs. The continuing utility of the NCLAN data to economists and policymakers is a strong testimonial to the success of this integrated research approach. The NCLAN approach may offer a valuable prototype in designing research efforts to address other bioeconomic issues.

ACKNOWLEDGEMENTS

The authors gratefully acknowledge the assistance of the NCLAN research management committee and researchers in providing critical data for this paper. Constructive comments from Tom Crocker and two anonymous reviewers on earlier versions of this chapter are appreciated. We are particularly indebted to Scott Johnson for developing numerous algorithms and other procedures used in implementing the O_3 analysis.

REFERENCES

Adams, R. M. and T. D. Crocker. (1982). Dose–response information and environmental damage assessments: An economic perspective. *J. Air Pollut. Control Assoc.*, **32**, 1062–7.

Adams, R. M. and T. D. Crocker. (1984). Economically relevant ecosystem response estimation and the value of information: Acid disposition. In *Economic Perspectives on Acid Deposition Control*, ed. by T. D. Crocker, 35–64. Ann Arbor Science, Butterworth.

Adams, R. M. and B. A. McCarl. (1985). Assessing the benefits of alternative oxidant standards on agriculture: The role of response information. *J. Environ. Econ. Managemt*, **12**, 264–76.

Adams, R. M., S. A. Hamilton, and B. A. McCarl. (1984a). The economic effects of ozone on agriculture. Report No. EPA-600-3-84-090. Corvallis, OR, US Environmental Protection Agency, Environmental Research Laboratory.

Adams, R. M., T. D. Crocker, and R. W. Katz. (1984b). Assessing the adequacy of natural science information: A Bayesian approach. *Rev. Econ. Stat.*, **66**, 568–75.

Adams, R. M., S. A. Hamilton, and B. A. McCarl. (1986). The benefits of pollution control: The case of ozone and U.S. agriculture. *Am. J. Agric. Econ.*, **68**, 886–93.

Brown, D. and M. Smith. (1984). Crop substitution in the estimation of economic benefits due to ozone reductions. *J. Environ. Econ. Managemt*, **11**, 327–46.

Doorenbos, J., A. H. Kassam, C. L. M. Bentvelsen, V. Branscheid, J. M. G. A. Plusje, M. Smith, G. O. Uitenbogaard, and H. K. van der Wal. (1979). *Yield response to water*. FAO Irrigation and Drainage Paper, Food and Agriculture Organization of the United Nations, Rome.

Duloy, J. H. and R. D. Norton. (1973). CHAC: A programming model of Mexican agriculture. In *Multilevel planning: case studies in Mexico*, ed. by L. Goreux and A. Manne, 292–312. Amsterdam, North Holland.

Fajardo, D., B. A. McCarl, and R. Thompson. (1981). A multicommodity analysis of trade policy effects: The case of Nicaragua agriculture. *Am. J. Agric. Econ.*, **63**, 23–31.

Freeman, A. M., III. (1979). *The benefits of environmental improvement*. Baltimore, The John Hopkins University Press.

Garcia, P., B. L. Dixon, J. W. Mjelde, and R. M. Adams. (1986). Measuring the benefits of environmental change using a duality approach: The case of ozone and Illinois cash grain farms. *J. Environ. Econ. Managemt*, **13**, 69–80.

Hamilton, S. A., B. A. McCarl, and R. M. Adams. (1985). The effect of aggregate response assumptions on environmental impact analyses. *Am. J. Agric. Econ.*, **67**, 407–13.

Heady, E. O. and U. K. Srivastava. (1975). *Spatial sector programming models in agriculture*. Ames, Iowa State University Press.

Heagle, A. S., L. W. Kress, P. J. Temple, R. J. Kohut, J. E. Miller and H. E. Heggestad. (1988). Factors influencing ozone dose–yield response relationships in open-top field chamber studies. In *Assessment of Crop Loss from Air Pollutants*, ed. by W. W. Heck, D. T. Tingey and O. C. Taylor, 141–79. London, Elsevier Science Publishers.

Heck, W. W., O. C. Taylor, R. M. Adams, G. Bingham, J. Miller, E. Preston, and L. Weinstein. (1982). Assessment of crop loss from ozone. *J. Air Pollut. Control Assoc.*, **32**, 353–61.

Heck, W. W., R. M. Adams, W. W. Cure, A. S. Heagle, H. E. Heggestad, R. J. Kohut, L. W. Kress, J. O. Rawlings, and O. C. Taylor. (1983). A reassessment of crop loss from ozone. *Environ. Sci. Technol.*, **17**, 572A–81A.

Howitt, R. G. and C. Goodman. (1988). Economic impacts of regional ozone standards on agricultural crops. *Environ. Pollut.*, **53**, 387–95.

Howitt, R. E., T. W. Gossard, and R. M. Adams. (1984). Effects of alternative ozone levels and response data on economic assessments: The case of California crops. *J. Air Pollut. Control Assoc.*, **34**, 1122–7.

Just, R. E., D. L. Hueth, and A. Schmitz. (1982). *Applied welfare economics and public policy*. New York, Prentice-Hall.

Katz, G., P. J. Dawson, A. Bytnerowicz, J. Wolf, C. R. Thomson, and D. Olszyk. (1985). Effects of ozone or sulfur dioxide on growth and yield of rice. *Agric. Ecosyst. Environ.*, **14**, 103–17.

King, D. (1988). Modeling the impact of ozone × drought interactions on regional crop yields. *Environ. Pollut.*, **53**, 351–64.

Kopp, R. J., W. J. Vaughn, M. Hazilla, and R. Carson. (1985). Implications of environmental policy for U.S. agriculture: The case of ambient ozone standards. *J. Environ. Management.*, **20**, 321–31.

Lefohn, A. S., H. P. Knudsen, J. A. Logan, J. Simpson, and C. Bhumralkar. (1987). An evaluation of the kriging method to predict 7-h seasonal mean ozone concentrations for estimating crop losses. *J. Air Pollut. Control Assoc.,* **37,** 595–602.

McCarl, B. A. (1982). Cropping activities in agricultural sector models: A methodological proposal. *Am. J. Agric. Econ.,* **64,** 769–72.

McCurdy, T. (1987). Additional ozone air quality indicators in metropolitan areas. Ambient Standards Branch, Office of Air Quality Standards, EPA.

McGartland, A. M. (1987). The implications of ambient ozone standards for U.S. agriculture: A comment and some further evidence. *J. Environ. Managemt.,* **20,** 139–46.

Mjelde, J. W., R. M. Adams, B. L. Dixon, and P. Garcia. (1984). Using farmers' actions to measure crop loss due to air pollution. *J. Air Pollut. Control Assoc.,* **34,** 360–3.

Murtaugh, B. and M. Saunders. (1977). *MINOS: Users Guide.* Stanford University, Systems Operations Laboratory, Research Technical Report, No. 77-9.

OAQPS Staff Paper. (1986). Review of the national ambient air quality standards for ozone preliminary assessment of scientific and technical information. Research Triangle Park, NC, US EPA, Strategies and Air Standards Division, Office of Air Quality Planning and Standards.

Samuelson, P. A. (1952). Spatial price equilibrium and linear programming. *Am. Econ. Rev.,* **42,** 283–303.

Segerson, K. (1987). Economic impacts of ozone and acid rain: Discussion. *Am. J. Agric. Econ,* **69,** 970–1.

Shortle, J. S., M. Phillips, and J. W. Dunn. (1988). Economic assessment of crop damage due to air pollution: The role of quality effects. *Environ. Pollut.,* **53,** 377–85.

Takayama, T. and G. Judge. (1971). *Spatial and temporal price and allocation models.* Amsterdam, North Holland.

United States Department of Agriculture. (1984a). *Agricultural statistics 1984.* Washington, DC.

United States Department of Agriculture, Economic Research Service. (1984b). *Livestock and Meat Statistics, 1983.* Stat. Bul. No. 715, Washington, DC.

United States Department of Agriculture, Statistical Reporting Service. (1984c). *Usual planting and harvesting dates for US field crops.* Washington, DC, Agricultural Handbook Number 628.

White, K. (1978). A general computer program for econometric methods—SHAZAM. *Econometrica,* **46,** 239–40.

Willig, R. D. (1976). Consumers' surplus without apology. *Am. Econ. Rev.,* **66,** 589–97.

21

THE EUROPEAN OPEN-TOP CHAMBERS PROGRAMME: OBJECTIVES AND IMPLEMENTATION

P. MATHY

Commission of the European Communities, Brussels, Belgium

21.1 INTRODUCTION

In terms of the environment, the Commission of the European Communities implements two types of action: (1) *Action Programmes*, to protect the environment by adopting curative and preventive legal measures in the form of directives (Community Environment Policy), and (2) *Research Programmes*, to provide scientific information to supply the Community Environment Policy and to address longer-term environmental problems. Apart from these particular objectives, the general goal of the Community Research and Development Programme in the field of Environmental Protection is to enhance the coordination of national research activities and to increase the efficiency of the overall European research effort.

The current R & D Programme will be in effect from 1986 to 1990. The total budget available is 55 million ECU. Twenty per cent of this amount will be devoted to research on air quality. About 7·5 million will be available for research on the effects of air pollution on terrestrial and aquatic ecosystems, including agricultural ecosystems.

The Proceedings of workshops and Air Pollution Research reports listed as references provided much more valuable information for development and implications of the European Communities Research and Development Programme in the field of Environmental Protection than is described in this chapter.

21.2 IMPLEMENTATION OF COMMUNITY R & D PROGRAMME IN THE FIELD OF ENVIRONMENTAL PROTECTION

In the field of Air Quality, the Community R & D Programme on environmental protection encompasses two concerted actions and research contracts between the Commission and the European research institutions. The concerted actions cover two main items: the physicochemical behaviour of air pollutants (COST project 611) and the effects of air pollution on terrestrial and aquatic ecosystems (COST project 612).

The objective of the concerted actions is to coordinate the national research carried out in Member States as well as in countries associated with the Community by scientific cooperation agreements, such as Sweden, Norway, Switzerland, and Austria.

The objective is achieved by organising technical workshops and symposia and also by stimulating the development of joint coordinated research projects, thereby associating several scientific teams from different countries. This approach applies particularly well to multidisciplinary research involving complementary, ecological, and physiological studies.

Selected research projects that fill gaps in the national programmes are granted by the Community. The tendency is to support primarily the joint coordinated projects emanating from the Concerted Actions.

The Community finances 50% of these research projects and 50% is financed by the national or regional authorities which also participate in the defining of research objectives.

21.3 EUROPEAN OPEN-TOP CHAMBERS PROGRAMME

Within the framework of its R & D Programme on Environmental Protection, the European Community coordinates and finances several research projects with the objective of investigating the effects of air pollution on agricultural crops and forest trees. One of the main projects involves an open-top chambers network and is divided into two subprogrammes: one on forest trees and one on agricultural crops. Eight scientific institutions are involved in the agricultural crops subprogramme:

—The Imperial College of Science and Technology
—The University of Nottingham
—The Institute of Terrestrial Ecology in the United Kingdom
—The Centre Départmental d'Etude de l'Environnment des Pyrénées Atlantiques in France
—The Institute for Chemical Research in Belgium
—The University of Dublin in Ireland
—The Institut für Produktions- und Ökotoxikologie in the Federal Republic of Germany
—Risø Research Centre in Denmark.

Two scientific teams are also associated with this subprogramme without being funded by the Community: the Swedish Environmental Research Institute and the Eidgedom Forschungsanstalt für Agrikulturchemie und Umwelthygiene in Switzerland.

Eight scientific institutions are also participating in the forest trees subprogramme:

—The Universität Hohenheim in Federal Republic of Germany
—The Fraunhofer-Institut für Umweltchemie und Ökotoxikologie in Federal Republic of Germany
—The Universität Essen in Federal Republic of Germany
—The Institut National de la Recherche Agronomique in France
—The Faculté des Sciences Agronomiques in Belgium
—The Risø Research Institute in Denmark
—The Institute of Terrestrial Ecology in the United Kingdom.

21.4 EUROPEAN COMMUNITY OBJECTIVE

One of the main features of the European open-top chambers programme comes from the fact that when it was initiated by the Commission, many valuable experiments with open-top chambers were already under way in several European countries. The general objective was to investigate the effects of air pollution on agricultural or forest plants. However, the specific objectives, the biological investigations and even the type of chambers differed from one site to another. Furthermore, the microclimate and the pollution climate were recorded in different ways. Finally, the skills, the degree of expertise of the research staff and the stage of development of experimental facilities differed from one site to another.

The objective of the Commission was to improve the existing network and enhance its efficiency in generating scientific information. It is important to emphasise that the Commission's primary objective is not to assess the agricultural crop losses at the Community level. This differentiates the European Open-Top Chambers Programme from the US National Crop Loss Assessment Programme. Nevertheless, the need to generate sufficient information in order to improve the existing air quality standards is taken into account in the experiment design. The concepts of "critical loads of pollutants" and "thresholds of injury" will probably be given more attention in the future.

However, for the time being, the emphasis is placed on the coordination of existing experiments to investigate the mechanisms of injury in relation to the variety of pollution climates which characterise Europe and the influence of interacting factors.

Thus, the core of the Commission's objective is to achieve a compromise which would be summarised in the following way:

—to enhance the coherence of the existing European open-top chambers network by introducing common monitoring protocols to allow comparisons between sites, particularly regarding the chamber effects and pollution climates;
—to generate sufficient information needed to improve Environment Policy;
—to avoid setting up a stringent system which would not preserve the

advantages of the existing diversity, primarily in regard to the biological investigations;

—to preserve and enhance scientific creativity by allowing project leaders to design their own experiments within their own field of interest. This has to be done within certain protocols, in order to generate comparable data.

21.5 ACHIEVEMENTS

21.5.1 Biological investigations: common protocols and coordination of experiments

Regarding the forest trees subprogramme, the purpose, for the time being, is to determine the effects of air pollution concentrations occurring in forest sites, remote from industrial and urban areas.

It is very unlikely that either macroscopic investigations or a mere observation of tree growth and development over a short period of time (4–5 years) would succeed in achieving that purpose. Furthermore, before now, no biological parameter (or set of parameters) could be used reliably as a specific indicator of air pollution effects.

This is why we encourage the *diversity of experiments* and the *diversity of investigated biological parameters,* assuming that this could increase the chances of:

—providing the scientific evidence for air pollution effects; and
—understanding the mechanisms of injury.

The investigations are concerned with anatomical as well as with biochemical and physiological parameters. The project leaders are free to design their experiments, but must consult each other in order to function in a coordinated and complementary way. Three treatments are compulsory: filtered air, non-filtered air in chambers, and open-field. Obviously, additional treatments are allowed.

Regarding the agricultural crop subprogramme, there are two levels of coordination. The first level is the more stringent one whereby all teams must record plant growth and development in filtered and non-filtered air and in an open-field plot. This is done according to a *common protocol* (see Appendix), to determine any effect of specific pollution climates on plant growth and development.

The second level concerns the most advanced teams who are examining mechanisms. To do so, they may develop additional treatments, such as fumigations or experiments with interacting factors. At this level, and similarly with the forest trees subprogramme, the teams are free to design these additional experiments according to their own field of interest.

21.5.2 Record of pollution climates and chamber effect: common protocols

Whether the subprogramme is forest trees or agricultural crops, a *common protocol* is applied with regard to the record of the microclimate and the pollution climate inside and outside the chambers (see Appendix). The Commission is aware of the importance of collecting comparable data from the different sites, as far as environmental variables are concerned.

The first reason is because there is a need to know more about the European pollution climates at rural sites, which are likely to be very diverse, changing in time and space. The Commission intends to map these pollution climates and the open-top chambers network offers an opportunity to collect the information needed to accomplish this mapping.

The second reason is that the chambers, being different from one site to another, may modify the microclimate to different extents. This is important when evaluating the chamber effect, in interpreting results on each site and for comparing results between sites.

The Commission intends to build up a common database and designate experts to analyse the results.

21.5.3 Exposure–effects relationships

The Commission encourages the development of exposure–response relationships in regard to agricultural crops.

The most advanced European teams are already able to establish these relationships for several crops. Information is available concerning the critical loads of air pollutants below which undesirable effects are unlikely to occur. To some extent, crop losses can already be estimated and possibly predicted. Some estimates and predictions are currently being carried out in Europe, based upon the conditions of particular sites.

Nevertheless, this information, particularly exposure–response relationships, is valid in well defined conditions. In view of the fact that plant response to pollutants is highly dependent on many external and internal factors, it is not possible, for the time being, to generalise the use of these relationships (in European conditions) on a Community scale.

21.6 CONCLUSIONS

A reliable assessment and prediction of crop losses requires a better knowledge of mechanisms of injury in various pollution climates, taking into account the influence of many interacting ecophysiological factors which modify the response of plants to air pollutants.

The main topics to be addressed by the European Research Programme regarding the effects of air pollution on plants in the next few years will be to: identify specific indicators of air pollution effects, characterise European pollution climates, and determine interactions between air pollution and other biotic and abiotic stresses. These items will be developed within the framework of an integrated ecophysiological approach, taking into account the possible indirect effects of air pollution via soils.

Obviously, the open-top chamber methodology cannot answer every question raised in that field. The Commission is convinced that the open-top chamber network has to be complemented with other experimental facilities like closed chambers, open field fumigation systems, lysimetry, catchment methodologies, etc.

Increasingly, the challenge is to combine the advantages of these facilities, to interrelate the skills of many laboratories, and to coordinate complementary, well-designed experiments.

In the future, the Commission of the European Community intends to emphasise this coordination. A mechanism for common evaluation of all data should be established.

Within that perspective, there is room to strengthen the cooperation between American and European scientists, and my personal opinion is that joint research projects, with teams from both sides of the Atlantic, should be set up.

REFERENCES

Mathy, P. Indirect effects of air pollution on forest trees—root–rhizosphere interaction. In Proceedings of a workshop held in Jülich, Federal Republic of Germany, 5–6 December 1985. Brussels, Commission of the European Communities.

Mathy, P. How are the effects of air pollutants on agricultural crops influenced by the interaction with other limiting factors? In Proceedings of a workshop held in Risø, Denmark, 23–25 March 1986. Brussels, Commission of the European Communities.

Mathy, P. Microclimate and plant growth in open-top chambers. In Proceedings of a workshop held in Freiburg and Hohenheim, Federal Republic of Germany, September 1986. Air Pollution Research Reports (No. 5) EUR 11.257 EN. Brussels, Commission of the European Communities.

Mathy, P. Direct effects of dry and wet deposition on forest ecosystems—in particular canopy interactions. In Proceedings of a workshop held in Lökeborg, Sweden, 19–23 October 1986. Air Pollution Research Reports (No. 4.) EUR 11.264 EN. Brussels, Commission of the European Communities.

Mathy, P. Definition and characterisation of European pollution climates. In Proceedings of a workshop held in Bern, Switzerland, 27–30 April 1987. Air Pollution Research Reports (No. 6). Brussels, Commission of the European Communities.

Mathy, P. Air pollution and ecosystems. In Proceedings of an international symposium, held in Grenoble, France, 18–22 May 1987. Air Pollution Research Reports (No. 7) EUR 11.244. Brussels, Commission of the European Communities.

APPENDIX: EUROPEAN OPEN-TOP CHAMBERS PROGRAMME AND PROTOCOLS APPLICABLE TO AGRICULTURAL CROPS

A.1 Hypothesis to test

Do ambient concentrations of *atmospheric pollutants* in important agricultural areas decrease crop yield?

A.2 Means to test hypothesis

A.2.1 General concept

The hypothesis will be tested by means of *open-top* chambers installed in sites remote from industrial and urban areas; the chosen sites will represent a range of ecological and air pollution environments. These sites will be characterised by relatively small average concentrations of air pollutants, as compared to those occurring in areas where visible injury is undoubtedly attributable to air pollution. Plants will be grown in a set of open-top chambers with filtered atmospheres. Comparison will be made with similar plants grown in both chambers with unfiltered atmospheres as well as adjacent field plots.

The contractor will take steps to ensure the necessary comparability of data generated at the above mentioned sites.

A.2.2 Protocols

A.2.2.1 Choice of crops. It is strongly recommended that spring-sown crops be used (cereals or legumes).

A.2.2.2 Management of experiments

(a) *Type of chamber:* To minimize "chamber effects", it is recommended that all teams work with similar chambers.
(b) *Type of filters:* carbon
(c) Air (filtered and non-filtered) in chambers should be distributed from a perforated annulus adjusted to be 10 to 15 cm above developing crops.
(d) Crop management should be consistent with good husbandry practices:

—*Pests and pathogens* will be controlled by an adequate programme of *preventive* sprays.
—*Adequate surface irrigation* will be provided to avoid undue water stress in the chambers and field plots.
—*Fertilisers* will be applied according to proper crop husbandry procedures.

(e) *Rates of air change* (throughout day and night) should be managed so as to:

—minimise boundary layer resistances (at least 3 air changes per min); and
—restrict air temperature differences to 2°C (when comparing inside chamber temperature with the outside temperature).

(f) The chambers should be sited in field crops, using on-site field soils.

A.2.2.3 Biological assessments (as judged appropriate for different crops)

(a) *Non-destructive measurements and observations on growing crops:* in addition to an early season baseline, the following growth measurements and phenological observations will be made:

—successive measurements of height;
—successive measurements of leaf number (living and dead);
—successive counts of numbers of branches (e.g. tillers);

—onset of flowering (ear emergence);

—date of maturity;

—successive assessments of fruit set and development;

—successive records of the incidence of pests and pathogens;

—occurrence of other damage (e.g. necrotic spots), other than that attributable to pests and pathogens; and

—photographic records of crop development.

(b) *Destructive samples*—at least two samples will be taken, one at the onset of flowering and the other at final harvest. (The final harvest will be done for each treatment separately as the crop matures, when an attempt will be made to estimate root protection).

Where appropriate and depending on the crop, the following attributes will be measured:

(a) *Quantitative assessment*

—numbers of plants per unit area at the beginning and at the end of study;

—numbers of stems per plant;

—number and areas of leaves, living and senescent (judged by coloration);

—numbers of inflorescences;

—numbers of fruit and/or seeds;

—fresh and dry weights of different plant components.

(b) *Qualitative assessment*—quality of harvest should be assessed.

(c) The above assessments are regarded as minimal; other observations are to be encouraged, but they should be non-destructive until the final harvest.

A.2.2.4 Monitoring atmospheric pollutants (also applicable to forest tree subprogramme):

(a) SO_2, NO, NO_2 and O_3 will be monitored according to EPA and VDI approved methods and the use of ethylene will be avoided.

(b) Ambient atmospheres and those within chambers (with and without filters) will be monitored. Measurements will be done 10 cm above the developing crop.

(c) "Continuous" measurements will be made in one chamber of each treatment and occasional checks will be made in replicate chambers. Air should be sampled from a perforated tube fitted with a glass fibre or equivalent filter.

(d) Frequency of measurements: it is recommended that at least one cycle of measurements be made per hour.

(e) Calibration

—In addition to continuously monitoring the ambient atmosphere, a series of 'zero-air' samples should be taken daily.

—Instruments must be calibrated weekly, with independent calibrations every 6 months.

A.2.2.5 Monitoring the climate both in and outside of chambers (also applicable to forest tree subprogramme):

(a) It is assumed that the experimental sites will be near weather stations and will routinely supply the following data: wind speed and direction, daily maximum and minimum temperatures, rainfall, and humidity.

(b) Additionally, it is strongly recommended that the following measurements be made within chambers and in field plots:

—continuous records of air temperature at crop canopy levels;
—continuous integrated radiation assessments using tube solarimeters;
—continuous wet and dry bulk records; and
—continuous soil temperatures at a depth of 5 cm.

A.2.2.6 Experimental design and sampling:

(a) At least five replicate chambers are needed to detect effects of 5 to 10%.

(b) To minimise errors, early season growth measurements will be taken so that later covariance analyses can be made.

22

POLICY IMPLICATIONS FROM CROP LOSS ASSESSMENT RESEARCH—A UNITED KINGDOM PERSPECTIVE

R. B. WILSON and A. C. SINFIELD

Department of the Environment, London, UK

22.1 INTRODUCTION

Research on crop loss assessment in the United Kingdom (UK) forms part of an integrated research programme to evaluate the impact of air pollution on the natural and man-made environment. The programme is designed to identify critical target areas and provide guidance to policy makers on future pollution control options in terms of both benefits and costs.

A knowledge of pollution climate is all important in identifying sensitive species or areas at risk. Further, it is useful to first consider the UK pollution climate as part of a larger European system and to make some general comparisons with the situation in the eastern United States. Table 22.1 (Unsworth *et al.*, 1988) indicates that SO_2 and NO_x concentrations in agricultural areas of the UK typically average between 0·002 and 0·015 ppm annually. Occasionally, values that are an order of magnitude larger than these are experienced for periods of a few hours and may persist for a few days if the air is stagnant. The close proximity of industrial and urban areas to agricultural areas in a small country like the UK is responsible for these relatively large concentrations and there are relatively high levels of dry deposition of SO_2 and NO_x over much of southern and central England. More remote areas of the northern and western UK receive less dry deposition, but, due to greater rainfall amount, wet deposition is proportionally larger and rainfall averages about pH 4·0 (Barrett *et al.*, 1987).

Although there are only limited O_3 records, it appears that episodes of O_3 exceeding 0·008 ppm occur for only a few hours per year and are probably more common in the southern than northern UK (Derwent *et al.*, 1987).

There are some similarities between pollution climates in the rural UK and

515

TABLE 22.1

A comparison of rural pollutant concentrations in the United Kingdom, Central and Southern Germany, and Eastern United States

Concentrations	United Kingdom		Central Germany	Southern Germany[a]	Eastern United States
	Rural	Remote rural			
Annual mean SO$_2$ in ppm	0·002–0·015	0·002	0·005–0·017	<0·004	<0·006
Annual mean NO$_2$ in ppm	0·002–0·015	0·002	0·006–0·012	<0·004	<0·004
Annual dry deposition	2–4 g S m^{-2} 1–2 g N m^{-2}	0·4 g S m^{-2} 0·2 g N m^{-2}	2–4 g S m^{-2} 1 g N m^{-2}	1 g S m^{-2} 0·5 g N m^{-2}	0·5–1 g S m^{-2} 0·3–0·5 g N m^{-2}
Annual wet deposition	0·8 g S m^{-2} 0·4 g N m^{-2}	2 g S m^{-2} 1 g N m^{-2}	1–2 g S m^{-2} 0·5–1 g N m^{-2}	1 g S m^{-2} 0·5 g N m^{-2}	0·5–1 g N m^{-2} 0·3–0·5 g N m^{-2}
Hours per year O$_3$ > 0·07 ppm	60–150	5–20	150–300	>250	~550
Rainfall pH	4·1–4·7	4·0	4·0–4·6	4·0–4·6	4·1–4·3

[a] At high altitudes in Southern Germany, wet deposition of sulphate and nitrate may be three times higher than that at lower altitude sites, and concentrations of O$_3$ tend to exceed 0·07 ppm during more than 350 h a year.

in central Germany but southern Germany appears to have a much higher frequency of O$_3$ episodes. Similarly, there are probably much higher levels of dry deposition of SO$_2$ and NO$_x$ over agricultural regions in the UK than in the eastern United States, where it appears O$_3$ episodes are more frequent.

Consequently, the complex mixtures of pollutants that occur in the UK and elsewhere in Europe necessitate different research and policy approaches than those undertaken in the US NCLAN programme, which considers O$_3$ to be the dominant pollutant. Understanding these differences is necessary in assessing the implications of control strategies, and extrapolation of other countries' strategies must be viewed with some suspicion. For this reason, there has been an emphasis in the UK on supporting research aimed at understanding the physiological and biochemical mechanisms by which crops respond to the mixtures of pollutants *encountered in the pollution climates in which the crops are growing*.

The references listed and those in the Bibliography provided valuable information in preparation of this paper which describes policy implications from crop loss assessment research in the UK.

22.2 UNITED KINGDOM CROP LOSS ASSESSMENT PROGRAMME

The main objective of this programme is to assess potential economic losses of crops in the UK due to air pollutants. Initially, the programme concentrated on the impacts of the gaseous SO$_2$ and NO$_x$ on overwintering crops such as winter barley, winter wheat, and pasture grasses. It has been the deliberate policy of the programme to support the development of several different

TABLE 22.2

Examples of work being carried out at various United Kingdom research establishments

Closed chamber and glass house	University of Lancaster
	University of Newcastle
	Imperial College
	Central Electricity Research Laboratory
Wind tunnel	Grasslands Research Institute
	Imperial College
Open-top chamber	Institute of Terrestrial Ecology
	University of Nottingham
	Imperial College
	Central Electricity Research Laboratory
Chamberless field exposure systems	University of Nottingham
	University of Newcastle
	Central Electricity Research Laboratory
Field observation and transect studies	Imperial College
Plant biochemical and physiological response	Institute of Terrestrial Ecology
	University of Newcastle
	Imperial College
	Central Electricity Research Laboratory
Acid rain simulation studies	Institute of Terrestrial Ecology
	Imperial College
	Central Electricity Research Laboratory

approaches to achieve these objectives. Examples of the type of work being carried out are shown in Table 22.2.

Although the UK programme initially concentrated on the impacts of gaseous SO_2 and NO_x to overwintering crops, it has since become apparent from the mechanistic work with mixtures of pollutants that both O_3 and acid rain may also play significant roles in plant damage processes. It has also become apparent that other stresses such as drought, cold, light intensity, and pest infestation also play important roles.

Consequently, the UK programme on the assessment of crop loss from air pollutants has been expanded to take these factors into account.

The UK programme has been in operation since 1978, but only during the past 5 years have sufficient resources been made available to enable an integrated approach to crop damage assessment to be undertaken. The programme is now in what could be called its second phase and is scheduled for completion by 1990–91. It is hoped that at that time the UK will be in a position to define dose–response relationships relating air pollutant damage for each pollutant individually and more realistically for combined pollutant mixtures for a number of economically significant crops in the UK so that:

(1) The importance of crops in the overall damage assessment model compared to other impact areas (e.g. forests, freshwaters, buildings and materials) can be ascertained.
(2) The economic significance of air pollutant crop damage can be assessed.
(3) Appropriate policy decisions can be formulated.

Thus, it is apparent that the UK has built up a number of centres of expertise to evaluate the impact of air pollutants on crop (Table 22.2). The agency coordinating this work is the UK Department of the Environment which has also been responsible for setting up a number of peer review groups comprised of experts from the academic world, research institutes and industry to report on air pollution matters relevant to the UK. A list of the various publications produced by these groups is given in the Appendix. These references cover subjects such as acid deposition, acid waters, photoxidants, terrestrial responses and materials and buildings.

Because of the expertise on crop loss assessment that has been built up in the UK, it has been invited to take lead responsibility for this issue in both the European Economic community (EEC) and the United Nations Economic Commission for Europe (UN ECE) International Collaborative Programme on Crop Damage throughout the ECE Region.

The UK Department of the Environment has also devoted considerable resources to the development of an analytical framework for the integrated economic and scientific evaluation of the various available technological and policy options for reducing the impact of air pollution in the environment. This model is intended to assess the costs and benefits to the UK, and to the rest of Europe, of reducing SO_2 and NO_x emissions, both nationally and regionally.

The philosophy behind the development of the model has not necessarily been to produce an 'objective' numerical result with all costs and benefits fully expressed in monetary terms. Rather it has been an exploration of the extent to which a system can be constructed that enables analysts and decision makers to think through the implications of an option in a consistent and coherent manner. This is of particular importance in the environment field where subjective opinion and uncertainty is as likely as objective fact, but where all need to be included in an evaluation.

The model has been developed on the top-down approach, initially using highly simplified relationships to evaluate the reduction in damage to human health, lakes and aquatic systems, forests, crops, and buildings and materials. In the case of crops, rye grass (a significant permanent pasture species) is used as the general indicator of damage. The damage function currently used is of the form that combines SO_2 and NO_x effects on crops additively. Since there is so much uncertainty about the damage function, the various parameters involved (thresholds, slopes and additive factors) are described probabilistically in the model with low, model and high values.

As UK research progresses, more crop species will be added including overwintering and summer crop species leading to the derivation of more complex damage functions.

The model has not yet reached an operational state as it still requires further data input. However, the model's potential has been recognised to the extent that proposals have been made to extend it to utilise emission and damage scenarios on a European scale, so that it can be used for both national and regional policy evaluation. In particular, the UK is leading the work of a group of economic experts in support of the United Nations Economic Commission for Europe Convention on Long Range Transboundary Air Pollution.

From the strength gained by international collaboration in the UN ECE and European Community Programmes, it is hoped that a form of European NCLAN might emerge permitting exchange of information, expertise and co-ordinated research between the many countries comprising the UN ECE region. The ultimate objective of this programme is to achieve an overall assessment of the impact of air pollutants on the most economically important crops throughout the UN ECE region. Attainment of this will provide better scientific input to discussions on pollutant reduction policies being formulated on national, regional and pan European scales.

REFERENCES

Barrett, C. F. *et al.* (1987). *Acid deposition in the United Kingdom 1981–1985.* Second Report of the UK Review Group on Acid Rain, May 1987. London, Department of the Environment.

Derwent, R. G. *et al.* (1987). *Ozone in the United Kingdom.* Interim Report of the UK Photochemical Oxidants Group, February 1987. London, Department of the Environment.

Unsworth, M. H. *et al.* (1988). *Effects of acid deposition on the terrestrial environment in the United Kingdom.* UK Terrestrial Effects Review Group, September 1988. London, Department of the Environment.

BIBLIOGRAPHY

Barrett, C. F. *et al.* (1983). *Acid deposition in the United Kingdom*: UK Review Group on Acid Rain, December 1983. Ruislip, Middlesex, UK, DoE Publications.

Barrett, C. F. *et al.* (1987). *Acid deposition in the United Kingdom 1981–1985.* Second Report of the UK Review Group on Acid Rain, May 1987. Ruislip, Middlesex, UK, DoE Publications.

Derwent, R. G. *et al.* (1987). Interim Report of the UK Photochemical Oxidants Review Group, February 1987. Ruislip, Middlesex, UK, DoE Publication.

Everett, L. H. *et al. Effects of acid deposition on buildings and materials in the United Kingdom.* First Report of the UK Building Review Group (in press). London, UK, HMSO.

Unsworth, M. H. *et al.* (1988). *Effects of acid deposition on the terrestrial environment in the United Kingdom.* UK Terrestrial Effects Review Group, September 1988. London, Department of the Environment.

Warren, S. C. *et al.* (1986). *Acidity in UK Freshwaters.* (1986) First Report of the UK Acid Waters Review Group, April 1986.

23

POLICY IMPLICATIONS FROM CROP LOSS ASSESSMENT RESEARCH: THE U.S. PERSPECTIVE

BRUCE C. JORDAN, ALLEN C. BASALA, PAMELA M. JOHNSON,
MICHAEL H. JONES and BRUCE MADARIAGA

*US Environmental Protection Agency, Research Triangle Park, North
Carolina, USA*

23.1 INTRODUCTION

Crop loss assessment research provides important information for our regulatory efforts to protect public welfare. The process of transforming crop loss assessments from air pollution research into rulemaking is somewhat complex. The process involves an assessment of biological, economic, and environmental factors, as well as a consideration of agricultural policies. This paper concerns the policy process and problems of putting crop loss research into a framework conducive for evaluating alternative levels and forms of US secondary national ambient air quality standards (SNAAQS). The goals of the paper are to (1) provide a summary of the statutory authorities for developing these standards, (2) highlight the task of developing secondary standards, (3) describe the process for incorporating effects data and analyses into the regulatory process, and (4) provide an illustrative example of how crop assessment data may be used in the ozone standard review.

This paper is organized as follows. Section 23.2 provides background information regarding the goal of the Clean Air Act and the standard development process. Section 23.3 describes analytical and policy problems that must be addressed to crop loss information for standard setting. Section 23.4 provides an illustrative benefit analysis using National Crop Loss Assessment Network (NCLAN) data and discusses future directions for enhancing the credibility and expanding the coverage of crop loss research and benefit analysis. Lastly, the paper is summarized in Section 23.5.

23.2 NATIONAL STANDARDS AND THE POLICY PROCESS

23.2.1 Background

On February 19, 1987, Environmental Protection Agency (EPA) Administrator Lee Thomas testified before the United States House Subcommittee on Health and the Environment and characterized the national ambient air quality standards (NAAQS) and their attainment as a major environmental priority. (Testimony of Lee M. Thomas, Administrator, US Environmental Protection Agency before the Subcommittee on Health and the Environment, Committee Energy and Commerce, United States House of Representatives, February 19, 1987.) Attainment and maintenance of the NAAQS is a multibillion dollar effort and probably the most resource-intensive of any environmental regulatory program in the United States. Because these standards are the markers that drive NAAQS control programs, the process for developing and updating these standards is of extreme importance. Indicative of national interest in this process, the United States Congress signaled its resolve to be informed on the need for evolution or change in the way standards are developed and reviewed by commissioning the General Accounting Office (GAO) to assess the standard development process (US GAO, 1986).

EPA's charter for conducting NAAQS development and reviews is defined in the Clean Air Act and related amendments passed by the United States Congress. (The Clean Air Act as Amended (42 USC 1857 et seq.). The standard development and review process includes the following: (1) assessment of scientific information, (2) generation of a consensus within the scientific community on the veracity of this assessment, (3) an exchange of views and information with the public sector following proposal of a standard, and (4) if necessary, promulgation of a final rule. It is in this process, particularly steps 1 and 2, where scientific evidence such as that obtained from NCLAN plays a key role in establishing national priorities.

The goals of the Clean Air Act are to protect the public health and welfare and to enhance the quality of the Nation's air. Under the Act, the Federal government is responsible for establishing, on a nationwide basis, ambient air quality standards at a level adequate to protect the public health and welfare. To provide for attainment of these standards, States are responsible for specifying emission limitations and other programs for individual sources through State implementation plans (SIPs). Two categories of standards are called for in the Clean Air Act: (1) primary or health-based standards and (2) secondary or welfare-based standards that address environmental damage not included in the public health category.

The Clean Air Act, as amended, contains a number of provisions that make explicit mention of public "welfare" (e.g., new source performance standards and prevention of significant deterioration of air quality). The central provisions involving protection of public welfare, however, are generally regarded as those giving authority and guidance for listing and developing scientific criteria for certain ambient air pollutants that may endanger public health or welfare (Section 108), setting and revising NAAQS for those pollutants (Section 109), and implementing those standards (principally Section 110).

Under Section 109(b)(2), SNAAQS must be adequate to protect the public welfare from "any known or anticipated adverse effects" associated with the presence of an ambient air pollutant listed under Section 108 (i.e., a criteria pollutant). Welfare effects, which are defined in Section 302(h) of the Act, include, but are not limited to, effects on "soils, water, crops, vegetation, man-made materials, animals, wildlife, weather, visibility, climate, damage to and deterioration of property, and hazards to transportation, as well as effects on economic values and on personal comfort and well-being". Based on this definition, it appears that "welfare" in the Act has almost the same meaning as the term used by economists, with the exception that most direct health aspects are excluded. In specifying a level or levels for secondary standards, the Administrator must judge at what point "known or anticipated" welfare effects become "adverse" and base this judgement on the scientific criteria for welfare effects. The Act requires EPA to prepare a criteria document which contains the "latest scientific knowledge useful in indicating the kind and extent of all identifiable effects on public health or welfare that may be expected from the presence of such pollutant in the ambient air". Unlike primary standards, the Act does not call for a "margin of safety" in setting secondary standards.

23.2.2 Standard development process

In 1975, EPA developed a plan for reviewing the NAAQS in a phased process over a 5-year period beginning with the photochemical oxidant standard (now ozone). The Congress added authority to this process by making it a legal requirement of the 1977 Clean Air Act Amendments. Figure 23.1 illustrates the various steps in this process.

The main responsibility for production of the criteria document rests with the Environmental Criteria and Assessment Office (ECAO) in EPA's Office of Research and Development (ORD). ECAO coordinates the process of accumulating and analyzing literature and writing initial rough drafts of document chapters.

Once the criteria document has been reviewed by the public and the Clean Air Science Advisory Committee (CASAC) and the document is nearing its final form, the Agency staff prepares a staff paper that evaluates the key studies in the criteria document and identifies critical elements to be considered in the review of the standard. The paper identifies and evaluates those studies that the staff believes should be used in making the necessary scientific judgements on the level at which adverse effects signal a danger to public welfare. While the paper does not present a judgement on what concentration level should be established for the standard, the staff paper will normally present staff recommendations on the range of standards that appear reasonable, given existing scientific knowledge. The paper helps bridge the gap between the science contained in the criteria documents and the judgement required of the Administrator in setting ambient standards.

The staff paper is reviewed externally by the public and the CASAC. A

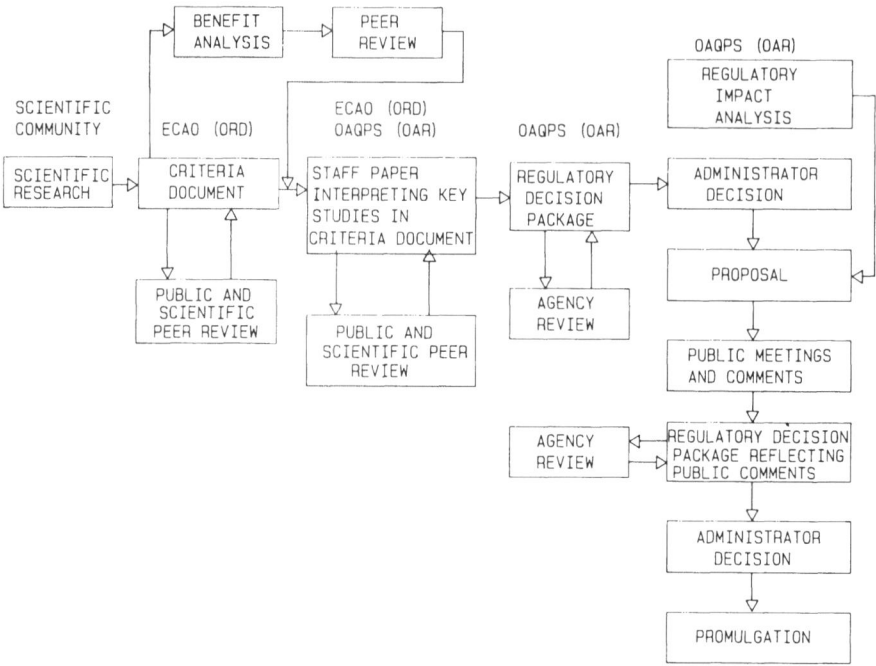

Fig. 23.1. Standard setting process for national ambient air quality standards. ORD, Office of Research and Development; ECAO, Environmental Criteria and Assessment Office; OAR, Office of Air and Radiation; OAQPS, Office of Air Quality Planning and Standards.

public meeting is held with the CASAC to receive their comments and the comments of the public. Once the paper has been reviewed by the CASAC, the scientific judgments made in the paper form the basis for the staff's recommendation to the Administrator on any revisions to the standard. Once an Agency position is established, an intra-Agency work group advises and assists the lead office in preparing a proposed regulation. The Steering Committee initially reviews the regulation. This committee is comprised of the six Assistant Administrators' staffs. Following Steering Committee review, proposed regulations are reviewed by all Assistant Administrators, General Counsel, and chief Staff Office directors, signed by the Administrator, and subsequently published. Once the Administrator has the benefit of public comment, a final regulation is published.

The impetus for conducting cost and economic analyses comes from a growing concern that governmental regulation of the private sector may stiffle economic productivity. Obviously this concern, which has been developing for a decade or more, applies to all types of regulations and not just ambient air quality standards. As a result, a series of Executive Orders (EO) issued by the past several administrations have mandated cost-impact analyses for all "major" regulations (generally defined as a regulation likely to result in expenditures over $100 million, although other economic factors are considered as well). In the Ford Administration, Inflation Impact Analyses were required. Under the Carter Administration, Executive Order 12044 required cost and economic analyses. These analyses were then reviewed by the Regulatory Analysis and Review Group (RARG) headed by the Chairman of the Council of Economic Advisors. In 1981, the Reagan Administration issued EO 12291, which required a full cost-benefit analysis for all major regulations. (Office of the President "Executive Order 12291". 46 FEDERAL REGISTER 13193 (February 17, 1981).) The cost-benefit analysis is the core of what is known as a Regulatory Impact Analysis (RIA) and is reviewed by the Office of Management and Budget (OMB). EPA's interpretation of the Clean Air Act has been that costs incurred by industry and the public to attain an ambient air quality standard are not to be considered in setting such standards, although these costs may be considered to some extent by State air pollution control agencies in implementing attainment measures. These analyses are prepared to help inform the public, Congress, and others in government.

23.3 LEVEL OF THE STANDARD: KEY ANALYTICAL AND POLICY ISSUES

The key decision in specifying a level for secondary standards is to determine at what point "known or anticipated" effects on public welfare become "adverse", based on the scientific criteria. The staff paper analysis does not make this judgement, but does attempt to identify potentially adverse effects and to bound ranges of interest for standard levels when the data are relatively clear. Judgements in the staff papers have been based directly on the effects information contained in the criteria document and on past decisions on secondary standards.

A number of suggestions have been advanced to define adverse effects for several welfare effects categories. In the area of vegetation effects for SNAAQS, adverse effects have largely been determined by defining a level of crop loss which is considered to be significant. More recently, in the ozone standard review, the CASAC advised the Agency to consider economic measures of adversity, such as changes in producer and consumer surplus, in assessing the level which is to be considered adverse. An illustration of EPA's assessment of economic benefits associated with lower ambient ozone concentrations is presented in the next section. While some economists have called for analyses providing quantitative estimates of benefits associated with more stringent SNAAQS, others have suggested that "public welfare" be interpreted in a broad economic context, and that projected costs of control be deducted from projected benefits of lower pollution levels to estimate net welfare benefits. While such analyses may be required under Executive Order 12291, legal questions may be raised with respect to their use in setting SNAAQS. It is becoming clear, however, that quantification of benefits for alternative standard levels is an important element of the secondary standard review.

Once reliable crop loss information has been developed, benefits analysis can be employed to help determine an appropriate level for the standard.

However, several key analytical and policy issues also need to be considered for the benefits analysis and for standard setting in general. These issues include: (1) averaging time and form of the standard, (2) coordination of US agricultural and environmental policies, (3) incomplete information, (4) integrating economic, environmental and biological information, and (5) distributional issues. The rest of this section describes the requirements and factors to be considered in addressing these issues.

23.3.1 Averaging time and form of the standards

The decision on averaging time for SNAAQS is driven by the effects information, but with practical constraints. If adequate protection can be provided, it is desirable to make the averaging times for primary and secondary standards identical. If it is necessary to use a different averaging time, the number of different averaging times should be minimized. In general, longer averaging times are preferred when they are protective because they provide additional stability for control programs.

Determining the relative protection offered by different averaging times usually involves statistical analyses of measured or modeled air quality data. While current data may suggest a different conclusion, such an analysis was instrumental in the 1979 revision of the ozone SNAAQS. The analysis related ozone effects on vegetation and crops to the seasonal mean of daily maximum 7-h ozone levels. The analysis then found that meeting the primary 1-h standard provided a reasonable degree of assurance that seasonal mean 7-h concentrations would not reach levels likely to produce significant reductions in growth or yield.

Serious examination of alternative forms of all NAAQS is continuing. The general approach is shifting from the original deterministic form (e.g., a 1-h level not to be exceeded more than once per year) to a statistical form (e.g., 1-h level with no more than one expected exceedance per year) as was determined for ozone in 1979 and PM_{10} in 1987. The statistical form can offer equivalent protection and should facilitate more stable and reliable implementation programs that recognize the stochastic nature of air quality and emissions (Biller and Feagans, 1981).

In the case of ozone, substantial uncertainty exists with regard to the role of various exposure parameters, particularly averaging time and dynamics (i.e., constant or variable), as well as concentration. Which exposure components actually cause plant response, however, remains unclear. Although many studies tend to report exposures as mean ozone concentrations, the averaging times may be peak hourly, daily, weekly, monthly, or seasonal means; number of hours above selected concentration; or number of hours above selected concentration intervals. None of these statistics has adequately characterized the relationship between concentration, exposure duration, interval between exposures, and plant responses. The lack of correlation between exposure statistics and ambient air measurements has posed a major problem for those trying to assess the effects of ozone exposure.

The implication inherent in the use of mean ozone concentrations is that all ozone concentrations are equally effective in causing plant responses. Use of the mean statistic minimizes the importance of peak concentrations by treating low-level long-term exposures the same as high concentration short-term exposures. Studies with beans and tobacco (Menser and Hodges, 1972), however, have shown that a given dose (concentration × time) distributed over a short period induced more injury than did the same dose distributed over a longer period. Moreover, concentration was shown to be approximately twice as important as time of exposure with respect to causing foliar injury in tobacco plants. Thus, a judgement must be made as to whether greater protection of plants from O_3 exposure is provided by limiting short-term peak exposures than by limiting long-term average exposures.

Not only are concentration and time important, but the dynamic nature of the O_3 exposure is also important, i.e., whether the exposure is at a constant or variable concentration. Several studies (Musselman *et al.*, 1983; Hogsett *et al.*, 1985) have demonstrated that although constant concentrations cause the same type of plant response as as variable concentrations at equivalent doses, the constant concentrations had less effect on plant-growth responses. Initial studies have compared the response of alfalfa to daily peak and episodic O_3 exposure profiles that had the equivalent total O_3 dose over the growing season; alfalfa yield was reduced to a greater extent in the episodic than the daily peak exposure (Hogsett *et al.*, 1985). The plants that displayed the greater growth reduction (in the episodic exposure) were exposed to a significantly lower 7-h seasonal mean concentration, which supports the concern that the 7-h seasonal mean may not properly consider peak concentrations.

23.3.2 Coordination of US agricultural and environmental policies

Crop loss assessment research provides a means to quantify the physical gains in agricultural output that may result from improved air quality. Physical gains, however, do not necessarily imply economic gains. Physical gains of valueless goods do not provide economic benefits. Indeed, debate has arisen regarding the value of additional US agricultural production that might result from improved air quality in light of current crop surpluses.

Two national models currently exist that have used NCLAN crop-response data, along with adequate aerometric and economic information, to estimate the agricultural benefits associated with the control of ozone. Using a 200 region producer supply model, Kopp *et al.* (1985) estimated national benefits of changing to a standard of 0·09 ppm from the current 0·12 ppm at more than one billion dollars. Adams *et al.* (1984) obtained national benefit estimates of similar size using a mathematical programming model. Both of these studies measured benefits as the change in the sum of producer and consumer surplus resulting from the supply shift after an ozone reduction. However, neither of these studies considered the effects of US agricultural subsidy programs on their benefit measures.

McGartland (1987) countered that agricultural benefit measures based on the assumption of free market conditions were grossly misleading. Agricultural subsidy programs distort many US crop markets and induce excessive agricultural production. This means that the societal cost of producing additional crop units exceeds the consumption value of these units. Air quality improvements would reduce the cost of current production, but they would also induce additional excess production. Further, under current programs for several major crops, additional output would have to be purchased and stored (at some cost) by the government. Thus, given current subsidy programs, the net effect on agricultural sector benefits from an air quality improvement could be negative.

McGartland's observations illustrate the need for environmental policy makers to recognize and account for ongoing agricultural policies. However, McGartland ignored the possibility that agricultural policy makers may recognize and account for ongoing environmental policies. For example, air quality improvements will likely reduce farmers' average costs and induce farmers to expand production. Since prices received by farmers are generally fixed by subsidy policies, farmer profits will likely rise, perhaps substantially, after air quality improvements. This would allow agricultural policy makers to reduce agricultural subsidies while maintaining farmer incomes at the level established before air quality control measures were put into effect. Reduced subsidies would discourage excessive production. Hence, additional crop surpluses and subsidy payments could be avoided through compensatory agricultural policy.

Other compensatory policies could be employed by agricultural policy makers that would allow the agricultural sector to benefit from cleaner air. For example, output controls could be imposed to maintain output at the level established before air quality control measures were put into effect. In this case, crop surpluses and subsidy payments would not change after air quality improvements were employed, and the benefits from improved air quality could be calculated simply as the cost savings on the fixed quantity produced. Alternatively, output could be controlled indirectly via the use of input controls, though this may cause some inefficiency in input mix.

The point of this discussion is that yield increases induced by air quality improvements need not cause additional crop surpluses as long as these yield increases are recognized and accounted for in agricultural policies. The major question then becomes, "Will agricultural policies actually adjust to improved crop yields?".

Precise prediction of future agricultural policy is, of course, not possible. However, tangible evidence that agricultural policy makers will respond to crop surplus and subsidy payment increases can be found by examining current law. For example, the Food Security Act of 1985 allows the Secretary of Agriculture to reduce wheat subsidy rates in response to surplus increases and specifies that other crop target prices decline through the life of the act. More notable, this act explicitly requires that crop acreage restrictions be increased in response to crop surplus increases. Thus, it seems reasonable to expect that compensatory agricultural policy will maintain production near current levels and that the benefits of improved air quality in the agricultural sector will be positive and substantial even when the effect of US agricultural subsidy programs are considered.

This discussion suggests two distinct policy contributions from assessments of crop loss from air pollutants. First and foremost, this information aids environmental policy makers in their pollution control decisions. In addition, this information can help agricultural policy makers adjust agricultural policies to expected changes in crop production induced by air quality control programs.

23.3.3 Incomplete information

The balancing that some call for in comparing the incremental benefits and costs of an ozone SNAAQS is a challenge. For although the cost characterization is fairly complete, the benefit characterization is much more limited. Without extending the benefits information, regulatory decision makers would be comparing a fairly complete assessment of costs with a partial assessment of the benefits.

Whether one extends the benefits characterization or not seems to depend on the effects endpoint under consideration. In the case of potential human carcinogens, experimental data on small populations of laboratory animals are extended to all segments of the human population. However, in the area of agriculture, some researchers state emphatically the empirical results for cultivar x in one area cannot be extended to other cultivars of the same crop or other regions.

If economic considerations are to have more of a role in the regulatory decision-making process and benefit-cost considerations are to be environmentally neutral, we simply must develop and implement approaches which address these research gaps. In the near term we need to employ the judgements of plant physiologists and others regarding extrapolating results to other cultivars and regions. In the longer term we need to incorporate the requirement for more comprehensive treatment of benefits in our experimental design.

23.3.4 Integrating economic, environmental, and biological information

Willingness-to-pay simulations require several features if they are to prove useful in the regulatory decision-making process. Predictive powers alone are not sufficient to guarantee their use in the regulatory decision-making process. We are asking regulatory decision makers to draw cause–effect inferences from correlative and other analytical procedures. To foster adoption of such inferences, simulation models must accommodate (1) biological hypotheses and empirical results regarding concentration/response relationships; (2) behavioral response hypotheses and empirical relationships on the part of the demanders, suppliers, and regulators of agricultural markets; and (3) descriptors regarding the imperfect nature of the information used in the simulations.

Further, economic and environmental/biological information must be developed in a coordinated manner. For example, endpoints studied by biologists need to be amenable to economic valuation. Endpoints such as photosynthesis rates might be important to the biologist, but crop yield and perhaps foliar injury for ornamental crops are the important endpoints for the economist. Thus, the need for coordinated research is apparent.

23.3.5 Distributional issues

SNAAQS must be met in a reasonable period of time. The proposed time periods are set forth in the State implementation plans. As a consequence, although national assessments of the agricultural sector benefits may prove useful in establishing SNAAQS, regional assessments may prove more useful in developing reasonable attainment time proposals.

Other distributional considerations may also prove useful to regulatory decision makers. For example, how are the benefits of cleaner air that accrue to the agricultural sector shared among the farmer, consumer, and taxpayer? What does cleaner air do for the marginal farm? What crops are better served by cleaner air?

23.4 ILLUSTRATIVE AGRICULTURAL BENEFIT ANALYSIS FOR OZONE CONTROL

The scope of welfare effects considerations for alternative ozone SNAAQS will likely be broader than selected agricultural commodities. Effects on materials, silviculture, horticulture, ecosystems, and worker productivity may also warrant attention. The focus of this section, however, is on selected agricultural commodities and the effects of ozone on crop yield. The agricultural component is likely to be the primary focus of the welfare benefits analysis for alternative ozone SNAAQS. More effects information is known about this component than about any other, and effects from ozone can be observed in agricultural markets whereas effects on other components are often nonmarket in nature and thus harder to measure. Furthermore, initial benefits analysis efforts indicate that agriculture may represent the largest benefit category.

To characterize adversity, it is not sufficient to know the physical yield decrement resulting from ozone exposure. A 10% loss of crop A may be much more or less important than a 10% loss of crop B depending on the value attached to these crops by society. Thus, the need for benefit analysis to weight effects by their relative per-unit value is apparent.

The purpose of this section is to show how NCLAN data can be used in a benefits analysis and to provide illustrative estimates of the value of ambient ozone reduction to the US agricultural sector. The damage function approach to agricultural benefit measurement is outlined, and benefit estimates of a 10% reduction in ozone via the soybean, corn, wheat, and cotton markets are provided.

23.4.1 The damage function approach

The damage function approach to estimating agricultural benefits is basically a four-step procedure:

(1) estimate baseline rural ozone levels;
(2) estimate the change in baseline concentrations after an environmental policy is implemented;
(3) estimate the change in crop yields resulting from the change in ozone concentrations; and
(4) estimate the change in economic benefits resulting from the change in crop yields.

An application of each of these steps is discussed in turn.

Ideally, baseline rural ozone concentrations could be measured. Unfortunately, ozone is only sparsely monitored in rural areas and thus baseline rural ozone levels must be estimated. The estimation technique employed in this analysis is known as "kriging". The kriging method interpolates rural ozone levels from available monitoring data. Concentration estimates for 1978 for each of the crop-producing counties were generated using this approach. Use of the 1978 data may result in the overestimation of effects, since current O_3 levels are probably somewhat lower than they were in 1978. Future analysis will use more recent kriged data.

Step 2 involves the estimation of concentration changes after an ozone control policy is implemented. Ozone control policies are primarily enforced in urban areas. Thus, the effect of urban control efforts on rural areas need to be estimated. However, until the EPA's Regional Oxidant Model (ROM) is completed, no satisfactory photochemical dispersion model exists to make this link. Hence, for this illustrative analysis, all baseline rural ozone concentrations are simply reduced by 10%.

The third step involves the estimation and use of dose–response information. NCLAN crop and region-specific dose–response functions were used to link ozone concentration changes to changes in crop yield. These functions were estimated from open-top chamber studies (Heck *et al.*, 1980, 1981) using a very flexible Weibull specification. The ozone exposure statistic employed was the 7-h seasonal mean statistic. Shortcomings with this statistic were addressed in Section 23.3.

Finally, Step 4 involves the link between crop yield changes and economic benefits. To estimate this link, this analysis employed the Regional Model Farm agricultural benefits model developed by Resources for the Future (Kopp *et al.*, 1984). This model used 1978 farm-budget data from the Federal Enterprize Data System (FEDS) to estimate average costs for corn, soybeans, wheat, and cotton. By assuming that average costs equaled marginal costs within each FEDS region, regional crop supply curves were constructed. These curves were then aggregated to obtain national supply curves. Demand curves were estimated using demand elasticity estimates for the four crops.

After a 10% reduction in ozone, the national supply curves for the four crops shift outward in accordance with the estimated dose–response function. The change in producer and consumer surpluses resulting from the ozone improvement can then be calculated from areas between the pre-control and post-control supply curves. However, as discussed in Section 23.3, step 4 requires that ongoing and future agricultural policies be considered.

In this analysis, a pure cost savings measure of agricultural benefits is employed to calculate benefits of ozone reduction associated with production of corn, wheat, and cotton. (See Madariaga, B. (1987). Implications of ambient ozone standards for U.S. agriculture: a comment and some further evidence. Unpublished paper, Research Triangle Park, NC, US EPA, OAQPS.) This measure is obtained by assuming that crop output is constant and then calculating the areas between the pre-control and post-control supply curves. The assumption that crop output will remain constant after an air quality improvement is based on the expectation that additional production will be countered by output-restricting agricultural policies such as specified in the acreage restriction provisions of the 1985 Food Security Act. Since soybean prices are not currently supported via agricultural policy, benefits associated with soybeans were calculated as the simple sum of producer and consumer surpluses without compensatory agricultural policies.

To account for other future policy changes already specified in the 1985 Food Security Act, the Food and Agricultural Policy Research Institute (FAPRI) econometric models were employed (Food and Agricultural Policy Institute, 1987). These models predict future agricultural production by focusing on participation in crop programs. The models account for future changes in loan rates and other policy parameters as specified in the current law.

23.4.2 Estimation results

Illustrative benefit estimates for the four crops after a 10% reduction in rural ambient ozone are presented in Table 23.1. Though illustrative, these estimates are the same order of magnitude as the previously generated (Kopp et al., 1985; Adams et al., 1984) estimates. For the hypothesized 10% reduction, total estimated benefits equaled approximately 1·5 billion dollars

TABLE 23.1

US welfare benefits of a 10% reduction in rural ambient ozone in 1986–1990 necessary under the Food Security Act of 1985 (millions of 1986 dollars)

Crop	1986	1987	1988	1989	1990
Soybeans	294	310	308	303	290
Corn	411	362	343	351	351
Wheat	376	340	328	334	328
Cotton	435	451	513	500	501
Total	1 506	1 463	1 492	1 488	1 470

per year. Total agricultural sector benefits would be larger since these estimates account for only the four crops.

Two other points are notable in Table 23.1. First, despite soybean sensitivity to ozone, estimated soybean benefits were somewhat less than the estimated benefits from the other three crops. This is in part due to the different measure used to calculate soybean benefits. Second, the temporal considerations are interesting. The yearly total benefit estimates are relatively stable over time, though changes in policy parameters affected the individual crop benefit estimates.

23.4.3 Future directions

Despite tremendous strides in crop loss research and valuation procedures, there is always room for improvement. Several avenues can be pursued to enhance the credibility and expand the coverage of our assessment of crop loss from air pollutants and valuation estimates. The following improvements can be made to aerometric, biological, and economic data and models.

Uncertainty regarding crop loss assessment and benefit estimates can be reduced in various ways. Improved and more recent kriged data, or preferably ROM model data, are needed to better characterize baseline and rural ambient air quality after policy implementation. Biological research is needed to assess the interaction of crop loss from air pollution with other factors such as drought and temperature. Several improvements in economic models are needed, such as improved crop demand curve estimates and accountability of producer mitigating behavior. In many cases, sensitivity analysis on uncertain parameters can help characterize the extent of uncertainty regarding the crop effects and benefits estimates.

Another way to enhance the credibility of effects and valuation estimates would be to employ alternative analytical methods. Statistical multivariate field studies or subjective probability encoding can be employed to produce crop loss effects information and perhaps, valuation estimates. Unlike experimental dose–response methods, both of these approaches may be able to account for mitigating behavior by farmers.

To expand coverage of the assessment and valuation of crop loss, at least two directions can be taken: expansion to other crop types and cultivars, and inclusion of effects on crop quality.

Credible dose–response information for crops and crop cultivars not studied by NCLAN is limited. A more comprehensive assessment of the effects of air pollutants on agriculture will require additional dose–response for nonNCLAN crops and for alternative cultivars. Furthermore, the effects of air pollutants on horticultural crops need to be researched. Rather than yield, foliary injury may be the most important endpoint to study for some of these crops.

Too little is known about the assessment and valuation of crop quality effects. NCLAN dose–response data are useful for assessing total agricultural yield changes from quality changes, but not yield quality effects. There are various features of crop quality that may be of value, such as appearance, taste, and nutritional composition. Identified examples of crop quality effects caused by ozone include reduced oil in soybean seeds (Kress and Miller, 1983), reduced nutritional content in alfalfa (Neely *et al.*, 1977), and adverse quality effects in potatoes when used to make potato chips (Pell *et al.*, 1980). Valuation of such quality effects has not yet been attempted. In future work, it may be possible to statistically relate crop prices (or perhaps USDA grade levels) to ambient air quality levels across the United States. Estimates of the differences in crop prices associated with air quality differences could then be used to adjust valuation measures based solely on quantity of yield.

23.5 SUMMARY

This paper has described the generic process of assessing crop loss (from air pollution) research in the standard-setting process. Part of that rulemaking process is responding to the advice of the CASAC. To respond to the CASAC's request for an assessment of the level and distribution of economic benefits, several analytical and policy issues have to be addressed. The paper identifies these issues as well as the requirements and factors to be considered in addressing these issues. It also prescribes ways to enhance the credibility and expand the coverage of future economic assessments.

Crop loss assessment research has been central to the process of assessing the need for as well as level and form of a SNAAQS for ozone. Without that research, the Agency would not now be considering a SNAAQS for ozone. However, this process is not yet complete. The Agency plans to revise the current Staff Paper in accordance with comments from the CASAC and the public. Once the CASAC has endorsed the staff paper, it becomes the basis for staff recommendations for a SNAAQS to the EPA Administrator. The staff paper serves to bridge the gap between science contained in the criteria document and judgements that are ultimately required of the Administrator in setting ambient standards. Consequently, although crop loss research has been central to the rulemaking process to date, its ultimate impact on the SNAAQS for ozone is not yet known.

REFERENCES

Adams, R. M., S. A. Hamilton, and B. A. McCarl. (1984). *The economic effects of ozone on agriculture.* Corvallis, OR, US Environmental Protection Agency, Report No. EPA-600/3-84-090.

Biller, W. F. and T. B. Feagans. (1981). Statistical forms of national ambient air quality standards. Paper presented at the 1981 Environmetrics conference, 8–10 April, Alexandria, VA. Society for Industrial and Applied Mathematics, Philadelphia, Pennsylvania.

Food and Agricultural Policy Research Institute (FAPRI). (1987). *Comparative analysis of selected policy options for U.S. agriculture.*, FAPRI Staff Report 1–87. University of Missouri-Columbia and University of Iowa.

Heck, W. W., O. C. Taylor, R. M. Adams, G. Bingham, J. E. Miller, and L. H. Weinstein. (1980). *National crop loss assessment network 1980 annual report.*

Corvallis, OR, Corvallis Environmental Research Laboratory, US EPA, Research Triangle Park, NC.

Heck, W. W., O. C. Taylor, R. M. Adams, G. Bingham, J. E. Miller, and L. H. Weinstein. (1981). *National crop loss assessment network 1980 annual report.* Corvallis, OR, Corvallis Environmental Research Laboratory, US EPA, Research Triangle Park, NC.

Hogsett, W. E., D. T. Tingey, and S. R. Holman. (1985). A programmable exposure control system for determination of the effects of pollutant exposure regimes on plant growth. *Atmos. Environ.*, **19**, 1135–45.

Kopp, R. J., W. J. Vaughan, and M. Hazilla. (1984). *Agricultural sector benefits analysis for ozone: methods evaluation and demonstration.* Final report submitted to OAQPS. Research Triangle Park, NC, US EPA.

Kopp, R. J., W. J. Vaughan, M. Hazilla, and R. Carson. (1985). Implication of environmental policy for U.S. agriculture: the case of ambient ozone standards. *J. Environ. Managemt*, **20**, 321–31.

Kress, L. W. and J. M. Kelly. (1983). Impact of ozone on soybean yield. *J. Environ. Qual.*, **12**, 139–46.

Kress, L. W. and Miller, J. E. (1983). Impact of ozone on soybean yield. *J. Environ. Qual.*, **12**, 276–81.

McGartland, A. (1987). Implications of ambient ozone standards for U.S. agriculture: a comment and some further evidence. *J. Environ. Managemt*, **24**, 139–46.

Menser, H. A. and G. H. Hodges. (1972). Oxidant injury to shade tobacco cultivars developed in Connecticut for weather fleck resistance. *Agron. J.*, **64**, 189–92.

Musselman, R. C., R. J. Oshima, and R. E. Gallawan. (1983). Significance of pollutant concentration distribution in the response of red kidney beans to ozone. *J. Am. Soc. Hortic. Sci.*, **108**, 347–51.

Neely, G. E., D. T. Tingey, and R. G. Wilhour. (1977). Effects of ozone and sulfur dioxide singly and in combination on yield, quality and N-fixation of alfalfa. In *Proceedings of the international conference on photochemical oxidant pollutant and its control: Volume II*, US Environmental Protection Agency, EPA report no. EPA-600/3-77-0016b. Raleigh, NC, Sept., edited by B. Dimitriades, pp. 663–73.

Pell, E. J., W. C. Weissberger, and J. J. Speroni. (1980). Impact of ozone on quantity and quality of greenhouse-grown potato plants. *Environ. Sci. Technol.*, **14**, 568–71.

US General Accounting Office, *EPA's standard-setting process should be more timely and better planned.* Washington, D.C., US GAO, December 1986.

LIST OF FIRST AUTHORS AND CHAIRMEN

RICHARD M. ADAMS
Department of Agricultural and
Resource Economics, Oregon State
University, Corvallis, Oregon 97333,
USA

A. PAUL ALTSHULLER
US EPA; Mail Drop 59, Research
Triangle Park, North Carolina 27711,
USA

ROBERT G. AMUNDSON
Boyce Thompson Institute, Tower
Road, Ithaca, New York 14853, USA

ALLEN S. HEAGLE
Air Quality Research Program,
USDA-ARS, Department of Plant
Pathology, North Carolina State
University, 1509 Varsity Drive,
Raleigh, North Carolina 27606, USA

ROBERT L. HEATH
Department of Botany and Plant
Sciences, University of California,
Riverside, California 92521, USA

WALTER W. HECK
Air Quality Research Program,
USDA-ARS, Department of Botany,
North Carolina State University,
1509 Varsity Drive, Raleigh,
North Carolina 27606, USA

HOWARD E. HEGGESTAD
3112 Casteleigh Road, Silver Spring,
Maryland. 20904, USA

BRUCE B. HICKS
Environmental Research
Laboratories, NOAA/ARATDD, PO
Box E, Oak Ridge, Tennessee 37831,
USA

WILLIAM E. HOGSETT
US EPA/CERL,
200 Southwest 35th Street, Corvallis,
Oregon 97333, USA

DONALD A. HOLT
Illinois Agricultural Experiment
Station, 1301 W. Gregory Drive,
Urbana, Illinois 61801, USA

PATRICIA M. IRVING
NAPAP, 722 Jackson Place,
Washington, DC 20503, USA

BRUCE C. JORDAN
US EPA, Chief Ambient Standards
Branch, Mail Drop 12, RTP, North
Carolina 27711, USA

ROBERT J. KOHUT
Boyce Thompson Institute for Plant
Research, Cornell University, Tower
Road, Ithaca, New York 14853, USA

H. PETER KNUDSEN
Montana College of Mineral Science
and Technology, West Park Street,
Butte, Montana 59701, USA

LANCE W. KRESS
USDA-Forest Service, PO Box
12254, Research Triangle Park,
North Carolina 22709, USA

JOHN A. LAURENCE
Boyce Thompson Institute for Plant
Research, Cornell University, Tower
Road, Ithaca, New York 14853, USA

SHARON K. LEDUC
NOAA/NESDIS, University of
Missouri, CIAM, Federal Building,
Room 200, 608 East Cherry Street,
Columbia, Missouri 65201, USA

ALLEN S. LEFOHN
ASL and Associates, 111 N. Last
Chance Gulch, Room 4A, Helena,
Montana 59601, USA

ROBERT J. LUXMOORE
Oak Ridge National Laboratory, PO
Box X, Oak Ridge, Tennessee 37831-
6038, USA

WILLIAM J. MANNING
Department of Plant Pathology,
University of Massachusetts, Fernald
Hall, Amherst, Massachusetts 01002,
USA

TERRY A. MANSFIELD
Department of Biological Sciences,
University of Lancaster, Bailrigg,
Lancaster, LA1 4YG, UK

PIERRE MATHY
Directorate-General for Science, Re-
search and Development Commission
of the European Communities, Rue de
la Loi 200, B-1049 Brussels, Belgium

ANDREW R. MCLEOD
Central Electricity Research
Laboratory, Kelvin Avenue,
Leatherhead, Surrey KT22 7SE, UK

JOSEPH E. MILLER
Air Quality Research Programs,
USDA-ARS, Crop Science
Department, 1509 Varsity Drive,
North Carolina State University,
Raleigh, North Carolina 27606, USA

THOMAS J. MOSER
Northrop Services, Inc., 200 SW 35th
Street, Corvallis, Oregon 97333, USA

DOUGLAS P. ORMROD
Department of Horticultural Science,
University of Guelph, Guelph,
Ontario, Canada N1G 2W1

ERIC M. PRESTON
US EPA/CERL, 200 SW 35th Street,
Corvallis, Oregon 97333, USA

JOHN O. RAWLINGS
Statistics Department, 604-F Cox
Hall, Box 8203, North Carolina State
University, Raleigh, North Carolina
27695, USA

RICHARD A. REINERT
Air Quality Research Program,
USDA-ARS, Plant Pathology
Department, North Carolina State
University, 1509 Varsity Drive,
Raleigh, North Carolina 27606, USA

VICTOR C. RUNECKLES
Department of Plant Science,
University of British Columbia,
Vancouver, British Columbia,
Canada V6T 2A2

GEORGE E. TAYLOR, JR
Oak Ridge National Laboratory,
Environmental Sciences Division, PO
Box X, Oak Ridge,
Tennessee 37830, USA

O. CLIFTON TAYLOR
Statewide Air Pollution Research
Center, University of California,
Riverside, California 92521,
USA

DAVID T. TINGEY
US EPA/CERL, 200 SW 35th Street,
Corvallis, Oregon 97330, USA

MICHAEL H. UNSWORTH
University of Nottingham School of
Agriculture, Sutton Bonington,
Loughborough, Leicestershire,
LE12 5RD, UK

LEONARD H. WEINSTEIN
Boyce Thompson Institute for Plant
Research, Cornell University, Tower
Road, Ithaca, New York 14853, USA

RAYMOND G. WILHOUR
US EPA, Gulf Breeze Environmental
Research Laboratory, Sabine Island,
Gulf Breeze, Florida 32651, USA

ROBERT B. WILSON
Department of the Environment,
Room 511, Becket House, Lambeth
Palace Road, London SE1 7ER, UK

LIST OF NCLAN PUBLICATIONS

Adams, R. M. (1983). Issues in assessing the economic benefits of ambient ozone control: Some examples from agriculture. *Environ. Internat.*, **9**, 539–48.

Adams, R. M. (1986). Agriculture, forestry, and related benefits of air pollution control: A review and some observations. *Am. J. Agric. Econ.*, **86**, 464–72.

Adams, R. M. and T. D. Crocker. (1982). Dose–response information and environmental damage assessments: An economic perspective. *J. Air Pollut. Contr. Assoc.*, **32**, 1062–67.

Adams, R. M. and B. A. McCarl. (1985). Assessing the benefits of alternative ozone standards on agriculture: The role of response information. *J. Environ. Econ. Managmt*, **12**, 264–76.

Adams, R. M., S. A. Hamilton, and B. A. McCarl. (1984). The economic effects of ozone on agriculture. Corvallis, OR, US Environmental Protection Agency, Environmental Research Laboratory. EPA-600/3-84-090. September 1984, 175 pp.

Adams, R. M., S. A. Hamilton, and B. A. McCarl. (1985). An assessment of the economic effects of ozone pollution on U.S. Agriculture. *J. Air Pollut. Contr. Assoc.*, **35**, 938–43.

Adams, R. M., S. A. Hamilton, and B. A. McCarl. (1986). The benefits of air pollution control: The case of ozone and U.S. agriculture. *Am. J. Agric. Econ.*, **68**, 886–93.

Amundson, R. G. (1983). Yield reduction of soybean due to exposure to sulfur dioxide and nitrogen dioxide and combinations. *J. Environ. Qual.*, **12**, 454–59.

Amundson, R. G., R. M. Raba, A. W. Schoettle, and P. B. Reich. (1986). Response of soybean to low concentrations of ozone: II. Effects on growth, biomass allocation, and flowering. *J. Environ. Qual.*, **15**, 161–67.

Amundson, R. G., R. J. Kohut, A. W. Schoettle, R. M. Raba, and P. B. Reich. (1987). Correlative reductions in whole-plant photosynthesis and yield of winter wheat caused by ozone. *Phytopathology*, **77**, 75–79.

Atkinson, S. E., R. M. Adams, and T. D. Crocker. (1985). Optimal measurements of factors affecting crop production: Maximum likelihood methods. *Am. J. Agric. Econ.*, **67**, 413–18.

Brown, D. J. and J. Pheasant. (1983). An assessment of economic damages from ozone to a set of representative midwest farms. Dept. of Agricultural Economics, Agricultural Experimental Station Bulletin No. 435. West Lafayette, IN, Purdue University, 36 pp.

Cure, W. W., J. S. Sanders, and A. S. Heagle. (1986). Crop yield response predicted with different characterizations of the same ozone treatments. *J. Environ. Qual.*, **15**, 251–54.

Dixon, B. L., P. Garcia, J. W. Mjelde, and R. M. Adams. (1984). Estimation of the cost of ambient ozone on Illinois cash grain farms: An application of duality. Illinois Agricultural Economics Staff Paper, No. 84-E-276. University of Illinois, Urbana. 64 pp.

Dixon, B. L., P. Garcia, R. M. Adams, and J. W. Mjelde. (1985). Combining economic and biological data to estimate the impact of pollution on crop production. *Western J. Agric. Econ.*, **9**, 293–302.

Flagler, R. B., R. P. Patterson, A. S. Heagle, and W. W. Heck. (1987). Ozone and soil moisture deficit effects on nitrogen metabolism of soybean. *Crop Sci.*, **27**, 1177–84.

Garcia, P., B. L. Dixon, J. Mjelde, and R. M. Adams. (1986). Measuring the benefits of environmental change using a duality approach: The Case of Ozone and Illinois Cash Grain farms. *J. Environ. Econ. and Manag.*, **13**, 69–80.

Hamilton, S. A., B. A. McCarl, and R. M. Adams. (1985). The effect of aggregate response assumptions on environmental impact analysis. *Amer. J. Agric. Econ.*, **67**, 407–13.

Heagle, A. S., W. W. Heck, J. O. Rawlings, and R. B. Philbeck. (1983). Effects of chronic doses of ozone and sulfur dioxide on injury and yield of soybeans in open-top field chambers. *Crop Sci.*, **23**, 1184–91.

Heagle, A. S., M. B. Letchworth, and C. Mitchell. (1983). Injury and yield responses of peanuts to chronic doses of ozone in open-top chambers. *Phytopathology*, **73**, 551–55.

Heagle, A. S., W. W. Cure, and J. O. Rawlings. (1985). Response of turnips to chronic doses of ozone in open-top field chambers. *Environ. Pollut.*, *(Series A)*, **38**, 305–19.

Heagle, A. S., W. W. Heck, V. M. Lesser, J. O. Rawlings, and F. Mowry. (1986). Injury and yield response of cotton to chronic doses of ozone and sulfur dioxide. *J. Environ. Qual.*, **15**(4), 375–82.

Heagle, A. S., V. M. Lesser, J. O. Rawlings, W. W. Heck, and R. B. Philbeck. (1986). Response of soybeans to chronic doses of ozone applied as constant or proportional additions to ambient air. *Phytopathology*, **76**, 51–56.

Heagle, A. S., R. B. Flagler, R. P. Patterson, V. M. Lesser, S. R. Shafer, and W. W. Heck. (1987). Injury and yield response of soybean to chronic doses of ozone and soil moisture deficit. *Crop Sci.*, **27**, 1016–24.

Heagle, A. S., W. W. Heck, V. M. Lesser, and J. O. Rawlings. (1987). Effects of daily ozone exposure duration and concentration fluctuation on yield of tobacco. *Phytopathology*, **77**, 856–62.

Heck, W. W., O. C. Taylor, R. Adams, G. Bingham, J. Miller, E. Preston, and L. Weinstein. (1982). Assessment of crop loss from ozone. *J. Air Pollut. Contr. Assoc.*, **32**, 353–61.

Heck, W. W., R. M. Adams, W. W. Cure, A. S. Heagle, H. E. Heggestad, R. J. Kohut, L. W. Kress, J. O. Rawlings, and O. C. Taylor. (1983). A reassessment of crop loss from ozone. *Environ. Sci. Technol.*, **17**, 573A-580A.

Heck, W. W., W. W. Cure, J. O. Rawlings, L. J. Zaragoza, A. S. Heagle, H. E. Heggestad, R. J. Kohut, L. W. Kress, and P. J. Temple. (1984). Assessing impacts of ozone on agricultural crops: I. Overview. *J. Air Pollut. Contr. Assoc.*, **34**, 729–35.

Heck, W. W., W. W. Cure, J. O. Rawlings, L. J. Zaragoza, A. S. Heagle, H. E. Heggestad, R. J. Kohut, L. W. Kress, and P. J. Temple. (1984). Assessing impacts on ozone on agricultural crops: II. Crop yield functions and alternative exposure statistics. *J. Air Pollut. Contr. Assoc.*, **34**, 810–17.

Heggestad, H. E., J. H. Bennett, E. H. Lee, and L. W. Douglas. (1986). Effects of increasing doses of sulfur dioxide and ambient ozone on tomatoes: Plant growth, leaf injury, elemental composition, fruit yields and quality. *Phytopathology*, **76**, 1338–44.

Heggestad, H. E., T. J. Gish, E. H. Lee, J. H. Bennett, and L. W. Douglas. (1985). Interactions of soil moisture stress and ambient ozone on growth and yield of soybeans. *Phytopathology*, **75**, 472–77.

Heggestad, H. E., E. L. Anderson, T. J. Gish, and E. H. Lee. (1988). Effects of ozone and soil water deficit on roots and shoots of field-grown soybeans. *J. Environ. Pollut.*, **50**, 259–78.

Howitt, R. E., T. W. Gossard, and R. M. Adams. (1984). Effects of alternative ozone concentrations and response data on economic assessments: The case of California crops. *J. Air Pollut. Contr. Assoc.*, **34**, 1122–27.

Howitt, R. E., T. W. Gossard, and R. M. Adams. (1985). The economic effects of air pollution on annual crops. *Calif. Agric.*, **39**, 22–24.

Irving, P. M. and J. E. Miller. (1984). Synergistic effects on field grown soybeans from combinations of SO_2 and NO_2. *Can. J. Bot.*, **62**, 840–46.

King, D. A. (1987). A model for predicting the influence of moisture stress on crop losses caused by ozone. *Ecol. Model.*, **35**, 29–44.

King, D. A. and W. L. Nelson. (1987). Assessing the impacts of soil moisture stress on regional soybean yield and its sensitivity to ozone. *Agric., Ecosys., Environ.*, **20**, 23–35.

King, D. A., J. R. Kercher, and G. E. Bingham. (1983). Modeling the effects of air pollutants on soybean yield. In *Analysis of ecological systems: State of the art in ecological modeling*, ed. by W. K. Lauenroth, G. V. Skogerboe, and M. Flug, 545–52. New York, Elsevier Scientific Publishing Co.

Kohut, R. J. (1986). The National Crop Loss Assessment Network (NCLAN): An update of research results and a program review. In *Evaluation of the scientific basis for ozone/oxidant standards*, ed. by S. D. Lee, 132–43. Pittsburgh, PA, Air Pollution Control Association.

Kohut, R. J. and J. A. Laurence. (1983). Yield response of Red Kidney Bean *Phaseolus vulgaris* to incremental ozone concentrations in the field. *Environ. Pollut. (Series A)*, **32**, 233–40.

Kohut, R. J., R. G. Amundson, and John A. Laurence. (1986). Evaluation of growth and yield of soybean exposed to ozone in the field. *Environ. Pollut.*, **41**, 219–34.

Kohut, R. J., R. G. Amundson, J. A. Laurence, L. Colavito, P. van Leuken, and P. King. (1987). Effects of ozone and sulfur dioxide on yield of winter wheat. *Phytopathology*, **77**, 71–74.

Kress, L. W. and J. E. Miller. (1983). Impact of ozone on soybean yield. *J. Environ. Qual.*, **12**, 276–81.

Kress, L. W. and J. E. Miller. (1985). Impact of ozone on field corn yield. *Can. J. Bot.*, **63**, 2408–15.

Kress, L. W. and J. E. Miller. (1985). Impact of ozone on grain sorghum yield. *Water, Air, Soil Pollut.*, **25**, 377–90.

Kress, L. W., J. E. Miller, and H. J. Smith. (1985). Impact of ozone on winter wheat yield. *Environ. Expt. Bot.*, **25**, 211–28.

Kress, L. W., J. E. Miller, H. J. Smith, and J. O. Rawlings. (1986). Impact of ozone and sulfur dioxide on soybean yield. *Environ. Pollut.*, **41**, 105–23.

Lefohn, A. S., W. E. Hogsett, and D. T. Tingey. (1986). A method for developing ozone exposures that mimic ambient conditions in agricultural areas. *Atmos. Environ.*, **20**, 361–66.

Lefohn, A. S., W. E. Hogsett, and D. T. Tingey. (1987). The development of sulfur dioxide and ozone exposure profiles that mimic ambient conditions in the rural southeastern United States. *Atmos. Environ.*, **21**, 659–70.

Lefohn, A. S., H. P. Knudsen, J. Logan, J. Simpson, and C. Bhumralkar. (1987). An evaluation of the Kriging Method to predict 7-h seasonal mean ozone concentrations for estimating crop losses. *J. Air Pollut. Contr. Assoc.*, **37**, 595–602.

Lefohn, A. S., J. A. Laurence, and R. J. Kohut. (1988). A comparison of indices that describe the relationship between exposure to ozone and reduction in the yield of agricultural crops. *Atmos. Environ.*, **22**, 1229–41.

Lefohn, A. S., C. E. David, C. K. Jones, D. T. Tingey, and W. E. Hogsett. (1987). Co-occurrence patterns of gaseous air pollutant pairs at different minimum concentrations in the United States. *Atmos. Environ.*, **21**, 2435–44.

McCarl, B. A., D. Brown, R. M. Adams, and J. Pheasant. (1986). Linking farm and sector models in spatial equilibrium analysis: An application to ozone standards as they effect corn belt agriculture. In *Quantitative methods for market oriented economic analysis over space and time*, ed. by W. Labys, T. Takayama and N. Uri. Greenwich, Conn., JAI Press.

Mjelde, J. W., R. M. Adams, B. L. Dixon, and P. Garcia. (1984). Using farmers' actions to measure crop loss due to air pollution. *J. Air Pollut. Contr. Assoc.*, **34**, 360–64.

Rawlings, J. O. and W. W. Cure. (1985). The Weibull function as a dose–response model to describe ozone effects on crop yields. *Crop Sci.*, **25**, 807–14.

Rawlings, J. O., V. M. Lesser, A. S. Heagle, and W. W. Heck. (1988). Alternative ozone dose metrics to characterize ozone impact on crop yield loss. *J. Environ. Qual.*, **17**, 285–91.

Reich, P. B. and R. G. Amundson. (1984). Low level O_3 and/or SO_2 exposure causes a linear decline in soybean yield. *Environ. Pollut.*, **34**, 345–55.

Reich, P. B. and R. G. Amundson. (1985). Ambient levels of ozone reduce net photosynthesis in tree and crop species. *Science*, **230**, 566–70.

Reich, P. B., A. W. Schoettle, and R. G. Amundson. (1985). Effects of low concentrations of ozone, leaf age and water stress on leaf diffusive conductance and water use efficiency in soybean. *Physiol. Plant.*, **63**, 58–64.

Reich, P. B., A. W. Schoettle, R. M. Raba, and R. G. Amundson. (1986). Response of soybean to low concentrations of ozone: I. Reductions in leaf and whole plant net photosynthesis and leaf chlorophyll content. *J. Environ. Qual.*, **15**, 31–36.

Rodecap, K. D. and D. T. Tingey. (1986). Ozone-induced ethylene release from leaf surfaces. *Plant Sci.*, **44**, 73–76.

Snow, M. D. and D. T. Tingey. (1985). Evaluation of a system for the imposition of plant water stress. *Plant Physiol.*, **77**, 602–7.

Temple, P. J. (1986). Stomatal conductance and transpirational responses of field-grown cotton to ozone. *Plant, Cell, Environ.*, **9**, 315–21.

Temple, P. J. and L. F. Benoit. (1988). Effects of ozone and water stress on canopy temperature, water use, and water use efficiency of alfalfa. *Agron. J.*, **80**, 439–46.

Temple, P. J., O. C. Taylor, and L. F. Benoit. (1985). Effects of ozone on yield of two field-grown barley cultivars. *Environ. Pollut. (Series A)*, **39**, 217–25.

Temple, P. J., O. C. Taylor, and L. Benoit. (1985). Cotton yield responses to ozone as mediated by soil moisture and evapotranspiration. *J. Environ. Qual.*, **14**, 55–60.

Temple, P. J., O. C. Taylor, and L. F. Benoit. (1986). Yield response of head lettuce (*Lactuca sativa* L.) to ozone. *Environ. Expt. Bot.*, **26**, 53–58.

Temple, P. J., R. W. Lennox, A. Bytnerowicz, and O. C. Taylor. (1987). Interactive effects of simulated acidic fog and ozone on field-grown alfalfa. *Environ. Expt. Bot.*, **27**, 409–17.

Temple, P. J., L. F. Benoit, R. W. Lennox, C. A. Reagan, and O. C. Taylor. (1988). Combined effects of ozone and water stress on alfalfa growth and yield. *J. Environ. Qual.*, **17**, 108–13.

Tingey, D. T. (1985). The impact of ozone on vegetation. In *Perspectives in environmental botany*, ed. by D. N. Rao, K. J. Ahmad, M. Yunus and S. N. Singh. Luchnow, India, Print House.

Tingey, D. T. (1986). The impact of ozone on agriculture and its consequences. In *Acidification and its policy implications,* ed. by T. Schneider, 53–63. Amsterdam, Elsevier Science Publishers B.V.

Tingey, D. T. and W. E. Hogsett. (1985). Water stress reduces ozone injury via a stomatal mechanism. *Plant Physiol.,* **77,** 944–47.

Tingey, D. T., K. D. Rodecap, E. H. Lee, T. J. Moser, and W. E. Hogsett. (1986). Ozone alters the concentrations of nutrients in bean tissue. *Angewandte Botanik,* **60,** 481–93.

Unsworth, M. H., A. S. Heagle, and W. W. Heck. (1984). Gas exchange in open-top field chambers I. Measurement and analysis of atmospheric resistances to gas exchange. *Atmos. Environ.,* **18,** 373–80.

Unsworth, M. H., A. S. Heagle, and W. W. Heck. (1984). Gas exchange in open-top field chambers II. Resistance to ozone uptake by soybeans. *Atmos. Environ.,* **18,** 381–85.

Unsworth, M. H., V. M. Lesser, and A. S. Heagle. (1984). Radiation interception and the growth of soybeans exposed to ozone in open-top field chambers. *J. Appl. Ecol.,* **21,** 1059–79.

INDEX